Edward R. Corino

# HEAT AND MASS TRANSFER

# McGRAW-HILL SERIES IN MECHANICAL ENGINEERING

ROBERT M. DRAKE, JR., AND STEPHEN J. KLINE, *Consulting Editors*

BEGGS · *Mechanism*

CAMBEL AND JENNINGS · *Gas Dynamics*

DURELLI, PHILLIPS, AND TSAO · *Introduction to the Theoretical and Experimental Analysis of Stress and Strain*

ECKERT AND DRAKE · *Heat and Mass Transfer*

HAM, CRANE, AND ROGERS · *Mechanics of Machinery*

HARTMAN · *Dynamics of Machinery*

HINZE · *Turbulence*

JACOBSEN AND AYRE · *Engineering Vibrations*

PHELAN · *Fundamentals of Mechanical Design*

SABERSKY · *Elements of Engineering Thermodynamics*

SHIGLEY · *Machine Design*

SHIGLEY · *Kinematic Analysis of Mechanisms*

SPALDING AND COLE · *Engineering Thermodynamics*

STOECKER · *Refrigeration and Air Conditioning*

---

The Series was established in 1954 under the Consulting Editorship of Richard G. Folsom, who continued in this capacity until he assumed the Presidency of Rensselaer Polytechnic Institute in 1958.

# Heat and Mass Transfer

by E. R. G. ECKERT
*Professor of Mechanical Engineering
and Director of the Heat Transfer Laboratory
University of Minnesota*

WITH PART A, HEAT CONDUCTION
AND APPENDIX OF PROPERTY VALUES
by **ROBERT M. DRAKE, Jr.**
*Professor and Chairman
Mechanical Engineering Department Princeton University*

*Second Edition of*
INTRODUCTION TO THE TRANSFER OF HEAT AND MASS

McGRAW-HILL BOOK COMPANY, INC.

New York  Toronto  London

1959

HEAT AND MASS TRANSFER. Copyright © 1959 by the McGraw-Hill Book Company, Inc. Copyright, 1950, by the McGraw-Hill Book Company, Inc. Printed in the United States of America. All rights reserved. This book, or parts thereof, may not be reproduced in any form without permission of the publishers. *Library of Congress Catalog Card Number* 58-9999

II

THE MAPLE PRESS COMPANY, YORK, PA.

# PREFACE

The goal at which this book is aimed is to introduce the reader to the basic physical phenomena involved in transport processes in general and the transfer of heat and mass in particular, and to acquaint him with the laws and analytical methods which have been developed for their description. The intent was the same in the first edition of this book; however, the tremendous amount of research and the accumulation of knowledge during the past eight years has made certain shifts in the approach to the above goal necessary.

In the first edition, an attempt was made to present this material in such a way that a reader with an education obtained through a course of study in a normal engineering curriculum could follow the derivations without undue difficulty. The many favorable comments that I received on this point made it appear desirable to maintain this mode of presentation. This, however, made it necessary to restrict the discussion to selected topics. The choice of these topics was made with the object of considering as much of the important heat-transfer phenomena as our present reasonably good understanding of their nature and laws makes possible. There is still a considerable number of heat-transfer situations of which our present-day understanding is rather poor. Generally speaking, they are usually situations in which the surface or interphase area through which the heat flows is not well-defined. Because of our lack of understanding, knowledge of such processes is essentially restricted to empirical relationships that can be found in the various reference books on heat transfer.

The topics were also generally selected with the view in mind of presenting as much physical information as possible without excessive mathematical manipulation. The main problem with which a worker in engineering research is faced is the task of translating a physical process into mathematical language. This is a task which he invariably must do by himself, whereas he can obtain the help of a mathematician if he does not master the tools for obtaining a solution to the equations in which he has formulated his problem. In some of the new chapters this rule could not be followed completely. There the presentation is somewhat more sophisticated and condensed, requiring a higher knowledge in mathematics than the rest. Such chapters were included because they are essential to an understanding of specific phases of heat or mass transfer;

however, they may be skipped without impairing the reading of the rest of the text. Sections 3-8, 4-1, 4-2, 5-1, 5-2, 6-3, 6-6, 7-2, 7-6, 10-3, 11-3, and 16-2 belong to this category.

No attempt was made to be complete in the references or even to include all the important ones. Such an endeavour would be futile indeed, since approximately 1,000 papers per year have been published recently in the area of heat and mass transfer. The purpose of the references in the book is to acknowledge original contributions to the topics which are discussed and to indicate sources for more information. Readers who are interested in a fairly complete coverage of recent literature may want to look up the surveys on heat and mass transfer which are printed yearly in the magazines *Industrial and Engineering Chemistry* and *Mechanical Engineering*.

Problems were also added in this edition. They are designed to test whether the reader or student has grasped the derivations presented in the preceding chapter. It may be necessary to point out that some problems do not have a unique answer; the same thing is true in most problems encountered in engineering practice. No analytic solution is obtainable in some of them, and in others the reader may have to go to numerical calculations at some point, depending on his familiarity with analytical methods. The intent is again to stress the fact that many problems encountered in engineering practice will be of such a nature.

I have received a large number of comments and suggestions since the first edition appeared on the market, and I want to thank all my colleagues and friends for them. They were incorporated as far as possible into the new edition. Professor R. M. Drake, Jr., this time contributed Part A, "Heat Conduction," and Sec. 10-3, "Heat Transfer in Gases at Low Densities," in addition to the "Appendix of Property Values," and in this way helped essentially to round off the coverage of the text. I have to thank Mr. Norman C. Stirling for his careful proofreading of the manuscript and for many helpful suggestions. Thanks are certainly due also to my wife, Josefine, who not only kept away from me a good deal of the everyday chores and disturbances, but also helped in the proofreading.

Those familiar with the first edition will see that the major part of it was rewritten completely and that this edition can be considered in this respect practically a new book. This was felt necessary for the presentation of a science which has made such rapid progress during recent years. It is hoped that, in this way, the text has been brought up to date completely and that it will find as favorable an acceptance among those studying or working in the heat- and mass-transfer field as did the first edition.

*E. R. G. Eckert*

# CONTENTS

*Preface*   v

*Nomenclature*   xi

**Chapter 1. Introduction**   1
  1-1. Dimensions and units   1
  1-2. The different modes of heat transfer   7
  1-3. Thermal conductivity, film, and over-all heat-transfer coefficients   8
  1-4. Parallel flow, counterflow, crossflow   13

## PART A   HEAT CONDUCTION

**Chapter 2. Theory of Heat Conduction and the Heat-conduction Equation**   25
  2-1. Concept of heat conduction   25
  2-2. The fundamental law of heat conduction   27
  2-3. The heat-conduction equation   30
      Heat-conduction equation for isotropic materials   30
      Heat-conduction equation for nonisotropic materials   32

**Chapter 3. Steady Heat Conduction**   36
  3-1. Simple conduction equation solutions for steady conditions   36
  3-2. Heat convection from bounding surfaces   38
  3-3. Critical thickness of insulation   39
  3-4. The thin rod   39
  3-5. Finned heating surfaces   43
  3-6. The wall with heat sources   58
  3-7. The buried cable   60
  3-8. Two-dimensional steady heat conduction   64

**Chapter 4. Unsteady Heat Conduction**   76
  4-1. Transient heat conduction   76
  4-2. Periodic heat conduction   97

**Chapter 5. Heat Conduction with Moving Boundaries**   110
  5-1. Heat conduction in melting or solidification   110
  5-2. Moving heat sources   113

## PART B   HEAT TRANSFER BY CONVECTION

The various types of heat transfer   121

**Chapter 6. Flow along Surfaces and in Channels**   124
  6-1. Boundary layer and turbulence   124
  6-2. The momentum equation of the boundary layer   131

|   |   |
|---|---|
| 6-3. The laminar-flow boundary-layer equation | 134 |
| 6-4. The plane plate in longitudinal flow | 138 |
| 6-5. Pressure gradients along a surface | 148 |
| 6-6. Exact solutions of the laminar boundary-layer equations for a flat plate | 150 |
| 6-7. The flow in a tube | 153 |
| 6-8. The cylinder in crossflow | 160 |
| 6-9. Flow around rotationally symmetric objects | 163 |

**Chapter 7. Forced Convection in Laminar Flow** — 167

|   |   |
|---|---|
| 7-1. The heat-flow equation of the boundary layer | 167 |
| 7-2. Laminar boundary-layer energy equation | 169 |
| 7-3. The plane plate in longitudinal flow | 173 |
| 7-4. The plane plate with arbitrarily varying wall temperature | 178 |
| 7-5. Cylindrical bodies in flow normal to their axes | 184 |
| 7-6. Exact solutions of the laminar-boundary-layer energy equation | 187 |
| 7-7. Flow through a tube | 190 |

**Chapter 8. Forced Convection in Turbulent Flow** — 201

|   |   |
|---|---|
| 8-1. Analogy between momentum and heat transfer | 201 |
| 8-2. Flow in a tube | 206 |
| 8-3. The plane plate in longitudinal flow | 215 |
| 8-4. Recent developments in the theory of turbulent heat transfer | 218 |

**Chapter 9. Forced Convection in Separated Flow** — 229

|   |   |
|---|---|
| 9-1. Dimensional analysis of heat transfer | 229 |
| 9-2. Tubes and tube bundles in crossflow | 239 |
| 9-3. Spheres and packed beds | 248 |

**Chapter 10. Special Heat-transfer Processes** — 255

|   |   |
|---|---|
| 10-1. Heat transfer at high velocities | 255 |
| 10-2. Heat transfer in gases at high velocities | 261 |
| 10-3. Heat transfer in rarefied gases | 272 |
| 10-4. Heat transfer in liquid metals | 298 |
| 10-5. Transpiration and film cooling | 301 |

**Chapter 11. Free Convection** — 311

|   |   |
|---|---|
| 11-1. Laminar heat transfer on a vertical plate and horizontal tube | 312 |
| 11-2. Turbulent heat transfer on a vertical plate | 322 |
| 11-3. Derivation of the boundary-layer equations | 326 |
| 11-4. Free convection in a fluid enclosed between two plane walls | 328 |
| 11-5. Mixed free and forced convection | 331 |

**Chapter 12. Condensation and Evaporation** — 334

|   |   |
|---|---|
| 12-1. Condensation | 334 |
| 12-2. Evaporation | 340 |

## PART C  THERMAL RADIATION

|   |   |
|---|---|
| Basic concepts | 353 |

**Chapter 13. Thermal Radiation Properties** — 355

|   |   |
|---|---|
| 13-1. The black body | 360 |

13-2. Solids and liquid bodies ... 371
13-3. Gases ... 381

**Chapter 14. Heat Exchange by Radiation** ... 395

14-1. Black bodies ... 395
14-2. Solid, liquid, and gaseous bodies ... 403
14-3. Radiative heat exchange inside an enclosure ... 407
14-4. Radiation of flames ... 419
14-5. The heat-transfer coefficient for radiation ... 422
14-6. Radiation error in temperature measurements ... 426
14-7. Pyrometry ... 429
14-8. Solar radiation ... 432

## PART D   THE TRANSFER OF MASS

**Chapter 15. Relations for Two-component Mixtures** ... 440

15-1. Basic equations for two-component gas mixtures ... 440
15-2. Basic equations and $s$-$i$ diagram for humid air ... 441

**Chapter 16. Mass Transfer** ... 449

16-1. Diffusion ... 449
16-2. Laminar boundary layer on a flat plate with mass and heat transfer ... 456
16-3. The integrated boundary-layer equations of mass transfer ... 463
16-4. Similarity relations for mass transfer ... 469
16-5. The evaporation of water into air ... 476

## PART E   HEAT EXCHANGERS

**Chapter 17. Heat-Exchanger Calculations** ... 480

17-1. The transfer-type heat exchanger ... 480
17-2. The storage-type heat exchanger ... 482

**Appendix of Property Values** ... 493

*Name Index* ... 523

*Subject Index* ... 527

# NOMENCLATURE

$a$    acceleration; accommodation coefficient (p. 281)
$b$    wall thickness; fin thickness (p. 44)
$c$    specific heat
$c_p$    specific heat at constant pressure
$d$    diameter
$d_h$    hydraulic diameter
$f$    friction coefficient for flow through a tube (p. 155); dimensionless stream function
$f_c$    drag coefficient of a cylinder (p. 162)
$f_p$    friction coefficient of a plane plate (p. 141)
$f_s$    velocity slip coefficient (p. 281)
$g$    acceleration of gravity
$h$    film heat-transfer coefficient (p. 11)
$i$    enthalpy
$i_{fg}$    heat of evaporation
$j$    degrees of freedom
$k$    thermal conductivity
$l$    length
$m$    mass
$\dot{m}$    mass rate of flow
$n$    normal; harmonic number (p. 98); number of molecules (p. 273)
$p$    pressure
$q$    heat flow per unit time and area (specific rate of heat flow)
$r$    radius, recovery factor (p. 256)
$s$    thickness; distance; molecular speed ratio (p. 293)
$t$    temperature
$t_B$    bulk temperature (p. 192)
$u$    internal energy
$u, v, w$    velocity components
$v$    specific volume
$v^*$    shearing-stress velocity (p. 156)
$x, y, z$    coordinates
$x_0$    wavelength (p. 103)

## NOMENCLATURE

$A$    area
$C$    constant; circumference
$E$    voltage; energy
$F$    force
$J$    turbulence intensity (p. 127); electric current; integral
$L$    length
$L_e$    entrance length (p. 131)
$M$    molecular weight
$Q$    heat flow per unit time
$R$    resistance; gas constant
$R_c$    thermal-conduction resistance (p. 10)
$R_t$    thermal-convection resistance (p. 12)
$T$    absolute temperature
$U$    internal energy; over-all transfer coefficient (p. 13)
$V$    volume; velocity
$\dot{V}$    volume flow
$W$    weight; work

$\alpha$    angle; thermal diffusivity (p. 31); parameter for properties (p. 237)
$\beta$    angle; parameter for stream velocity (p. 153); expansion coefficient
$\gamma$    parameter for wall temperature (p. 188); ratio of specific heats
$\delta$    boundary-layer thickness (p. 126); phase lag (p. 98)
$\delta^*$    boundary-layer displacement thickness (p. 133)
$\delta_i$    boundary-layer momentum thickness (p. 133)
$\delta_t$    thermal boundary-layer thickness (p. 167)
$\epsilon$    porosity
$\epsilon_m$    eddy diffusivity for momentum (p. 219)
$\epsilon_q$    eddy diffusivity for heat (p. 219)
$\zeta$    ratio of thermal and flow boundary-layer thicknesses
$\eta$    dimensionless wall distance (p. 150); fin effectiveness (p. 49)
$\theta$    temperature difference; parameter describing temperature (p. 187)
**k**    Boltzmann constant (p. 292)
$\lambda$    form parameter for velocity profile (p. 149); eigenvalue (p. 80); mean-free-path length (p. 272)
$\mu$    dynamic viscosity
$\nu$    kinematic viscosity
$\rho$    density; radius; electric resistivity
$\sigma$    surface tension
$\tau$    shearing stress; time
$\tau_0$    period

$\phi$ heat dissipation; heat source, fin effectiveness
$\psi$ stream function (p. 150)

## DIMENSIONLESS VALUES

Bi Biot number (p. 48)
E dissipation number (p. 233)
Fo Fourier number (p. 77)
Gr Grashof number for heat transfer (p. 236)
$Gr_D$ Grashof number for mass transfer (p. 474)
Le Lewis number (p. 459)
Ma Mach number (p. 238)
Nu Nusselt number (p. 176)
Pe Peclet number (p. 178)
Pr Prandtl number (p. 174)
$Pr_t$ turbulent Prandtl number (p. 220)
Re Reynolds number (p. 128)
Sc Schmidt number (p. 458)
St Stanton number (p. 207)

### Indices

$b$ values on the border between laminar sublayer and turbulent boundary layer
$cr$ critical values for transition to turbulent flow
$d$ based on diameter $d$
$dy$ dynamic values (dynamic pressure, etc.)
$e$ exit
$f$ fluid
$i$ inlet; based on enthalpy
$l$ liquid
$m$ mean value
$p$ partial
$r$ recovery
$s$ values in the stream outside the boundary layer
$st$ static values (static pressure, etc.)
$t$ turbulent
$v$ vapor
$w$ values on the wall surface
$x$ based on length $x$
$B$ bulk
$M$ arithmetic mean value
$T$ total values (total pressure, etc.)
$0$ values at a reference point
$\infty$ values at great distance from a body

|   |   |
|---|---|
| – | mean value |
| * | reference |
| ′ | fluctuating value, dimensionless |

## ADDITIONALLY FOR THERMAL RADIATION

| | |
|---|---|
| $a$ | absorption coefficient (p. 383) |
| $c$ | velocity of light |
| $e$ | emissive power (p. 357) |
| $h$ | Planck's quantum (p. 370) |
| $i$ | radiation intensity (p. 361) |
| $\kappa$ | gas constant per molecule |
| $n$ | refraction index |
| $p_r$ | radiation pressure (p. 360) |
| $u$ | radiation density (p. 360) |
| $B$ | total radiation leaving an interphase (p. 405) |
| $E$ | error |
| $F$ | shape factor (p. 396) |
| $H$ | irradiation = total radiation approaching an interphase (p. 409) |
| $L_e$ | equivalent radius (p. 391) |
| $\alpha$ | absorptivity (p. 357) |
| $\beta$ | angle |
| $\epsilon$ | emissivity (p. 359) |
| $\eta$ | efficiency |
| $\lambda$ | wavelength |
| $\nu$ | frequency |
| $\rho$ | reflectivity (p. 357) |
| $\sigma$ | Boltzmann constant (p. 365) |
| $\tau$ | transmissivity (p. 357) |
| $\omega$ | solid angle |

### Indices

| | |
|---|---|
| $b$ | black |
| $c$ | convection; color |
| $e$ | equivalent |
| $g$ | gas; green |
| $i$ | impinging |
| $n$ | normal to surface |
| $r$ | radiation; red |
| $s$ | surface |
| $\beta$ | under an angle |
| $\lambda$ | monochromatic |
| $c$ | concentration |

## ADDITIONALLY FOR MASS TRANSFER

- $h_D$ mass-transfer coefficient (p. 460)
- $i$ enthalpy
- $n$ number of molecules
- $r$ relative humidity (p. 442)
- $s$ specific humidity (p. 441)
- $w$ mass ratio
- $C_p$ specific heat per unit volume
- $D$ diffusion coefficient
- $N$ number of moles
- $\mathcal{R}$ universal gas constant
- $\delta_D$ diffusion boundary-layer thickness
- $\sigma$ evaporation coefficient (p. 476)
- $\varphi$ parameter describing mass fraction (p. 458)

### Indices

- $a$ air
- $s$ saturation
- $v$ vapor

## ADDITIONALLY FOR HEAT-EXCHANGER CALCULATIONS

- $\epsilon$ effectiveness (p. 481)
- $\eta$ dimensionless parameter for time (p. 486)
- $\xi$ dimensionless parameter for length (p. 486)

### Indices

- $l$ large
- $s$ small
- $C$ cooling
- $H$ heating
- $L$ per unit length
- $P$ period
- $S$ solid

CHAPTER 1

# INTRODUCTION

**1-1. Dimensions and Units.** The laws of physics and also those of heat transfer are most effectively presented in mathematical language. The equations which are used in physics, however, differ from the ones which we use in mathematics in the respect that they deal with physical quantities. Such quantities have *dimensions* (length, velocity, energy) and are measured in certain *units* (ft, ft/sec, ft lb$_f$). A physical quantity is known only when the unit in which it is measured is stated and a number is given indicating how often the unit is contained in the quantity. Each symbol in a physical equation stands, therefore, for the product of a number with a unit ($V = 3$ ft/sec). We distinguish between basic units (ft, lb, sec) and derived units. The latter are composed of groups of basic units (ft/sec, ft lb$_f$). The concepts of dimension and unit should be clearly distinguished. A physical quantity has only one dimension but can be expressed in different units. The dimension length, for instance, can be measured in feet, inches, or miles. It is possible to convert each of the different units for the same dimension into any one of the others. For instance,

$$1 \text{ hr} = 3{,}600 \text{ sec}$$
$$1 \text{ Btu} = 778.26 \text{ ft lb}_f$$

It is never possible to convert one dimension into another. It is of advantage to keep the number of basic units small. Different systems of units are distinguished depending on which units are used as basic ones.

The physical system of units is based on three units: the centimeter (cm) as unit for the dimension length, the gram (g) as unit for the dimension mass, and the second (sec) as unit for the dimension time. Sometimes the degree centigrade (°C)[1] is used as a fourth basic unit. All other units are formed from these basic ones as derived units. The number of basic units is in no way prescribed to us by nature. The degree centigrade (C), for instance, can be expressed by the basic units cm, g, with

---

[1] The notation °C is often used. In derived units it may, however, lead to errors, since the symbol ° for degree may be interpreted as a power (ft$^2$°C). Therefore, the letter C only will be used as symbol for degree centigrade (and correspondingly F, K, R for degree Fahrenheit, Kelvin, Rankine).

the use of the laws of thermodynamics. The number of basic units can in principle be reduced still more. It would, for instance, be possible to eliminate the gram (g) as a basic unit for the mass by the use of the general gravitational law. The force with which two masses $m_1$ and $m_2$ at a distance $r$ attract each other is expressed by Newton's gravitational law, usually written

$$F = K \frac{m_1 m_2}{r^2} \tag{1-1}$$

This expression contains a universal constant $K$ which in the physical system of units has the value $K = 6.685 \times 10^{-8}$ cm$^3$/g sec$^2$. The constant is omitted when the equation is used to derive a unit for the mass, and the equation is written as

$$F = \frac{m_1 m_2}{r^2}$$

To eliminate the force, this equation is combined with Newton's law, written, for instance, for the mass $m_1$ subject to an acceleration $a$

$$F = m_1 a$$

Combination of both equations results in

$$m_2 = ar^2$$

It can be seen that the right-hand side of the equation has the dimension cm$^3$/sec$^2$. Since different terms appearing in a physical equation must have the same dimension, we have obtained in this way a dimension in which mass can be measured. It is customary, however, to retain the gram as the basic unit for the mass in addition to the others mentioned above.

The engineering system of units, which in this country is based on English units, has grown under the influence of engineering needs and has actually never developed into such a clear and universally accepted system as the physical system of units. For instance, there is no complete agreement as to the number and the kind of basic units used. Usually the engineering system is defined in such a way that basic units are used for the dimensions length (ft), force (lb$_f$), time (sec), and temperature (F). Other units, however, are used continuously besides the ones listed above (in., hr, R). Moreover, quite often in the engineering literature the pound (lb) is used to represent a mass and a force simultaneously.

In 1954, the Tenth General Conference for Units and Weights adopted internationally a new system called the mksa or Giorgi system which is based on the following units:

m = meter (for length)
kg = kilogram (for mass)
s = second (for time)
A = ampere (for electric current)
°K = degree Kelvin (for temperature)
cd = candela (for light flux)

In this system energy is measured in joules (1 J = 1 m² kg/sec²) or in kilojoules (1 kJ = 1000 J) and power in watts (1 W = 1 m² kg/sec³) or kilowatt (1 kW = 1,000 W).

The following relation exists:

$$1 \text{ kJ} = 0.23844 \text{ kcal}\dagger = 0.94621 \text{ Btu}\dagger$$

This system has many advantages. It combines, for instance, in a natural way electric units with those used in mechanics and thermodynamics. A number of countries have already adopted it by law. It is to be hoped that it will also be used more and more in the United States.

Because of the diversified applications of heat transfer, papers on this subject are found in magazines on physics as well as on engineering. As a consequence, the necessity arises for everyone working in this field continuously to convert values from one system of units to the other. The possibility of errors in such calculations, which is great, can be substantially reduced if the following rules are adhered to.

1. Always remember that a physical quantity is described only when the unit is stated as well as the number indicating how often the unit is contained in the quantity. If, therefore, a physical equation is used to make numerical calculations, then for each symbol not only a number but also the corresponding unit should be introduced. Some examples in the following paragraphs will indicate the application of this rule.

2. It has to be kept in mind that only quantities of the same kind, which means with the same dimension, can be added, subtracted, or compared. As a consequence, all terms in a physical equation must have the same dimension.

3. The pound as unit of mass or unit of force has to be clearly distinguished. In particular it has to be clarified in each instance whether property values (density $\rho$, specific heat $c$, etc.) are based on mass or weight. In this country they are usually based on mass, whereas on the European continent an engineering system is preferred which bases properties on a unit standard weight, which is defined as the force which a mass of 1 kg experiences under the influence of the standard gravitational acceleration. In this text the pound as unit for a force will be written $lb_f$ and as unit for a mass lb.

† International Steam Table values.

The whole problem of systems of units loses much of its significance when the above rules are consistently applied, and conversion from one system to another becomes almost automatic. The following examples will explain these rules.

1. If a substance expands reversibly under constant pressure $p$ by the volume $\Delta V$, then the first law of thermodynamics relates the heat $Q$ added in this process to the change in internal energy $\Delta U$ and the mechanical work performed in the expansion by the following equation:

$$Q = \Delta U + \frac{p\,\Delta V}{\mathbf{J}} \tag{1-2}$$

$\mathbf{J}$ is the mechanical equivalent of heat. Assume that the values for the different quantities appearing on the right-hand side of the equation are known ($\Delta U = 3$ Btu, $p = 1{,}000$ lb$_f$/ft$^2$, $\Delta V = 2$ ft$^3$) and that the heat added in this process is to be calculated. Remembering that $\mathbf{J} = 778.26$ ft lb$_f$/Btu and introducing the values for the symbols on the right-hand side Eq. (1-2) lead to the following expression:

$$Q = 3 \text{ Btu} + \frac{1{,}000 \text{ lb}_f/\text{ft}^2 \times 2 \text{ ft}^3}{778.26 \text{ ft lb}_f/\text{Btu}} = 5.57 \text{ Btu}$$

Actually the factor $\mathbf{J}$ appearing in the first law of thermodynamics as written above is only a conversion factor from one unit for the dimension work into another, and it is not necessary to have a special symbol for it in a physical equation. We usually do not indicate on other physical quantities what unit will be used. The expression for the first law, in the process considered here, can then be written in the following way:

$$Q = \Delta U + p\,\Delta V \tag{1-3}$$

We shall show that the calculation procedure which applies the rules stated above leads to exactly the same result for the heat $Q$ added in the process. Introducing the prescribed values into the right-hand side of the equation, the following expression is obtained:

$$Q = 3 \text{ Btu} + 1{,}000 \text{ lb}_f/\text{ft}^2 \times 2 \text{ ft}^3 = 3 \text{ Btu} + 2{,}000 \text{ ft lb}_f$$

This time the two terms on the right-hand side of the equation appear in different units. In order to add the terms, one term must be converted to the units of the other. Since it is customary to express heat in Btu, the conversion will be made on the second term. Introducing the conversion factor from ft lb$_f$ into Btu, which was given before, the calculation can be continued: 1 ft lb$_f$ = 1/778.26 Btu;

$$Q = 3 \text{ Btu} + 2{,}000 \times \frac{1}{778.26} \text{ Btu} = 5.57 \text{ Btu}$$

## INTRODUCTION

2. In flow of an inviscid, incompressible fluid Bernoulli's equation holds:

$$p_T = p_{st} + \frac{\rho v^2}{2} \tag{1-4}$$

This equation states that the sum of the static pressure $p_{st}$ and the dynamic pressure $\rho v^2/2$ at each location in the fluid is equal to the total pressure $p_T$. Let us assume that the following values are prescribed:

$$p_{st} = 2 \text{ lb}_f/\text{ft}^2 \qquad v = 50 \text{ fps} \qquad \rho = 0.1 \text{ lb}/\text{ft}^3$$

and that the total pressure is to be calculated. The calculation proceeds in the same way as in the first example, introducing the values for the symbols on the right-hand side of the equation:

$$p_T = 2 \text{ lb}_f/\text{ft}^2 + 0.1 \frac{\text{lb}}{\text{ft}^3} \times \frac{2{,}500}{2} \frac{\text{ft}^2}{\text{sec}^2} = 2 \text{ lb}_f/\text{ft}^2 + 125 \text{ lb}/\text{ft sec}^2$$

Again, it is seen that the two terms on the right-hand side, which must have the same dimension, are expressed in different units. Therefore, one of the units has to be converted into the other. For this conversion we have to remember the definition of the pound force: One pound force is that force which the mass of one pound exerts in a gravitational field with the standard gravitational acceleration 32.174 ft/sec². This definition can be written as an equation:

$$1 \text{ lb}_f = 1 \text{ lb} \times 32.174 \text{ ft/sec}^2 = 32.174 \text{ lb ft/sec}^2$$

The equation is used in the above expression to replace the lb:

$$p_T = 2 \text{ lb}_f/\text{ft}^2 + \frac{125}{32.174} \frac{1}{\text{ft sec}^2} \times \frac{\text{lb}_f \text{ sec}^2}{\text{ft}} = 5.89 \text{ lb}_f/\text{ft}^2$$

Bernoulli's equation is sometimes found in the literature in the following form:

$$p_T = p_{st} + \rho \frac{v^2}{2g_c} \tag{1-5}$$

The symbol $g_c$ in this equation is the conversion factor

$$g_c = 32.174 \text{ lb ft/lb}_f \text{ sec}^2$$

The calculation is made in the same way as above with the following results:

$$p_T = 2 \text{ lb}_f/\text{ft}^2 + 0.1 \text{ lb}/\text{ft}^3 \times \frac{2{,}500}{2} \frac{\text{ft}^2}{\text{sec}^2} \times \frac{1}{32.174}$$

$$\frac{\text{lb}_f \text{ sec}^2}{\text{lb ft}} = 5.89 \text{ lb}_f/\text{ft}^2$$

It can again be seen that either one of the two Eqs. (1-4) or (1-5) for Bernoulli's law leads to the same value for the total pressure.

3. A third example deals with temperature measurements in high-speed gas flow. Later on in this book it will be explained that two different kinds of temperature can be distinguished in such a flow: a total temperature $T_T$ and a static temperature $T_{st}$. Both are connected by the following equation when the specific heat $c_p$ is considered constant:

$$T_T = T_{st} + \frac{v^2}{2c_p}$$

In this equation, the specific heat $c_p$ at constant pressure appears in the second term on the right-hand side. In order to make calculations properly, it has to be known whether this specific heat is the heat per pound-mass and per degree Fahrenheit or the specific heat per pound-force (weight) and degree Fahrenheit. In the above equation the specific heat is based on a unit mass. The calculation proceeds in the following way when again the values for the terms on the right-hand side are prescribed: $T_{st} = 100$ F, $v = 1{,}000$ fps, $c_p = 0.240$ Btu/lb F

$$T_T = 100 \text{ F} + \frac{10^6}{2} \frac{\text{ft}^2}{\text{sec}^2} \times \frac{1}{0.240} \frac{\text{lb F}}{\text{Btu}}$$
$$= 100 \text{ F} + 2.08 \times 10^6 \text{ lb ft}^2 \text{ F/sec}^2 \text{ Btu}$$

Obviously the second term must have the dimension of temperature. This suggests replacement of the lb in the numerator by $\text{lb}_f$ to obtain an expression for the mechanical work:

$$T_T = 100 \text{ F} + 2.08 \times 10^6 \frac{\text{ft}^2 \text{ F}}{\text{sec}^2 \text{ Btu}} \times \frac{1}{32.174} \frac{\text{lb}_f \text{ sec}^2}{\text{ft}}$$
$$= 100 \text{ F} + 6.46 \times 10^4 \text{ lb}_f \text{ ft F/Btu}$$

Finally the Btu can be converted to ft $\text{lb}_f$:

$$T_T = 100 \text{ F} + 6.46 \times 10^4 \frac{\text{lb}_f \text{ ft F}}{\text{Btu}} \times \frac{1}{778.26} \frac{\text{Btu}}{\text{ft lb}_f}$$
$$= 100 \text{ F} + 83 \text{ F} = 183 \text{ F}$$

It can be easily shown that the equation above for the total temperature changes to

$$T_T = T_{st} + \frac{v^2}{2gc_p}$$

with $g$ indicating standard gravitational acceleration when the specific heat is based on unit weight.

It has been demonstrated by these examples that conversion factors like the mechanical equivalent of heat $\mathbf{J}$ and the gravitational conversion factor $g_c$ can be omitted from physical equations when numerical calculations are made with the utilization of the three stated rules. In the equations in this book, therefore, such conversion factors are not given. In addition, all property values like density, specific heat, etc., are consistently based on the unit mass, which is the more satisfactory dimension for measuring a certain amount of material.

**1-2. The Different Modes of Heat Transfer.** In the course of time, temperature differences in a body are reduced by heat flowing from regions of higher temperature to those of lower temperature. This process takes place in all substances which are found in nature, in liquids and gases as well as in solid bodies. Knowledge of the laws governing this process is of great importance throughout the entire field of engineering, since this knowledge affords means for channeling the flow of heat in a desirable manner.

For instance, one may be given the task of improving upon the flow of heat by any possible means. In the development of heat engines we are confronted with this task again and again. According to thermodynamic theory, such a heat engine consists in principle of two heat reservoirs at different temperatures, with the engine that performs the work placed between them. The working medium frequently changes in the course of such a process. The heat must then be exchanged between the individual media using the smallest possible temperature drop. In steam power plants the heat is contained initially as internal energy in the combustion gases. In the steam boiler the heat is transferred to the steam. In the condenser the steam gives off its heat to the cooling water, and the cooling water while passing through the cooling tower transfers this heat to the air. In internal-combustion engines this type of heat exchange does not exist, since the heat is produced by combustion directly in the working medium. Some of this heat is converted to work, and the waste heat as well as the waste gases are jointly exhausted. Nevertheless, a complete understanding of heat transmission is also of great importance in the design of such engines, since here, with the use of air or water coolants, the cylinder walls must be kept at temperatures which can be safely withstood by the material. Therefore the admissible work output of a cylinder of a high-performance engine is determined essentially by cooling problems. For gas turbines, now entering at an increasing rate a stage of economical and reliable industrial application, the mastery of heat-transmission problems is likewise of fundamental significance.

On the other hand, we often face the task of preventing an undesirable amount of heat transfer in so far as is possible. This can be done by applying an insulating layer of a material with a low coefficient of thermal conductivity. The prevention of heat losses is also of far-reaching influence on the efficiency of thermal processes. According to the second law of thermodynamics, a quantity of heat can be the more completely converted to work the higher the temperature. Any decrease of the temperature without performance of work is thus undesirable.

In a solid body the flow of heat is the result of the transfer of internal energy from one molecule to another. This process is called *conduction*. The same process takes place within liquids and gases. In these substances, however, the molecules are no longer confined to a certain point but constantly change their position, even if the substance is in a state of rest. The heat transfer by this process is also included in the term *heat transfer by conduction*.

In addition there is still another mode of heat transfer in liquids and gases. In such a medium, motions of a macroscopic nature may exist and heat may be transported from one point to another by being carried along as internal energy with the flowing medium. This process is called *heat transfer by convection*.

A third mode of heat transfer is the result of *radiation*. Solid bodies, as well as liquids and gases, are capable of radiating thermal energy in the form of electromagnetic waves and of picking up such radiant energy by absorption.

In industrial processes all three mechanisms often participate simultaneously in the transmission of heat. For a study of heat-transfer processes, however, it is necessary to distinguish clearly among the various modes, since they are subject to different laws. We shall consider the processes of heat conduction, convection, and radiation separately in the following chapters of this book.

**1-3. Thermal Conductivity, Film, and Over-all Heat-transfer Coefficients.** In engineering computations we are often interested in knowing the quantity of heat being exchanged per unit of time between two liquids or two gases at different temperatures if the substances are separated by a wall. According to the statements made in Sec. 1-2 it has already become clear that in this case a superposition of diverse processes occurs. First, the heat in one gas (or one liquid) has to be brought to the separating wall. Subsequently the heat must pass through the separating wall, and finally, from the other wall surface it flows into the cooler gas (or the cooler liquid). This section deals with the computation of this heat flow for the simplest case, namely, a plane wall with temperatures constant with respect to time, in order to introduce the fundamental laws of heat transmission. The individual processes will be thoroughly investigated in the following chapters.

Thus, we consider first a plane wall with thickness $b$, both surfaces of which are kept at different but constant temperatures $t_{w1}$ and $t_{w2}$ in steady state (Fig. 1-1). The quantity of heat which by this temperature difference is caused to flow through the area $A$ of the wall per unit of time is called the *rate of heat flow* and will be designated by $Q$.[1] For this quantity of heat, Fourier's law[2] is valid:

$$Q = \frac{k}{b} A(t_{w1} - t_{w2}) \tag{1-6}$$

where $k$, the *thermal conductivity*, is a property of the substance of which the wall consists. Restrictions to the validity of this equation will be discussed in the section on heat conduction. From Eq. (1-6) its dimension can easily be derived:

$$k = \frac{Qb}{A(t_{w1} - t_{w2})} \quad \text{Btu/hr ft F}$$

The amount of heat penetrating a unit area of the surface per unit of time is called the *specific rate of heat flow*. For this, therefore, the following equation is valid:

$$q = \frac{Q}{A} = \frac{k}{b}(t_{w1} - t_{w2}) \tag{1-7}$$

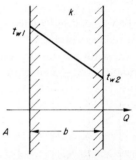

Fig. 1-1. Steady heat conduction through a plane wall.

As illustrated in Fig. 1-1, the temperature within the wall decreases linearly from the value $t_{w1}$ to $t_{w2}$ if the thermal conductivity is independent of the temperature. The thermal conductivities of a number of substances are indicated in the Appendix. There it appears that, among solid bodies, metals have the highest values for thermal conductivities. For instance, iron has a thermal conductivity of approximately 30 Btu/hr ft F, copper of approximately 200 Btu/hr ft F. Metal alloys have thermal-conductivity values substantially lower than pure substances. Stainless steel, for instance, has a value for thermal conductivity of about 9 Btu/hr ft F. Nonmetallic substances have thermal conductivity values of approximately 0.03 to 2 Btu/hr ft F. That of gases is one-tenth as large. Gases, therefore, have the lowest thermal-conductivity values of all substances. The low thermal conductivity of heat-insulating materials (diatomaceous earth, slag wool, peat, cork) is accounted for by their porosity. The flow of heat in these materials, therefore, is essentially a process of heat conduction through the air contained in the pores. The solid substance has only the task of preventing this air from getting into motion as a result of the temperature differences and in this way of avoiding the transport of additional heat by convection. Fourier's law for the heat-conduction process has a striking similarity to Ohm's law for

---

[1] Differing from thermodynamics, where $Q$ denotes an amount of heat without reference to time.

[2] Jean Baptiste Joseph Fourier (1768–1830).

electric currents. This can readily be seen if Eq. (1-6) is stated in the form

$$t_{w1} - t_{w2} = \frac{b}{kA} Q \qquad (1\text{-}8)$$

and compared with Ohm's law:

$$E = RI \qquad (1\text{-}9)$$

The heat flow $Q$ corresponds here to the electric current $I$. The driving force for the heat flow is the temperature difference $(t_{w1} - t_{w2})$; this corresponds to the driving force for the electric current, or the impressed voltage $E$. The expression $b/kA$ is called *the thermal resistance of the heat-conducting process* and is designated by $R_c$ (it corresponds to Ohm's resistance $R$), or

$$R_c = \frac{b}{kA} \qquad (1\text{-}10)$$

The inverse value of thermal conductivity, i.e., the specific resistance to heat conduction, corresponds to the specific resistance in electrical engineering

$$\frac{1}{k} = R_c \frac{A}{b}$$

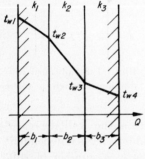

FIG. 1-2. Steady heat conduction through a composite plane wall.

Assuming a plane wall to be composed of several layers, e.g., three, of different materials with thermal conductivities $k_1, k_2, k_3$ and designating the temperatures at the interfaces by $t_{w2}$ and $t_{w3}$ (Fig. 1-2), we can write Eq. (1-8) for each layer:

$$t_{w1} - t_{w2} = \frac{b_1}{k_1 A} Q$$

$$t_{w2} - t_{w3} = \frac{b_2}{k_2 A} Q$$

$$t_{w3} - t_{w4} = \frac{b_3}{k_3 A} Q$$

By adding all the equations we obtain the expression

$$t_{w1} - t_{w4} = \left( \frac{b_1}{k_1 A} + \frac{b_2}{k_2 A} + \frac{b_3}{k_3 A} \right) Q = (R_{c1} + R_{c2} + R_{c3})Q \qquad (1\text{-}11)$$

With this equation the heat flow $Q$ can be calculated from the temperatures of the two surfaces $t_{w1}$ and $t_{w4}$. The resistance to heat conduction of the composite wall equals the sum of the thermal resistances of the single layers. In this case, therefore, the same law can be applied as is used in electrical engineering for resistors connected in series.

In heat exchangers, we are mostly concerned with walls separating liquids or gases from each other. In these cases we do not know the temperatures of both surfaces of the separating walls, but only the temperatures of the liquids on both sides of the wall. In Fig. 1-3 these temperatures are indicated as $t_1$ and $t_2$. By measuring the temperature field in the liquids one obtains the curves shown. The temperature gradient is confined to a relatively narrow layer of thickness $\delta$ quite close to the wall, whereas at a greater distance from the wall in most cases only small temperature differences exist. Simplified, the temperature curve can be replaced by the dashed broken line. This can be explained by assuming that a thin film of liquid (of the thickness $\delta'$) adheres to the wall whereas outside this film all temperature differences vanish as a result of mixing motions of the liquid. This picture oversimplifies the actual process considerably as we shall see later, but it brings out the salient features. Within the film the heat transfer takes place by conduction, as in a solid wall. The temperature in the film, therefore, is again linear, and the flow of heat follows from Eq. (1-6), into which the thermal conductivity $k$ of the liquid or of the gas and the thickness of the film $\delta'$ are to be inserted.

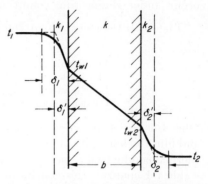

Fig. 1-3. Steady film heat transfer and heat conduction in a plane wall.

Thus, for the heat flow to the wall, the equation

$$Q = \frac{k}{\delta'} A (t - t_w) \qquad (1\text{-}12)$$

is obtained. From this equation the rate of heat flow $Q$ can be calculated as soon as the film thickness $\delta'$ is known. The latter, however, depends to a very great extent upon the external flow conditions, for instance, upon the velocity with which the liquid flows along the wall, upon the shape of the wall, upon the structure of the wall surface, and similar factors. In engineering practice it has become customary to calculate with the expression $k/\delta'$ rather than directly with the film thickness $\delta'$. This value is called *film heat-transfer coefficient* and is designated by the letter $h$. Thus the expression

$$Q = hA(t - t_w) \qquad (1\text{-}13)$$

as formulated by Isaac Newton (1643–1727) is obtained. For some time after its introduction the heat-transfer coefficient $h$ was considered a property of the flowing liquid or gas. It was not until the more recent developments in the theory of heat transmission that the intricate relations became apparent which determine its value. In this book a chapter of greater length is devoted to the evaluation of the heat-transfer coefficients. The orders of magnitude of the heat-transfer coefficients as they occur under engineering conditions have been compiled in Table 1-1.

TABLE 1-1. ORDER OF MAGNITUDE OF FILM HEAT-TRANSFER COEFFICIENTS

$Btu/hr\,ft^2\,F$

| | |
|---|---|
| Flowing air | 2–50 |
| Flowing water | 100–1000 |
| Boiling water | 500–1000 |
| Condensing water vapor | 1000–5000 |

Since the heat-transfer coefficient is the quotient of thermal conductivity and film thickness, it is to be expected that gases, because of their smaller thermal conductivity, have smaller heat-transfer coefficients than liquids. Table 1-1 confirms this conclusion.

Applying Eq. (1-13) to the two surfaces in Fig. 1-3 produces

$$Q = h_1 A (t_1 - t_{w1})$$
$$Q = h_2 A (t_{w2} - t_2)$$

Both these heat quantities must be equal in steady state. The equations, too, can be put into the form corresponding to Ohm's law:

$$t_1 - t_{w1} = \frac{1}{h_1 A} Q$$
$$t_{w2} - t_2 = \frac{1}{h_2 A} Q \qquad (1\text{-}14)$$

The value $1/hA$ is called *thermal resistance* $R_t$ of the film heat-transfer process

$$R_t = \frac{1}{hA} \qquad (1\text{-}15)$$

By writing down Eqs. (1-8) and (1-14) and summing all the equations a relation is obtained between the heat flow through the separating wall and the two temperatures $t_1$ and $t_2$ (Fig. 1-3):

$$t_1 - t_{w1} = \frac{1}{h_1 A} Q$$
$$t_{w1} - t_{w2} = \frac{b}{kA} Q$$
$$t_{w2} - t_2 = \frac{1}{h_2 A} Q$$

$$\overline{t_1 - t_2 = \left(\frac{1}{h_1 A} + \frac{b}{kA} + \frac{1}{h_2 A}\right) Q = (R_{t1} + R_c + R_{t2})Q = R_o Q \qquad (1\text{-}16)}$$

INTRODUCTION

The sum of the individual resistances is the thermal resistance $R_o$ of the over-all heat-transfer process. For this process also, the same law applies as for electric resistances connected in series. Instead of the resistances, technical calculations in most cases employ the thermal conductivities and heat-transfer coefficients. For these, the following relations are valid, which result readily from Eq. (1-16):

$$Q = UA(t_1 - t_2) \qquad (1\text{-}17)$$

$$\frac{1}{U} = \frac{1}{h_1} + \frac{b}{k} + \frac{1}{h_2} \qquad (1\text{-}18)$$

The value $U$ is called the *over-all heat-transfer coefficient*. It has the dimension Btu/hr ft² F. Considering a wall consisting of several layers with different thermal conductivities $k_i$ and thicknesses $b_i$ the middle term in Eq. (1-18) is replaced by the sum $\Sigma b_i/k_i$.

**Example 1-1.** A plane iron wall, 0.5 in. thick, is washed by air on both sides, the heat-transfer coefficients $h_1 = h_2$ being 2 Btu/hr ft² F (Table 1-1). The values of the over-all thermal resistance and of the over-all heat-transfer coefficient are to be calculated. The heat conductivity of the wall is $k = 30$ Btu/hr ft F. According to Eq. (1-16)

$$R_o = \frac{1}{A}\left(\frac{1}{h_1} + \frac{b}{k} + \frac{1}{h_2}\right) = \frac{1}{A}\left(\frac{1}{2} + \frac{1}{24 \times 30} + \frac{1}{2}\right) = \frac{721}{720}\frac{1}{A} \qquad \text{hr F/Btu}$$

The over-all heat-transfer coefficient is $U = 1/AR_o = 1$ Btu/hr ft² F. The thermal resistance to heat conduction of the iron wall is seen to be relatively negligible for this particular heat passage. It would be pointless to consider decreasing its resistance if a substantial increase in the heat flow is desired. To accomplish this, the heat-transfer coefficient must be enlarged.

**Example 1-2.** An iron wall, 0.5 in. thick, and an aluminum wall, 1.0 in. thick, are laid one upon the other, leaving an air space of 0.0005 in. between them. The thermal resistance of this composite wall is to be calculated. According to Eq. (1-11) and Tables A-1 and A-4 (Appendix),

$$R_o = \frac{1}{A}\left(\frac{b_1}{k_1} + \frac{b_2}{k_2} + \frac{b_3}{k_3}\right) = \frac{1}{A}\left(\frac{0.5}{12 \times 30} + \frac{0.0005}{12 \times 0.015} + \frac{1}{12 \times 118}\right)$$

$$= \frac{1}{A}\left(\frac{1}{720} + \frac{1}{360} + \frac{1}{1{,}416}\right) \qquad \text{hr F/Btu}$$

The thermal resistance of the composite wall above is therefore influenced essentially by the air space, which, however small, cannot be avoided even by the most careful construction. The resistance without an air space would be

$$\frac{1}{A}\left(\frac{1}{720} + \frac{1}{1{,}416}\right) \qquad \text{hr F/Btu}$$

which is less than one-half the above.

**1-4. Parallel Flow, Counterflow, Crossflow.** The preceding section treated heat exchange between two gases or liquids on the assumption that the temperatures on both sides of the heating surface were constant over the surface $A$. In reality, the temperatures of the two liquids usually vary, while passing along the heating surface, as a result of the heat exchange. Hence, in order to apply the formulas of the previous section, a mean temperature difference is to be inserted into these formulas. This mean temperature difference will now be calculated. The first case to be considered shows both liquids flowing past the heating surface in the same direction (Fig. 1-4); this is called *parallel flow*. Plotted also in Fig. 1-4 are the temperatures of both liquids as they flow along the heating surface $A$. Now consider a surface element of the area $dA$. The temperatures of the liquids at this point are $t_1$ and $t_2$. The temperature difference between the two liquids is to be denoted by $\Delta t$. Hence, the equation

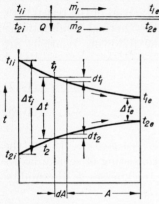

FIG. 1-4. Over-all heat transfer in parallel flow.

$$\Delta t = t_1 - t_2 \tag{1-19}$$

is valid. The heat flow $dQ$ through the area $dA$ is given by Eq. (1-17); i.e., with the symbols applied here,

$$dQ = U\, dA\, \Delta t \tag{1-20}$$

The over-all heat-transfer coefficient $U$ is assumed constant along the heating surface. Because of the heat exchange, the hotter liquid cools by the amount $dt_1$. For this the relation

$$dQ = -m_1 c_1\, dt_1 \tag{1-21}$$

is valid if $m_1$ is the mass of liquid flowing along the heating surface per unit of time (the mass rate of flow) and $c_1$ is its specific heat. The factor $m_1 c_1$ represents the water equivalent or heat capacity of the mass rate of liquid flow. According to the equation

$$dQ = m_2 c_2\, dt_2 \tag{1-22}$$

the colder fluid is heated in the same manner by the amount $dt_2$; $m_2$ and $c_2$ are the mass rate of flow and the specific heat of the second fluid, respectively. By differentiating Eq. (1-19), one obtains

$$d\,(\Delta t) = dt_1 - dt_2 \tag{1-23}$$

When the temperature differentials from Eqs. (1-21) and (1-22) are substituted herein,

## INTRODUCTION

$$d(\Delta t) = -\left(\frac{1}{m_1 c_1} + \frac{1}{m_2 c_2}\right) dQ = -\mu \, dQ \tag{1-24}$$

if the symbol $\mu$ is chosen instead of the expression in parentheses. Equation (1-24) can be integrated immediately. Expressing the temperature difference at the beginning (intake) of the heating surface $A$ by $\Delta t_i$ and at the end by $\Delta t_e$,

$$\Delta t_i - \Delta t_e = \mu Q \tag{1-25}$$

Inserting the heat flow $dQ$ from Eq. (1-20) into Eq. (1-24), the relation

$$\frac{d(\Delta t)}{\Delta t} = -\mu U \, dA \tag{1-26}$$

is obtained, and integration across the entire heating surface $A$, with consideration of the boundary condition ($A = 0$ for $\Delta t = \Delta t_i$), gives

$$\ln \frac{\Delta t_e}{\Delta t_i} = -\mu U A \tag{1-27}$$

The temperature difference at the end of the heating surface, therefore, can be calculated with the aid of the equation

$$\Delta t_e = \Delta t_i e^{-\mu U A} \tag{1-28}$$

By replacing the value $\mu$ in Eq. (1-25) with that from Eq. (1-27) we obtain for the heat flow through the entire heating surface $A$ the relation

$$Q = UA \frac{\Delta t_i - \Delta t_e}{\ln \Delta t_i/\Delta t_e} \tag{1-29}$$

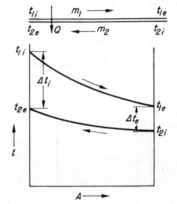

FIG. 1-5. Over-all heat transfer in counterflow.

The fraction on the right-hand side of the equation represents the mean temperature difference $\Delta t_m$ which we sought to find through the above calculation. The rate of heat flow, therefore, can be calculated with the aid of the following formulas:

$$Q = UA \, \Delta t_m \tag{1-30}$$

$$\Delta t_m = \frac{\Delta t_i - \Delta t_e}{\ln \Delta t_i/\Delta t_e} \tag{1-31}$$

Another group of heat exchangers is built in such a way that the two liquids flow in opposite directions, as shown in Fig. 1-5. This arrangement is called *counterflow*. In this case, calculation of the mean temperature difference is accomplished in the same manner as in the case of

TABLE 1-2. Ratio $a$ of the Logarithmetic Mean Temperature Difference $\Delta t_m$ to the Arithmetic Mean $\Delta t_M$ for Parallel Flow and Counterflow*

$$\Delta t_m = a \, \Delta t_M$$

| $\dfrac{\Delta t_i}{\Delta t_e}$ | $a$ | $\dfrac{\Delta t_i}{\Delta t_e}$ | $a$ | $\dfrac{\Delta t_i}{\Delta t_e}$ | $a$ | $\dfrac{\Delta t_i}{\Delta t_e}$ | $a$ |
|---|---|---|---|---|---|---|---|
| 1.0 | 1.000 | 2.0 | 0.962 | 3.0 | 0.910 | 6.0 | 0.798 |
| 1.2 | 0.998 | 2.2 | 0.952 | 3.5 | 0.889 | 7.0 | 0.770 |
| 1.4 | 0.991 | 2.4 | 0.942 | 4.0 | 0.867 | 8.0 | 0.748 |
| 1.6 | 0.981 | 2.6 | 0.928 | 4.5 | 0.846 | 9.0 | 0.729 |
| 1.8 | 0.971 | 2.8 | 0.918 | 5.0 | 0.829 | 10.0 | 0.710 |

* The values in the table are valid also for the reciprocal values $\Delta t_e/\Delta t_i$.

parallel flow, except that in Eq. (1-22) a minus sign appears, since the cooler liquid likewise cools off when proceeding along the heating surface $A$ in a positive direction. The value $\mu$ therefore is

$$\mu = \frac{1}{m_1 c_1} - \frac{1}{m_2 c_2} \tag{1-32}$$

Formulas (1-28) to (1-31) of the preceding paragraph remain unchanged and apply also to counterflow.

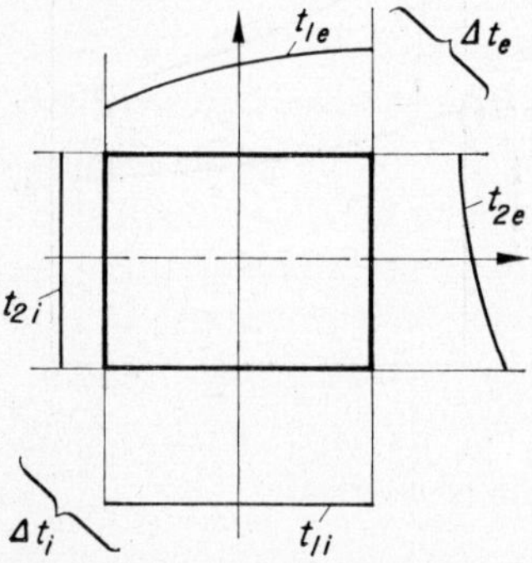

The mean temperature difference calculated according to Eq. (1-31) is always smaller than the arithmetic mean value $\Delta t_M$ of the initial and final temperature differences:

$$\Delta t_M = \frac{\Delta t_i + \Delta t_e}{2} \tag{1-33}$$

Fig. 1-6. Over-all heat transfer in crossflow.

The ratio $a$ of the logarithmic mean of Eq. (1-31) to the arithmetic mean is dependent on the ratio $\Delta t_i/\Delta t_e$ of the temperature differences as shown in Table 1-2. This table can be employed for a simpler determination of the logarithmic mean value by calculating the arithmetic mean and then multiplying this value by the factor $a$ indicated in the table.

A third method of guiding liquids in a heat exchanger is to cause the two liquids to flow at right angles to each other. This is called *crossflow* and is illustrated in Fig. 1-6. As shown there, one liquid flows in front of the heating surface, the other one behind it. The velocities are assumed locally uniform. The calculation of the mean temperature difference is considerably more difficult in this case than in the case of parallel flow or counterflow. Such a calculation was carried out by W. Nusselt.[1]

As shown in Fig. 1-6, the temperatures of the two liquids when leaving the heating surface are not uniform across the cross section of the outlet. The mean outlet temperatures are now expressed by the symbol $t_e$. The value

$$\Delta t_i = t_{1i} - t_{2i} \tag{1-34}$$

is to be understood as the initial temperature difference. In this equation $t_{1i}$ represents the inlet temperature of the hot fluid and $t_{2i}$ represents the inlet temperature of the cold fluid. Accordingly, the final temperature difference is given by the expression

$$\Delta t_e = t_{1e} - t_{2e} \tag{1-35}$$

with $t_{1e}$, the mean outlet temperature of the substance giving off heat and $t_{2e}$, the mean outlet temperature of the substance absorbing heat. The mean temperature difference in this case depends not only on the ratio $\Delta t_i/\Delta t_e$ but also on the ratio $m_1c_1/m_2c_2$ of the heat capacities of the two liquids. From Table 1-3, calculated on the basis of the results of Nusselt's investigation, the ratio of the mean temperature difference as defined by Eq. (1-30) $\Delta t_m$ to the arithmetic mean $\Delta t_M$ can be found, and therewith the mean temperature difference can be determined as was done earlier in the case of parallel flow and counterflow.

The smallest heating surface for a given rate of heat flow and at given initial and final temperature differences is obtained with the liquids flowing in counterflow. Parallel flow requires the largest heating surface. Moreover, counterflow is more favorable than parallel flow, because in the case of counterflow the outlet temperature $t_{2e}$ of the heated liquid can be raised to a temperature higher than the outlet temperature $t_{1e}$ of the heating liquid (Fig. 1-5). The same result can be attained with crossflow (in the region of the negative values of $\Delta t_e/\Delta t_i$ in Table 1-3). With respect to the size of the heating surface, crossflow lies between the two other possibilities.

The difference in the heating surfaces is greatest when the heat capacities of both liquids have the same magnitude. When the heat capacity of one liquid is greater, the difference in the heating surfaces becomes smaller. If the heat capacity of one liquid becomes infinite, there no longer exists any difference in the size of the heating surfaces. This case is illustrated in Fig. 1-7. Here the temperature of the one liquid remains constant along the heating surface, and really, in this case one cannot speak of parallel flow, counterflow, or crossflow any more. Practically,

[1] W. Nusselt, *Z. Ver. deut. Ingr.*, **55**:2091 (1911), and *Forsch. Gebiete Ingenieurw.*, **1**:417 (1930).

TABLE 1-3. RATIO $a$ OF THE TRUE MEAN TEMPERATURE DIFFERENCE $\Delta t_m$ TO THE ARITHMETIC MEAN $\Delta t_M$ FOR CROSSFLOW (FROM VALUES CALCULATED BY W. NUSSELT)*

$\Delta t_m = a\, \Delta t_M$

| $\Delta t_e/\Delta t_i$ \\ $m_1c_1/m_2c_2$ | 0.9 | 0.8 | 0.7 | 0.6 | 0.5 | 0.4 | 0.3 | 0.2 | 0.1 |
|---|---|---|---|---|---|---|---|---|---|
| 0 | 0.998 | 0.993 | 0.986 | 0.978 | 0.962 | 0.939 | 0.902 | 0.836 | 0.707 |
| 0.2 | 0.998 | 0.993 | 0.987 | 0.981 | 0.971 | 0.958 | 0.936 | 0.902 | 0.848 |
| 0.5 | 0.998 | 0.993 | 0.988 | 0.982 | 0.977 | 0.968 | 0.945 | 0.935 | 0.913 |
| 1 | 0.998 | 0.993 | 0.989 | 0.983 | 0.978 | 0.974 | 0.961 | 0.948 | 0.933 |

| $\Delta t_e/\Delta t_i$ \\ $m_1c_1/m_2c_2$ | 0 | $-0.1$ | $-0.2$ | $-0.3$ | $-0.4$ | $-0.5$ | $-0.6$ | $-0.7$ | $-0.8$ |
|---|---|---|---|---|---|---|---|---|---|
| 0 | 0 | | | | | | | | |
| 0.2 | 0.762 | 0.630 | | | | | | | |
| 0.5 | 0.874 | 0.819 | 0.750 | 0.632 | 0.519 | | | | |
| 1 | 0.912 | 0.876 | 0.835 | 0.766 | 0.710 | 0.618 | 0.500 | 0.380 | 0.220 |

* The values in this table are valid also for the reciprocal values $m_2c_2/m_1c_1$.

the temperature curve shown in Fig. 1-7, for instance, occurs with evaporation or condensation of a liquid.

If the heat capacity is the same for both liquids in case of counterflow, the initial temperature difference $\Delta t_i$ becomes, according to Eq. (1-25),

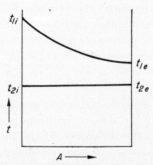

FIG. 1-7. Heat transfer with infinite heat capacity of one fluid.

equal to the final temperature difference $\Delta t_e$, as Eq. (1-32) in this case gives $\mu = 0$. The logarithmic and the arithmetic mean likewise have the same magnitude:

$$\Delta t_m = \Delta t_M = \Delta t_i \qquad (1\text{-}36)$$

INTRODUCTION 19

In this case, the temperatures change linearly along the heating surface. Hence, temperature curves result as shown in Fig. 1-8.

The heating surface need not have the shape of a plane wall as shown in Figs. 1-4 to 1-8. The cooling of liquids and gases while passing through a bundle of tubes (Fig. 1-9) can be calculated with the previously derived formulas. These relations, however, are strictly valid only when the number of tube rows is very large, though for engineering purposes the same formulas are usually applied even if there is a small number of tube rows.

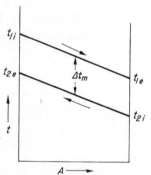

FIG. 1-8. Counterflow heat transfer with equal heat capacity of both fluids.

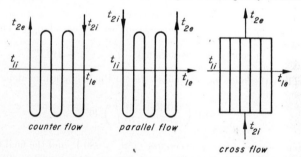

FIG. 1-9. Counterflow, parallel flow, and crossflow on tube bundles.

In the preceding calculations, the over-all heat-transfer coefficient $U$ was assumed constant along the heating surface. As this frequently is not the case, it is necessary to divide the entire heating surface into sections. The over-all heat-transfer coefficient varies over each of these sections, but so slightly that it can be assumed to be constant. Thus, these individual sections can be calculated with the given formulas.

Heat exchange in parallel flow and counterflow has also been treated for the case in which the over-all heat-transfer coefficient varies linearly with temperature.[1] The formula

[1] J. Nikuradse, *Proc. Intern. Congr. Appl. Mech.*, 3rd Congr., **1**:239 (Stockholm, 1950).

$$Q = A \frac{U_e \, \Delta t_i - U_i \, \Delta t_e}{\ln U_e \, \Delta t_i / U_i \, \Delta t_e}$$

now takes the place of Eqs. (1-30) and (1-31). Alternatively, we can use an average over-all heat-transfer coefficient for such calculations. The reference temperature for this can be taken from diagrams.[1]

Besides the combinations dealt with in this chapter and represented in Fig. 1-9, others, such as mixed parallel flow and counterflow, are encountered in practice. The temperature differences to be inserted into Eq. (1-30) for a large number of these combinations were treated in an extensive investigation and plotted in diagrams by Bowman, Mueller, and Nagle.[2] Some additional information will also be found in a later chapter on heat-exchanger calculations.

**Example 1-3.** $m_1 = 100$ lb/hr of water is to be heated from 50 to 170 F with flue gases having an initial temperature of 330 F. The mass rate of flow of the flue gases is $m_2 = 400$ lb/hr; their specific heat $c_p = 0.25$ Btu/lb F; the over-all heat-transfer coefficient $U = 20$ Btu/hr ft$^2$ F. The size of the heating surface $A$ is to be calculated for parallel flow, counterflow, and crossflow.

First, the cooling of the flue gases can be found for each case with the aid of a heat balance. The heat absorption of the water is

$$100(170 - 50) = 12,000 \text{ Btu/hr}$$

When there are no heat losses, the heat delivery of the flue gases must be equal to heat absorbed by the water. From this relation we can calculate the final temperature of the gases:

$$12,000 = 400 \times 0.25(330 - t_e) \qquad t_e = 210 \text{ F}$$

With *parallel* flow the following scheme is valid for the temperatures along the heating surface:

$$280 \text{ F} \left( \begin{array}{c} 330 \text{ F} \rightarrow 210 \text{ F} \\ 50 \text{ F} \rightarrow 170 \text{ F} \end{array} \right) 40 \text{ F}$$

The initial temperature difference therefore is $\Delta t_i = 280$ F, and the final temperature difference $\Delta t_e = 40$ F. The arithmetic mean becomes $\Delta t_M = (280 + 40)/2 = 160$ F, and the ratio $\Delta t_i / \Delta t_e = 280/40 = 7$. The value $a = 0.77$ is taken from Table 1-2. Thus we obtain the logarithmic mean temperature difference $\Delta t_m = 0.77 \times 160 = 123$ F. From Eq. (1-30) it follows that

$$A = \frac{12,000}{20 \times 123} = 4.88 \text{ ft}^2$$

For *counterflow* we obtain from the scheme

$$160 \text{ F} \left( \begin{array}{c} 330 \text{ F} \rightarrow 210 \text{ F} \\ 170 \text{ F} \leftarrow 50 \text{ F} \end{array} \right) 160 \text{ F}$$

[1] A. P. Colburn, *Ind. Eng. Chem.*, **25**:873–877 (1933).
[2] R. A. Bowman, A. C. Mueller, and W. M. Nagle, *Trans. ASME*, **62**:283–294 (1940).

where $\Delta t_i = \Delta t_e = 160$ F. Thus $\Delta t_i/\Delta t_e = 1$; therefore $a = 1$. Hence, the logarithmic mean is $\Delta t_m = 160$ F and

$$A = \frac{12{,}000}{20 \times 160} = 3.75 \text{ ft}^2$$

According to the following scheme, for *crossflow* there is

$$\Delta t_i = 280 \text{ F} \qquad \Delta t_e = 40 \text{ F} \qquad \Delta t_M = \frac{280 + 40}{2} = 160 \text{ F}$$

$$\frac{\Delta t_e}{\Delta t_i} = 0.143 \qquad \text{and} \qquad \frac{m_1 c_1}{m_2 c_2} = 1$$

```
              330°
      280°/    |
              |
       50° ———+———> 170°
              |
              | /40°
              v
             210°
```

It follows by interpolation from Table 1-3 that $a = 0.939$ and therewith $\Delta t_m = 0.939 \times 160 = 150$ F. The size of the heating surface thereby becomes

$$A = \frac{12{,}000}{20 \times 150} = 4 \text{ ft}^2$$

The results show that counterflow requires the smallest and parallel flow the largest heating surface. The size of the heating surface for parallel flow increases to a very considerable extent with greater heating of the water. At a water outlet temperature of 190 F, the final temperature difference will be $\Delta t_e = 0$ and therefore the heating surface will be infinite. Hence, it follows that an increase above this temperature cannot be obtained by parallel flow, but only by counterflow or crossflow.

## PROBLEMS

**1-1.** It has been shown in this chapter that the thermal resistance of a composite wall can be calculated in the same way as electric resistances in series. Demonstrate that a similar analogy exists between the electric resistance of resistors in parallel and the thermal resistance of a wall of thickness $b$ composed of various materials such that the surface area is subdivided into parts $A_1, A_2, \ldots, A_n$ and the material is uniform throughout the thickness $b$ behind each part.

**1-2.** 20 lb/sec of water is heated from 80 to 212 F, then evaporated and superheated to 300 F in a counterflow heat exchanger, all processes being at a pressure of 14.7 psia. Combustion gases at a flow rate of 100 lb/sec are used as the heating medium with an entering temperature of 1200 F. If the over-all heat-transfer coefficient is constant over the whole heat-transfer area and has a value of 35 Btu/hr ft² F, find the necessary total area of the heat exchanger. (Use properties of air for the combustion gases.)

**1-3.** Set up the differential equations by which the temperature field in a heat exchanger with crossflow can be calculated, and state the boundary conditions. How do the differential equations change when one fluid is constantly mixed within each cross section normal to the flow direction?

PART A

# HEAT CONDUCTION

In the next four chapters we shall examine the physical phenomena, the fundamental laws, the physical properties, and the characteristic mathematical formulations that are significant in the process of heat conduction in matter. The flow of heat in solids takes place exclusively by the conduction process, while in liquids and gases the processes of conduction, convection, and radiation occur simultaneously. In what follows, therefore, the major interest will be with solids, and the results of the problems considered in the chapters on heat conduction can be applied to all solid bodies. In certain specific cases, where heat exchange by convection is prevented and exchange by radiation is minimized, the principles of heat conduction can be applied to liquids and gases as well.

The treatment of heat conduction in solids has been separated into four specific topics: (1) theory of heat conduction and the heat-conduction equations, (2) steady heat conduction, (3) unsteady heat conduction, and (4) heat conduction with moving boundaries. These subdivisions are necessarily arbitrary but serve as an aid in the presentation of the material.

Heat conduction from the macroscopic, phenomenological point can be easily understood without a companion understanding of the microscopic notions of the mechanism of heat conduction in solids as proposed by the theories of solid-state physics. This is not to say that the theories of solid-state physics are unimportant, for many advances of the future will be in that area. An acceptable understanding of the science of heat and heat transfer requires complete familiarity with both the microscopic and macroscopic points of view. In the chapters that follow, heat conduction will be treated in the macroscopic, phenomenological sense. In such a treatment it is well to realize that although the physical notions are simple, the mathematical techniques which are required to obtain usable results, i.e., temperature distributions, heat rates, temperature-time histories etc., are generally complex and in many cases difficult. The mathematical complexities have not been avoided; however, an attempt was made to keep the presentation from becoming a mathematical treatise. For more complicated problems and problems with diverse boundary conditions, the reader is referred to the excellent book by Carslaw and Jaeger.[1]

With an exception or two, the thermophysical properties which occur in the heat-conduction problems have been considered to be independent of temperature. Such practice not only simplifies the mathematical treatment but also is a reasonable approximation in many physical problems where the temperature variation is not large. In problems involving chemical reactions or phase changes, the neglect of temperature-dependent properties may be a serious omission. Therefore, each problem must be carefully considered physically before the assumption of constant properties is applied.

Certain heat-conduction problems involve the convection (or radiation) mode of heat transfer, usually in the statement of some boundary condition. In the consideration of heat-conduction problems in which heat convection plays a part, it will be assumed that the heat-transfer coefficients are known. The nature of these heat-transfer coefficients and the methods for their determination are considerations for the chapters of this book on convection heat transfer.

[1] H. S. Carslaw and J. C. Jaeger, "Conduction of Heat in Solids," Oxford University Press, London, 1947.

CHAPTER 2

# THEORY OF HEAT CONDUCTION AND HEAT-CONDUCTION EQUATION

**2-1. Concept of Heat Conduction.** The currently accepted theory of heat is closely associated with the internal energy of matter which in thermodynamics is referred to as the energy related to the physical and chemical state of the body—the orientation and motion of the molecules and atoms within the body. Although incomplete, the dynamic theory of heat permits some important conclusions to be drawn which are quite generally confirmed by experiment:

1. Since heat as energy is associated with translational, rotational, and vibrational motions of the molecules, atoms, and their components, heat transfer by conduction must be strictly related to these motions.

2. Increased temperature increases the intensity and frequency of molecular and atomic motions; therefore, the conduction of heat should increase with increasing temperature.[1]

In accordance with the theories of heat and structure of matter, it is generally accepted today that heat conduction in amorphous solids, liquids, and gases is the result of direct transfer of molecular (or atomic) motions from molecule to molecule at the contact areas. This type of heat transfer is frequently visualized as a process of heat diffusion. In more organized matter, for example in crystals, the atomic motions are converted into a vibratory motion of the whole crystal lattice.

*Solids.* The current theory of heat conduction in solids makes a clear distinction between dielectric (nonmetallic) and metallic substances. In the dielectric, the transfer of heat is accomplished by means of lattice waves produced by atomic motions. In metals and other solid electrical conductors the heat carriers comprise the lattice waves and the free

---

[1] There are exceptions to this statement:
1. Thermal conductivity and, thus, heat conduction of dielectric crystals and metals decrease from a maximum value at very low temperatures toward the melting point.
2. Thermal conductivity of most liquids decreases with temperature; however, in some liquids such as water it increases with temperature in one temperature range and decreases with temperature in another. See Appendix.

electrons; thus these substances in general have the larger thermal conductivities.

*Liquids.* The process of phase change from the solid to the liquid state suggests a change from a relatively orderly state of molecular arrangement to a more disorderly state. This phase change produces some important changes in the molecular structure. It loosens the molecular bonds, destroys the order of the solid state, and creates the possibility of thermal movement of the molecules. These changes lead one to the conclusion that a liquid might be quite similar to a gas in that the molecules in the liquid are completely random but oriented at smaller average intermolecular distances than gaseous molecules. This conclusion is not entirely warranted, since near the melting point the liquid and solid states cannot be very different and, consequently, there can be little possibility for free molecular movement. For this reason modern theories propose that the structure of liquids is more like that of solids than that of gases and retains a semblance of order to a large extent. This notion is substantiated by experimental efforts such as X-ray diffraction measurements of liquids.

*Gases.* The evaporation process loosens the intermolecular bonds which existed in the liquid state and increases the intermolecular distance to such a degree that the molecules are free to move in any direction, the only obstacles in their path being other molecules with which they may collide. The gaseous state has, at least far from any boundary surface, a completely random molecular distribution. In this instance, all the properties and behavior of gases can be adequately explained in terms of the kinetic theory of gases. In other words, the heat conduction in gases may be compared with the processes of molecular diffusion from hotter to colder regions but is restricted to the exchange of location and energy of molecules.

It can be seen that there is still some conjecture about the physical mechanism of thermal-energy transfer; however, in any event and for any theory, the energy in transition is referred to as *heat* and the process of the energy transition is known as *conduction*.

Even though the actual mechanism of heat conduction is imperfectly understood, the hypothesis upon which the science of heat conduction is founded was based on experimental observations. Subsequent use of this hypothesis as a basis for mathematical analysis to obtain results which have been experimentally verified is sufficient to establish the particular law which is characteristic of the transfer itself. The basic law so established is entirely consistent with classical thermodynamics. The concern here will be less with the physical mechanism of the heat-conduction process and more with the application of the basic law of heat conduction to heat-transfer systems.

## 2-2. The Fundamental Law of Heat Conduction.

To be consistent thermodynamically is to require, by virtue of the second law of thermodynamics, that heat will be transferred from one body to another body (or from one part of a body to another part of the same body) only when the bodies are at different temperatures and that the heat will flow from the highest temperature to the lowest temperature. This states specifically that a temperature gradient exists and that the heat flows in the direction of decreasing temperature. The first law of thermodynamics (i.e., conservation of energy) states that the flowing thermal energy is conserved in the absence of heat sources or sinks. To this end a solid may have a temperature distribution which is dependent upon the space coordinates and time of observation:

$$t = f(x,y,z,\tau)$$

We may suppose that within this solid is a surface such that, when observed at a certain time, each point on it has an identical temperature. Such a surface is called an *isothermal surface*. We can further visualize other isothermal surfaces within this body which differ from one another by being hotter or colder by amounts $\pm \delta t$, respectively. These isothermal surfaces never intersect because no point in this solid can exist at two different temperatures at the same time. The solid is thus visualized as being composed of a number of arbitrarily thin isothermal shells that, of course, vary with time.

In that which follows, unless stated differently, we shall consider only *isotropic solids*, that is, solids whose properties and constitution in the neighborhood of any point are invariant with the direction from the point. In such a case, and because of the symmetry involved, the heat flow at a point is along a path perpendicular to the isothermal surface through the point. For a nonisotropic solid the heat-flow direction is not necessarily in a direction perpendicular to the isothermal surface through the point. This situation will be discussed in a subsequent paragraph.

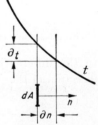

Fig. 2-1. Heat-conduction system.

The mathematical formulation of the law of heat conduction can be expressed as

$$Q = \frac{dQ_i}{d\tau} = -kA\frac{\partial t}{\partial n} \qquad (2\text{-}1)$$

Equation (2-1) can be interpreted with the aid of Fig. 2-1.

The heat flux $Q/A$ flows along the normal $n$ to the area $A$ in the direction of the decreasing temperature, i.e., the negative thermal gradient. The negative sign in Eq. (2-1) indicates that the heat flow is in the direction of the negative gradient and serves to make the heat flux positive in this sense. The proportionality factor is $k$, the thermal conductivity, and is a property of the material through which the heat flows.

Equation (2-1) can be rewritten in the case of an infinitesimal area as

$$dQ = -k\, dA\, \frac{\partial t}{\partial n} \qquad (2\text{-}2)$$

Equations (2-1) and (2-2) are generally attributed to the French mathematician *Jean Baptiste Fourier* and in his honor are designated the *Fourier heat-conduction equations*.

The heat flow per hour per unit area across any surface is called the *heat flux* $q$ and has units of Btu/hr ft$^2$. The heat flux is a vector; that is, it must be specified as to magnitude and direction.

The heat flux can be calculated for any point in reference to any arbitrary direction through the point if the area normal to this direction is considered.

In Fig. 2-2 are shown the isotherms $t$ and $t + dt$ in a body. The normal to these isotherms is designated by the axis $n$, which is also the normal to the differential area $dA$. The heat flux can be calculated in the direction of the normal and in the direction $s$ as shown below:

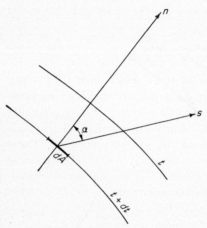

FIG. 2-2. Direction of heat flow.

$$q_n = \frac{dQ}{dA} = -k\frac{\partial t}{\partial n}$$

$$q_s = \frac{dQ}{dA\, \cos\alpha} = -k\frac{\partial t}{\partial s}$$

It is easily shown that $n = s\cos\alpha$. Therefore

$$q_s = -k\,\frac{\partial t}{\partial n}\cos\alpha \qquad (2\text{-}3)$$

Or in other words, $q_s$ is a component of the heat-flux vector $q_n$. It can be seen also from Eq. (2-3) that the greatest heat flux is that which is calculated along the normal to the isothermal surfaces. In particular, if the component fluxes are related to the planes containing the $x$, $y$, $z$ coordinate system, the fluxes are

$$q_x = -k\frac{\partial t}{\partial x} \qquad q_y = -k\frac{\partial t}{\partial y} \qquad q_z = -k\frac{\partial t}{\partial z} \tag{2-4}$$

The fluxes shown in Eq. (2-4) are components of the heat-flux vector

$$\mathbf{q} = \mathbf{i}q_x + \mathbf{j}q_y + \mathbf{k}q_z$$

*Thermal Conductivity.* It should be noted here that the thermal conductivity $k$ is not necessarily a constant but, in fact, is a function of the temperature for all phases and in liquids and gases depends also upon the pressure, especially when near the critical state. The thermal conductivity in wood and crystals also varies markedly in direction. The thermal conductivity in wood across the grain as compared with along the grain varies by a factor of 2 to 4.

The dependence of thermal conductivity on temperature for small, select temperature ranges can be acceptably expressed in a linear form:

$$k = k_0(1 \pm at) \tag{2-5}$$

where $k_0$ is the value of the thermal conductivity at some reference condition and $a$ is the temperature coefficient and is positive or negative depending upon the material in question. Figure 2-3 shows the effect on the temperature gradient in a body as a result of the positive or negative characteristic of $a$.

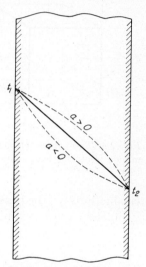

FIG. 2-3. Temperature gradient in a simple slab.

It can be readily seen that a linear temperature gradient exists only when the thermal conductivity is a constant.

It is interesting to note that the Fourier equation for heat conduction is exactly analogous to Ohm's law for an electrical conductor. Ohm's law for a conductor of any shape can be expressed as

$$dI = -\lambda \, dA \, \frac{\partial E}{\partial n} \tag{2-6}$$

In Eq. (2-6) the electrical current $I$ corresponds to the heat flow $Q$, the electrical potential $E$ corresponds to the thermal potential $t$, and the electrical conductivity $\lambda$ ($\lambda = 1/\rho$ where $\rho$ is the electrical resistivity) corresponds to the thermal conductivity $k$. Since Eqs. (2-2) and (2-6) have the same form, the temperature field within a heated body and the field of electrical potential in bodies of the same shape correspond to each other provided the temperature distribution on the surface corresponds to the surface distribution of the electrical potential. This analogy enables one to study heat-conduction problems in accurate detail by means of similar electrical models.

**2-3. The Heat-conduction Equation.** The following section will discuss the establishment of the heat-conduction equation in differential form in the rectangular coordinate system. The differential form of the heat-conduction equation is most advantageous.

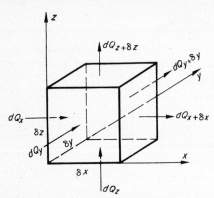

FIG. 2-4. The volume element for determination of the heat-conduction equation.

*Heat-conduction Equation for Isotropic Materials.* Consider an infinitesimal space lattice of dimensions $\delta x$, $\delta y$, and $\delta z$ which is oriented into a three-dimensional coordinate system $x$, $y$, and $z$ as in Fig. 2-4. The considerations here will include the nonsteady condition of temperature variation with time $\tau$.

According to the Fourier heat-conduction law, the heat flowing into the leftmost face of the lattice in the $x$ direction can be expressed as

$$dQ_x = -k \, \delta y \, \delta z \, \frac{\partial t}{\partial x}$$

The value of the heat flow out of the right face of the lattice can be obtained by expanding $dQ_x$ in a Taylor series and retaining only the first two terms as a reasonable approximation:

$$dQ_{x+\delta x} = dQ_x + \frac{\partial}{\partial x}(dQ_x)\,\delta x + \cdots$$

The net heat flow by conduction in the $x$ direction is therefore

$$dQ_x - dQ_{x+\delta x} = \frac{\partial}{\partial x}\left(k\,\frac{\partial t}{\partial x}\right)\delta x \, \delta y \, \delta z \tag{2-7}$$

Two more equations similar to Eq. (2-7) for the $y$ and $z$ directions can be written in the same way:

$$dQ_y - dQ_{y+\delta y} = \frac{\partial}{\partial y}\left(k\,\frac{\partial t}{\partial y}\right)\delta x \, \delta y \, \delta z$$

$$dQ_z - dQ_{z+\delta z} = \frac{\partial}{\partial z}\left(k\,\frac{\partial t}{\partial z}\right)\delta x \, \delta y \, \delta z$$

The sum of the net quantities of heat is the heat which must be stored in the lattice:

$$\left[\frac{\partial}{\partial x}\left(k\,\frac{\partial t}{\partial x}\right) + \frac{\partial}{\partial y}\left(k\,\frac{\partial t}{\partial y}\right) + \frac{\partial}{\partial z}\left(k\,\frac{\partial t}{\partial z}\right)\right]\delta x \, \delta y \, \delta z \tag{2-8}$$

If per unit of time and space the heat quantity $Q'(x,y,z,\tau)$ is generated, then the generation of heat in the lattice is

$$Q' \, \delta x \, \delta y \, \delta z \qquad (2\text{-}9)$$

The heat remaining in the lattice owing to conduction [Eq. (2-8)] and the heat generated within the lattice [Eq. (2-9)] together serve to increase the internal energy of the volume element. Such an increase in the internal energy is reflected in the time rate of change in the heat capacity of the volume element and can be written

$$c\rho \, \delta x \, \delta y \, \delta z \, \frac{\partial t}{\partial \tau} \qquad (2\text{-}10)$$

where $c$ = specific heat
$\rho$ = density
$\tau$ = time

An energy balance can be made on the volume element to equate the time rate of change of heat storage to the net heat flowing into the element owing to conduction and to that heat generated within the element to yield the expression

$$c\rho \frac{\partial t}{\partial \tau} = \frac{\partial}{\partial x}\left(k \frac{\partial t}{\partial x}\right) + \frac{\partial}{\partial y}\left(k \frac{\partial t}{\partial y}\right) + \frac{\partial}{\partial z}\left(k \frac{\partial t}{\partial z}\right) + Q' \qquad (2\text{-}11)$$

It should be noted here that $k = k(x,y,z,t)$, $c = c(x,y,z,t)$, and

$$\rho = \rho(x,y,z,t)$$

so that Eq. (2-11) is valid for isotropic, heterogeneous media.

If the heat-source term can be omitted for the situation where the body is free of sources, Eq. (2-11) can be written in terms of the heat-flux components as

$$-c\rho \frac{\partial t}{\partial \tau} = \frac{\partial}{\partial x}(q_x) + \frac{\partial}{\partial y}(q_y) + \frac{\partial}{\partial z}(q_z) \qquad (2\text{-}12)$$

This formulation is somewhat more general and will be of use in the section on nonisotropic materials.

Equation (2-11) can be further simplified for an isotropic, homogeneous material where $k$, the thermal conductivity, can be considered a constant; thus

$$\frac{\partial t}{\partial \tau} = \frac{k}{\rho c}\left[\frac{\partial^2 t}{\partial x^2} + \frac{\partial^2 t}{\partial y^2} + \frac{\partial^2 t}{\partial z^2}\right] + \frac{Q'}{\rho c} = \alpha \nabla^2 t + \frac{Q'}{\rho c} \qquad (2\text{-}13)$$

The group $k/\rho c$ has the dimensions of length squared per time, is referred to as the thermal diffusivity $\alpha$, and is a property of the conducting material.

HEAT-CONDUCTION EQUATION IN CYLINDRICAL COORDINATE SYSTEM. By means of a transformation of coordinates, Eq. (2-13) can be expressed in a form better suited for cylindrical systems. Thus from Fig. 2-5 for $x = r \cos \theta$, $y = r \sin \theta$, and $z = z$,

$$\frac{\partial t}{\partial \tau} = \alpha \left( \frac{\partial^2 t}{\partial r^2} + \frac{1}{r} \frac{\partial t}{\partial r} + \frac{1}{r^2} \frac{\partial^2 t}{\partial \theta^2} + \frac{\partial^2 t}{\partial z^2} \right) + \frac{Q'}{\rho c} \tag{2-14}$$

HEAT-CONDUCTION EQUATION IN SPHERICAL COORDINATE SYSTEM. A similar transformation yields for the spherical system (Fig. 2-6) for $x = r \sin \psi \cos \phi$, $y = r \sin \psi \sin \phi$, and $z = r \cos \psi$,

$$\frac{\partial t}{\partial \tau} = \alpha \left[ \frac{1}{r} \frac{\partial^2 (rt)}{\partial r^2} + \frac{1}{r^2 \sin \phi} \frac{\partial}{\partial \phi} \left( \sin \phi \frac{\partial t}{\partial \phi} \right) + \frac{1}{r^2 \sin^2 \phi} \frac{\partial^2 t}{\partial \psi^2} \right] + \frac{Q'}{\rho c} \tag{2-15}$$

*Heat-conduction Equation for Nonisotropic Materials.* In the preceding section the heat-conduction equation for isotropic media was derived.

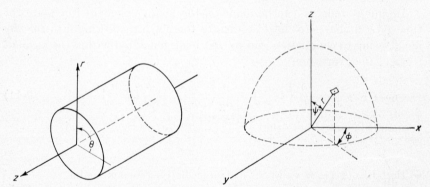

FIG. 2-5. Cylindrical coordinate system.   FIG. 2-6. Spherical coordinate system.

Certain technically important materials and laminates have thermal conductivities that vary markedly, depending upon the direction of heat flow through the body. Included in this category of materials are crystalline substances, woods, laminated plastics, and laminated metals such as are used in transformer cores and plywood. The heat-conduction equation requires certain modifications before these nonisotropic materials can be adequately treated. The general treatment[1] is quite complex and somewhat beyond the scope of this book; however, the fundamental notions in terms of the two-dimensional problem will be considered here.

The thermal conductivity in the two-dimensional case is so oriented that maximum and minimum values occur along the preferred axes, which are called the *principal axes*. Other directions through the body yield values of the thermal conductivity which are between the principal

[1] H. S. Carslaw and J. C. Jaeger, "Conduction of Heat in Solids," p. 27, Oxford University Press, London, 1947.

THEORY OF HEAT CONDUCTION 33

values and are distributed as an ellipse, with the principal values making the major and minor axes.

Consider a material, as shown in Fig. 2-7, where the body is oriented in the $(x,y)$ coordinate system which makes an angle $\beta$ with the principal axis of thermal conductivity of the material. The coordinate system $(\xi,\eta)$ is aligned with the principal axes of conductivity. We can consider

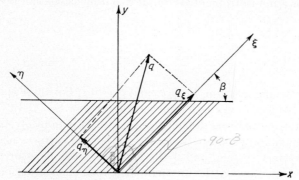

FIG. 2-7. Heat flux in nonisotropic medium.

the heat flux through the body to be made up of the component heat fluxes in the direction of the coordinates $\xi$ and $\eta$:

$$q_\xi = -k_\xi \frac{\partial t}{\partial \xi} \qquad q_\eta = -k_\eta \frac{\partial t}{\partial \eta}$$

The fluxes in the $x$ and $y$ directions are therefore

$$q_x = q_\xi \cos \beta - q_\eta \sin \beta = -k_\xi \cos \beta \frac{\partial t}{\partial \xi} + k_\eta \sin \beta \frac{\partial t}{\partial \eta}$$
$$q_y = q_\xi \sin \beta + q_\eta \cos \beta = -k_\xi \sin \beta \frac{\partial t}{\partial \xi} - k_\eta \cos \beta \frac{\partial t}{\partial \eta} \qquad (2\text{-}16)$$

The temperature gradient can be transformed into gradients in $x$ and $y$ by the following relations:

$$\frac{\partial t}{\partial \xi} = \frac{\partial t}{\partial x}\frac{\partial x}{\partial \xi} + \frac{\partial t}{\partial y}\frac{\partial y}{\partial \xi}$$
$$\frac{\partial t}{\partial \eta} = \frac{\partial t}{\partial x}\frac{\partial x}{\partial \eta} + \frac{\partial t}{\partial y}\frac{\partial y}{\partial \eta}$$

and by the geometry of the figure,

$$y = \xi \sin \beta = \eta \cos \beta \qquad x = \xi \cos \beta = -\eta \sin \beta$$

Substituting these values into Eqs. (2-16) and rearranging, we obtain for the heat fluxes

$$q_x = -(k_\xi \cos^2\beta + k_\eta \sin^2\beta)\frac{\partial t}{\partial x} - (k_\xi - k_\eta)\cos\beta\sin\beta\frac{\partial t}{\partial y}$$
$$q_y = -(k_\xi - k_\eta)\sin\beta\cos\beta\frac{\partial t}{\partial x} - (k_\xi \sin^2\beta + k_\eta \cos^2\beta)\frac{\partial t}{\partial y} \quad (2\text{-}17)$$

With the use of the general form of the conduction equation as shown in Eq. (2-12) and the heat fluxes as given in Eq. (2-17), the heat-conduction equation for a nonisotropic material in two dimensions can be expressed:

$$\rho c \frac{\partial t}{\partial \tau} = (k_\xi \cos^2\beta + k_\eta \sin^2\beta)\frac{\partial^2 t}{\partial x^2} + (k_\xi \sin^2\beta + k_\eta \cos^2\beta)\frac{\partial^2 t}{\partial y^2}$$
$$+ (k_\xi - k_\eta)\sin 2\beta \frac{\partial^2 t}{\partial x\,\partial y} \quad (2\text{-}18)$$

For an isotropic medium, $k_\xi = k_\eta$ and $\beta = 0$. For these specifications Eq. (2-18) reduces to the two-dimensional form of Eq. (2-13).

It is interesting to recognize here that if a sheet of nonisotropic material such as a laminate is held between the isothermal surfaces of a thermal-conductivity test system, and if the laminated sample is prepared such that its principal axes make an angle of $\beta$ with the isothermal surfaces, the thermal conductivities that are measured (depending on whether the measurements were made in the $x$ or $y$ direction) are given by the expressions

$$k_x = k_\xi \cos^2\beta + k_\eta \sin^2\beta$$
$$k_y = k_\xi \sin^2\beta + k_\eta \cos^2\beta$$

If the geometrical axes of the nonisotropic body are oriented with the principal axes of the thermal conductivities, then a simplification of Eq. (2-18) gives

$$\rho c \frac{\partial t}{\partial \tau} = k_\xi \frac{\partial^2 t}{\partial \xi^2} + k_\eta \frac{\partial^2 t}{\partial \eta^2} \quad (2\text{-}19)$$

$\beta = 0$

In the case of a medium such as wood which has different conductivities along the grain $z$, across the grain $r$, and circumferentially $\theta$, we can use the idea of Eq. (2-19) in Eq. (2-14) by orienting the $z$ axis along the center line of the tree and neglecting the heat source term; thus in cylindrical coordinates,

$$\rho c \frac{\partial t}{\partial \tau} = \frac{k_r}{r}\frac{\partial}{\partial r}\left(r\frac{\partial t}{\partial r}\right) + \frac{k_\theta}{r^2}\frac{\partial^2 t}{\partial \theta^2} + k_z\frac{\partial^2 t}{\partial z^2} \quad (2\text{-}20)$$

**Example 2-1.** A slab of laminated material is being used in a thermal-conductivity test. The laminations make an angle $\beta$ with the smooth surfaces of the sample (Fig. 2-8). The surfaces $A$ are kept at constant but different temperatures and thus are isothermal surfaces. It is required to calculate the angle that the heat-flux vector makes with the normal $n$ to the isothermal surfaces. From Fig. 2-8 it may be seen that for $k_\xi > k_\eta$,

$$\tan\gamma = \frac{q_\eta}{q_\xi} = \frac{k_\eta}{k_\xi}\frac{\partial t/\partial \eta}{\partial t/\partial \xi}$$

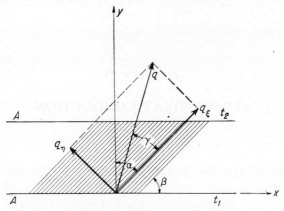

FIG. 2-8. Heat-flux vector in a laminated material.

but from earlier calculations,

$$\frac{\partial t}{\partial \eta} = -\sin \beta \frac{\partial t}{\partial x} + \cos \beta \frac{\partial t}{\partial y}$$

$$\frac{\partial t}{\partial \xi} = \cos \beta \frac{\partial t}{\partial x} + \sin \beta \frac{\partial t}{\partial y}$$

thus, 
$$\tan \gamma = \frac{k_\eta [\cos \beta (\partial t/\partial y) - \sin \beta (\partial t/\partial x)]}{k_\xi [\cos \beta (\partial t/\partial x) + \sin \beta (\partial t/\partial y)]} \qquad (a)$$

In the system we are considering here, however, the bounding surfaces are kept isothermal; thus the coordinate axis $y$ is in the direction of the normal $n$, and $x$ lies in the isothermal surface. Thus $\partial t/\partial x = 0$, and expression (a) becomes

$$\tan \gamma = \frac{k_\eta}{k_\xi} \cot \beta = \frac{k_\eta}{k_\xi} \tan \alpha$$

Therefore $\gamma < \alpha$ and the heat-flux vector is not normal to the isothermal surface as it would be if the material were isotropic. If $k_\xi = 2k_\eta$, as is possible in wood, and if $\beta = 45°$, then

$$\tan \gamma = \tfrac{1}{2}$$
$$\gamma = 26.6°$$

## PROBLEMS

**2-1.** Suggest a method for the measurement of the thermal conductivity of liquid metals at high temperatures. Indicate the essential components of the apparatus by a sketch, and estimate the error of the proposed measurements.

**2-2.** Making use of the cylindrical coordinate system and a small volume element in that coordinate system, develop Eq. (2-14) in the manner of the development of Eq. (2-11).

**2-3.** Making use of the spherical coordinate system and a small volume element in that coordinate system, develop Eq. (2-15) in the manner of the development of Eq. (2-11).

**2-4.** Develop the heat-conduction equation for a nonisotropic medium in three dimensions where the principal conductivities $k_\xi \neq k_\eta \neq k_\zeta$.

CHAPTER 3

# STEADY HEAT CONDUCTION

**3-1. Simple Conduction Equation Solutions for Steady Conditions.**
Whenever the geometrical form of the conducting body is simple and the flow of heat is in one direction only and is invariant with time, the heat-conduction equation can be greatly simplified. Some very important practical cases of heat-conduction problems fall into these simple categories, for example, the slab, the hollow tube, and the hollow sphere.

*The Slab.* In the case of the slab of constant thermal conductivity and a thickness $l$ (Fig. 3-1) which extends to infinity in the other dimensions so that in effect the heat flow in the region considered is truly one-dimensional, it is convenient to treat the system in the rectangular coordinate system. If there are no heat sources (or sinks) in the slab and the heat flow is steady and one-dimensional, Eq. (2-13) becomes

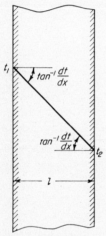

Fig. 3-1. Steady heat conduction in a slab.

$$\frac{d^2 t}{dx^2} = 0 \tag{3-1}$$

Equation (3-1) is readily solved to yield

$$t = C_1 x + C_2$$

The constants $C_1$ and $C_2$ can be evaluated from the boundary conditions which prescribe the temperature of the surfaces $x = 0$ and $x = l$. Application of these conditions results in an expression for the temperature distribution in the slab:

$$\frac{t - t_1}{t_2 - t_1} = \frac{x}{l} \tag{3-2}$$

The heat flow through the slab can be obtained from the Fourier conduction law:

$$Q = -kA \frac{dt}{dx} = -kA \frac{t_2 - t_1}{l} = \frac{t_1 - t_2}{l/kA} \tag{3-3}$$

STEADY HEAT CONDUCTION 37

Again it is well to note here the similarity of Eq. (3-3) to the usual statement of Ohm's law. The term $l/kA$ is the equivalent of the electrical resistance and is called appropriately the *thermal resistance*.

*The Tube.* The tube (Fig. 3-2) or the hollow cylinder is more conveniently treated from the cylindrical coordinate system. A case in point is one of steady state, constant properties, and absence of heat sinks or sources. With these restrictions and another which limits the heat flow to the radial direction only (this presupposes the tube to be very long axially), Eq. (2-14) reduces to the ordinary differential equation

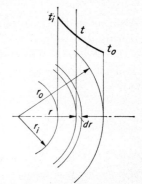

$$\frac{d^2t}{dr^2} + \frac{1}{r}\frac{dt}{dr} = 0 \qquad (3\text{-}4)$$

the solution of which becomes

$$t = C_1 \ln r + C_2$$

The appropriate boundary conditions are those specifying the temperatures $t_i$ and $t_o$ at the radii $r_i$ and $r_o$, from which the constants $C_1$ and $C_2$ can be evaluated to yield the expression for the radial temperature distribution in the tube

FIG. 3-2. Steady heat conduction in a thick-walled tube.

$$\frac{t - t_o}{t_i - t_o} = \frac{\ln (r/r_o)}{\ln (r_i/r_o)} \qquad (3\text{-}5)$$

The heat flow across the tube wall is determined from the Fourier conduction law; however, the area normal to the heat-flow vector now changes with radius and must be properly accounted for. Thus

$$Q = -kA(r)\frac{dt}{dr} = -k(2\pi rL)\frac{dt}{dr} \qquad (3\text{-}6)$$

where $L$ is the axial length of the tube. Differentiating Eq. (3-5) once and using that result in Eq. (3-6) gives the specification of the heat flow

$$Q = \frac{t_i - t_o}{(1/2\pi kL) \ln (r_o/r_i)} \qquad (3\text{-}7)$$

Equation (3-7) has a form similar to Eq. (3-3) for the slab, except that here the thermal resistance is

$$\frac{1}{2\pi kL} \ln \frac{r_o}{r_i}$$

*The Composite Tube* (Fig. 3-3). As in the case of the composite slab, it is possible to treat a composite tube. Physically this may be a pipe with covers of various sorts of insulation. The treatment is that of resistances in series as for the composite slab (Chap. 1):

$$Q = \frac{t_1 - t_4}{\frac{1}{2\pi k_1 L}\ln\frac{r_2}{r_1} + \frac{1}{2\pi k_2 L}\ln\frac{r_3}{r_2} + \frac{1}{2\pi k_3 L}\ln\frac{r_4}{r_3}} \qquad (3\text{-}8)$$

Also, as in the case of the slab, the intermediate temperatures can be readily determined.

*The Sphere.* The temperature distribution and heat flow through the walls of a hollow sphere can be calculated from Eq. (2-15) and the Fourier conduction law by using assumptions similar to those for the slab and tube. The temperature distribution so obtained is

$$\frac{t - t_0}{t_i - t_0} = \frac{(1/r) - (1/r_0)}{(1/r_i) - (1/r_0)} \qquad (3\text{-}9)$$

The heat flow is thence found to be

$$Q = \frac{t_i - t_0}{(1/4\pi k)[(1/r_i) - (1/r_0)]} \qquad (3\text{-}10)$$

The *composite sphere* case can be readily synthesized from Eq. (3-10) in the manner used to obtain Eq. (3-8).

**3-2. Heat Convection from Bounding Surfaces.** Although the study of heat conduction has as its primary objective the specification of the temperature distribution within the conducting body and the heat-flow rate within the body, surface conditions are most important. In determining the temperature distribution, the boundary conditions can be either a specification of the surface temperatures directly or a specification for the heat rate to or from the surface.

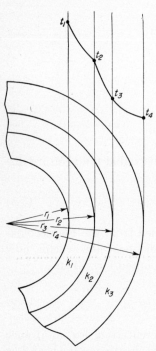

FIG. 3-3. Steady heat conduction in a composite tube.

In the case where the heat flow is desired, the convection conditions on the surfaces often are in control; i.e., heat-flow resistance in the conducting body is much smaller than the convection resistance at the surface, with the result that conditions on the surface, not those within the body itself, regulate the heat flow.

The general problem of convective heat transfer is rather complicated, depending on several variables, and as such will be treated in some detail in later sections of this text. However, certain similarities exist between the gross convection problem and pure conduction as has been discussed in Chap. 1 in regard to the formulation of Newton's law of cooling. We shall use this statement of the magnitude of convective heat transfer to establish boundary conditions for the conduction problems that will be treated in this chapter.

**3-3. Critical Thickness of Insulation.** In the case of small-diameter, circular tubes the application of insulating material to the outer surface may in special instances *increase* the heat loss from the surface. This interesting phenomenon can be treated as follows. The heat loss from a tube with insulation to a surrounding fluid is given by

$$Q = \frac{2\pi L(t_i - t_0)}{(1/k)\ln(r_0/r_i) + (1/h_0 r_0)} \tag{3-11}$$

The heat rate will be a maximum when the denominator becomes a minimum. The minimum value for the denominator can be calculated by taking the derivative of the denominator with respect to $r_0$ while $r_i$ is held as a constant parameter and setting the result equal to zero. This gives

$$\frac{1}{kr_0} - \frac{1}{h_0 r_0^2} = 0 \tag{3-12}$$

from which

$$r_{0\,\text{crit}} = \frac{k}{h_0} \tag{3-13}$$

It can be seen that this result is independent of $r_i$. The heat-transfer coefficient was considered constant in this calculation. This approximation is reasonable for many practical cases where the variation in $r_0$ is small.

Physically the result of Eq. (3-12) can be explained in the following way. The term $(1/k)\ln(r_0/r_i)$ is the thermal resistance due to the insulation; the term $1/h_0 r_0$ is the thermal resistance due to the film of fluid. The former increases with $r_0$, while the latter decreases with increasing $r_0$. At the critical radius $r_{0\,\text{crit}}$, the rate of increase of the insulation resistance is equal to the rate of decrease in the film resistance, thus giving the minimum value of the sum of the resistances as shown by Eq. (3-12). The conclusion is that tubes whose outside radii (in this case $r_i$) are smaller than the critical radius $r_{0\,\text{crit}}$, as calculated here, can have their heat losses increased by adding insulation up to the value of the critical thickness. This usually requires small tube radii, relatively large thermal conductivities in the insulation, and small heat-transfer coefficients. A practical application is the problem of insulating electrical wires where the objective would be the provision of adequate electrical insulation at the same time providing for maximum wire cooling.

**3-4. The Thin Rod.** Another simple but important solution of Eq. (2-13) is that for the physical system of a thin rod which is transferring heat at its surface to a surrounding fluid and which is connected at its base to a heated wall. The system is shown in Fig. 3-4. The temperature of the base at the wall is $t_1$; the cross-sectional area of the rod, $A$; its circumference, $C$; and its length, $l$. The convection on the surface is supposed to provide a constant value of the heat-transfer coefficient over the entire surface. The area $A$ and circumference $C$ are constant along the length of the rod. If the diameter of the rod is small as compared with its length, and if the convection film essentially controls the heat flow, there will be no radial temperature distribution in the rod but there will be a large axial temperature distribution. This, then, is a case of one-dimensional steady heat conduction in the rod. The fact that the heat being conducted along the rod from the base is being lost to the surrounding fluid by convection suggests that the problem can be solved by reducing Eq. (2-13) to the terms describing axial conduction and a distributed heat sink which is equal to the convection loss. Such a reduction becomes

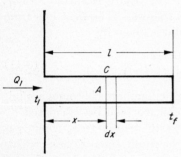

Fig. 3-4. Steady heat conduction in a thin rod.

$$\frac{d^2t}{dx^2} = \frac{Q'}{k} \tag{3-14}$$

The heat sink $Q'$ must be evaluated in terms of the convection loss, where $Q'$ is the heat sink per unit volume:

$$\frac{d^2t}{dx^2} = \frac{hC\,\delta x\,(t - t_f)}{kA\,\delta x} \tag{3-15}$$

Equation (3.15) can be readily recognized as the same result that would be obtained by equating the net heat conduction in a volume to the convection loss by that same volume. If the symbol $\vartheta = t - t_f$ is used for the temperature difference, Eq. (3-15) becomes

$$\frac{d^2\vartheta}{dx^2} - \frac{hC}{kA}\vartheta = 0 \tag{3-16}$$

Equation (3-16) can be solved by routine methods[1] to yield a general solution in the form

$$\vartheta = C_1 e^{mx} + C_2 e^{-mx} \tag{3-17}$$

where

$$m = \sqrt{\frac{hC}{kA}}$$

[1] I. S. Sokolnikoff and E. S. Sokolnikoff, "Higher Mathematics for Engineers and Physicists," 2d ed., McGraw-Hill Book Company, Inc., New York, 1941.

STEADY HEAT CONDUCTION

The constants $C_1$ and $C_2$ must be calculated with the aid of suitable boundary conditions. The end of the rod fastened to the wall is at the wall temperature $t_1$. The other end may transmit heat to the surrounding fluid. Mathematically this end-loss process can be expressed by

$$-kA\left(\frac{d\vartheta}{dx}\right)_{x=l} = hA\vartheta_{x=l} \qquad (3\text{-}18)$$

If the rod is long and thin as we have so far allowed, the heat loss from the end may be zero. This condition can be written as

$q_{end} = 0$ ∴ $\left(\frac{d\vartheta}{dx}\right)_{x=l} = 0$  $\vartheta_1 = t_1 - t_f$ (3-19)
$\vartheta_2 = t_2 - t_f$

Since no transfer $q = 0$, $hA\frac{d\vartheta}{dx}$

Using the boundary condition in Eq. (3-19) for the moment and introducing it into Eq. (3-17) with the wall condition gives

$$\vartheta_1 = t_1 - t_f = C_1 + C_2 \qquad (x=0)$$
$$\left(\frac{d\vartheta}{dx}\right)_{x=l} = 0 = mC_1 e^{ml} - mC_2 e^{-ml} \qquad x=l \qquad (3\text{-}20)$$

The solution of Eqs. (3-20) for $C_1$ and $C_2$ and the introduction of these values into Eq. (3-17) gives

$$\frac{\vartheta}{\vartheta_1} = \frac{e^{m(l-x)} + e^{-m(l-x)}}{e^{ml} + e^{-ml}} = \frac{\cosh m(l-x)}{\cosh ml} \qquad (3\text{-}21)$$

The excess temperature* of the end of the rod ($x = l$) is

$$\vartheta_2 = \frac{\vartheta_1}{\cosh ml} \qquad (3\text{-}22)$$

The heat flow through the base of the rod ($x = 0$) is

$$Q_1 = -kA\left(\frac{d\vartheta}{dx}\right)_{x=0} = mkA\vartheta_1\left(\frac{\sinh m(l-x)}{\cosh ml}\right)_{x=0} = \sqrt{hCkA}\,\vartheta_1 \tanh ml \qquad (3\text{-}23)$$

same for fin

The two functions $1/\cosh ml$ and $\tanh ml$ are given in Fig. 3-5 and in Table 3-1. As can be seen in Fig. 3-5, when the length $l$ is increasing beginning at zero, the heat flow increases rapidly at first but the incremental increase becomes smaller and smaller and finally the heat flow

TABLE 3-1. CALCULATION FUNCTIONS FOR HEAT CONDUCTION IN A ROD

| $ml$ | 0 | 0.5 | 1 | 1.5 | 2 | 3 | 4 | 5 | 6 |
|---|---|---|---|---|---|---|---|---|---|
| $\cosh ml$ | 1 | 1.1276 | 1.543 | 2.352 | 3.762 | 10.07 | 27.31 | 74.21 | 201.7 |
| $\tanh ml$ | 0 | 0.4621 | 0.7616 | 0.9052 | 0.9640 | 0.9951 | 0.9993 | 0.9999 | 1 |

*excess temp.— the end of the rod is presumed to be $t_f$, but if rod is not sufficiently long the end will have a temp other than this — the difference between this temp and $t_f$ is called excess temp.

approaches an asymptotic value. The excess temperature at the end of a very long rod is zero.

The solution of Eq. (3-16) with the boundary condition expressed by Eq. (3-18) (i.e., the case of heat loss from the end of the rod) becomes somewhat lengthy but results in the expression for the temperature distribution along the rod

$$\frac{\vartheta}{\vartheta_1} = \frac{\cosh m(l - x) + (h_2/mk) \sinh m(l - x)}{\cosh ml + (h_2/mk) \sinh ml} \qquad (3\text{-}24)$$

The temperature excess at the end of the rod ($x = l$) becomes

$$\frac{\vartheta_2}{\vartheta_1} = \frac{1}{\cosh ml + (h_2/mk) \sinh ml} \qquad (3\text{-}25)$$

The heat flow through the base of the rod ($x = 0$) becomes

$$Q_1 = mkA\vartheta_1 \frac{(h_2/mk) + \tanh ml}{1 + (h_2/mk) \tanh ml} \qquad (3\text{-}26)$$

In the last three equations the value $h_2$ is the heat-transfer coefficient at the end of the rod; this value is generally different from the heat-transfer coefficient along the rod surface. Equation (3-26) will reduce to Eq. (3-23) for small enough values of $h_2/mk$.

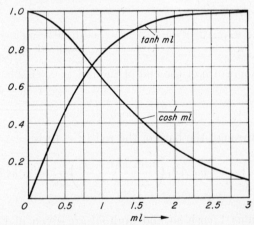

FIG. 3-5. Functions for determination of heat flow and temperature distribution in a thin rod.

The following calculated example has an important technical application. The temperature of a fluid flowing in a tube is usually measured by a thermometer or thermocouple put into a well which is welded into the tube wall as shown in Fig. 3-6. If the fluid temperature differs greatly from the outside temperature, then the tube wall has a lower temperature than the gas and heat flows by conduction from the well to the tube wall. The end of the well where the thermometer bulb or thermocouple junction is placed may become colder than the fluid, and the indicated temperature will not be true fluid temperature. This error can be calculated by Eq. (3-22) or (3-25), whichever is deemed appropriate. Figure 3-5 and Table 3-1 give the necessary length of the tube when the error must be confined within a certain limit.

**Example 3-1.** In a tube of 3.6-in. diameter in which flows superheated steam, there is placed an iron well (Fig. 3-6) with a diameter $d = 0.6$ in. for a thermometer. The steam has a pressure of 14.26 lb/in.$^2$ and a temperature of 600 F. The flow velocity is 65 fps. The length of the well, which gives an error in temperature measurement of less than 0.5 per cent of the difference between the gas temperature and the tube-wall temperature, is to be determined. The film heat-transfer coefficient between the steam and the tube wall is $h = 18.5$ Btu/hr ft$^2$ F as calculated. If the wall thickness of the well is $s = 0.036$ in., the cross section for the heat flow in the well is $A = d\pi s$, and the circumference is $C = \pi d$, then

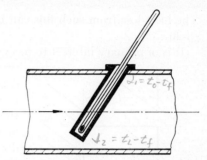

Fig. 3-6. Temperature measurement in flow in a tube.

$$m = \sqrt{\frac{hC}{kA}} = \sqrt{\frac{h}{ks}} = \sqrt{\frac{18.5}{32 \times 0.003}} = 13.7/\text{ft}$$

With the signs used in this paragraph, the error is $\vartheta_2/\vartheta_1 = 0.005$. For this value, the product $ml$ must be 6 as is found in Table 3-1. Therefore, the length of the well $l = 6/m = 6/13.7 = 0.44$ ft $= 5.28$ in. As this length is greater than the tube diameter, it is necessary to locate the well obliquely in the tube (Fig. 3-6).

Heat radiation between the end of the well and the tube wall may cause an additional error in temperature measurement. This will be dealt with in a later chapter.

**3-5. Finned Heating Surfaces.** In Chap. 1 it was shown that the overall heat-transfer resistance of a plane wall is determined principally by the greatest single resistance. If this resistance is one of the convection resistances, the heat flow through the wall can be increased by putting fins on the surface where this large resistance occurs. Such finned surfaces are widely used, for example, in economizers in steam power plants, convectors for steam and hot-water heating systems, or electrical transformers; for the cylinders of aircraft engines; etc.

*The Rectangular Fin.* The following calculations are made for the plane finned surface as the simplest case. As long as the height of the fins on a tube is comparatively small with respect to the tube diameter, the formulas derived from the plane wall can also be used for the tube.

For fins of constant thickness with the height $l$, the formulas derived in Sec. 3-4 hold true. Using the symbols in Fig. 3-7, the cross-sectional area of the fin is $bL$ and the circumference $C$ is $2L$ when $b$ is small compared with $L$. The value $m_r$ therefore, is derived by introducing the above values into the expression

$$m = \sqrt{\frac{hC}{kA}} = \sqrt{\frac{2h}{kb}} \tag{3-27}$$

The heat loss from such fins can be calculated from either Eq. (3-23) or (3-26).

It is of primary interest to recognize the conditions for which the finned surface has advantages over the unfinned surface. The answer to this question depends on whether the price, the weight, or the space needed is of the most importance. So first we shall answer the question, "Under what circumstances can the heat flow through a wall be increased at all by fins?" Obviously, fins are advantageous for the case where the heat flow through a fin increases with increased fin height. If the converse is true, it is advantageous to make the fins shorter until they are actually omitted. The limit of the conditions for which fins are advantageous is given by the expression

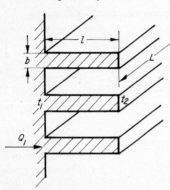

Fig. 3-7. Heating surface with rectangular fins.

$$\frac{dQ_1}{dl} = 0 \tag{3-28}$$

Using Eq. (3-26) and considering the factors $k$, $A$, $m$, and $\vartheta_1$ as constants, it suffices to differentiate only the fraction in Eq. (3-26). The resulting differentiation becomes zero when the numerator is zero or the denominator infinite. The last possibility results in a trivial statement only, for when $k = 0$, Eq. (3-28) is fulfilled. Consider then the numerator

$$\left(1 + \frac{h_2}{km} \tanh ml\right) \frac{m}{\cosh^2 ml} - \left(\frac{h_2}{km} + \tanh ml\right) \frac{h_2/k}{\cosh^2 ml} = 0$$

This equation can be simplified to the expression

$$m - \frac{h_2{}^2}{k^2 m} = 0$$

Introducing Eq. (3-27) gives

$$\frac{2k}{hb} = 1$$

or rewriting,

$$\frac{1}{h} = \frac{b/2}{k} \tag{3-29}$$

The left-hand expression is the film heat-transfer resistance and the right-hand term is the thermal-conduction resistance of a plane wall of thickness equal to one-half the fin thickness. When both resistances have the same magnitude, the limit is reached beyond which fins are useless. It should be borne in mind, however, that as fins become shorter, the heat flow becomes two-dimensional and therefore differs from the assumptions made in the derivation of Eq. (3-26). The heat-flow lines and isotherms in such an actual short fin are as shown in Fig. 3-8. The numerical value of Eq. (3-29) is influenced by this two-dimensional effect. We may be sure, however, that it is advantageous to use finned surfaces as soon as the condition

$$\frac{2k}{hb} > 5 \qquad (3\text{-}30)$$

is fulfilled.

Fig. 3-8. Temperature distribution in a short rectangular fin.

Figure 3-9[1] represents a comparison of the heat flow through the base of three fins, (a) the short fin, (b) the long fin with insulated end, and (c) the long fin with heat loss from the end. This figure shows the fin lengths that can be treated as long fins with the subsequent use of the simpler one-dimensional heat-flow analysis.

**Example 3-2.** It is to be determined when iron fins of 0.12-in. thickness on a heating surface are advantageous. The thermal conductivity of iron is found in the Appendix. We take the mean value $k = 33$ Btu/hr ft F. If the heat is transferred to air, the heat-transfer coefficient is 2 to 20 Btu/hr ft² F. We insert the higher value and find

$$\frac{2k}{hb} = 330$$

For heat transfer to air (or other gases), fins are therefore advantageous. If the heat is transferred to water, the heat-transfer coefficients lie between the limits 100 and 1000 Btu/hr ft F. If we again take the higher value, the characteristic value becomes

$$\frac{2k}{hb} = 6.6$$

As this value is very low, fins materially result in no increase in heat transfer. The effectiveness of fins for liquids can be increased by using thinner fins of a metal with higher thermal conductivity. But practically, the thickness cannot be reduced appreciably below 0.12 in., and the conductivity of copper, the best commercial conduction metal, is only five times that of iron. Therefore, fins provide no great improvement in heat transfer to liquids.

When fins are advantageous, the heat transfer is increased by placing them as near to each other as practicable. There is a limit, however, to the distance between two fins, for the heat-transfer coefficients decrease when the boundary layers which occur on the surfaces of the fins mutually influence each other. The distance between two

[1] Private communication from Dr. D. L. Doughty.

fins must not be appreciably smaller, therefore, than twice the boundary-layer thicknesses. Boundary-layer-thickness calculations will be considered in another section of this book; however, as an order of magnitude, air flowing along a plate 1 ft long with

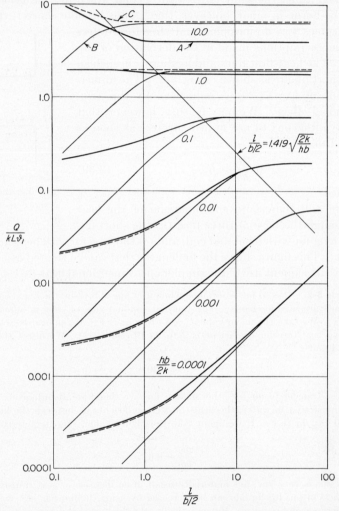

Fig. 3-9. Performance of short and long fins. (*A*) Short fins. (*B*) Long fins, insulated end. (*C*) Long fin, heat loss from end. (*Courtesy of Dr. D. L. Doughty.*)

50-fps velocity creates a boundary layer which is approximately 0.1 in. thick. The low velocities which occur in free convection, for instance, in a household convector, result in larger boundary layers about 0.5 in. thick.

*The Rectangular Fin of Minimum Weight.* For the design of cooling devices on vehicles, especially aircraft, the problem of exchanging the greatest amount of heat with the least amount of weight in the heat exchanger is of paramount importance. This problem of minimum-weight–maximum-heat transfer will be discussed for finned surfaces. The weight of one fin (Fig. 3-7) is

$$W = \rho L b l = \rho L A_1 \qquad (3\text{-}31)$$

where $\rho$ is the density of the material for the fin and $A_1$ is the area of the fin cross section normal to $L$. The length $L$ is fixed at a given dimension, whereas the two dimensions $b$ and $l$ are to be changed so as to give maximum heat flow for a given area $A_1$. In Eq. (3-23), for example, we insert $m = \sqrt{2h/kb}$ and $A = bL$. (Note the difference between $A$ and $A_1$. $A$ is measured normal to the direction of heat flow, and $A_1$ is measured in the plane of the drawing in Fig. 3-7 to be perpendicular to dimension $L$.) If the height $l$ of the fin is expressed in terms of the area $A_1$ ($l = A_1/b$), then

$$Q_1 = L \sqrt{2hkb}\, \vartheta_1 \tanh\left(\sqrt{\frac{2h}{kb}} \frac{A_1}{b}\right) \qquad (3\text{-}32)$$

The expression in Eq. (3-32) becomes a maximum for $dQ_1/db = 0$; thus by differentiating and setting the result equal to zero, we get

$$\frac{1}{2\sqrt{b}} \tanh\left(\sqrt{\frac{2h}{kb}} \frac{A_1}{b}\right) - \sqrt{b}\, \sqrt{\frac{2h}{k}}\, A_1 \frac{3}{2}\, b^{-5/2}\, \frac{1}{\cosh^2\left[\sqrt{2h/kb}(A_1/b)\right]} = 0 \qquad (3\text{-}33)$$

Introducing the abbreviation

$$\sqrt{\frac{2h}{kb}} \frac{A_1}{b} = u \qquad (3\text{-}34)$$

gives
$$\tanh u = \frac{3u}{\cosh^2 u}$$

This transcendental equation must be solved numerically or graphically, for example, by plotting both sides against $u$ and thus determining the point of intersection of the two curves. Thus, the value $u = 1.419$ that satisfies the equation can be obtained. The maximum heat flow through a fin of given weight is obtained, therefore, when the following equation is fulfilled:

$$\frac{l}{b/2} = 1.419 \sqrt{\frac{2k}{hb}} \qquad (3\text{-}35)$$

Here it can be seen that the ratio of height to half this fin thickness depends on the same characteristic value found in Eq. (3-29). The grouping $h\dfrac{b}{2}\bigg/k$ occurs so often in heat-conduction problems with convection boundary conditions that the *general form* of the group has been given a name—the *Biot modulus*. It is dimensionless and is similar to the Nusselt number, which will be encountered in the study of heat convection. However, there is an important difference: The thermal conductivity in the Biot modulus refers to the conducting body, while in the Nusselt number the thermal conductivity is that of the convecting fluid.

The temperature excess at the end of the optimum fin over the surrounding fluid is

$$\vartheta_2 = \frac{\vartheta_1}{\cosh u} = 0.457 \vartheta_1 \qquad (3\text{-}36)$$

Equation (3-36) provides a test as to whether or not the optimum height is obtained by measurement of $\vartheta_1$ and $\vartheta_2$.

When the heating surface has no fins, the area equivalent to the fin base transmits the heat flow $Q' = hbL\vartheta_1$. The ratio of the heat flow through the fins $Q_1$ to the heat flow $Q'$ without the fins is, for the best fin according to Eq. (3-32),

$$\frac{Q_1}{Q'} = \sqrt{\frac{2k}{hb}} \tanh u = 0.889 \sqrt{\frac{2k}{hb}} \qquad (3\text{-}37)$$

Equation (3-37) makes it possible to determine the heat-flow increase through the wall as a result of the addition of fins for the system considered.

**Example 3-3.** For an iron fin 0.12 in. thick, the characteristic value $2k/hb$ was determined to be 330. When it is introduced into Eq. (3-35), the optimum ratio $l/(b/2)$ becomes 25.8 for the heat transfer to air and 3.64 for the heat transfer to water. An aluminum fin 0.04 in. thick, with a thermal-conductivity value of $k = 120$, has an optimum ratio $l/(b/2) = 85.1$ for air and 12.0 for water. It can be seen that the fin must be thicker when the ratio of the film resistance to the conduction resistance becomes smaller. Iron fins for heat transfer to water are of little use.

The steel fins of air-cooled cylinder barrels for reciprocating aeroengines are approximately 0.04 in. thick and 0.8 in. high with 0.16-in spacing. The cylinder head is generally equipped with aluminum fins about 0.06 in. thick and 1.4 in. high with 0.2-in. spacing. These dimensions are very close to the optimum values according to Eq. (3-35). The radiators of liquid-cooled aeroengines also have very thin fins (approximately 0.004 to 0.008 in. thick, 0.2 in. high). Stationary heat-transfer equipment has fins which are generally thicker than the optimum value, as the saving of weight is not so important there. Finned tubes of economizers for steam boilers have fins 0.16 to 0.25 in. thick, 0.6 to 1 in. high, with 0.6- to 0.8-in. spacing; finned tubes for warm-water heating have fins 0.04 to 0.1 in. thick, 1 to 1.6 in. high, with 0.4- to 1-in. spacing.

*Fin Effectiveness.* Equation (3-37) is one expression for "fin effectiveness." Since the reference heat-flow system is somewhat arbitrary, two definitions for fin effectiveness need to be discussed. These are (1) effectiveness relative to the base area without the fin and (2) effectiveness relative to a similar fin of infinite thermal conductivity. It is clear that

Eq. (3-37) belongs to type (1). As a means of illustration, both effectivenesses are shown below for the simple rectangular fin.

Infinitely long fin:

$$\eta = \frac{\sqrt{hCkA}\ \vartheta_1}{hA\vartheta_1} = \sqrt{\frac{kC}{hA}} \tag{3-38}$$

$$\phi = \frac{\sqrt{hCkA}\ \vartheta_1}{hCl\vartheta_1} = \frac{1}{ml} \tag{3-39}$$

Finite fin:

$$\eta = \frac{\sqrt{hCkA}\ \vartheta_1 \tanh ml}{hA\vartheta_1} = \sqrt{\frac{kC}{hA}}\ \tanh ml \tag{3-40}$$

$$\phi = \frac{\sqrt{hCkA}\ \vartheta_1 \tanh ml}{hCl\vartheta_1} = \frac{\tanh ml}{ml} \tag{3-41}$$

In Eqs. (3-38) through (3-41), $\eta$ is the effectiveness relative to the base area and $\phi$ is the effectiveness relative to a similar fin of infinite thermal conductivity.

*Straight Fin of Triangular Profile.* In determining the optimum fin, the question arises as to whether or not weight advantage can be gained by using a shape other than rectangular for the fin cross section that has been considered thus far. In the discussion that follows, a straight fin of triangular cross section will be considered. Such a fin is shown in Fig. 3-10. The mathematical treatment in this case is similar to the case of the fin of rectangular cross section except that the area normal to the heat flow is a function of the distance along the fin, decreasing as the fin length increases. Thus, when written for constant thermal conductivity and heat-transfer coefficient the differential equation becomes

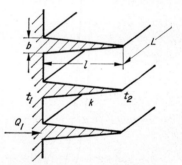

Fig. 3-10. Fins of triangular cross section.

$$\frac{d}{dx}\left[A(x)\frac{dt}{dx}\right] = \frac{hC}{k}(t - t_f) \tag{3-42}$$

The area $A(x)$ can be expressed directly in terms of the fin-length ratio $x/l$.

$$A(x) = bL\frac{x}{l}$$

and for the case where $L \gg b$, the periphery becomes

$$C = 2L$$

If these values are substituted into Eq. (3-42), the following differential equation is the result:

$$\frac{d^2\vartheta}{dx^2} + \frac{1}{x}\frac{d\vartheta}{dx} - \beta\frac{\vartheta}{x} = 0 \tag{3-43}$$

where $\vartheta$ is the temperature excess over the surrounding fluid temperature as before and

$$\beta = \frac{2hl}{kb}$$

Equation (3-43) is a modified Bessel equation[1] and has the solution

$$\vartheta = AI_0(2\sqrt{\beta x}) + BK_0(2\sqrt{\beta x}) \tag{3-44}$$

The functions $I_0(\alpha)$ and $K_0(\alpha)$ are graphed in Fig. 3-11 and are given in tabular form in Table 3-2.

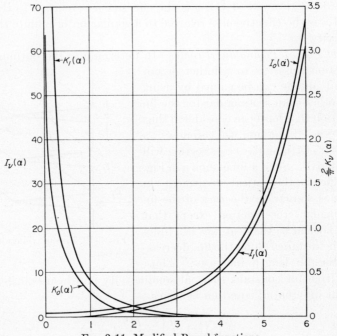

FIG. 3-11. Modified Bessel functions.

It can be seen that $K_0(2\sqrt{\beta x})$ has the value infinity at the tip of the fin ($x = 0$), and as the temperature is physically not infinite at this location, the coefficient $B$ for this term must be zero. The remaining boundary condition relates to the temperature at the base of the fin:

$$\vartheta = \vartheta_1 \quad \text{for } x = l$$

[1] See, for example, N. W. MacLachlan, "Bessel Functions for Engineers," Oxford University Press, London, 1934.

STEADY HEAT CONDUCTION

TABLE 3-2. SELECTED MAGNITUDES OF BESSEL FUNCTIONS*

| $\alpha$ | $I_0(\alpha)$ | $I_1(\alpha)$ | $\dfrac{2}{\pi} K_0(\alpha)$ | $\dfrac{2}{\pi} K_1(\alpha)$ |
|---|---|---|---|---|
| 0.0 | 1.0000 | 0.0000 | | |
| 0.2 | 1.01003 | 0.1005 | 1. 116 | 3. 040 |
| 0.4 | 1.04040 | 0.2040 | 0.7 095 | 1. 391 |
| 0.6 | 1.09205 | 0.3137 | 0.4 950 | 0.8 294 |
| 0.8 | 1.1665 | 0.4329 | 0.3 599 | 0.5 486 |
| 1.0 | 1.2661 | 0.5652 | 0.2 680 | 0.3 832 |
| 1.2 | 1.3937 | 0.7147 | 0.2 028 | 0.2 767 |
| 1.4 | 1.5534 | 0.8861 | 0.15 512 | 0.2 043 |
| 1.6 | 1.7500 | 1.0848 | 0.11 966 | 0.15 319 |
| 1.8 | 1.9896 | 1.3172 | 0.09 290 | 0.11 626 |
| 2.0 | 2.2796 | 1.5906 | 0.07 251 | 0.08 904 |
| 2.2 | 2.6291 | 1.9141 | 0.05 683 | 0.06 869 |
| 2.4 | 3.0493 | 2.2981 | 0.04 470 | 0.05 330 |
| 2.6 | 3.5533 | 2.7554 | 0.03 527 | 0.04 156 |
| 2.8 | 4.1573 | 3.3011 | 0.02 790 | 0.03 254 |
| 3.0 | 4.8808 | 3.9534 | 0.02 212 | 0.02 556 |
| 3.2 | 5.7472 | 4.7343 | 0.017 568 | 0.02 014 |
| 3.4 | 6.7848 | 5.6701 | 0.013 979 | 0.015 915 |
| 3.6 | 8.0277 | 6.7927 | 0.011 141 | 0.012 602 |
| 3.8 | 9.5169 | 8.1404 | $0.0^2 8$ 891† | $0.0^2 9$ 999 |
| 4.0 | 11.3019 | 9.7595 | $0.0^2 7$ 105 | $0.0^2 7$ 947 |
| 4.2 | 13.4425 | 11.7056 | $0.0^2 5$ 684 | $0.0^2 6$ 327 |
| 4.4 | 16.0104 | 14.0462 | $0.0^2 4$ 551 | $0.0^2 5$ 044 |
| 4.6 | 19.0926 | 16.8626 | $0.0^2 3$ 648 | $0.0^2 4$ 027 |
| 4.8 | 22.7937 | 20.2528 | $0.0^2 2$ 927 | $0.0^2 3$ 218 |
| 5.0 | 27.2399 | 24.3356 | $0.0^2 2$ 352 | $0.0^2 2$ 575 |
| 5.2 | 32.5836 | 29.2543 | $0.0^2 1$ 888 | $0.0^2 2$ 062 |
| 5.4 | 39.0088 | 35.1821 | $0.0^2 15$ 181 | $0.0^2 16$ 531 |
| 5.6 | 46.7376 | 42.3283 | $0.0^2 12$ 214 | $0.0^2 13$ 262 |
| 5.8 | 56.0381 | 50.9462 | $0.0^3 9$ 832 | $0.0^2 10$ 648 |
| 6.0 | 67.2344 | 61.3419 | $0.0^3 7$ 920 | $0.0^3 8$ 556 |
| 6.2 | 80.72 | 73.89 | $0.0^3 6$ 382 | $0.0^3 6$ 879 |
| 6.4 | 96.98 | 89.03 | $0.0^3 5$ 146 | $0.0^3 5$ 534 |
| 6.6 | 116.54 | 107.30 | $0.0^3 4$ 151 | $0.0^3 4$ 455 |
| 6.8 | 140.14 | 129.38 | $0.0^3 3$ 350 | $0.0^3 3$ 588 |
| 7.0 | 168.6 | 156.04 | $0.0^3 2$ 704 | $0.0^3 2$ 891 |
| 7.2 | 202.9 | 188.3 | $0.0^3 2$ 184 | $0.0^3 2$ 331 |
| 7.4 | 244.3 | 227.2 | $0.0^3 17$ 646 | $0.0^3 1$ 880 |
| 7.6 | 294.3 | 274.2 | $0.0^3 14$ 262 | $0.0^3 15$ 172 |
| 7.8 | 354.7 | 331.1 | $0.0^3 11$ 530 | $0.0^3 12$ 248 |
| 8.0 | 427.6 | 399.9 | $0.0^4 9$ 325 | $0.0^4 9$ 891 |
| 8.2 | 515.6 | 483.0 | $0.0^4 7$ 543 | $0.0^4 7$ 991 |
| 8.4 | 621.9 | 583.7 | $0.0^4 6$ 104 | $0.0^4 6$ 458 |
| 8.6 | 750.5 | 705.4 | $0.0^4 4$ 941 | $0.0^4 5$ 220 |
| 8.8 | 905.8 | 852.7 | $0.0^4 4$ 000 | $0.0^4 4$ 221 |
| 9.0 | 1,093.6 | 1,030.9 | $0.0^4 3$ 239 | $0.0^4 3$ 415 |
| 9.2 | 1,320.7 | 1,246.7 | $0.0^4 2$ 624 | $0.0^4 2$ 763 |
| 9.4 | 1,595.3 | 1,507.9 | $0.0^4 2$ 126 | $0.0^4 2$ 236 |
| 9.6 | 1,927 | 1,824 | $0.0^4 17$ 226 | $0.0^4 1$ 810 |
| 9.8 | 2,329 | 2,207 | $0.0^4 13$ 962 | $0.0^4 14$ 658 |
| 10.0 | | | $0.0^4 11$ 319 | $0.0^4 11$ 872 |

* L. M. K. Boelter, V. H. Cherry, H. A. Johnson, and R. C. Martinelli, "Heat Transfer Notes," University of California Press, Berkeley, Calif., 1948.
† $0.0^2 8$ 891 = 0.008891.

Introduction of this boundary condition into Eq. (3-44) enables the determination of the remaining constant:

$$A = \frac{\vartheta_1}{I_0(2\sqrt{\beta l})}$$

Thus the final equation for the temperature distribution becomes

$$\frac{\vartheta}{\vartheta_1} = \frac{I_0(2\sqrt{\beta x})}{I_0(2\sqrt{\beta l})} \qquad (3\text{-}45)$$

The heat rate can be determined from the Fourier conduction law and the first derivative of Eq. (3-45), remembering that $dI_0(\alpha)/d\alpha = I_1(\alpha)$.

$$Q_1 = L\sqrt{2hkb}\,\frac{I_1(2\sqrt{\beta l})}{I_0(2\sqrt{\beta l})} \qquad (3\text{-}46)$$

If the heat flow into the fin of triangular cross section is optimized in order to determine the best ratio of height $l$ to base $b$, the following expression results:

$$\frac{l}{b/2} = 1.309\sqrt{\frac{2k}{hb}} \qquad (3\text{-}47)$$

The temperature excess at the tip of the fin is

$$\vartheta_2 = 0.277\vartheta_1 \qquad (3\text{-}48)$$

The ratio of the thickness of the triangular fin to the thickness of the rectangular fin with equal heat flow is 1.31; the ratio of the cross-sectional areas is 1:1.44. Therefore, the weight saved by using the triangular fin is 44 per cent.

*Fin of Minimum Weight.*[1,2] It is of further interest to determine the optimum shape of a fin having the smallest weight for a given heat flow. It is evident that in such a fin every part should be utilized to the same degree, or that the specific rate of heat flow should be constant throughout the fin. The proof of this fact was presented by E. Schmidt.[3] The heat-flow lines in such a fin must have shapes as indicated in Fig. 3-12. The heat-flow lines are equally spaced and parallel to the fin axis. With a constant specific rate of heat flow the temperature decreases linearly along any flow line from the value $t_1$ at the root of the fin. The tip of the fin will approach the value of the fluid surrounding the fin only if the

---

[1] E. Schmidt, *Z. Ver. deut. Ingr.*, **70**:885 (1926). Calculations were made for pin fins on a heating surface by R. Focke, *Forsch. Gebiete Ingenieurw.*, **13**:34–42 (1942).

[2] Experimental investigation on different fin arrangements by R. H. Norris and W. A. Spofford, *Trans. ASME*, **64**:489–496 (1942).

[3] Schmidt, *loc. cit.*

angle $\alpha$ goes to zero. For a finite value of $\alpha$ there will be a temperature discontinuity between the fin tip and the surrounding fluid. The temperatures within the fin are indicated in the upper part of Fig. 3-12. At a distance $x$ from the fin tip the temperature is $t$. Because of the linearity of the temperature distribution, the difference between the temperature $t$ and the fluid temperature $t_f$ can be expressed as

$$t - t_f = \frac{x}{l}(t_1 - t_f)$$

The constant specific rate of heat flow in the direction of the fin axis may be $q$(Btu/hr ft$^2$). Consider a surface element of the fin at a distance $x$. The element may be inclined toward the fin axis by the angle $\alpha$. The specific rate of heat flow through this element is $q \sin \alpha$, and since the heat flowing through this surface element has to be transferred to the surrounding fluid, the equation

$$q \sin \alpha = h(t - t_f) = h\frac{x}{l}(t_1 - t_f)$$

is valid. This equation determines the angle $\alpha$ of the surface element as a function of the distance $x$:

$$\sin \alpha = \frac{h\vartheta_1}{ql}x$$

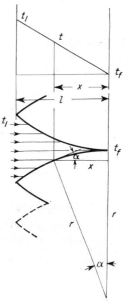

Fig. 3-12. Fin with smallest weight.

The fraction on the right-hand side of the equation has a constant value. The contour of a fin defined in this way is a circle, for a circle is described by the equation $\sin \alpha = x/r$ which corresponds exactly to the above equation. The radius $r$ of the circle has the value $ql/h\vartheta_1$. F. Weinig first pointed out the fact that the surface contours of fins of minimum weight are circles. It is not necessary to give the fin an infinitely thin edge as shown in Fig. 3-12. Any circular segment can be used to build up a contour for a fin of constant heat-flow distribution. Circular arc fins having finite angles at the tip are shown by the dashed lines in Fig. 3-12. The difference in weight between a fin in the shape of a circular arc and one of a triangular shape, however, is very small. As the triangular shape is much easier to manufacture, it may, for practical purposes, be regarded as the best form.

*Fin Arrangement.* The cross-sectional area $A_1$ necessary for a given heat flow in the instance of a rectangular fin is derived by combining

HEAT CONDUCTION

TABLE 3-3. COMPARISON OF FIN MATERIALS

| Material | Thermal conductivity $k$, Btu/hr ft F | Density $\rho$, lb/ft$^3$ | $\rho/k$, $\dfrac{\text{hr lb F}}{\text{Btu/ft}^2}$ | $\dfrac{\rho/k}{(\rho/k)_{\text{Al}}}$ |
|---|---|---|---|---|
| Copper.......................... | 220 | 560 | 2.55 | 1.96 |
| Aluminum, pure................ | 130 | 170 | 1.30 | 1 |
| Aluminum alloy................ | 90 | 166 | 1.85 | 1.42 |
| Magnesium, pure.............. | 100 | 110 | 1.31 | 1.01 |
| Steel........................... | 32 | 490 | 15.3 | 11.8 |
| Steel, stainless................ | 8 | 490 | 61.2 | 47.1 |

Eqs. (3-32) and (3-34) and solving for $A_1$:

$$A_1 = \left(\frac{Q_1}{\vartheta_1}\right)^3 \frac{1}{L^3} \frac{u}{\tanh^3 u} \frac{1}{4h^2 k} = \frac{2.109}{4L^3 h^2 k}\left(\frac{Q_1}{\vartheta_1}\right)^3 \qquad (3\text{-}49)$$

This equation shows that it is advantageous to make the fins as small as possible. To double the heat flow, the area of one fin must be increased eightfold, whereas it is sufficient to use two fins of the original size.

Equation (3-49) allows for the comparison of different materials for fins. The cross-sectional area $A_1$ is inversely proportional to the thermal conductivity $k$. The mass is proportional, therefore, to $\rho/k$. Table 3-3 lists the quotient of the density $\rho$ to the thermal conductivity $k$ for different materials. It can be seen that by the use of aluminum, for example, a weight savings of 50 per cent can be realized as compared with the use of copper. Iron fins have a tenfold weight, and iron alloy fins a fiftyfold weight. It is of no benefit, therefore, to make the fins on air-cooled cylinders of aircraft engines of copper instead of aluminum, whereas the use of aluminum instead of iron or steel promises great advantages.

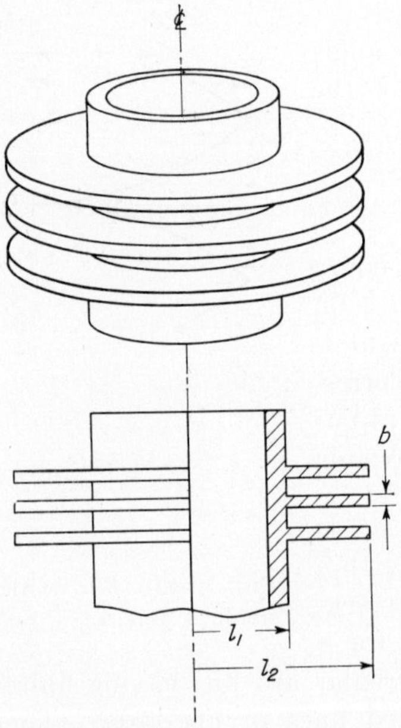

FIG. 3-13. Surface with circumferential fins.

*Cylindrical Fins.* Fins which are arranged around tubes are called cylindrical fins and are quite important from an engineering point of view. Such a fin system is shown in Fig. 3-13. Here again the treatment is substantially the same as for the rectangular fin except that the area must

# STEADY HEAT CONDUCTION

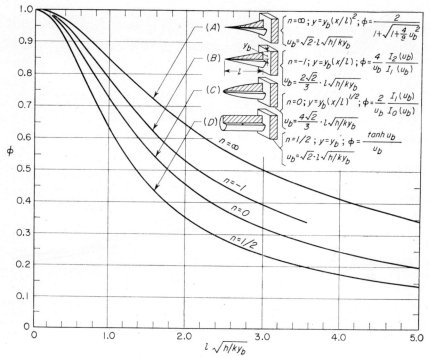

FIG. 3-14. Performance of pin fins. (*Courtesy of K. A. Gardner, Trans. ASME, 1945.*)

be allowed to vary with the radius. The area normal to the heat flux vector can be written as

$$A = 2\pi r b$$

and the periphery can be expressed as

$$C = 4\pi r$$

These values, when introduced into Eq. (3-42) where $x$ is replaced by the radius $r$, yield the differential equation

$$\frac{d^2\vartheta}{dr^2} + \frac{1}{r}\frac{d\vartheta}{dr} - \frac{2h}{kb}\vartheta = 0 \tag{3-50}$$

The appropriate boundary conditions for Eq. (3-50) may be the following:

$$r = l_1 \qquad \vartheta = \vartheta_1$$
$$r = l_2 \qquad \frac{d\vartheta}{dr} = 0$$

The solution of Eq. (3-50) is obtained in the form of modified Bessel functions:

$$\frac{\vartheta}{\vartheta_1} = \frac{K_1(l_2\sqrt{\beta})I_0(r\sqrt{\beta}) + I_1(l_2\sqrt{\beta})K_0(r\sqrt{\beta})}{K_1(l_2\sqrt{\beta})I_0(l_1\sqrt{\beta}) + I_1(l_2\sqrt{\beta})K_0(l_1\sqrt{\beta})} \tag{3-51}$$

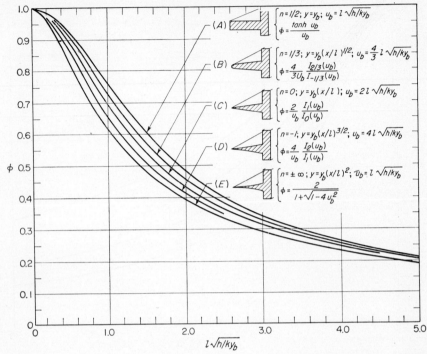

Fig. 3-15. Performance of straight fins. (*Courtesy of K. A. Gardner, Trans. ASME,* 1945.)

The heat flow into the base becomes

$$Q_1 = 2\pi l_1 bk \sqrt{\beta}\, \vartheta_1 \left[ \frac{I_1(l_2\sqrt{\beta})K_1(l_1\sqrt{\beta}) - K_1(l_2\sqrt{\beta})I_1(l_1\sqrt{\beta})}{I_1(l_2\sqrt{\beta})K_0(l_1\sqrt{\beta}) + K_1(l_2\sqrt{\beta})I_0(l_1\sqrt{\beta})} \right] \quad (3\text{-}52)$$

where
$$\beta = \frac{2h}{kb}$$

K. A. Gardner[1] has generalized the relationships for extended-surface heat-exchanger systems to include not only the examples given in the preceding sections but also several other shapes. These results are useful in design and are presented in Figs. 3-14 to 3-17 in terms of the effectiveness based on a similar fin of infinite conductivity. Additional details of solutions for extended-surface heat-transfer devices are presented by P. J. Schneider.[2]

[1] K. A. Gardner, *Trans. ASME*, November, 1945, pp. 621–631.
[2] P. J. Schneider, "Conduction Heat Transfer," Addison-Wesley Publishing Company, Reading, Mass., 1955.

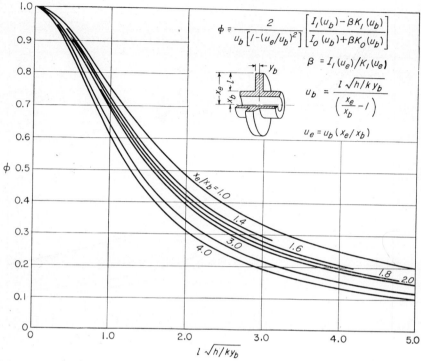

FIG. 3-16. Performance of circumferential fins of rectangular cross section. (*Courtesy of K. A. Gardner, Trans. ASME, 1945.*)

**Example 3-4.** If it is desirable to increase the heat rejection from a cylinder barrel of an air-cooled engine by adding fins to enlarge the heat-transfer area, the usual type of fin is the cylindrical fin as shown in Fig. 3-13. We may compare the heat rejection for a finned barrel with one without fins. Equation (3-52) gives the heat loss for one fin of this type. Suppose the specifications require the fins to be 1 in. long and 0.090 in. thick and separated by a width 0.180 in. The heat-transfer coefficient may be taken as $h = 50$ Btu/hr ft F due to forced flow over the fin. $k = 28$ Btu/hr ft F from the Appendix. Thus, if the cylinder wall is 250 F and the air temperature is 70 F,

$$l_1 = \tfrac{3}{12} \text{ ft} \qquad l_2 = \tfrac{4}{12} \text{ ft}$$
$$\vartheta_1 = 250 - 70 = 180 \text{ F}$$
$$\beta = \frac{2h}{kb} = \frac{2 \times 50 \times 12}{28 \times 0.090} = 476 \qquad \sqrt{\beta} = 21.8$$
$$l_1 \sqrt{\beta} = \frac{65.4}{12} = 5.45$$
$$l_2 \sqrt{\beta} = \frac{21.8 \times 4}{12} = 7.26$$

From Table 3-2

$$I_1(7.26) = 188 \qquad K_1(7.26) = 0$$
$$I_1(5.45) = 37.2 \qquad K_1(5.45) = 0.0024$$
$$I_0(5.45) = 41.0 \qquad K_0(5.45) = 0.0022$$

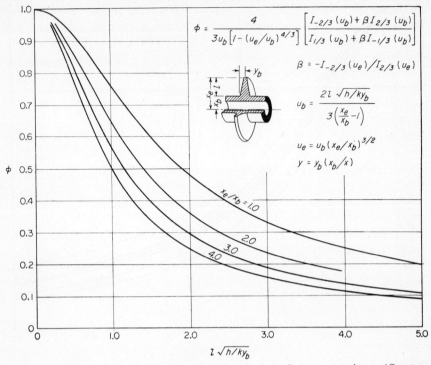

Fig. 3-17. Performance of circumferential fins of triangular cross section. (*Courtesy of K. A. Gardner, Trans. ASME, 1945.*)

Thus $\quad Q_1 = \dfrac{(6.28)(0.090)(28)(21.8)(180)}{144} \dfrac{(188)(0.0024)}{(188)(0.0022)} = 1390$ Btu/hr

For a bare tube

$$Q_B = \dfrac{(0.270)(6.28)(3)(50)(180)}{144} = 318 \text{ Btu/hr}$$

$$\dfrac{Q_1}{Q_B} = \dfrac{1390}{318} = 4.4$$

## 3-6. The Wall with Heat Sources.

The previous considerations must be extended when heat is generated in any manner, that is to say, heat sources prevail in the heat-conducting body. An explicit case of heat sources in a conducting body is an electrical conductor in which the dissipated electrical energy is transformed into heat. The calculations of the temperatures which originate by this heat generation are of specific interest in the design of electrical machines and apparatus. Other problem areas involving heat-source considerations lie in the chemical and nuclear fields. For example, special refrigeration systems must be designed in order to prevent production of inadmissibly high temperatures by the generation of heat during the setting of concrete. The flow losses in fluids are also transformed into heat. The temperatures caused by the dissipation of mechanical energy into heat in the case of high-speed gas flow (frictional heating) in aircraft currently is of the order of several hundreds of degrees and increases with the square of the Mach number. Considerable temperature increases occur from frictional heating in the oil films used for lubrication of fast-running bearings.

In this section, as an example, the treatment will be limited to solid bodies, and for simplicity the initial consideration will involve the plane wall or slab (Fig. 3-18). Uniformly distributed heat sources exist in the wall, and therefore a quantity of heat $Q'$ per unit volume and time is generated. Hence $Q'$ is independent of the space coordinate $x$. On each of the exposed surfaces the slab is bounded by a circulating fluid of temperature $t_f$. The film heat-transfer coefficient for each surface is $h$. If the thermal conductivity is considered constant and the conditions are steady with time, Eq. (2-13) reduces to

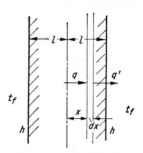

FIG. 3-18. The plane wall with heat sources.

$$\frac{d^2\vartheta}{dx^2} + \frac{Q'}{k} = 0 \qquad (3\text{-}53)$$

wherein $\vartheta$ is again the excess temperature.

The boundary conditions for the determination of the two constants that will appear in the solution of Eq. (3-53) can be expressed as follows:

$$\left(\frac{d\vartheta}{dx}\right)_{x=0} = 0$$

$$-k\left(\frac{\partial\vartheta}{\partial x}\right)_{x=l} = h\vartheta_{x=l}$$

The solution of Eq. (3-53) is

$$\vartheta = -\frac{Q'}{2k}x^2 + C_1 x + C_2 \qquad (3\text{-}54)$$

By evaluating the constants in Eq. (3-54)

$$\vartheta = \frac{Q'}{2k}(l^2 - x^2) + \frac{Q'l}{h} \qquad (3\text{-}55)$$

Hence, the isotherms in a plate with uniform heat generation are parabolas, as shown in Fig. 3-19. The maximum excess temperature is obtained from Eq. (3-55) when $x = 0$, thus

$$\vartheta_0 = \frac{Q'}{2k}l^2 + \frac{Q'l}{h} \qquad (3\text{-}56)$$

The temperature gradient on the surface can be found easily by the method given in Fig. 3-19. If we plot the distance $k/h$ from the surface, add the temperature $t_f$ as ordinate, and connect point 1, which is obtained in this manner, with point 2 as defined by the surface temperature, the direction of the straight line connecting both points yields the temperature gradient at the wall. This result can be obtained directly by the solution of the second part of the boundary conditions for the gradient at the wall:

$$-\left(\frac{d\vartheta}{dx}\right)_{x=l} = \frac{\vartheta_{x=l}}{k/h} \qquad (3\text{-}57)$$

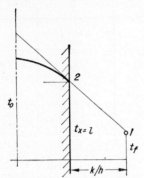

Fig. 3-19. Temperature distribution in the plane wall with heat sources.

**Example 3-5.** A cylindrical transformer coil made of insulated copper wire has an inside diameter of 6.6 in. = 0.55 ft and an outside diameter of 9.6 in. = 0.8 ft. The fraction $\phi = 0.6$ of the total cross section of the coil is copper, and the rest insulation. The density of the current in the conductors is $j = 1,300$ amp/in.$^2$; the specific resistance of copper is $\rho = 9.5 \times 10^{-6}$ ohm in.$^2$/ft. Hence the generation of heat per unit of space in the coil is $r = \phi j^2 \rho = 0.6 \times 1.69 \times 10^6$ amp$^2$/in.$^4$ $\times 9.5 \times 10^{-6}$ ohm in.$^2$/ft $\times 144$ in.$^2$/ft$^2$ $\times 3.412$ Btu/watthour = 4730 Btu/hr ft$^3$. The heat-transfer coefficient on both surfaces of the coil, which are cooled by air at 70 F, is $h = 4$ Btu/hr ft$^2$ F; the thermal conductivity of the coil is $k = 0.2$ Btu/hr ft F (mica, glue). If we consider the coil in a first approximation as a plane wall with the thickness $2l = 1.5$ in. = 0.125 ft, we obtain the highest temperature prevailing in it according to Eq. (3-56):

$$\vartheta_0 = \frac{4730 \times 0.0039}{2 \times 0.2} + \frac{4730 \times 0.0625}{4} = 120.1 \text{ F}$$

Thus the temperature $t_0$ at the center of the spool is $120.1 + 70 = 190$ F. Accurate calculation of the thermal field in the coil when considered as a hollow cylinder yields the value 191 F. Consequently, it can be seen that the calculation, as for a plane wall, yields useful results even on rather thick-walled coils.

**3-7. The Buried Cable.** The use of heat sources (and sinks) makes it possible to analyze certain types of steady-state systems that are intractable by other means. Such a case is the buried cable. A cable is supposed to be buried in homogeneous surroundings such as firmly packed soil, and the cable is generating heat by some means, perhaps electrically (or in the case of a pipe carrying some fluid, perhaps chemically), such that the specific heat rate is constant. The medium surrounding the cable terminates in a level surface (Fig. 3-20) at a distance $a$ above the center line of the cable. The temperature of the cable surface is taken to be uniformly $t_0$, and the surface temperature uniformly $t_s$. The excess temperature at any point $P(x,y)$ is then

$$\vartheta = t - t_s$$

and the temperature excess at the cable surface is
$$\vartheta_0 = t_0 - t_s$$
It is required to determine the temperature distribution around the cable within the soil.

To effect a solution, the cable is treated as a heat source causing heat to be conducted radially, the cable being assumed very long in the direction perpendicular to the plane of the paper. Now it is clear that if the

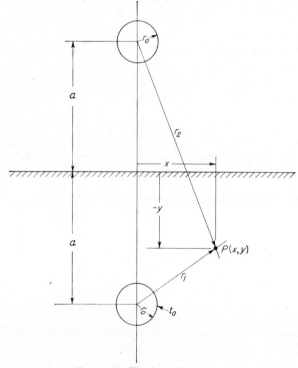

FIG. 3-20. The buried cable.

soil extended infinitely in the $x$ and $y$ directions, the isotherms would be circles oriented concentrically about the cable center and there would be no linear constant-temperature surface as is required by the proposed problem. However, if we imagine that a mirror-image system exists as shown in Fig. 3-20 except that the upper fictitious cable is taken, not as a heat source, but as a heat sink, at points equidistant from the source and sink the temperature will be just between the two. This system will yield an isotherm along $y = 0$.

Consider the cable. We know that the cable will have a heat flow
$$Q = \frac{\vartheta_0 - \vartheta_1}{(1/2\pi kL) \ln (r_1/r_0)} \tag{3-58}$$

and that due to the cable alone the temperature excess at point $P(x,y)$ will be

$$\vartheta_1 = \vartheta_0 - \frac{Q}{2\pi kL} \ln \frac{r_1}{r_0} \tag{3-59}$$

If the fictitious cable is to act as a heat sink, it must have the temperature $-\vartheta_0$ at the surface but must absorb the same amount of heat that is generated by the real cable. Thus its effect on the point $P(x,y)$ is

$$\vartheta_2 = -\vartheta_0 + \frac{Q}{2\pi kL} \ln \frac{r_2}{r_0} \tag{3-60}$$

Since the heat-conduction equation is linear, one solution can be superimposed upon another solution to yield a third solution. Therefore, Eqs. (3-59) and (3-60) can be added together to give a specification of the temperature excess at $P(x,y)$ as a result of both source and sink:

$$\vartheta = \frac{Q}{2\pi kL} \ln \frac{r_2}{r_1} \tag{3-61}$$

Now the radii $r_1$ and $r_2$ can be written in terms of $x$, $y$, and $a$ to read

$$r_1^2 = x^2 + (y+a)^2 \quad \text{and} \quad r_2^2 = x^2 + (y-a)^2$$

When these values are introduced into Eq. (3-61),

$$\vartheta = \frac{Q}{4\pi kL} \ln \left[ \frac{x^2 + (y-a)^2}{x^2 + (y+a)^2} \right] \tag{3-62}$$

Equation (3-62) can likewise be written

$$C = e^{(4\pi kL/Q)\vartheta} = \frac{x^2 + (y-a)^2}{x^2 + (y+a)^2} \tag{3-63}$$

The equations for the isotherms can be obtained by holding the left-hand term a constant which we can call $C$. Thus when an equation in $x$ and $y$ parametric in $C$ is solved,

$$x^2 + \left(y - \frac{1+C}{1-C} a\right)^2 = \frac{4Ca^2}{(1-C)^2} \tag{3-64}$$

Equation (3-64) represents a family of circles with origins at $x = 0$; $y = a(1+C)/(1-C)$, with radii $r = 2a\sqrt{C}/(1-C)$. The circle representing the $\vartheta = 0$ isotherm is the one in which $C = 1$ with a radius of infinity, therefore, a straight line. The center of this circle is located at negative infinity on the $y$ axis.

The heat transfer from a cable such as that considered above can be obtained for any depth of submergence. It must be recognized that the actual line heat source will be higher than the center line of the cable (which in reality is one of the isotherms) (Fig. 3-21) because the value $(1+C)/(1-C) >$ unity, since $C > 0$; thus the centers of the isotherm

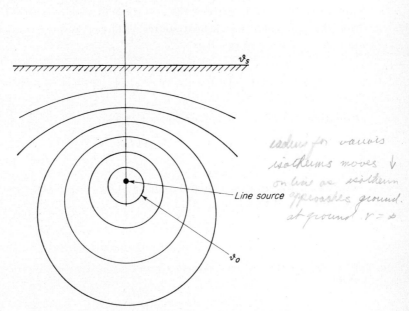

FIG. 3-21. Temperature distribution about the buried cable.

circles move downward as the radii increase. If $D$ is the cable diameter and $N$ is the depth of the cable center line below the surface,

$$N = \frac{1+C}{1-C} a$$

$$\frac{D}{2} = \frac{2a\sqrt{C}}{1-C} = \frac{2\sqrt{C}\,N}{1+C}$$

Equation (3-64) can be written explicitly in $C$:

$$C^2 + \left(2 - 16\frac{N^2}{D^2}\right) C + 1 = 0 \tag{3-65}$$

The quadratic equation in $C$ can be solved by the usual methods:

$$C = \left[8\left(\frac{N}{D}\right)^2 - 1\right] \pm 4\frac{N}{D}\sqrt{4\left(\frac{N}{D}\right)^2 - 1} = e^{(4\pi kL/Q)\vartheta} \tag{3-66}$$

The question of the correct sign for Eq. (3-66) must be determined by the physics of the system. Suppose $N/D$ is very large; therefore

$$C = 8\left(\frac{N}{D}\right)^2 \pm 8\left(\frac{N}{D}\right)^2$$

If the negative sign is used, $C = 0$, which indicates that

$$e^{(4\pi kL/Q)\vartheta} = 0$$

and, thus, that $\vartheta$ is negative. This statement violates the initial assumption that the cable was a heat source. Thus, the positive sign must be retained. The final result is

$$C_0 = \left[8\left(\frac{N}{D}\right)^2 - 1\right] + 4\frac{N}{D}\sqrt{4\left(\frac{N}{D}\right)^2 - 1} = e^{(4\pi kL/Q)\vartheta_0} \quad (3\text{-}67)$$

Equation (3-67) can be solved for $Q$ to yield

$$Q = \frac{4\pi kL\vartheta_0}{\ln\{[8(N/D)^2 - 1] + 4(N/D)\sqrt{4(N/D)^2 - 1}\}} \quad (3\text{-}68)$$

where $\vartheta_0$ is the temperature difference and the rest of the equation makes up the thermal resistance.[1]

$$R = \frac{\ln\{[8(N/D)^2 - 1] + 4(N/D)\sqrt{4(N/D)^2 - 1}\}}{4\pi kL} \quad (3\text{-}69)$$

If $N/D \gg 1$, Eq. (3-68) reduces to

$$Q = \frac{2\pi kL\vartheta_0}{\ln 4(N/D)} \quad (3\text{-}70)$$

It is useful to note that Eq. (3-68) is also the solution for the heat flow through a cylinder whose bore is also circular but eccentric with the outer wall. It is required, however, that the surfaces be maintained at constant temperature and that each represent an isotherm of the buried-cable problem.

**3-8. Two-dimensional Steady Heat Conduction.** The heat conduction equation for an isotropic, homogeneous material in two dimensions in which the temperature distribution is invariant with time and in which there are no heat sources can be reduced to a statement of the Laplace equation:

$$\frac{\partial^2 t}{\partial x^2} + \frac{\partial^2 t}{\partial y^2} = 0 \quad (3\text{-}71)$$

This equation can be interpreted in consideration of the functions of a complex variable.[2] Consider Fig. 3-22, which represents the $z$ plane where $z = x + iy$. On this plane are drawn traces $\phi$ and $\psi$ which are related to $z$ by the expression

$$\phi + i\psi = f(x + iy) = f(z)$$

[1] Thermal resistances for other buried geometries calculated by A. L. London are given in W. H. McAdams, "Heat Transmission," 3d ed., McGraw-Hill Book Company, Inc., New York, 1954.

[2] See, for example, I. S. and E. S. Sokolnikoff, "Higher Mathematics for Engineers and Physicists," p. 448, McGraw-Hill Book Company, Inc., New York, 1941.

$\phi$ and $\psi$ are conjugate functions of a complex variable. Therefore the following relations hold:

$$\frac{\partial \phi}{\partial x} + i\frac{\partial \psi}{\partial x} = f'(z) \qquad \frac{\partial \phi}{\partial y} + i\frac{\partial \psi}{\partial y} = if'(z)$$

and from these relations we can write, by separating and equating the real and imaginary parts, the Cauchy-Riemann relations

$$\begin{aligned}\frac{\partial \phi}{\partial y} &= -\frac{\partial \psi}{\partial x} \\ \frac{\partial \phi}{\partial x} &= \frac{\partial \psi}{\partial y}\end{aligned} \qquad (3\text{-}72)$$

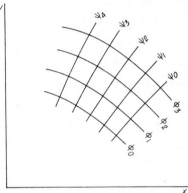

FIG. 3-22. Potential function and stream function in complex domain.

The traces $\phi = \text{const}$ and $\psi = \text{const}$ are thus *orthogonal*, i.e., perpendicular to each other at the points of intersection. Another property of conjugate functions is that each satisfies the Laplace equation. This property can be demonstrated by partially differentiating Eqs. (3-72) first by $x$ and then by $y$:

and

$$\frac{\partial^2 \phi}{\partial y\, \partial x} = -\frac{\partial^2 \psi}{\partial x^2} \qquad \frac{\partial^2 \phi}{\partial x\, \partial y} = \frac{\partial^2 \psi}{\partial y^2}$$

$$\frac{\partial^2 \phi}{\partial y^2} = -\frac{\partial^2 \psi}{\partial x\, \partial y} \qquad \frac{\partial^2 \phi}{\partial x^2} = \frac{\partial^2 \psi}{\partial x\, \partial y}$$

From these relations we obtain the two Laplace equations

$$\begin{aligned}\frac{\partial^2 \phi}{\partial x^2} + \frac{\partial^2 \phi}{\partial y^2} &= 0 \\ \frac{\partial^2 \psi}{\partial x^2} + \frac{\partial^2 \psi}{\partial y^2} &= 0\end{aligned} \qquad (3\text{-}73)$$

If we consider $\phi = t$ the temperature, then $\psi$ must be a path of constant heat flow and therefore related to $Q$.[1] Consider these values in the light of Fig. 3-23, which shows a coordinate system $s$, $n$ so oriented that $n$ is the normal to an isotherm and $s$ is tangent to the isotherm at the

---

[1] The conjugate functions in the Laplace equation are commonly called the *potential function* and the *stream function*. These correspond in here to the *temperature* and the *heat flow*, respectively.

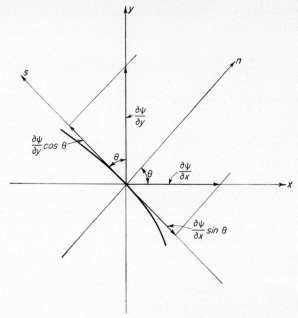

Fig. 3-23. Relations among components of the stream-function derivatives.

same point. For the heat flux along $n$, we can write

$$-k\frac{\partial \phi}{\partial n} = -k\cos\theta\,\frac{\partial \phi}{\partial x} - k\sin\theta\,\frac{\partial \phi}{\partial y}$$

When the conditions in Eq. (3-72) are used, this becomes

$$-k\frac{\partial \phi}{\partial n} = -k\cos\theta\,\frac{\partial \psi}{\partial y} + k\sin\theta\,\frac{\partial \psi}{\partial x} = -k\frac{\partial \psi}{\partial s} \qquad (3\text{-}74)$$

[Equation (3-74) is shown graphically in Fig. 3-23.] Therefore,

$$\frac{\partial \phi}{\partial n} = \frac{\partial \psi}{\partial s} \qquad (3\text{-}75)$$

In particular, if the net of lines of constant temperature and lines related to the heat flow are taken in a fine mesh but still in finite increments, the following approximation of Eq. (3-75) can be employed:

$$\frac{\Delta \phi}{\Delta n} = \frac{\Delta \psi}{\Delta s} \qquad (3\text{-}76)$$

Equation (3-76) is exactly true when the rates of change of $\phi$ and $\psi$ are constant, but regardless, it follows from Eq. (3-76) that the plane will be divided into curvilinear squares[1] if $\Delta\phi = \Delta\psi$ and the increments are small enough. Further, the smaller the increments, the more nearly the curvilinear squares will approach geometrical squares. Equation (3-76) has an important application to a graphical technique to be employed in the next section.

The nature of $\psi$, the stream function, can be expressed if we calculate the heat flow through an elemental area (Fig. 3-24).

$$dQ_x = -kL \frac{\partial t}{\partial x} dy = -kL \frac{\partial \phi}{\partial x} dy \qquad (3\text{-}77)$$

where $L$ is the dimension normal to the plane of the paper. Integrating Eq. (3-76) across a height $(y_2 - y_1)$ we obtain, with the aid of Eq. (3-72) again,

$$Q_x = -kL \int_{y_1}^{y_2} \frac{\partial \phi}{\partial x} dy =$$
$$-kL \int_{y_1}^{y_2} \frac{\partial \psi}{\partial y} dy = -kL(\psi_2 - \psi_1)$$

Thus $\quad \dfrac{Q_x}{kL} = -\Delta\psi \qquad (3\text{-}78)$

or the stream function $\psi$ becomes, in form,

$$\psi = \frac{Q}{kL} \qquad (3\text{-}79)$$

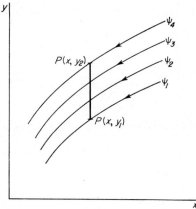

FIG. 3-24. The stream function.

Thus the stream function $\psi$ is the total rate of heat flow per unit thermal conductivity per unit depth. If $\psi$ is constant, then the rate of heat flow $Q$ is constant and the surfaces represented by the lines $\psi_1$ and $\psi_2$ are those which bound a *heat-flow tube* in which $Q$ is constant.

*The Flux Plot.* The ideas discussed in the last few paragraphs can be applied to effect a solution of two-dimensional, steady-state heat-conduction problems by the technique known as *flux-plotting*. In its crudest form, this method is essentially a freehand sketching of the flow lines (streamlines) and potential lines, carefully obeying the rules relating to orthogonality and curvilinear squares, to construct a *flow net* or *flux plot*. This method will be illustrated. Consider Fig. 3-25, which represents a parallelepiped, the upper and lower surfaces of which are at constant

---

[1] Curvilinear squares closely approximate true squares only in the limit where the number of squares approaches infinity. In finite form curvilinear squares may be considered for constructional purposes to have sides of lengths such that the average lengths of the sides opposite each other are equal. The internal angles are, of course, $\pi/2$ in each case.

temperatures $t_1$ and $t_2$, respectively, and the other surfaces are perfectly insulated. The upper surface can be ruled off to make an arbitrary number of heat-flow lanes (stream tubes). Lines representing isotherms can be scribed at intervals $\Delta y$ maintaining $\Delta y = \Delta x$ and the condition of perpendicularity of the isotherms and heat-flux lines at their intersections. The heat flow through any one of the flow tubes is

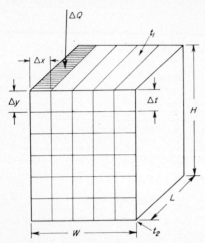

FIG. 3-25. Steady heat flow in a parallelepiped by flux plot.

$$\Delta Q = -kL \, \Delta x \, \frac{\Delta t}{\Delta y} \quad (3\text{-}80)$$

Equation (3-80) can be rewritten to be analogous to Eq. (3-78) for the condition $\Delta y = \Delta x$:

$$-\frac{\Delta Q}{kL} = \Delta t \quad (3\text{-}81)$$

The construction can be continued until the flow field is completed. The result is $N_C$ *flow tubes* and $N_R$ *temperature increments* $\Delta t$. The total heat flow in the body is obviously the sum of the incremental heat flows across the total temperature difference:

$$\frac{Q}{kL} = \frac{N_C}{N_R} \frac{\Delta x}{\Delta y} (t_1 - t_2) \quad (3\text{-}82)$$

where $t_1 > t_2$. In the construction, $\Delta x = \Delta y$; thus

$$\frac{Q}{kL} = \frac{N_C}{N_R} (t_1 - t_2) \quad (3\text{-}83)$$

The ratio $N_C/N_R = \xi$ is the *shape factor*[1] of the body. The shape factor is purely geometrical and can be determined once and for all for any given system. In the body in Fig. 3-25 the shape factor can be calculated by counting the rows and columns:

$$\xi = \frac{N_C}{N_R} = \frac{5}{6} \quad (3\text{-}84)$$

It can easily be shown that if the heat-flow lines are interchanged with the isothermals (this corresponds to heat flow horizontally across the body rather than vertically), the shape factor $\xi'$ for this alternate problem is

---

[1] The term *shape factor* occurs in several different usages in the study of heat transfer, each one having a different meaning which the reader will distinguish for himself. The property in common for all shape factors is that they are geometrical and thus can be determined once and for all for the bodies in question.

reciprocally related to the shape factor of Eq. (3-84):

$$\xi' = \frac{1}{\xi}$$

The system considered was a very simple one and as such has a place only as an illustrative example. The method described here, however, is not restricted to simple systems and is often quite useful for rapid results where extreme accuracy is not paramount.

**Example 3-6.** A more complex system can be treated which represents heat flow in the wall of a square duct whose surface temperatures are constant. Since the duct is symmetrical, only one corner need be treated. The isotherms and heat-flow lines are sketched in, attention being paid to the requirements of the construction of the resulting curvilinear squares. The result after several adjustments is shown in Fig. 3-26.

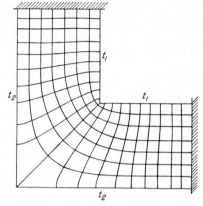

FIG. 3-26. Steady heat flow in an L-shaped corner by flux plot.

The shape factor can be readily determined by noting that the heat-flow lanes number 20 and the temperature increments number 7. The shape factor becomes

$$\xi = \frac{N_C}{N_R} = \frac{20}{7} = 2.85$$

and the heat transfer by conduction is

$$Q = 2.85kL(t_1 - t_2)$$

The heat flow can be evaluated on the basis of the material from which the duct wall is made, its length, and the temperature difference across its wall.[1]

*The Relaxation Method.* Solution of the conduction equation in two (or three) dimensions with or without heat sources can be accomplished by means of a numerical method developed by Southwell[2-4] called the

---

[1] There are a number of methods based on heat-conduction analogies with electrical conduction, fluid flow, etc., which can be used to obtain a flux plot of the desired shape or for gross determination of the shape factor. The more useful of these systems are discussed in P. J. Schneider, "Conduction Heat Transfer," chap. 13, Addison-Wesley Publishing Company, Reading, Mass., 1955.

[2] R. V. Southwell, "Relaxation Methods in Engineering Science," Oxford University Press, New York, 1940; "Relaxation Methods in Theoretical Physics" Oxford University Press, New York, 1946.

[3] H. W. Emmons, The Numerical Solution of Heat-conduction Problems, *Trans. ASME,* **65**(6): 607–612 (1943).

[4] G. M. Dusinberre, "Numerical Analysis of Heat Flow," McGraw-Hill Book Company, Inc., New York, 1949.

relaxation method. The method, of course, has general application to the solution of the Laplace-Poisson equation but the discussion of it here will be based on the heat-conduction system.

In steady state, with no internal heat sources, the Laplace equation is the differential equation whose solution with appropriate boundary conditions results in the temperature distribution in the body:

$$\frac{\partial^2 t}{\partial x^2} + \frac{\partial^2 t}{\partial y^2} = 0 \qquad (3\text{-}85)$$

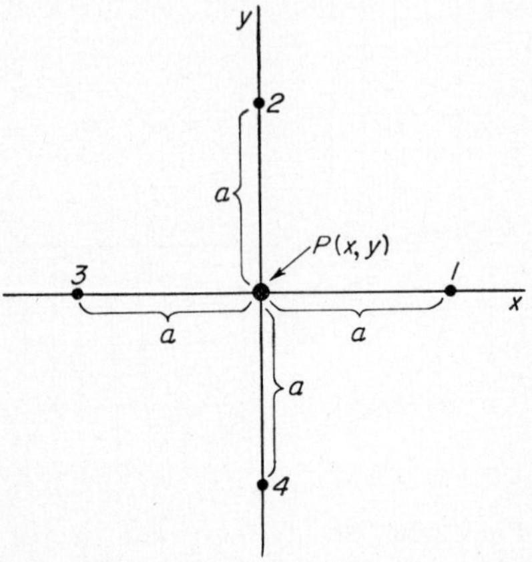

Consider a point in a body $P(x,y)$ through which is passed a rectangular coordinate system (Fig. 3-27). Other points in this body may be located at small distances away from $P(x,y)$ such as $P(x + a, y)$ $P(x, y + a)$, $P(x - a, y)$, $P(x, y - a)$. They are numbered 1, 2, 3, 4, respectively, for future convenience. If the temperature at point $P(x,y)$ is known, then by means of a Taylor series the temperatures at the surrounding points can be approximated to any desired degree of accuracy. For example, denoting the point $P(x,y)$ as 0,

Fig. 3-27. Coordinate system for the relaxation method.

$$t_1 = t_0 + \left(\frac{\partial t}{\partial x}\right)_0 a + \left(\frac{\partial^2 t}{\partial x^2}\right)_0 \frac{a^2}{2!} + \left(\frac{\partial^3 t}{\partial x^3}\right)_0 \frac{a^3}{3!} + \cdots \qquad (3\text{-}86)$$

and

$$t_3 = t_0 - \left(\frac{\partial t}{\partial x}\right)_0 a + \left(\frac{\partial^2 t}{\partial x^2}\right)_0 \frac{a^2}{2!} - \left(\frac{\partial^3 t}{\partial x^3}\right)_0 \frac{a^3}{3!} + \cdots \qquad (3\text{-}87)$$

If Eqs. (3-86) and (3-87) are added together, we obtain

$$t_1 + t_3 = 2t_0 + \left(\frac{\partial^2 t}{\partial x^2}\right)_0 a^2 + O(a^4) \qquad (3\text{-}88)$$

where $O(a^4)$ indicates a remainder of terms *of the order of* $a^4$.

Equation (3-88) can be rewritten explicitly in terms of $(\partial^2 t/\partial x^2)_0$:

$$\left(\frac{\partial^2 t}{\partial x^2}\right)_0 = \frac{t_1 + t_3 - 2t_0}{a^2} + O(a^4)$$

Similar equations which result in a statement for $(\partial^2 t/\partial y^2)_0$ can be written:

$$\left(\frac{\partial^2 t}{\partial y^2}\right)_0 = \frac{t_2 + t_4 - 2t_0}{a^2} + O(a^4)$$

STEADY HEAT CONDUCTION 71

The sum of these second derivatives constitutes a solution of the Laplace equation accurate to order $a^4$ of the neglected terms:

$$\left(\frac{\partial^2 t}{\partial x^2}\right)_0 + \left(\frac{\partial^2 t}{\partial y^2}\right)_0 = \frac{t_1 + t_2 + t_3 + t_4 - 4t_0}{a^2} = 0 \qquad (3\text{-}89)$$

Equation (3-89) results in an expression for the temperature $t_0$ in terms of the surrounding points at equilibrium:

$$t_0 = \frac{t_1 + t_2 + t_3 + t_4}{4} \qquad (3\text{-}90)$$

The *objective of the method* is to establish a temperature field in the body such that each point satisfies Eq. (3-90). To do this *systematically*, a *residual* $Q_0$ is defined and is written

$$Q_0 = t_1 + t_2 + t_3 + t_4 - 4t_0 \qquad (3\text{-}91)$$

and the *objective takes the form of making the residuals zero for each point in the body.* Regarding Fig. 3-27 again in light of Eq. (3-91), it can be seen that a unit temperature change in any one of the neighboring points can change the residual $Q_0$ by $\pm 1$ whereas a unit change in the temperature of the point itself changes the residual $Q_0$ by $\mp 4$. This, then, is the key to the procedure. Figure 3-28 depicts the relaxation pattern to be used for convenience. The following simple example will suffice to clarify the process.

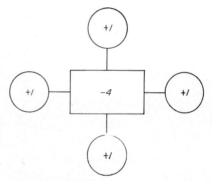

Fig. 3-28. Relaxation pattern for the Laplace equation.

**Example 3-7.** As an example to demonstrate this method consider the L-shaped corner. Such a corner is shown in Fig. 3-29. The temperatures at the boundaries are fixed, as shown in larger sized type. The initial temperature distribution is guessed at any values that might appear reasonable and compatible with the boundary conditions. These are shown to the right of the nodal points. The next step is to calculate the residuals for each of the nodal points by using the appropriate relaxation pattern, which in this case is that for the Laplace equation, since no heat sources are present. The residuals are shown to the left of the nodal points. Now attacking the largest residuals we go from point to point, changing the temperatures and adjusting the residuals. It is better to overshoot somewhat than to try to adjust the point exactly. Like any other trial-and-error process one gets better with practice. On the figure the first few calculations have been made to illustrate the methods. Only one-half of the corner has been computed because of the symmetry of the system.

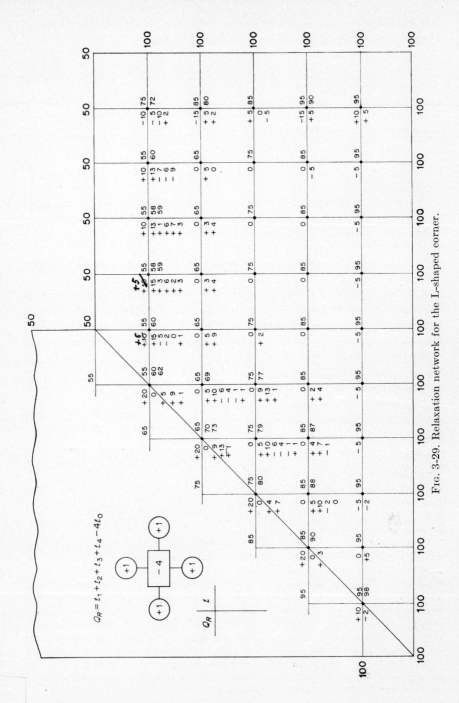

FIG. 3-29. Relaxation network for the L-shaped corner.

If the body under consideration has distributed heat sources, the conduction equation is a form of the Poisson equation

$$\frac{\partial^2 t}{\partial x^2} + \frac{\partial^2 t}{\partial y^2} = \frac{Q'(x,y)}{k} = f(x,y) \tag{3-92}$$

Again, using the expressions for the partial derivatives,

$$\frac{t_1 + t_2 + t_3 + t_4 - 4t_0}{a^2} + O(a^4) = f(x,y) \tag{3-93}$$

The residual $Q_0$ in this case, neglecting terms of fourth or higher order, becomes

$$Q_0 = t_1 + t_2 + t_3 + t_4 - 4t_0 - a^2 f(x,y) \tag{3-94}$$

In the case of Eq. (3-94) the relaxation pattern must be changed to account for the heat-generation term $a^2 f(x,y)$ as shown in Fig. 3-30. It is clear that the heat sources can be arbitrarily distributed throughout the body, the distribution being accommodated pointwise in the computation procedure. Thus the heat source is simply a modification of the effect of a change at zero point.

The relaxation method can be applied to cylindrical coordinate systems by developing the relaxation pattern within the cylindrical system; however, the equations corresponding to Eqs. (3-91) and (3-94) are not easy to use. A change of variable, however, will reduce the cylindrical coordinate system to a rectangular coordinate system in the new variable. The steady heat-conduction equation in the two dimensions $r$ and $\theta$ becomes

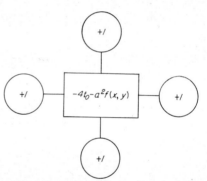

FIG. 3-30. Relaxation pattern for the Laplace-Poisson equation.

$$\frac{\partial^2 t}{\partial r^2} + \frac{1}{r}\frac{\partial t}{\partial r} + \frac{1}{r^2}\frac{\partial^2 t}{\partial \theta^2} = 0 \tag{3-95}$$

Let $\zeta = \ln r$, $\eta = \theta$; thus,

$$\frac{\partial^2 t}{\partial \theta^2} = \frac{\partial^2 t}{\partial \eta^2}$$

$$\frac{\partial t}{\partial r} = \frac{1}{r}\frac{\partial t}{\partial \zeta}$$

$$\frac{\partial^2 t}{\partial r^2} = \frac{1}{r^2}\frac{\partial^2 t}{\partial \zeta^2} - \frac{1}{r^2}\frac{\partial t}{\partial \zeta}$$

74                     HEAT CONDUCTION

When these values are substituted into Eq. (3-95),

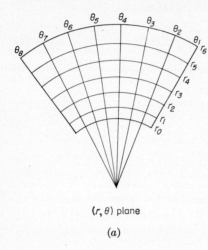

$$\frac{1}{r^2}\left(\frac{\partial^2 t}{\partial \zeta^2} - \frac{\partial t}{\partial \zeta}\right) + \frac{1}{r^2}\frac{\partial t}{\partial \zeta} + \frac{1}{r^2}\frac{\partial^2 t}{\partial \eta^2} = 0$$

which becomes

$$\frac{\partial^2 t}{\partial \zeta^2} + \frac{\partial^2 t}{\partial \eta^2} = 0 \qquad (3\text{-}96)$$

where now $t = t(\zeta,\eta)$ and the relaxation system of Eq. (3-91) applies. The new boundary conditions can be obtained from the relations $\zeta = \ln r$ and $\eta = \theta$. The results obtained can then be translated graphically from the $t(\zeta,\eta)$ system to the $t(r,\theta)$ system.

Consider Fig. 3-31. The (a) part of the figure shows a section of tube for which the temperature distribution is desired. The (b) part shows the tube section transformed into the $(\zeta,\eta)$ plane by the transformation $\zeta = \ln r; \eta = \theta$. The boundary conditions are shown here as constant surface temperatures; however, if the surface temperatures varied with either $r$ or $\theta$, or both, the particular value at the boundary point would be associated with the same point transformed into the $(\zeta,\eta)$ plane.

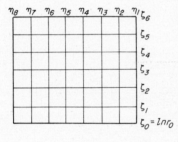

FIG. 3-31. The log transformation in cylindrical systems: (a) the tube in cylindrical coordinates; (b) the tube in transformed coordinates.

The rectangular relaxation net can be applied directly to the figure in the $(\zeta,\eta)$ plane to establish the temperatures at each point. Once this has been accomplished, the temperature values for those points are associated with the transformed point in the $(r,\theta)$ plane and the problem is complete.

## PROBLEMS

**3-1.** Develop an expression relating the temperature distribution with length in a conical fin whose base radius is $b$ and the included angle at the apex is $\alpha$. The temperature at the base can be considered constant.

**3-2.** Measurements made by means of an array of thermocouples inserted in a semi-infinite body along the direction of heat flow indicates that $dt/dx < 0$ and that $d^2t/dx^2 < 0$. What conclusions can be drawn from these experimental observations?

**3-3.** It is proposed to heat water with a 2-kw Calrod heater $\frac{3}{8}$ in. in diameter and 24 in. long, the heater being enclosed concentrically in a 0.625-in.-ID pipe. The heat-transfer coefficient due to the water flow over the heater is 300 Btu/hr ft² F, and the inlet water temperature is 140 F. Can a copper-sheathed heater be used safely in this application?

**3-4.** An 8-in.-OD pipe is buried 30 in. below the surface of the ground. Assuming a ground-surface temperature of 40 F and a pipe-wall temperature of 200 F, estimate the steady rate of heat transfer from the pipe to the ground surface by means of a flux plot. Compare this estimate with the calculated result.

**3-5.** How much Fiberglas or rock wool insulation is needed to guarantee that the temperature of the outer surface of a kitchen oven will not exceed 130 F? The maximum oven temperature to be maintained by means of a conventional thermostatic control is 550 F.

**3-6.** An exhaust stack thermocouple inserted in an 8-in.-long well made of 16-gauge $\frac{1}{4}$-in.-OD steel tubing indicates a temperature of 500 F. The stack surface temperature is measured at 400 F, and the heat-transfer coefficient is estimated at 15 Btu/hr ft² F. Estimate the temperature error and the corrected gas temperature.

**3-7.** The walls of a refrigerator are insulated with 3 in. of glass wool contained between the inner and outer metal walls. In the area near the evaporator the inner wall temperature is 20 F. Application of a heat meter to the outer steel wall indicates a heat flow of 15 Btu/hr ft² when the outside air temperature is 90 F. Is the insulation effective or not? If not, suggest reasons for the poor effectiveness.

**3-8.** Determine the steady-state temperature distribution in a bus bar of rectangular cross section ($\frac{1}{4}$ by 1 in.) when the length is very long compared with the other dimensions and the bar is carrying an electrical current of 100 amp. The bar is copper, and it is surrounded by flowing air such that the surface temperature remains at 65 F.

**3-9.** Calculate the heat transfer through the wall of a circular cylinder whose bore is eccentric. The outside diameter is 8 in., the inside diameter is 5 in., and the eccentricity is 1 in. The inner surface is maintained at 250 F, and the outer surface is at 85 F. The material from which the cylinder is made is transite.

**3-10.** Complete the relaxation problem started in Fig. 3-29.

**3-11.** A steel bar, 1 by $\frac{1}{4}$ in., is used as a hanger strap for a 2-in.-IPS insulated refrigeration line. The hanger extends 1 ft from the edge of $1\frac{1}{2}$-in. magnesia pipe insulation to the ceiling, the bar being uninsulated in this distance. The bar is fastened directly to the metal wall of the pipe. The temperature of the refrigerant within the pipe is 0 F, and the air temperature is 75 F. (a) Compare the heat loss through the hanger with that through the insulated section of the pipe. (b) How far up the hanger strip will frost form?

**3-12.** Determine the steady-state temperature distribution at any point $P(r,\theta)$ for a long rod of radius $R$, one-half of whose surface for $0 < \theta < \pi$ is kept at temperature $t_s$ while the other half $\pi < \theta < 2\pi$ is kept at zero.

CHAPTER 4

# UNSTEADY HEAT CONDUCTION

**4-1. Transient Heat Conduction.** The previous discussion was related to heat-conduction systems where the temperature ceased to vary with time. It was assumed physically that sufficient time had elapsed to permit the transient condition, brought about by the steady application of heat to some boundary of the system, to die out. In this section the heating or cooling of the body, the *transient state*, will be discussed.

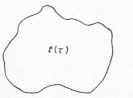

Fig. 4-1. The homogeneous billet.

*Solids of Infinite Thermal Conductivity.* These present the simplest cases. If the body has a large thermal conductivity and a corresponding low film heat-transfer coefficient, the heat flow to or from the body is controlled principally by the convection resistance and little or no temperature gradients exist in the body; i.e., the body is spacewise isothermal, the temperature varying only with time. Certain smaller bodies of finite thermal conductivity satisfy the above conditions. Consider the small billet shown in Fig. 4-1. This billet exists at some uniform temperature when at time zero it is placed in a fluid of a different temperature. The problem is to determine the change in temperature with time as a function of the system characteristics. A heat balance on the billet, equating the change in heat capacity with the convective heat loss, yields

$$\rho c V \frac{dt}{d\tau} = -hA(t - t_f) \tag{4-1}$$

In Eq. (4-1), $V$ is the volume of the billet and $A$ is the surface area. The appropriate initial condition is $t = t_0$ at $\tau = 0$. The surrounding fluid temperature $t_f$ is a constant. Equation (4-1) can be rewritten in the form of a differential equation for the excess temperature where $\vartheta = t - t_f$

$$\frac{d\vartheta}{\vartheta} = -\frac{hA}{\rho c V} d\tau \tag{4-2}$$

The appropriate condition for the evaluation of the single constant appearing in the solution of Eq. (4-2) is the initial condition, the temperature of the billet at time zero.

$$\tau = 0 \quad \vartheta = \vartheta_0$$

With the use of the above boundary condition the solution for Eq. (4-2) becomes

$$\frac{\vartheta}{\vartheta_0} = e^{-(hA/\rho cV)\tau} \qquad (4\text{-}3)$$

It is clear that a plot of Eq. (4-3) for $\ln \vartheta/\vartheta_0$ versus $\tau$ will yield a family of straight lines with $hA/\rho cV$ as a parameter. Equation (4-3) can be rewritten to advantage as follows:

$$\frac{\vartheta}{\vartheta_0} = e^{-(hL/k)(\alpha\tau/L^2)} \qquad (4\text{-}4)$$

where $hL/k$ = Biot modulus
$\alpha\tau/L^2$ = Fourier modulus
$\alpha$ = thermal diffusivity

The appropriate plot of Eq. (4-4) is $\ln \vartheta/\vartheta_0$ versus $\alpha\tau/L^2$ with $hL/k$ as a parameter (Fig. 4-2).

The value $L$ in the Biot and Fourier modulus is recognized as a significant dimension of the system and is derived as the ratio of the volume of the body to the surface area. Thus for simple geometrical shapes the value of $L$ can be readily obtained:

Sphere: $\qquad L = \dfrac{\frac{4}{3}\pi r_0^3}{4\pi r_0^2} = \dfrac{r_0}{3}$

Cylinder: $\qquad L = \dfrac{\pi r_0^2 l}{2\pi r_0 l} = \dfrac{r_0}{2} \quad l \gg r_0$

Cube: $\qquad L = \dfrac{l^3}{6l^2} = \dfrac{l}{6}$

The Biot and Fourier moduli are dimensionless and have the great advantage that when used in the manner of Eq. (4-4) the temperature-time histories for all bodies of infinite thermal conductivity can be reduced to a single universal plot for all values of the convection boundary condition. The electrical equivalent of the transient heat-transfer system of the small billet is the discharge of an electrical condenser in a circuit with a pure resistance. The electrical equation is

$$\frac{E}{E_0} = e^{-(\tau/RC)} \qquad (4\text{-}5)$$

where $E_0$ = potential at time zero and $E = E(\tau)$
$R$ = resistance
$C$ = electrical capacitance

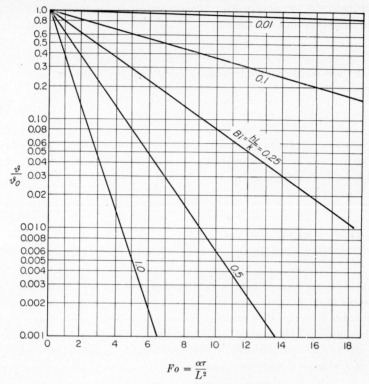

Fig. 4-2. Temperature-time history of the homogeneous billet.

This analogy led to the design of many computers for transient heat-conduction behavior based on electrical lumped $RC$ circuits.

**Example 4-1.** If it is intended to measure an unsteady temperature with a thermometer, it is very important to understand the speed with which the thermometer follows this process. The *half-value time* is the time within which the initial difference between the true temperature and the indicated temperature of the thermometer is reduced to half after a sudden change of the true temperature. This half-value time is to be determined for a mercury thermometer which is installed in an air stream. The mercury bulb may have a cylindrical shape with a 0.12-in. = 0.01-ft radius. The thermal conductivity of the mercury according to the Appendix is $k = 5$ Btu/hr ft F; its thermal diffusivity is $\alpha = 0.178$ ft²/hr. The thermal resistance of the thin glass wall is disregarded. The heat-transfer coefficient in the air stream may be

$$h = 10 \text{ Btu/hr ft}^2 \text{ F}$$

Hence the Biot modulus is $hL/k = (10 \times 0.01)/(5 \times 2) = 0.01$. The temperature ratio $t/t_0$ is 0.5 in Eq. (4-4) when the exponent has the numerical value 0.693. Thus the equation for the determination of the half-value time $\tau_H$ is $(\alpha \tau_H / L^2)(hL/k) = 0.693$. The characteristic value $\alpha \tau_H / L^2$ thus becomes

$$\frac{\alpha \tau_H}{L^2} = \frac{0.693}{0.01} = 69.3$$

and the half-value time,

$$\tau_H = \frac{1 \times 10^{-4} \times 69.3}{4 \times 0.178} \text{ hr} = 0.0098 \times 3{,}600 \text{ sec} = 35 \text{ sec}$$

Only for unsteady temperature changes, which are slower (with a sine-shaped temperature oscillation the period duration must be about tenfold), can we expect that the thermometer records the temperature trend accurately.[1]

*Solids of Infinite Conductivity, Boundary-condition Functions of the Time.* Consider the billet of Fig. 4-1, but imagine that the fluid temperature begins at time zero to change linearly with time; that is, the temperature of the fluid obeys the following expression:

$$t_f = B\tau \qquad (4\text{-}6)$$

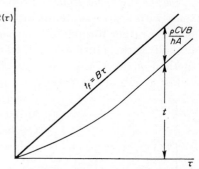

FIG. 4-3. Temperature-time history of a homogeneous billet subjected to linearly variable fluid temperatures.

Equation (4-1) can be rewritten to consider the variation of the fluid temperature:

$$\frac{dt}{d\tau} + \frac{hA}{\rho c V} t = \frac{hAB}{\rho c V} \tau \qquad (4\text{-}7)$$

The solution of Eq. (4-7) becomes

$$t = B\left(\tau - \frac{\rho c V}{hA}\right) + C_1 e^{-(hA\tau/\rho c V)} \qquad (4\text{-}8)$$

The constant $C_1$ can be evaluated from the initial condition which establishes the initial temperature at zero. The complete solution is therefore

$$t = B\tau - \frac{\rho c V}{hA} B[1 - e^{-(hA\tau/\rho c V)}] \qquad (4\text{-}9)$$

Equation (4-9) is demonstrated graphically in Fig. 4-3. It can be seen that the billet temperature always lags behind the fluid temperature. As soon as the initial transient dies out, the lag becomes a constant. This can be seen from Eq. (4-9) for $\tau$ very large.

**Example 4-2.** If the thermometer used in Example 4-1 was used to monitor the temperature of an oven for an ordinary household stove, it would be of interest to calculate the lag of the thermometer while the oven was heating at a rate of 400 F/hr.

---

[1] F. Lieneweg, *Wiss. Veröffentl. Siemens-Werken*, **16**:112–126 (1937); **17**:19–32 (1937); **19**:73–87 (1940); also a brief report on the subject in *Z. Ver. deut. Ingr.,* **85**: 272–273 (1941).

Suppose the heat-transfer coefficient $h = 2$ Btu/hr ft$^2$ F; in this case,

$$\frac{\rho c V}{hA} = \frac{r_0 k}{2\alpha h} = \frac{0.01 \times 5}{2 \times 0.178 \times 2} = 0.0705 \, \frac{1}{\text{hr}}$$

$$\Delta t_{lag} = 0.0705 \times 400 = 28.2 \text{ F}$$

*The Infinite Flat Plate.* Problems in transient heat conduction in systems with spacewise temperature distributions constitute the beginning of some rather formidable mathematical treatments. Fourier[1] occupied himself with solutions to this problem and developed the renowned method of the Fourier series in this respect.

For a slab of thickness $l$ in the $x$ direction and infinite in the $y$ and $z$ directions, insulated on the face $x = 0$ and losing heat by convection on the face $x = l$, the heat-conduction equation in the absence of heat sources reduces to

$$\frac{\partial \vartheta}{\partial \tau} = \alpha \frac{\partial^2 \vartheta}{\partial x^2} \qquad (4\text{-}10)$$

where $\vartheta$ is the excess temperature anywhere in the body. The slab can be represented as shown in Fig. 4-4. Equation (4-10) can be solved for this set of boundary conditions by the method of separation of variables. Assume a solution of the form

$$\vartheta = F(\tau)G(x)$$

where, as indicated, $F$ is a function of $\tau$ only and $G$ is a function of $x$ only:

$$\frac{\partial \vartheta}{\partial \tau} = G(x)F'(\tau)$$

$$\frac{\partial^2 \vartheta}{\partial x^2} = F(\tau)G''(x)$$

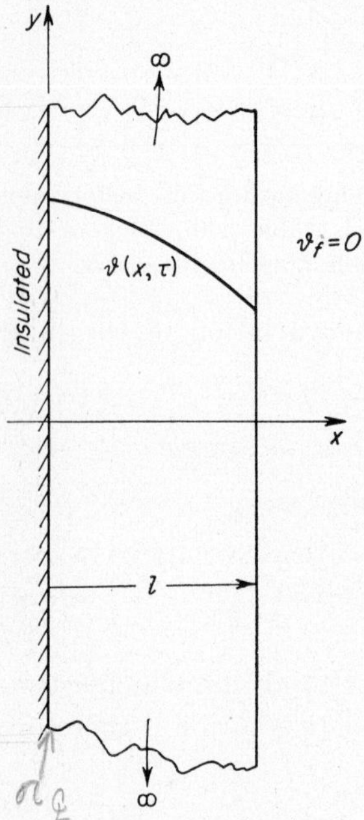

Fig. 4-4. Heat conduction in the infinite flat plate.

Substitution of these values in Eq. (4-10) gives

$$G(x)F'(\tau) = \alpha F(\tau)G''(x) \qquad (4\text{-}11)$$

whereby the variables are separable:

$$\frac{1}{\alpha} \frac{F'}{F} = \frac{G''}{G} = \pm\lambda^2 \qquad (4\text{-}12)$$

where $\lambda$ is a constant.[2]

[1] Jean Baptiste Fourier, "Théorie analytique de la chaleur, oeuvres de Fourier," Gauthier-Villars, Paris, 1822, English translation by Freeman, Cambridge, 1878.

[2] $\lambda$ is called an *eigenvalue*, and the problem an *eigenvalue problem*.

UNSTEADY HEAT CONDUCTION

Equation (4-12) presents two differential equations for solution instead of the one in Eq. (4-10), but these are ordinary linear equations and so are amenable to solution. The two equations are

$$F' \pm \lambda^2 \alpha F = 0 \tag{4-13}$$
$$G'' \pm \lambda^2 G = 0 \tag{4-14}$$

Solution of Eq. (4-13) yields

$$F = C_1 e^{\pm \lambda^2 \alpha \tau}$$

and since we know that the excess temperature will decrease with time, we accept the negative sign. Equation (4-14) with the negative $\lambda^2$ solves to yield

$$G = C_2 e^{+i\lambda x} + C_3 e^{-i\lambda x}$$

The complete solution is, of course, $\vartheta = F(\tau)G(x)$; therefore

$$\vartheta = e^{-\lambda^2 \alpha \tau}(A \cos \lambda x + B \sin \lambda x) \tag{4-15}$$

The boundary and initial conditions are

$$\frac{\partial \vartheta}{\partial x} = 0 \qquad x = 0 \tag{4-16}$$

$$\frac{\partial \vartheta}{\partial x} = -\frac{h}{k} \vartheta_l \qquad x = l \tag{4-16a}$$

$$\vartheta = \vartheta_0 \qquad \tau = 0 \tag{4-16b}$$

There are really three constants to be determined, $A$, $B$, and $\lambda$, and there are appropriately three conditions adequate for their determination. Equation (4-16) requires the derivative $\partial \vartheta / \partial x = 0$ at $x = 0$.

$$\frac{\partial \vartheta}{\partial x} = e^{-\lambda^2 \alpha \tau}(-A_1 \lambda \sin \lambda x + B\lambda \cos \lambda x) \tag{4-17}$$

It is clear from Eq. (4-17) that $B = 0$ in order that the temperature gradient $(\partial \vartheta / \partial x)_{x=0} = 0$. The solution now reads

$$\vartheta = A e^{-\lambda^2 \alpha \tau} \cos \lambda x \tag{4-18}$$

The second boundary condition [Eq. (4-16a)] can be used to obtain the $\lambda$.

$$\left(\frac{\partial \vartheta}{\partial x}\right)_{x=l} = -A_1 e^{-\lambda^2 \alpha \tau} \lambda \sin \lambda l = -\frac{h}{k} \vartheta_l \tag{4-19}$$

From Eq. (4-18)

$$\vartheta_l = A e^{-\lambda^2 \alpha \tau} \cos \lambda l \tag{4-20}$$

Combining Eqs. (4-19) and (4-20), we obtain

$$\cot \lambda l = \frac{\lambda k}{h} \tag{4-21}$$

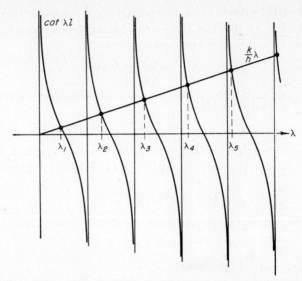

Fig. 4-5. Determination of eigenvalues for the infinite-flat-plate problem.

Equation (4-21) is a specification for $\lambda$, which can be obtained by plotting each side of the equation against $\lambda$ as in Fig. 4-5. From the intersections of the two functions as many values of $\lambda$ can be obtained as are necessary. The temperature distribution becomes

$$\vartheta = \sum_{n=1}^{\infty} A_n e^{-\lambda_n^2 \alpha \tau} \cos \lambda_n x \qquad (4\text{-}22)$$

where a value of $A_n$ for each $\lambda_n$ must now be determined. The initial condition is utilized for this purpose. Assume that $\vartheta_0$ can be expanded in an infinite series:

$$\vartheta_0 = A_1 \cos \lambda_1 x + A_2 \cos \lambda_2 x + A_3 \cos \lambda_3 x + \cdots \qquad (4\text{-}23)$$

If Eq. (4-23) is multiplied through by $\cos \lambda_n x\, dx$ and integrated from $x = 0$ to $x = l$ (assuming the integration permissible), it can be shown that

$$\int_0^l A_m \cos \lambda_n x \cos \lambda_m x\, dx = 0$$

if $m \neq n$. The remaining values are

$$\int_0^l \vartheta_0 \cos \lambda_n x\, dx = \int_0^l A_n \cos^2 \lambda_n x\, dx$$

and these can be evaluated to yield the constants $A_n$. In general, then,

$$A_n = \frac{2\vartheta_0 \sin \lambda_n l}{\lambda_n l + \sin \lambda_n l \cos \lambda_n l} \qquad (4\text{-}24)$$

UNSTEADY HEAT CONDUCTION

Now for convenience let $\lambda_n = \delta_n/l$, so that the temperature distribution is, finally,

$$\vartheta = \sum_{n=1}^{\infty} e^{-\delta_n^2(\alpha\tau/l^2)} \frac{2\vartheta_0 \sin \delta_n \cos (\delta_n x/l)}{\delta_n + \sin \delta_n \cos \delta_n} \tag{4-25}$$

The heat loss for the semi-infinite slab is obtained from the Fourier conduction law:

$$dQ_i = -kA \left(\frac{\partial \vartheta}{\partial x}\right)_{x=l} d\tau$$

$$\left(\frac{\partial \vartheta}{\partial x}\right)_{x=l} = -\frac{2\vartheta_0}{l} \sum_{n=1}^{\infty} e^{-\delta_n^2(\alpha\tau/l^2)} \frac{\delta_n \sin^2 \delta_n}{\delta_n + \sin \delta_n \cos \delta_n} \tag{4-26}$$

When Eq. (4-26) is substituted into the Fourier conduction law,

$$\frac{Q_i}{A} = \frac{2k\vartheta_0}{l} \int_0^\tau \sum_{n=1}^{\infty} e^{-\delta_n^2(\alpha\tau/l^2)} \frac{\delta_n \sin^2 \delta_n}{\delta_n + \sin \delta_n \cos \delta_n} d\tau \tag{4-27}$$

The integration of Eq. (4-27) yields the heat flow per unit area:

$$\frac{Q_i}{A} = \frac{2kl}{\alpha} \vartheta_0 \sum_{n=1}^{\infty} \frac{\sin^2 \delta_n \, [1 - e^{-\delta_n^2(\alpha\tau/l^2)}]}{\delta_n^2 + \delta_n \sin \delta_n \cos \delta_n} \tag{4-28}$$

Equation (4-21) involving the eigenvalue $\lambda_n$ can be written in terms of $\delta_n$:

$$\frac{hl}{k} = \delta_n \tan \delta_n \tag{4-29}$$

The result shows that

$$\delta_n = \phi \left(\frac{hl}{k}\right)$$

The results of Eqs. (4-25), (4-28), and (4-29) can be presented graphically in the following functional forms:

$$\frac{\vartheta}{\vartheta_0} = f_1 \left(\frac{hl}{k}, \frac{\alpha\tau}{l^2}, \frac{x}{l}\right)$$

$$\frac{\vartheta_c}{\vartheta_0} = f_2 \left(\frac{hl}{k}, \frac{\alpha\tau}{l^2}, 0\right)$$

$$\frac{\vartheta_l}{\vartheta_0} = f_3 \left(\frac{hl}{k}, \frac{\alpha\tau}{l^2}, 1\right)$$

$$\frac{Q_i}{Q_0} = f_4 \left(\frac{hl}{k}, \frac{\alpha\tau}{l^2}\right)$$

These results have been calculated and plotted for rapid use by Gröber,[1]

[1] H. Gröber and S. Erk, "Die Grundgesetze der Wärmeübertragung," Springer-Verlag, Berlin, (1933).

Gurney-Lurie,[1] Heisler,[2] and others. Solutions have been made for the similar cases of the cylinder and the sphere. The results of these three solutions are shown graphically in Figs. 4-6 to 4-8 for a limited range of the variables.

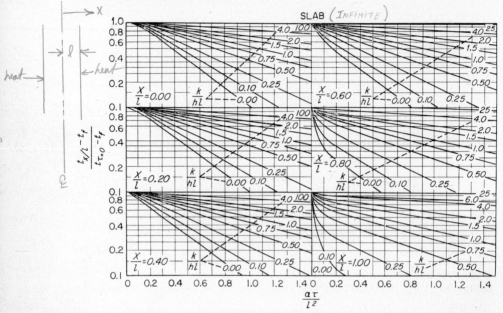

FIG. 4-6. Transient heat conduction in a slab. (*Courtesy of L. M. K. Boelter, H. A. Johnson, et al., "Heat Transfer Notes," University of California Press, Berkeley, Calif.*)

*Two- and Three-dimensional Solutions.* It is more often the case that transient heat-conduction problems concern finite bodies such as rectangular parallelepipeds or short cylinders. The solutions discussed so far cannot be used directly. However, a method due to A. B. Newman[3] enables the extension of the spacewise one-dimensional problems, for which solutions exist, to two- and three-space dimensions. The method for two dimensions is as follows, and it can be applied easily for three dimensions. Consider a long rectangular bar as shown in cross section Fig. 4-9. There is convection on the outer surfaces of the bar yielding heat-transfer coefficients $h$, which may be constant all around or equal on opposite sides. Only one quarter section of the bar need be considered because of the symmetry of the system.

The differential equation representing the system is

$$\frac{\partial \vartheta}{\partial \tau} = \alpha \left( \frac{\partial^2 \vartheta}{\partial x^2} + \frac{\partial^2 \vartheta}{\partial y^2} \right) \tag{4-30}$$

[1] H. P. Gurney and J. Lurie, *Ind. Eng. Chem.*, **15**:1170–1172 (1923).
[2] M. P. Heisler, *Trans. ASME*, **69**:227–236 (1947).
[3] A. B. Newman, *Trans. Am. Inst. Chem. Engrs.*, **24**:44 (1930).

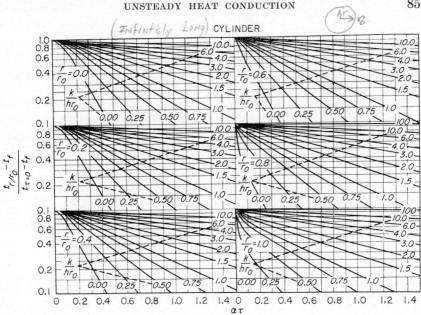

Fig. 4-7. Transient heat conduction in a cylinder. (*Courtesy of L. M. K. Boelter, H. A. Johnson, et al., "Heat Transfer Notes," University of California Press, Berkeley, Calif.*)

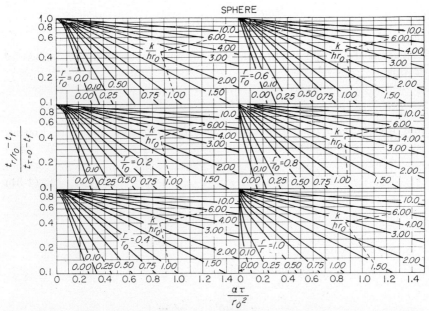

Fig. 4-8. Transient heat conduction in a sphere. (*Courtesy of L. M. K. Boelter, H. A. Johnson, et al., "Heat Transfer Notes," University of California Press, Berkeley, Calif.*)

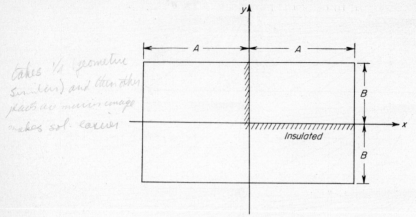

Fig. 4-9. Two-dimensional heat flow in an infinite rod.

The boundary conditions are as follows:

$$\frac{\partial \vartheta}{\partial x} = 0 \qquad x = 0$$

$$\frac{\partial \vartheta}{\partial y} = 0 \qquad y = 0$$

$$-k\left(\frac{\partial \vartheta}{\partial x}\right)_{x=A} = (h\vartheta)_A \qquad x = A$$

$$-k\left(\frac{\partial \vartheta}{\partial y}\right)_{y=B} = (h\vartheta)_B \qquad x = B$$

Equation (4-30) and the boundary conditions are made dimensionless by letting

$$x' = \frac{x}{A} \qquad y' = \frac{y}{B} \qquad \tau'_x = \frac{\alpha\tau}{A^2} \qquad \tau'_y = \frac{\alpha\tau}{B^2} \qquad \vartheta' = \frac{\vartheta}{\vartheta_0}$$

and taking note of the fact that the *dimensionless times* are Fourier numbers. Substitution results in

$$\frac{\partial \vartheta'}{\partial \tau'_x} + \left(\frac{A}{B}\right)^2 \frac{\partial \vartheta'}{\partial \tau'_y} = \frac{\partial^2 \vartheta'}{\partial x'^2} + \left(\frac{A}{B}\right)^2 \frac{\partial^2 \vartheta'}{\partial y'^2} \qquad (4\text{-}31)$$

$$\frac{\partial \vartheta'}{\partial x'} = 0 \qquad x' = 0$$

$$\frac{\partial \vartheta'}{\partial y'} = 0 \qquad y' = 0$$

$$\frac{\partial \vartheta'}{\partial x'} = \frac{hA}{k}\vartheta' \qquad x' = 1$$

$$\frac{\partial \vartheta'}{\partial y'} = \frac{hB}{k}\vartheta' \qquad y' = 1$$

A solution is assumed to be of the form $\vartheta' = \vartheta'_x \times \vartheta'_y$, where

$$\vartheta'_x = f_1\left(x', \frac{\alpha\tau}{A^2}, \frac{hA}{k}\right)$$

$$\vartheta'_y = f_2\left(y', \frac{\alpha\tau}{B^2}, \frac{hB}{k}\right)$$

Introduction of the assumed solution into Eq. (4-31) yields

$$\frac{1}{\vartheta'_x}\frac{\partial \vartheta'_x}{\partial \tau'_x} + \left(\frac{A}{B}\right)^2 \frac{1}{\vartheta'_y}\frac{\partial \vartheta'_y}{\partial \tau'_y} = \frac{1}{\vartheta'_x}\frac{\partial^2 \vartheta'_x}{\partial x'^2} + \left(\frac{A}{B}\right)^2 \frac{1}{\vartheta'_y}\frac{\partial^2 \vartheta'_y}{\partial y'^2} \qquad (4\text{-}32)$$

Equation (4-32) is of the form

$$\phi_x(x,\tau) = \phi_y(y,\tau) = 0\dagger$$

Thus, we have two equations:

$$\frac{\partial \vartheta'_x}{\partial \tau'_x} = \frac{\partial^2 \vartheta'_x}{\partial x'^2} \qquad \frac{\partial \vartheta'_y}{\partial \tau'_y} = \frac{\partial^2 \vartheta'_y}{\partial y'^2} \qquad (4\text{-}33)$$

Equations (4-33) are dimensionless forms of Eq. (4-10) which have solutions like Eqs. (4-25) and (4-28). Thus it has been shown that the product of the solution of two semi-infinite slab thicknesses $2A$ and $2B$ gives a solution for an infinite bar of cross section $2A$ by $2B$. It is easily seen that the parallelepiped of dimensions $2A$ by $2B$ by $2C$ has a solution which is the product of three semi-infinite slabs, dimensions $2A$, $2B$, and $2C$, respectively. Figure 4-10 shows thermal histories of bodies which

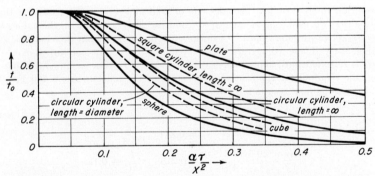

FIG. 4-10. Temperatures at the central axes of different shapes during the process of cooling. (*From H. Gröber and S. Erk, "Die Grundgesetze der Wärmeübertragung," p. 58, Fig. 28, Springer-Verlag, Berlin, 1933.*)

were initially held at an excess temperature $\vartheta_0$ and whose surfaces were

† The condition here requires that the two values equal a constant. However, either equation must apply in the absence of the other; thus the constant must be zero.

suddenly reduced to $\vartheta = 0$.[1] Many other solutions are discussed in the books by Carslaw and Jaeger,[2] Schneider,[3] and McAdams.[4]

**Example 4-3.** In the heat-treating of a steel billet shaped in the form of a rectangular parallelepiped measuring 2 by 2 by 4 ft, it is required that all parts of the billet be heated to just about 700 F but so that no part of the billet exceeds 750 F. To do this, the billet, initially at 70 F, is placed in an oven which provides an atmosphere of inert gas at 750 F and a circulation designed to give a heat-transfer coefficient $h = 100$ Btu/hr ft² F. How long must the billet remain in the oven for the center temperature to reach 700 F?

In order to use the existing solutions (which are applicable only for the case of solids existing initially at a uniform temperature, then at time zero being submerged in a fluid at zero temperature) we must state the billet problem in the same way. This is accomplished by inverting the boundary conditions and taking 750 F as the zero datum. This results in an equivalent problem statement that requires the billet to exist initially at $750 - 70 = 680$ F, submerged in a fluid of zero temperature at time zero, where now we wish to find the time it takes for the center temperature to reach $750 - 700 = 50$ F. From the Appendix, $k = 25$ Btu/hr ft² F/ft and the thermal diffusivity $\alpha = 0.570$ ft²/hr.

From the discussion of the Newman product solution it is required that

$$\left(\frac{t}{t_0}\right)_A \left(\frac{t}{t_0}\right)_B \left(\frac{t}{t_0}\right)_C = \frac{t}{t_0} = \frac{50}{680} = 0.0735$$

Since $A = B = 1$ ft and $C = 2$ ft, the expression above can become

$$\left(\frac{t}{t_0}\right)_A^2 \left(\frac{t}{t_0}\right)_C = 0.0735 \qquad (a)$$

$$\frac{hA}{k} = \frac{hB}{k} = \frac{(100)(1)}{25} = 4 \quad \text{and} \quad \frac{hC}{k} = \frac{(100)(2)}{25} = 8$$

The solution to the problem becomes one of trial and error. We assume a time $\tau$, calculate the two values $\alpha\tau/A^2$ and $\alpha\tau/C^2$, and use these values with the values $hA/k$ and $hC/k$ above to determine two values $(t/t_0)_A$ and $(t/t_0)_C$ from the chart for $x/l = 0$ in Fig. 4-6. These values are substituted into the expression (a) above. This process is repeated until expression (a) is satisfied. The time assumed for that case is the time required for the center temperature of the billet to reach 700 F. After several trials the final results are given below.

Assume that $\tau = 1.53$ hr

$$\frac{\alpha\tau}{A^2} = \frac{(0.570)(1.53)}{1} = 0.871 \qquad \frac{k}{hA} = 0.25$$

$$\frac{\alpha\tau}{C^2} = \frac{(0.570)(1.53)}{4} = 0.218 \qquad \frac{k}{hC} = 0.125$$

---

[1] H. Gröber and S. Erk, *op. cit.*

[2] H. S. Carslaw and J. C. Jaeger, "Conduction of Heat in Solids," Oxford University Press, New York, 1947.

[3] P. J. Schneider, "Conduction Heat Transfer," Addison-Wesley Publishing Company, Reading, Mass., 1955.

[4] W. H. McAdams, "Heat Transmission," 3d ed., McGraw-Hill Book Company, Inc., New York, 1954.

From Fig. 4-6 for $x/l = 0$ (the center of each slab) we read for the above parameters

$$\left(\frac{t}{t_0}\right)_A = 0.302$$

$$\left(\frac{t}{t_0}\right)_C = 0.805$$

Thus    $(0.302)^2(0.805) = 0.0735$

Thus the time required for the center temperature of the billet to reach 700 F under the conditions specified in the problem is $\tau = 1.53$ hr.

*The Infinite Solid.* We now consider the case of the infinite solid, a body extending in all directions to infinity. We wish to consider the nature of the temperature distribution in the solid for $\tau > 0$ which would result from a temperature distribution $t = f(x)$ at $\tau = 0$. The equation to solve is

$$\frac{\partial t}{\partial \tau} = \alpha \frac{\partial^2 t}{\partial x^2} \qquad (4\text{-}34)$$

Consider the expression

$$t = \frac{1}{\sqrt{\tau}} e^{-x^2/4\alpha\tau} \qquad (4\text{-}35)$$

which can be differentiated with respect to $\tau$ once and with respect to $x$ twice, substituted into Eq. (4-34), and shown to satisfy the equation exactly. The expression in Eq. (4-35) can be written in a slightly modified form which is still a solution of the differential equation:

$$t = \frac{f(\xi)}{2\sqrt{\pi\alpha\tau}} e^{-(x-\xi)^2/4\alpha\tau} \, d\xi \qquad (4\text{-}35a)$$

where $\xi$ is a parameter. This expression has the property that it is zero everywhere at time zero except at $x = \xi$, where it is finite. At increasing values of $\tau$ the temperature distribution over the space coordinate $x$ is as shown in Fig. 4-11. Equation (4-35a) can be interpreted physically as an instantaneous plane heat source at time zero located at $x = \xi$. The temperature at any location $x$ then is the result of the source at $\xi$. The strength of the source is $f(\xi)$, and the quantity of heat liberated is $f(\xi)\rho c$ per unit area of the plane; that is, $f(\xi)$ is the temperature to which the volume (unit area $\times$ $\delta\xi$) would be raised by the heat liberated by the source.

Since the conduction equation is linear, the sum of any number of particular solutions is also a solution; thus

$$t = \frac{1}{2\sqrt{\pi\alpha\tau}} \int_{-\infty}^{+\infty} f(\xi) e^{-(x-\xi)^2/4\alpha\tau} \, d\xi \qquad (4\text{-}36)$$

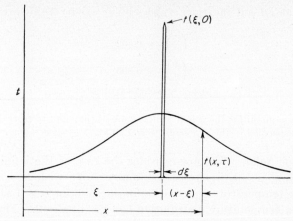

FIG. 4-11. The instantaneous heat source and ensuing temperature distribution in the infinite solid.

If a change of variable is made in Eq. (4-36), where

$$\xi = x + \beta \sqrt{4\alpha\tau}$$

then
$$t(x,\tau) = \frac{1}{2\sqrt{\pi\alpha\tau}} \int_{-\infty}^{+\infty} f(x + \beta \sqrt{4\alpha\tau}) e^{-\beta^2} \sqrt{4\alpha\tau} \, d\beta$$

or
$$t(x,\tau) = \frac{1}{\sqrt{\pi}} \int_{-\infty}^{+\infty} f(x + \beta \sqrt{4\alpha\tau}) e^{-\beta^2} \, d\beta$$

If in the above expression $\tau = 0$, then $f(x + \beta \sqrt{4\alpha\tau}) = f(x)$; therefore,

$$t(x,0) = \frac{2}{\sqrt{\pi}} f(x) \int_0^\infty e^{-\beta^2} \, d\beta = f(x)$$

Thus the solution proposed in Eq. (4-35a) satisfies both the differential equation and the initial condition and therefore constitutes a solution to the problem.

*The Semi-infinite Solid.* The semi-infinite solid may be a body bounded by a plane at $x = 0$ and extending to infinity in the positive $x$ direction. Suppose the body to have a temperature $t = f(x)$ at time zero. The temperature at the surface is zero for $\tau \gtrless 0$. The temperature distribution for this body can be deduced from the solution for the infinite body [Eq. (4-36)]. If we imagine the solid to be continued in the negative $x$ direction and the initial temperature distribution at $-\xi$ to be $-f(\xi)$, the initial temperature at $\xi$ being $f(\xi)$, the plane $x = 0$ will remain at the prescribed zero temperature (Fig. 4-12). Thus from Eq. (4-36),

$$t = \frac{1}{2\sqrt{\pi\alpha\tau}} \left\{ \int_0^\infty f(\xi) e^{-(x-\xi)^2/4\alpha\tau} \, d\xi + \int_{-\infty}^0 [-f(-\xi)] e^{-(x-\xi)^2/4\alpha\tau} \, d\xi \right\}$$

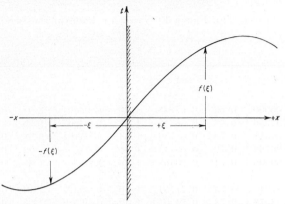

FIG. 4-12. Temperature-distribution model for the semi-infinite solid.

or rewriting,

$$t = \frac{1}{2\sqrt{\pi\alpha\tau}} \int_0^\infty f(\xi)[e^{-(x-\xi)^2/4\alpha\tau} - e^{-(x+\xi)^2/4\alpha\tau}]\, d\xi \tag{4-37}$$

When the initial temperature $t_0$ is a constant, Eq. (4-37) can be simplified by substituting for the first term in the integral $\xi = x + 2\beta\sqrt{\alpha\tau}$ and for the second term $\xi = -x + 2\beta\sqrt{\alpha\tau}$. We thus obtain

$$t = \frac{t_0}{\sqrt{\pi}} \int_{-x/2\sqrt{\alpha\tau}}^\infty e^{-\beta^2}\, d\beta - \frac{t_0}{\sqrt{\pi}} \int_{x/2\sqrt{\alpha\tau}}^\infty e^{-\beta^2}\, d\beta$$

$$= \frac{t_0}{\sqrt{\pi}} \int_{-x/2\sqrt{\alpha\tau}}^{x/2\sqrt{\alpha\tau}} e^{-\beta^2}\, d\beta = \frac{2t_0}{\sqrt{\pi}} \int_0^{x/2\sqrt{\alpha\tau}} e^{-\beta^2}\, d\beta \tag{4-38}$$

The definite integral in Eq. (4-38) has been tabulated, and a short tabulation is given in Table 4-1.

If we designate the definite integral in Eq. (4-38)

$$\operatorname{erf} z = \frac{2}{\sqrt{\pi}} \int_0^z e^{-\beta^2}\, d\beta$$

then the solution of the problem of the semi-infinite solid, whose surface is kept at zero temperature and whose initial temperature is $t_0$, is given by

$$\frac{t}{t_0} = \operatorname{erf} \frac{x}{2\sqrt{\alpha\tau}} \tag{4-39}$$

The rate of heat loss from the face $x = 0$ becomes

$$\frac{Q}{A} = -k\left(\frac{\partial t}{\partial x}\right)_{x=0} = -\frac{kt_0}{\sqrt{\pi\alpha\tau}} \tag{4-40}$$

Equation (4-36) can be applied to the case where two sheets of heavy material are welded together by infusing molten metal between them.

TABLE 4-1. THE ERROR FUNCTION OR PROBABILITY INTEGRAL*

$$\operatorname{erf} z = \frac{2}{\sqrt{\pi}} \int_0^z e^{-\beta^2}\, d\beta$$

| z | erf z | z | erf z | z | erf z | z | erf z |
|---|---|---|---|---|---|---|---|
| 0.00 | 0.00000 | 0.66 | 0.64938 | 1.30 | 0.93401 | 1.94 | 0.99392 |
| 0.02 | 0.02256 | 0.68 | 0.66378 | 1.32 | 0.93807 | 1.96 | 0.99443 |
| 0.04 | 0.04511 | 0.70 | 0.67780 | 1.34 | 0.94191 | 1.98 | 0.99489 |
| 0.06 | 0.06762 | 0.72 | 0.69143 | 1.36 | 0.94556 | 2.00 | 0.99532 |
| 0.08 | 0.09008 | 0.74 | 0.70468 | 1.38 | 0.94902 | 2.05 | 0.99626 |
| 0.10 | 0.11246 | 0.76 | 0.71754 | 1.40 | 0.95229 | 2.10 | 0.99702 |
| 0.12 | 0.13476 | 0.78 | 0.73001 | 1.42 | 0.95538 | 2.15 | 0.99764 |
| 0.14 | 0.15695 | 0.80 | 0.74210 | 1.44 | 0.95830 | 2.20 | 0.99814 |
| 0.16 | 0.17901 | 0.82 | 0.75381 | 1.46 | 0.96105 | 2.25 | 0.99854 |
| 0.18 | 0.20094 | 0.84 | 0.76514 | 1.48 | 0.96365 | 2.30 | 0.99886 |
| 0.20 | 0.22270 | 0.86 | 0.77610 | 1.50 | 0.96611 | 2.35 | 0.9991107 |
| 0.22 | 0.24430 | 0.88 | 0.78669 | 1.52 | 0.96841 | 2.40 | 0.9993115 |
| 0.24 | 0.26570 | 0.90 | 0.79691 | 1.54 | 0.97059 | 2.50 | 0.9995930 |
| 0.26 | 0.28690 | 0.92 | 0.80677 | 1.56 | 0.97263 | 2.60 | 0.9997640 |
| 0.28 | 0.30788 | 0.94 | 0.81627 | 1.58 | 0.97455 | 2.70 | 0.9998657 |
| 0.30 | 0.32863 | 0.96 | 0.82542 | 1.60 | 0.97635 | 2.80 | 0.9999250 |
| 0.32 | 0.34913 | 0.98 | 0.83423 | 1.62 | 0.97804 | 2.90 | 0.9999589 |
| 0.34 | 0.36936 | 1.00 | 0.84270 | 1.64 | 0.97962 | 3.00 | 0.9999779 |
| 0.36 | 0.38933 | 1.02 | 0.85084 | 1.66 | 0.98110 | 3.10 | 0.9999884 |
| 0.38 | 0.40901 | 1.04 | 0.85865 | 1.68 | 0.98249 | 3.20 | 0.9999940 |
| 0.40 | 0.42839 | 1.06 | 0.86614 | 1.70 | 0.98379 | 3.30 | 0.9999969 |
| 0.42 | 0.44747 | 1.08 | 0.87333 | 1.72 | 0.98500 | 3.40 | 0.9999985 |
| 0.44 | 0.46623 | 1.10 | 0.88020 | 1.74 | 0.98613 | 3.50 | 0.99999925691 |
| 0.46 | 0.48466 | 1.12 | 0.88679 | 1.76 | 0.98719 | 3.60 | 0.99999964414 |
| 0.48 | 0.50275 | 1.14 | 0.89308 | 1.78 | 0.98817 | 3.70 | 0.99999983285 |
| 0.50 | 0.52050 | 1.16 | 0.89910 | 1.80 | 0.98909 | 3.80 | 0.99999992300 |
| 0.52 | 0.53790 | 1.18 | 0.90484 | 1.82 | 0.98994 | 3.90 | 0.99999996521 |
| 0.54 | 0.55494 | 1.20 | 0.91031 | 1.84 | 0.99074 | 4.00 | 0.99999998458 |
| 0.56 | 0.57162 | 1.22 | 0.91553 | 1.86 | 0.99147 | 4.20 | 0.99999999714 |
| 0.58 | 0.58792 | 1.24 | 0.92051 | 1.88 | 0.99216 | 4.40 | 0.99999999951 |
| 0.60 | 0.60386 | 1.26 | 0.92524 | 1.90 | 0.99279 | 4.60 | 0.99999999992 |
| 0.62 | 0.61941 | 1.28 | 0.92973 | 1.92 | 0.99338 | 4.80 | 0.99999999999 |
| 0.64 | 0.63459 |  |  |  |  |  | 1.00 |

* L. M. K. Boelter, V. H. Cherry, H. A. Johnson, and R. C. Martinelli, "Heat Transfer Notes," University of California Press, Berkeley, Calif., 1948.

UNSTEADY HEAT CONDUCTION

If it is assumed that there are no complications because of the phase change and also that the exposed surface of the sheets loses heat at a much slower rate than the heat flows by conduction into the metal, the problem can be idealized into an initial temperature distribution as shown in Fig. 4-13. This temperature distribution indicates that $t = 0$ for

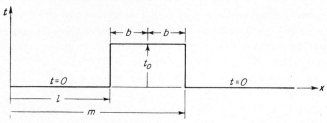

FIG. 4-13. Initial temperature distribution in infusion welding of two solid sheets.

$x < l$ and $x > m$, $t = t_0$ for $l < x < m$. The solution can be written by writing Eq. (4-36) sectionally. The integral from $l$ to $m$ is the only segment that gives a contribution:

$$t = \frac{1}{2\sqrt{\pi\alpha\tau}} \int_l^m t_0 e^{-(x-\xi)^2/4\alpha\tau} \, d\xi \tag{4-41}$$

Making the change of variable as before,

$$t = \frac{t_0}{\sqrt{\pi}} \int_{(l-x)/2\sqrt{\alpha\tau}}^{(m-x)/2\sqrt{\alpha\tau}} e^{-\beta^2} \, d\beta$$

or rewriting in terms of the error function we get

$$t = \frac{t_0}{2} \left( \operatorname{erf} \frac{m-x}{2\sqrt{\alpha\tau}} - \operatorname{erf} \frac{l-x}{2\sqrt{\alpha\tau}} \right) \tag{4-42}$$

Equation (4-42) is an adequate solution, but if $x = 0$ is equidistant from $l$ and $m$, $b = (m - l)/2$; then Eq. (4-42) becomes, in terms of $b$,

$$t = \frac{t_0}{2} \left( \operatorname{erf} \frac{b-x}{2\sqrt{\alpha\tau}} + \operatorname{erf} \frac{b+x}{2\sqrt{\alpha\tau}} \right) \tag{4-43}$$

*The Schmidt Graphical Method.* In many cases a quicker solution to the semi-infinite body and the slab can be obtained by the graphical approximation method proposed by E. Schmidt.[1] The heat-conduction equation (4-34) can be changed into an equation of finite differences by splitting the body depth (or thickness) into intervals $\Delta x$ and the time into

[1] E. Schmidt, in "Beiträge zur technischen Mechanik und technischen Physik," (Föppl-Festschrift), Springer-Verlag, Berlin, 1924; also E. Schmidt, "Einführung in die technische Thermodynamik," p. 262, Springer-Verlag, Berlin, 1936. (English translation by J. Kestin, Oxford University Press, New York.)

intervals $\Delta\tau$ and considering the temperature changes in these intervals (Fig. 4-14). Equation (4-34) can be written as a difference equation:

$$\frac{\Delta_\tau t}{\Delta\tau} = \alpha \frac{\Delta_x^2 t}{(\Delta x)^2} \tag{4-44}$$

The subscripts indicate that the time $\tau$ or the location $x$ is variable with the differential formation $\Delta t$. The length intervals and the time intervals can be numbered continuously. Thus the designation $n$ (Fig. 4-14) may refer to any arbitrary place in the plate, and the number $k$ may indicate any arbitrary instant in the time scale.

Hence, we can write the value $\Delta_\tau t$ in the form

$$\Delta_\tau t = t_{n,k+1} - t_{n,k}$$

and also the value

$$\Delta_x t = t_{n+1,k} - t_{n,k}$$

The expression $\Delta_x^2 t$ is the difference of two successive differences; hence we obtain

$$\Delta_x^2 t = (t_{n+1,k} - t_{n,k}) - (t_{n,k} - t_{n-1,k}) = t_{n+1,k} - 2t_{n,k} + t_{n-1,k}$$

Therewith, the difference equation (4-44) becomes

$$t_{n,k+1} - t_{n,k} = \alpha \frac{\Delta\tau}{\Delta x^2} (t_{n+1,k} - 2t_{n,k} + t_{n-1,k}) \tag{4-45}$$

From Eq. (4-45) the temperature throughout the wall at the time $(k+1)$ can be computed if the temperatures are known at the time $k$. By continued application of the equation, the development of the temperature field with time can be determined from a known initial temperature distribution. In lieu of the numerical calculations, the graphical solution illustrated in Fig. 4-14 can be used. The temperature field $t_k$, for the time $k$, is plotted in this figure. Now we connect the temperature points, which are two $\Delta x$ intervals apart, by a straight line, for example, in Fig. 4-14 the point $t_{n-1,k}$ on the vertical line $(n-1)$ with the point $t_{n+1,k}$ on the ordinate $(n+1)$. In this manner we arrive at the point of intersection $a$. The distance designated has the value

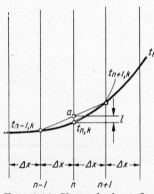

FIG. 4-14. Unsteady heat flow in a plane wall determined by the Schmidt method.

$$l = \frac{t_{n-1,k} + t_{n+1,k}}{2} - t_{n,k} = \frac{1}{2}(t_{n+1,k} - 2t_{n,k} + t_{n-1,k}) \tag{4-46}$$

Equation (4-46) agrees closely with the temperature difference defined by the right side of Eq. (4-45), the only difference being the occurrence of the coefficient ½ in one case and the coefficient $\alpha\,\Delta\tau/\Delta x^2$ in the other. However, we still have at our disposal the time increment $\Delta\tau$, which can always be arranged with an arbitrarily chosen space increment $\Delta x$ such that the condition $\alpha\,\Delta\tau/(\Delta x)^2 = \frac{1}{2}$ is fulfilled. By the procedure shown in Fig. 4-14 we have developed in point $a$ a point in the temperature field for the time $(k + 1)$ which is given by the expression

$$\Delta\tau = \frac{(\Delta x)^2}{2\alpha}$$

which expresses the time increment later than the time $k$. In the same manner by plotting additional straight lines, other points of the temperature field for the time trace $(k + 1)$ can be obtained, and thus the

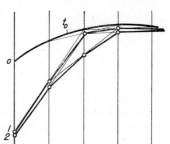

Fig. 4-15. Unsteady heat flow in a plane wall determined by the Schmidt method.

entire temperature field can be determined solely with a straightedge. In order to use this graphical solution, the temperature field in the wall must be known at some specific time. If, in addition, the change of the surface temperature with time is known, the temperature field in the wall can be plotted as shown in Fig. 4-15. Here the temperature $t_0$ for the time $\tau = 0$ and the temperatures of the surface (points 0, 1, 2, etc.) for successive time intervals are given.

Frequently, however, only the temperature $t_f$ of the fluid adjacent to the wall and the heat-transfer coefficient on the surface of the wall are known. For such a case, the subdivision into layers is more suitably accomplished in the manner of Fig. 4-16. In this case the temperature gradient on the surface for each instant is defined by the surface condition

$$-k\frac{dt}{dx} = h(t - t_f)$$

Graphically, this condition means that the tangent to the temperature curve at the surface must pass through a reference point whose distance from the wall is $k/h$ and whose ordinate is the fluid temperature $t_f$. This relationship has already been pointed out before. It can be used here in the following manner: We extend the given temperature distributions 1, 2, 3, etc., at the beginning of the equalization process (Fig. 4-16) by the straight line $(r, a)$ past the surface of the wall and start the process by connecting point 0 with 2, 1 with 3, etc., by straight lines. Thus a new temperature distribution 1', 2', 3', etc., is obtained. This line can be extended again by connection with the guide point $r$, thus yielding the

96    HEAT CONDUCTION

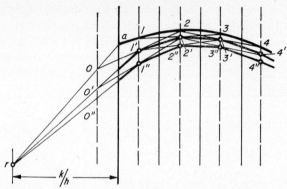

Fig. 4-16. Unsteady heat flow in a plane wall with convection on the exposed surfaces determined by the Schmidt method.

point $0'$. The process is repeated. If during the course of time the surrounding temperature $t_f$ or the heat-transfer coefficient $h$ changes, such changes can easily be incorporated in the process by appropriate displacement of the guide point $r$ in either the vertical or horizontal direction, respectively. This ease in handling the variable surface conditions is a marked advantage of the graphical procedure as compared with an analytical treatment of the differential equation, where a variable boundary condition leads to considerable difficulty mathematically. The Schmidt method has been extended by Nessi and Nissolle[1] to other body forms and to sectionally homogeneous systems.

**Example 4-4.** A living room with brick walls, which has initially an outside temperature of 30 F, is heated by some unit. It is to be determined how long it takes to establish a steady-state temperature distribution in the walls and in the room. The final temperature in the room is to be 70 F. The heat-transfer coefficient on the inside of the wall is $h_i = 1.2$ Btu/hr ft² F, the heat-transfer coefficient on the outside $h_a = 3.0$ Btu/hr ft F, the thermal diffusivity $\alpha = 0.012$ ft²/hr, and the thickness of the wall is 1.3 ft. The steady-state temperature in the wall is given by a straight line connecting the two points $a$ and $b$, which are separated from the two wall surfaces by $k/h_i$ and $k/h_a$, respectively, and have respective temperatures of 30 and 70 F. If we assume that the rate of heat transfer to the inside of the wall is constant, all temperature gradients in the wall during the heating process must have the same inclination to the inner surface. To start with the graphical method, the wall is divided into six layers, each with the thickness $\Delta x = 0.217$ ft. Then the time interval is given by

$$\Delta \tau = \frac{(\Delta x)^2}{2\alpha} = \frac{0.217^2}{2 \times 0.012} = 1.97 \text{ hr}$$

The construction of the temperature curves can be easily followed in Fig. 4-17. It can be seen that more than 80 hr, or 4 days, are needed before steady-state conditions

---

[1] A. Nessi and L. Nissolle, "Méthodes graphiques pour l'étude des installations de chauffrage et de réfrigeration en régime discontinu," Dunod, Paris, 1929. See also M. Jakob, "Heat Transfer," pp. 380–398, John Wiley & Sons, Inc., New York, 1949.

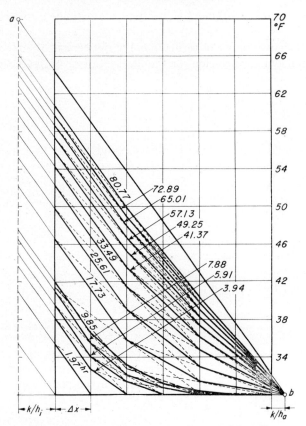

FIG. 4-17. Unsteady heat flow into a brick wall.

are reached. After the time 9.85 hr, the length interval was made $2\Delta x$; accordingly, the time interval now becomes $4 \times 1.97 = 7.89$ hr.

**4-2. Periodic Heat Conduction.** In many cases of practical interest, heat conduction occurs under conditions where the temperature (or heat-flux) boundary conditions change with time in a periodic manner. Examples of this phenomenon are evident in the cylinders of steam and internal-combustion engines, industrial processes where thermal cycling of the system is desirable from a controls point of view, and in many other instances. Such periodically transient systems can be treated by the Schmidt graphical method just described, but the method is apt to become graphically difficult. However, some analytical treatment is also possible, and by this means one can examine the nature of periodic heat conduction to an extent not possible graphically.

*The Solid of Infinite Thermal Conductivity, Surroundings Periodic in Time.* Consider the same little billet that was discussed in Sec. 4-1, but now consider that it is submerged in a fluid whose temperature is changing in a periodic manner. Any such periodic cycle can be represented by using a sufficient number of terms in a Fourier series;[1] therefore, we can write the following expression for the fluid-temperature variations:

$$t_f = \frac{a_0}{2} + \sum_{n=1}^{\infty} \left( a_n \cos \frac{2\pi n \tau}{\tau_0} + b_n \sin \frac{2\pi n \tau}{\tau_0} \right) \qquad (4\text{-}47)$$

where the $a_n$ and $b_n$ are Fourier coefficients and $\tau_0$ is the period of the temperature oscillation. When Eq. (4-47) is introduced into Eq. (4-1),

$$\frac{dt}{d\tau} + \frac{hA}{\rho cV} t = \frac{hA}{\rho cV} \left[ \frac{a_0}{2} + \sum_{n=1}^{\infty} \left( a_n \cos \frac{2\pi n \tau}{\tau_0} + b_n \sin \frac{2\pi n \tau}{\tau_0} \right) \right] \qquad (4\text{-}48)$$

which can be integrated[2] to yield

$$t = e^{-hA\tau/\rho cV} \left\{ \int e^{hA\tau/\rho cV} \frac{hA}{\rho cV} \left[ \frac{a_0}{2} \right. \right.$$
$$\left. \left. + \sum_{n=1}^{\infty} \left( a_n \cos \frac{2\pi n \tau}{\tau_0} + b_n \sin \frac{2\pi n \tau}{\tau_0} \right) \right] d\tau + C_1 \right\}$$

or when the operations are completed,[3]

$$t = \frac{a_0}{2} + \sum_{n=1}^{\infty} \frac{hA}{\rho cV} \left\{ \frac{a_n \cos\left[(2\pi n\tau/\tau_0) - \delta\right] + b_n \sin\left[(2\pi n\tau/\tau_0) - \delta\right]}{\sqrt{(hA/\rho cV)^2 + (2\pi n/\tau_0)^2}} \right\}$$
$$+ C_1 e^{-hA\tau/\rho cV} \qquad (4\text{-}49)$$

wherein $\delta = \tan^{-1}(2\pi n/\tau_0)(\rho cV/hA)$. The constant $C_1$ can be determined from the specification of the temperature at time zero; however, at long times the transient term goes to zero and the temperature oscillation becomes regular. The temperature-time history of the billet after the transient period can be expressed as

$$t = \frac{a_0}{2} + \sum_{n=1}^{\infty} \frac{a_n \cos\left[(2\pi n\tau/\tau_0) - \delta\right] + b_n \sin\left[(2\pi n\tau/\tau_0) - \delta\right]}{\sqrt{1 + \tan^2 \delta}} \qquad (4\text{-}50)$$

The conclusions to be drawn from Eq. (4-50) are that the temperature of the billet always lags behind the fluid temperature by the phase angle $\delta$

---

[1] R. V. Churchill, "Fourier Series and Boundary Value Problems," McGraw-Hill Book Company, Inc., New York, 1941.
[2] L. R. Ford, "Differential Equations," 2d ed., McGraw-Hill Book Company, Inc., New York, 1955.
[3] B. O. Pierce, "A Short Table of Integrals," 3d ed., Ginn & Company, Boston, 1929.

and the amplitude of the temperature variation is diminished by the factor

$$\frac{1}{\sqrt{1+\tan^2 \delta}}$$

**Example 4-5.** A resistance thermometer is used to measure the gas temperature in the cylinder of a machine running at 120 rpm. The output of the resistance thermometer is driving an oscilloscope. We desire to estimate the probable error in the measurement. Assuming the temperature oscillation to be sinusoidal, we can use the results of Eq. (4-50) with $n = 1$. The resistance bulb is roughly cylindrical in shape and is made of platinum, 0.020 in. in diameter. The thermal conductivity of platinum is $k = 40$ Btu/hr ft$^2$ F/ft, and the thermal diffusivity is $\alpha = 0.936$ ft$^2$/hr. The heat-transfer coefficient in the cylinder may be 50 Btu/hr ft$^2$ F. The significant dimension of the resistance bulb is $L = 0.020/4 = 0.005$ in. The machine makes $120 \times 60 = 7{,}200$ cycles/hr; thus

$$\tan \delta = \frac{2\pi}{\tau_0} \frac{\rho c V}{hA} = \frac{2\pi}{\tau_0} \frac{\rho c}{k} \frac{Lk}{h} = \frac{2\pi \times 7{,}200 \times 0.005 \times 40}{0.936 \times 12 \times 50} = 16.1$$

$$\tan^2 \delta = 259.$$

$$\text{Amplitude ratio} = \frac{1}{\sqrt{1 + 259.}} = 0.062$$

$$\delta = \tan^{-1} 16.1 = 86.5°$$

Therefore there is a 94 per cent temperature-reading error and a lag in the reading of 86.5° or almost one-quarter of a revolution.

*Semi-infinite Solid, Surface Temperature Periodic.* Consider again the semi-infinite solid extending from the surface $x = 0$ to infinity, where the temperature of the exposed surface at $x = 0$ is varying periodically with time. The differential equation for the system is

$$\frac{\partial t}{\partial \tau} = \alpha \frac{\partial^2 t}{\partial x^2}$$

and the initial and boundary conditions to be satisfied are

$$\tau = 0 \qquad t_i = t = 0$$
$$x = 0 \qquad t_0 = f(\tau)$$
$$x \to \infty \qquad t_\infty \ne \infty$$

A solution may be assumed to be of the form

$$t = F(\tau)G(x) \tag{4-51}$$

Since the temperature is to be periodic, it is necessary that the time $\tau$ and the space coordinate $x$ both appear in the argument of some trigonometric function. This can be forced by making the exponent for the solution for $F(\tau)$ imaginary. Differentiating Eq. (4-51) and separating the variables, we obtain

$$\frac{F'(\tau)}{\alpha F(\tau)} = \frac{G''(x)}{G(x)} = \pm i\lambda^2 \tag{4-52}$$

where $i = \sqrt{-1}$.

## HEAT CONDUCTION

As noted before in the treatment of product solutions, Eq. (4-52) is actually two equations:

$$F'(\tau) - (\pm i\lambda^2)\alpha F(\tau) = 0$$
$$G''(x) - (\pm i\lambda^2)G(x) = 0$$

from whose solutions comes

$$t = C e^{\pm \lambda^2 \alpha \tau} e^{\pm \lambda^2 \sqrt{\pm i} x} \tag{4-53}$$

Equation (4-53) actually results in four particular solutions:[1]

$$t_1 = C_1 \exp\left[-\sqrt{\tfrac{1}{2}}\,\lambda x + i(\lambda^2 \alpha \tau - \sqrt{\tfrac{1}{2}}\,\lambda x)\right]$$
$$t_2 = C_2 \exp\left[-\sqrt{\tfrac{1}{2}}\,\lambda x - i(\lambda^2 \alpha \tau - \sqrt{\tfrac{1}{2}}\,\lambda x)\right]$$
$$t_3 = C_3 \exp\left[\sqrt{\tfrac{1}{2}}\,\lambda x + i(\lambda^2 \alpha \tau - \sqrt{\tfrac{1}{2}}\,\lambda x)\right]$$
$$t_4 = C_4 \exp\left[\sqrt{\tfrac{1}{2}}\,\lambda x - i(\lambda^2 \alpha \tau - \sqrt{\tfrac{1}{2}}\,\lambda x)\right]$$

Two of these solutions are not physically admissible, since the temperature cannot be infinite for $x$ infinite. These two are discarded ($t_3$ and $t_4$) and the other two are added to obtain another particular solution:

$$t = e^{-\sqrt{\tfrac{1}{2}}\,\lambda x}\left[C_1 e^{+i(\lambda^2 \alpha \tau - \sqrt{\tfrac{1}{2}}\,\lambda x)} + C_2 e^{-i(\lambda^2 \alpha \tau - \sqrt{\tfrac{1}{2}}\,\lambda x)}\right] \tag{4-54}$$

Equation (4-54) can be written in terms of trigonometric functions as follows:

$$t = e^{-\sqrt{\tfrac{1}{2}}\,\lambda x}\left[A \cos(\lambda^2 \alpha \tau - \sqrt{\tfrac{1}{2}}\,\lambda x) + B \sin(\lambda^2 \alpha \tau - \sqrt{\tfrac{1}{2}}\,\lambda x)\right] \tag{4-55}$$

or in terms of a phase angle,

$$t = C e^{-\sqrt{\tfrac{1}{2}}\,\lambda x} \cos(\lambda^2 \alpha \tau - \sqrt{\tfrac{1}{2}}\,\lambda x - \delta) \tag{4-56}$$

where
$$\delta = \tan^{-1}\frac{B}{A} \quad \text{and} \quad C = \sqrt{A^2 + B^2}$$

It is necessary to determine the constants $A$, $B$, and $\lambda$ from the available boundary conditions. It is assumed that the function $t_0 = f(\tau)$ can be expressed by a Fourier series:

$$t_0 = \frac{a_0}{2} + \sum_{n=1}^{\infty}\left(a_n \cos\frac{2\pi n \tau}{\tau_0} + b_n \sin\frac{2\pi n \tau}{\tau_0}\right) \tag{4-57}$$

Equation (4-55) for $x = 0$ becomes

$$t_0 = A \cos \lambda^2 \alpha \tau + B \sin \lambda^2 \alpha \tau \tag{4-58}$$

[1] L. M. K. Boelter, V. H. Cherry, H. A. Johnson, and R. C. Martinelli, "Heat Transfer Notes," University of California Press, Berkeley, Calif., 1948, contains detailed solution.

# UNSTEADY HEAT CONDUCTION

Thus, comparing Eqs. (4-57) and (4-58), we see that

$$A = a_n \qquad B = b_n \qquad \lambda = \sqrt{\frac{2\pi n}{\alpha \tau_0}}$$

However, the constant term $a_0/2$ or the average temperature at $x = 0$ does not appear in Eq. (4-58). This term is actually the mean value of the oscillating temperature at $x = 0$ and is the result of the initial transient at time zero. We have developed such a step solution earlier which accounts for this variation:

$$t = \frac{a_0}{2}\left(1 - \text{erf}\,\frac{x}{2\sqrt{\alpha\tau}}\right) \qquad (4\text{-}59)$$

Equation (4-59) can be added to Eq. (4-56) for another particular solution:

$$t = \frac{a_0}{2}\left(1 - \text{erf}\,\frac{x}{2\sqrt{\alpha\tau}}\right) + \sum_{n=1}^{\infty} e^{-\sqrt{n\pi/\alpha\tau_0}\,x}\left[a_n \cos\left(\frac{2\pi n\tau}{\tau_0} - \sqrt{\frac{n\pi}{\alpha\tau_0}}\,x\right)\right.$$

$$\left. + b_n \sin\left(\frac{2\pi n\tau}{\tau_0} - \sqrt{\frac{n\pi}{\alpha\tau_0}}\,x\right)\right] \qquad (4\text{-}60)$$

That Eq. (4-60) satisfies the differential equation and its boundary conditions can be seen by evaluation at $x = 0$ and $x \to \infty$. Equation (4-60) is not satisfactory for the condition $t = 0$, $\tau = 0$ and cannot be used in that region of small time.[1] However, for $\tau$ large the erf $x/2\sqrt{\alpha\tau}$ goes to zero and the resulting expression adequately represents the periodic temperature distribution for the case where the initial transient has diminished.

$$t = \frac{a_0}{2} + \sum_{n=1}^{\infty} e^{-\sqrt{n\pi/\alpha\tau_0}\,x}\,C_n \cos\left(\frac{2\pi n\tau}{\tau_0} - \sqrt{\frac{n\pi}{\alpha\tau_0}}\,x - \delta_n\right) \qquad (4\text{-}61)$$

where $\qquad C_n = \sqrt{a_n^2 + b_n^2} \qquad$ and $\qquad \delta_n = \tan^{-1}\frac{b_n}{a_n}$

For the simpler case where the surface temperature is a cosine function

$$t_0 = t_{0M} \cos\frac{2\pi n\tau}{\tau_0}$$

and only the temperature oscillations about the mean temperature are considered; i.e., this is equivalent to taking the temperature datum as $a_0/2$. $t_{0M}$ is the maximum absolute value of the surface temperature variation. Equation (4-61) reduces to

$$t = t_{0M} e^{-\sqrt{n\pi/\alpha\tau_0}\,x} \cos\left(\frac{2\pi n\tau}{\tau_0} - \sqrt{\frac{n\pi}{\alpha\tau_0}}\,x\right) \qquad (4\text{-}62)$$

[1] The solution good for small $\tau$ is given in *ibid*.

The physical significance of the more complicated equations just written can be discussed in terms of the simpler Eq. (4-62). Consider Eq. (4-62) at a particular value of the space coordinate $x$ by assuming $x$ a constant. Since $\cos 2m\pi = +1$ when $m = 0, 1, 2, 3, \ldots$, $t$ will have maxima when

$$\frac{2\pi n \tau}{\tau_0} - \sqrt{\frac{n\pi}{\alpha \tau_0}} x = 2m\pi$$

or
$$\tau_{t\max} = m \frac{\tau_0}{n} + \frac{1}{2} \sqrt{\frac{\tau_0}{\alpha n \pi}} x$$

for any $x$.

For the surface $x = 0$, the temperature is

$$t_0 = t_{0M} \cos \frac{2\pi n \tau}{\tau_0}$$

and is a maximum when

$$\frac{2\pi n \tau}{\tau_0} = 2m\pi$$

or
$$\tau_{t_{0M}} = m \frac{\tau_0}{n} \qquad \text{at } x = 0$$

By comparing the values of time when the temperature maxima occur at a depth $x$ and at the surface, it can be seen that the oscillations have the

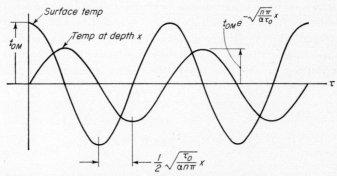

FIG. 4-18. Comparison of surface-temperature variation and temperature variation at depth $x$ during periodic heat conduction in semi-infinite body.

same period $\tau_0/n$ at each depth but the oscillations at depth $x$ lag those on the surface by $\frac{1}{2} \sqrt{\tau_0/\alpha n \pi}\, x$. Furthermore the surface amplitude is diminished at the depth $x$ by the factor

$$e^{-\sqrt{n\pi/\alpha \tau_0}\, x}$$

These physical features are shown in Fig. 4-18 and 4-19.

If it is recalled that the cosine is an even function, that is,

$$\cos(-\beta) = \cos \beta$$

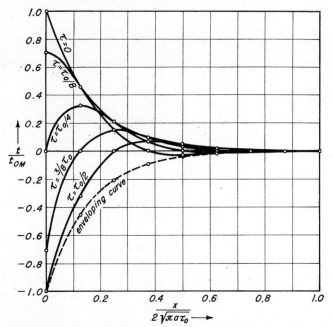

Fig. 4-19. Penetration of a temperature oscillation into an infinitely thick wall. (*From H. Gröber and S. Erk, "Die Grundgesetze der Wärmeübertragung," p. 75, Fig. 36, Springer-Verlag, Berlin, 1933.*)

Eq. (4-62) can be written as

$$t = t_{0M} e^{-\sqrt{n\pi/\alpha\tau_0}\,x} \cos\left(\sqrt{\frac{n\pi}{\alpha\tau_0}}\,x - \frac{2\pi n\tau}{\tau_0}\right) \qquad (4\text{-}63)$$

The term $t_{0M} \cos \sqrt{n\pi/\alpha\tau_0}\, x$ represents a cosine wave of amplitude $t_{0M}$ and wavelength $x_0$. The wavelength is equal to

$$x_0 = 2\sqrt{\frac{\pi\alpha\tau_0}{n}}$$

The term $t_{0M} \cos [\sqrt{n\pi/\alpha\tau_0}\, x - (2\pi n\tau/\tau_0)]$ represents the same wave as before, but moved in the positive $x$ direction by the amount $2\pi n\tau/\tau_0$. The velocity of wave propagation thus becomes $2\sqrt{\pi\alpha n/\tau_0}$. The advancing wave is decreasing in amplitude with increasing depth by the factor $e^{-\sqrt{n\pi/\alpha\tau_0}\,x}$. These characteristics are shown in Fig. 4-20. Equation (4-63) also shows that the higher the frequency (the larger $n$), the less the penetration of the thermal wave; i.e., high-frequency thermal oscillations are rapidly damped out in comparison with the fundamental or lower harmonic oscillations.

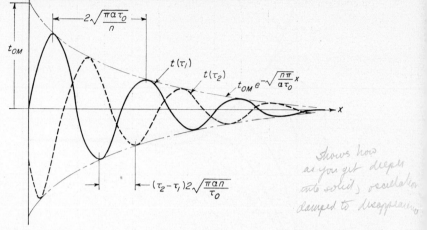

Fig. 4-20. Characteristics of a temperature oscillation penetrating an infinitely thick wall.

**Example 4-6.** It is to be determined to what depth the daily and yearly temperature fluctuations penetrate the ground. From the Appendix one can obtain the thermal diffusivity of claylike soil, 0.039 ft²/hr, and the thermal diffusivity of sandstone, 0.041 to 0.049 ft²/hr. Calculations are performed with the lower value 0.039. We infer from Fig. 4-19 that the oscillations have practically died off when $x/(2\sqrt{\pi\alpha\tau_0}) = 0.8$. For the daily fluctuations, $\tau_0 = 24$ hr. Thus we obtain

$$x = 1.6\sqrt{\pi 0.039 \times 24} \text{ ft} = 2.73 \text{ ft}$$

With the yearly fluctuation, the depth of penetration is then $\sqrt{365}$-fold, hence 52.2 ft.

After the initial transient effect has dissipated itself, the heat in a periodic system must necessarily flow into and out of the body periodically, since the temperature gradient at the surface is sometimes positive and other times negative. The heat absorbed or rejected can be deduced from the Fourier conduction equation

$$dQ_i = -kA \left(\frac{\partial t}{\partial x}\right)_{x=0} d\tau \qquad (4\text{-}64)$$

If Eq. (4-62) is differentiated, for example, with respect to $x$ and substituted into Eq. (4-64), we obtain

$$dQ_i = -kAt_{0M}\sqrt{\frac{n\pi}{\alpha\tau_0}}\left(\sin\frac{2\pi n\tau}{\tau_0} - \cos\frac{2\pi n\tau}{\tau_0}\right)d\tau$$

or

$$\frac{dQ_i}{d\tau} = -\sqrt{2}\,kAt_{0M}\sqrt{\frac{n\pi}{\alpha\tau_0}}\sin\left(\frac{2\pi n\tau}{\tau_0} - \frac{\pi}{4}\right)$$

which, when integrated between limits 1 and 2, becomes

$$Q = Q_i \bigg|_{\tau_1}^{\tau_2} = kAt_{0M}\sqrt{\frac{\tau_0}{2\pi n\alpha}}\cos\left(\frac{2\pi n\tau}{\tau_0} - \frac{\pi}{4}\right)\bigg|_{\tau_1}^{\tau_2} \qquad (4\text{-}65)$$

The relationships between the temperature maxima and minima and the heat-flow maxima and minima are shown in Table 4-2.

TABLE 4-2. PHASE RELATIONSHIPS IN PERIODIC HEAT FLOW*

| $t_{x=0}$ | $\tau$ | $Q$ | $\tau$ | $Q_i$ |
|---|---|---|---|---|
| Max | 0 | 0 | $\dfrac{1}{8}\dfrac{\tau_0}{n}$ | Max |
| 0 | $\dfrac{1}{4}\dfrac{\tau_0}{n}$ | Min | $\dfrac{3}{8}\dfrac{\tau_0}{n}$ | 0 |
| Min | $\dfrac{1}{2}\dfrac{\tau_0}{n}$ | 0 | $\dfrac{5}{8}\dfrac{\tau_0}{n}$ | Min |
| 0 | $\dfrac{3}{4}\dfrac{\tau_0}{n}$ | Max | $\dfrac{7}{8}\dfrac{\tau_0}{n}$ | 0 |
| Max | $\dfrac{\tau_0}{n}$ | 0 | $\dfrac{9}{8}\dfrac{\tau_0}{n}$ | Max |

* From L. M. K. Boelter, V. H. Cherry, H. A. Johnson, and R. C. Martinelli, "Heat Transfer Notes," University of California Press, Berkeley, Calif., 1948.

Using the values of time for $Q_\text{max}$ and $Q_\text{min}$ in Eq. (4-65) we can calculate the heat flowing into the solid during a half period:

$$Q_i \bigg|_{\frac{1}{8}(\tau_0/n)}^{\frac{5}{8}(\tau_0/n)} = kAt_{0M}\sqrt{\frac{2\tau_0}{\pi n\alpha}} \tag{4-66}$$

Equation (4-65) is shown graphically in Fig. 4-21.

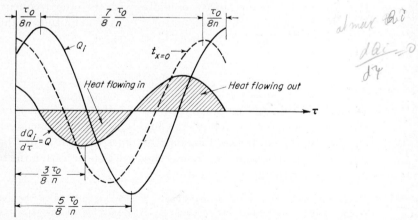

FIG. 4-21. Phase relationships in periodic heat flow. (*Courtesy of L. M. K. Boelter, H. A. Johnson, et al.,* "Heat Transfer Notes," *University of California Press, Berkeley, Calif., 1948.*)

*Semi-infinite Solid, Surrounding Fluid Temperature Periodic with Time.* If the fluid temperature adjacent to the exposed surface $x = 0$ varies periodically and yields a heat-transfer coefficient $h$, the solution follows in much the same way as in the preceding section except that now the boundary condition at $x = 0$ relates to the heat transfer by convection. The fluid temperature varies as

$$t_f = t_{fM} \cos \frac{2\pi n \tau}{\tau_0} \qquad (4\text{-}67)$$

and the boundary condition imposed on the surface to account for the heat convection is

$$k \left( \frac{\partial t}{\partial x} \right)_{x=0} = -h(t - t_f) \qquad (4\text{-}68)$$

A particular solution of the differential equation is given in Eq. (4-56), repeated here for convenience:

$$t = C_1 e^{-\sqrt{1/2}\, \lambda x} \cos (\lambda^2 \alpha \tau - \sqrt{1/2}\, \lambda x - \delta)$$

It is therefore necessary to determine the values of $C_1$, $\lambda$, and $\delta$ using Eqs. (4-56), (4-67), and (4-68) to obtain a solution of the proposed problem. Upon evaluation of these required constants, the solution becomes

$$t = t_{fM} \frac{e^{-\sqrt{\pi n/\alpha \tau_0}\, x}}{\sqrt{1 + 2\sqrt{(\pi n k^2/\alpha \tau_0 h^2)} + 2(\pi n k^2/\alpha \tau_0 h^2)}} \\ \cos \left[ \frac{2\pi n \tau}{\tau_0} - \sqrt{\frac{\pi n}{\alpha \tau_0}}\, x - \tan^{-1} \left( \frac{1}{1 + \sqrt{\alpha \tau_0 h^2 / \pi n k^2}} \right) \right] \qquad (4\text{-}69)$$

An examination of Eq. (4-69) for the surface conditions ($x = 0$) will show that the surface temperature $t_0$ oscillates with the same frequency $n/\tau_0$ as does the surrounding fluid temperatures but the amplitude of the surface-temperature oscillation is reduced by the factor

$$\frac{1}{\sqrt{1 + 2\sqrt{\pi n k^2/\alpha \tau_0 h^2} + 2(\pi n k^2/\alpha \tau_0 h^2)}} \qquad (4\text{-}70)$$

From the above equations it can also be seen that for a large Biot modulus the surface temperature follows the fluid temperature more closely and for smaller Biot moduli the difference becomes greater. Also higher frequency thermal oscillations are less effective in penetrating the body than are the lower frequency oscillations.

**Example 4-7.** A case of periodic heat transfer is exhibited in the cylinder of a reciprocating internal-combustion engine. It is possible to calculate the depth of penetration of the temperature oscillations into the cylinder wall. Assume that the engine operates at 2,000 rpm. If the engine is the two-stroke type, it will produce

2,000 peak temperatures per minute, or the period of one oscillation is

$$\tau_0 = \frac{1}{60 \times 2{,}000} = \frac{1}{12 \times 10^4} \text{ hr}$$

The thermal diffusivity of iron is $\alpha = 0.64$ ft²/hr. Thus again using the value $x/2 \sqrt{\pi \alpha \tau_0} = 0.8$ from Fig. 4-19 as a measure of the practical depth of penetration, we can calculate $x$:

$$x = 1.6 \sqrt{\frac{\pi \times 0.64}{12 \times 10^4}} = 6.54 \times 10^{-3} \text{ ft}$$

Thus the temperature fluctuations penetrate only to a depth of 0.08 in. into the cylinder wall.

Because of the finite heat-transfer coefficient, the surface temperature of the cylinder wall will fluctuate with a much smaller amplitude than that of the gas in the cylinder. The amplitude reduction can be calculated with the aid of Eq. (4-70). Take $n = 1$ for the fundamental wave, use from the Appendix $k = 30$ Btu/hr ft F, and estimate $h = 100$ Btu/hr ft² F; then

$$\frac{\pi k^2}{\alpha \tau_0 h^2} = \frac{\pi (30)^2 \times 12 \times 10^4}{0.64 \times 100^2} = 5390$$

$$(1 + 2\sqrt{5390} + 2 \times 5390)^{-\frac{1}{2}} = 0.0096$$

If the gas fluctuates through a double amplitude of 3000 F, then the cylinder wall surface temperature has a double amplitude of

$$3000 \times 0.0096 = 29 \text{ F}$$

According to tests made by A. Meier[1] with an Otto engine at $n = 2{,}000$ rpm, the fluctuation of the surface temperature was measured at approximately 20 F.

## PROBLEMS

**4-1.** A billet of steel in the form of a parallelepiped, with dimensions 4 by 4 by 10 ft, originally at a temperature of 500 F, is placed in a radiant furnace where the surface temperature is held at 2200 F. Specify the temperature of a point near a corner of the billet after an elapsed time of 25 min. The point in question is located 2 in. from one surface and 8 in. from each of the other surfaces.

**4-2.** The surface of a solid is exposed to quiet air, and the temperature throughout the solid is the same as that of the air. The surface of the solid is suddenly exposed to a radiant flux of $q = 400$ Btu/hr ft². Specify the variation of the temperature of the surface of the solid as a function of time. After 1 sec, the irradiation ceases. Specify the temperature variation of the surface with time from the moment when the irradiation ceases.

|       | $k$, Btu-ft/hr ft² F | $\rho$, lb/ft³ | $C_P$, Btu/lb F |
|-------|---------------------|----------------|-----------------|
| Solid | 0.10                | 34             | 0.50            |
| Air   | 0.015               | 0.075          | 0.24            |

[1] A. Meier, *Forsch. Gebiete Ingenieurw.*, **10**:41–54 (1939).

**4-3.** A slab of 15 per cent manganese-steel armor plate ($\alpha = 0.12 \text{ ft}^2/\text{hr}$; $k = 10$ Btu/hr ft F) 16 in. thick is taken from a soaking pit for surface-hardening operations. The plate is quenched in water at 212 F, yielding a heat-transfer coefficient of 2200 Btu/hr ft$^2$ F on the surface. How long will it take for the temperature of points in the slab $\frac{1}{4}$ in. away from the surface to reach a temperature of 700 F?

**4-4.** If it is specified that a rolled roast of beef is medium well-done when the center has attained the temperature of 160 F, estimate the time required to cook an 8-lb roast having a cylindrical form with the length equal to the diameter. The initial temperature of the meat is 50 F; the oven temperature is 350 F.

**4-5.** The heat-transfer coefficient for the convection heat transfer from a circular cylinder in crossflow is to be checked by observation of the temperature-time history of a 1-in.-diameter copper cylinder as it cools in an air stream.

*a.* Assuming an air temperature of 60 F, determine the maximum value of the heat-transfer coefficient for which the internal resistance of the cylinder can be neglected.

*b.* Plot the estimated temperature-time history of the cylinder from an initial cylinder temperature of 150 F for this heat-transfer coefficient.

**4-6.** Air flows in a duct. A thermocouple for measuring the air temperature is inserted into a well made of 16-gauge, $\frac{1}{4}$-in.-OD steel tubing, the well being arranged perpendicular to the direction of flow. The air temperature is 150 F, and the heat-transfer coefficient over the thermocouple well is 35 Btu/hr ft$^2$ F. A change of 5 F occurs instantaneously. Estimate the time required for the thermocouple to indicate a temperature change of 2 F, assuming that the thermocouple indication closely approximates the inside wall temperature of the well.

**4-7.** Consider a slice of bread inserted into an automatic electric toaster. The bread is heated from both sides by coils of wires which form a discontinuous surface having a uniform radiant emission of 500 Btu/hr ft$^2$. If free convection is present and is taken into account, estimate by means of a Schmidt plot the time required for the bread surface temperature to reach 350 F. How does this time compare with the usual toasting time of such units?

**4-8.** Derive the expression for the temperature-time history of a body of very large thermal conductivity which is suddenly immersed in a fluid bath which is caused to undergo a temperature variation of the form

$$t = t_0(1 + B \sin \omega_0 \tau)$$

Plot the bath- and body-temperature variations with time.

**4-9.** To what depth must a water pipe be buried in order to prevent freezing in a region having climatic conditions typical of the Minneapolis area? Give recommendations as to the depth of burial, considering the effect of soil and moisture conditions.

**4-10.** The temperature variation in a semi-infinite body due to a sinusoidal surface-temperature variation is expressed as

$$t = t_{0M} e^{-\sqrt{n\pi/\alpha\tau_0}\, x} \cos\left(\frac{2\pi n\tau}{\tau_0} - \sqrt{\frac{n\pi}{\alpha\tau_0}}\, x\right)$$

If damping is considered complete when the final double amplitude of the temperature wave is 5 per cent of that at the surface, how much deeper will the first harmonic penetrate into the body than the ninth harmonic?

**4-11.** A "fireproof" safe is to be made of sheet steel with asbestos board between the inner and outer shell. Estimate the thickness of asbestos required if 1-hr fire protection is to be anticipated on the basis of an outer temperature of 1500 F for this period, during which the inside surface temperature may not exceed 250 F. Compare the results obtained here with guarantees for commercial safe boxes of this type to determine if the guarantees on protection are reasonable.

**4-12.** Two semi-infinite bodies with different thermal properties initially existing at different but constant temperatures are suddenly placed in intimate contact with each other. Develop an expression for the temperature at the interface between the bodies.

**4-13.** Consider a slab of steel which is thick enough so that the semi-infinite solid solution will apply. The temperature at one surface varies periodically so that $t = 5 \sin (2\pi\tau/0.10) + 50$ and the other surface is at 50 F.

 a. Compare the rates of heat flow into the surfaces for the two cases.
 b. How thick must the slab be for the above conditions to apply?

CHAPTER 5

# HEAT CONDUCTION WITH MOVING BOUNDARIES

**5-1. Heat Conduction in Melting or Solidification.** Many problems arise in engineering where heat transfer is accompanied by a phase change in the conducting medium or where the chemical composition changes because of some chemical reaction which is propagated through the medium. Such phenomena as these are generally accompanied by the release or absorption of heat in the active zone. The energy so formed is commonly distributed in the system by the mechanism of heat conduction. Examples of such phenomena are the melting and solidification of solids, chemical reactions such as combustion, and penetration of frost into the ground. The essential and common feature in these systems is that an interface exists separating two regions of different thermophysical properties and this interface moves as some function of time. In addition, heat is liberated or absorbed at the interface. It is necessary for the solution of such problems to determine the way in which the interface will move. The problems so stated involve considerable difficulties and despite their obvious practical value have not received their share of attention.[1-4] The problem of solidification of a liquid will be considered here, with the realization that the other problems can be considered in a similar manner. In a problem such as this, we must consider the latent heat of fusion and at the same time account for the fact that properties such as thermal conductivity, thermal diffusivity, specific heats, and densities are different, sometimes markedly so, for the two phases. Consider Fig. 5-1, and suppose the substance fills the region $x > 0$ initially and is being frozen by the removal of heat at the exposed surface which is maintained at a constant temperature $T_1$. At any time $\tau$, the surface separating the liquid and solid phases is at $X(\tau)$. The bulk temperature of the fluid at a large distance away from the interface is $T_2$ and constant. Heat is conducted, therefore, from the liquid through the solid phase to

---

[1] H. S. Carslaw and J. C. Jaeger, "Conduction of Heat in Solids," pp. 71–74, Oxford University Press, London, 1947.

[2] H. Gröber and S. Erk, "Die Grundgesetze der Wärmeübertragung," 2d ed., pp. 118–121, Springer-Verlag, Berlin, 1933.

[3] Ingersoll and Zobel, "Mathematical Theory of Heat Conduction."

[4] J. H. Weiner, *Brit. J. Appl. Phys.*, **6**:361–363 (October, 1955).

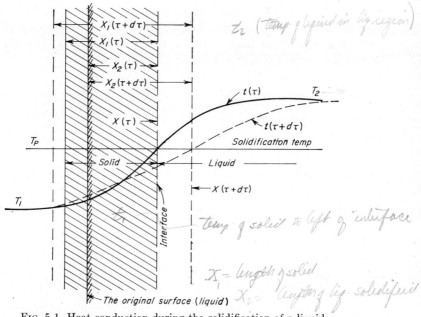

FIG. 5-1. Heat conduction during the solidification of a liquid.

the free surface. At the interface the system releases the latent heat of fusion.

At some time $\tau$, the region $x < X(\tau)$ consists of the solid phase with properties $k_1, a_1, \rho_1, c_1$, and if $t_1$ is the temperature within this solid phase, it must satisfy the expression

$$\frac{\partial^2 t_1}{\partial x_1^2} - \frac{1}{\alpha_1}\frac{\partial t_1}{\partial \tau} = 0 \qquad (5\text{-}1)$$

and $t_1 = T_1$ at $x = 0$.

The region $x > X(\tau)$ consists of the liquid phase with properties $k_2, a_2, \rho_2$, and $c_2$, and if $t_2$ is the temperature within this liquid phase (neglecting convection currents), it must satisfy the expression

$$\frac{\partial^2 t_2}{\partial x_2^2} - \frac{1}{\alpha_2}\frac{\partial t_2}{\partial \tau} = 0 \qquad (5\text{-}2)$$

and $t_2 \to T_2$ at $x \to \infty$.

In the case of the solidification of water to ice there is an increase in volume (decrease in density), and this effect can be taken into account by noting that the ice surface will move away from the original surface according to the density of each phase.[1] This can be expressed by the relationship

$$\frac{X_1}{X_2} = \frac{\rho_2}{\rho_1} = \beta \qquad (5\text{-}3)$$

[1] A. J. Rothmann private communication, 1953.

In addition we require at the interface, which is at the temperature necessary for phase change, that $t_1 = t_2 = T_p$ at $x_1 = X_1(\tau)$ or $x_2 = X_2(\tau)$.

If $Q_L$ is the latent heat of fusion of the solid, then, when the phase separation surface moves a distance $dx$, a quantity of heat

$$Q_L \rho_1 \frac{dX_1}{d\tau} = Q_L \rho_2 \frac{dX_2}{d\tau}$$

is liberated and must be conducted away. This requires that per unit area,

$$k_1 \left(\frac{\partial t_1}{\partial x_1}\right)_{x_1=X_1} - k_2 \left(\frac{\partial t_2}{\partial x_2}\right)_{x_2=X_2} = Q_L \rho_1 \frac{dX_1}{d\tau} = Q_L \rho_2 \frac{dX_2}{d\tau} \quad (5\text{-}4)$$

From this point on, the procedure follows the method due to Neumann.[1] Assume solutions of the form

$$\vartheta_1 = t_1 - T_p = (T_1 - T_p) + A \operatorname{erf} \frac{x_1}{2\sqrt{\alpha_1 \tau}} \quad (5\text{-}5)$$

$$\vartheta_2 = t_2 - T_p = (T_2 - T_p) + B \operatorname{erfc} \frac{x_2}{2\sqrt{\alpha_2 \tau}} \quad (5\text{-}6)$$

where $A$ and $B$ are constants, and thus Eqs. (5-5) and (5-6) satisfy Eqs. (5-1) and (5-2), respectively.[2] The condition relating to the interfacial temperature, i.e., $t_1 = t_2 = T_p$ at $x_1 = X_1(\tau)$ and $x_2 = X_2(\tau)$, results in

$$T_P - T_1 = A \operatorname{erf} \frac{X_1}{2\sqrt{\alpha_1 \tau}}$$
$$-(T_P - T_2) = B \operatorname{erfc} \frac{X_2}{2\sqrt{\alpha_2 \tau}} \quad (5\text{-}7)$$

Now since Eq. (5-7) must hold for all values of $X_1$ or $X_2$, these must be proportional to $\sqrt{\tau}$. Making use of Eq. (5-3), therefore, we have the relations

$$X_1 = K\beta \sqrt{\tau}$$
$$X_2 = K \sqrt{\tau} \quad (5\text{-}8)$$

where $K$ is a constant to be determined. When the results of Eqs. (5-5), (5-6), and (5-8) are used in Eq. (5-4), then

$$\frac{Ak_1}{\sqrt{\pi \alpha_1}} e^{-K^2\beta^2/4\alpha_1} - \frac{Bk_2}{\sqrt{\pi \alpha_2}} e^{-K^2/4\alpha_2} = \frac{Q_L \rho_1 K \beta}{2} \quad (5\text{-}9)$$

and when (5-7) is used in (5-9), then

$$\frac{(T_P - T_1)k_1 e^{-K^2\beta^2/4\alpha_1}}{\sqrt{\pi \alpha_1} \operatorname{erf}(K\beta/2\sqrt{\alpha_1})} - \frac{(T_2 - T_P)k_2 e^{-K^2/4\alpha_2}}{\sqrt{\pi \alpha_2} \operatorname{erfc}(K/2\sqrt{\alpha_2})} = \frac{Q_L \rho_1 K \beta}{2} \quad (5\text{-}10)$$

[1] Carslaw and Jaeger, op. cit., p. 72.
[2] erfc $z = (1 - \operatorname{erf} z)$

Equation (5-10) can be solved numerically to give the value $K$ in terms of $T_1$, $T_2$, and $T_P$ and the thermal properties of the material. When $K$ is known, $A$ and $B$ are found from Eqs. (5-7) and (5-8).

From Eq. (5-8) it is clear that for $\tau \to 0$, $X_1 \to 0$ and $X_2 \to 0$, and from Eq. (5-6) it follows that for $x_2 > 0$ and $\tau \to 0$, the temperature $t_2 = T_2$; thus the initial condition is that the region $x > 0$ at $\tau = 0$ is all liquid at temperature $T_2$. Equation (5-10) has been solved for $K$ for limited values of $T_1$ and $T_2$ around a solidification temperature of 32 F characteristic of water-ice systems for $\beta = 1$[1] and for $\beta = 1.09$[2], the density ratio of water to ice. An adequate approximation[2] for $K$ in the case of the water-ice system can be effected:

$$K_0 = \sqrt{\frac{2(T_p - T_1)k_1}{Q_L \rho_1}} \qquad \beta = 1$$

$$K_\beta = \sqrt{\frac{2(T_p - T_1)k_1}{Q_L \rho_1 \beta^2}} \qquad \beta > 1$$

Thus $K_0 = 1.09 K_\beta$ in the water-ice case. It can be seen, then, that a slower freezing time is calculated if the change in density in the ice phase from that of the water phase is considered.

**5-2. Moving Heat Sources.** Heat conduction in a body resulting from a moving heat source (or sink) is of great technical importance and has broad applications in the fields of arc welding, surface hardening, continuous casting or quenching, and cooling rotating systems with jets of coolant. Moving heat sources have been considered quite thoroughly by Rosenthal.[3] The analysis is based upon the notion that if a heat source is moved on a body of sufficiently large dimensions, a *quasi-steady state* occurs, *which is to say that the system appears to be steady state from the standpoint of the observer located in and traveling with the source*. For a three-dimensional system in the cartesian coordinate system, Eq. (2-13) is applicable if no heat sources are considered.

$$\frac{\partial t}{\partial \tau} = \alpha \left( \frac{\partial^2 t}{\partial x^2} + \frac{\partial^2 t}{\partial y^2} + \frac{\partial^2 t}{\partial z^2} \right)$$

If the properties of the material are considered to be constant, a quantity of heat at rate $Q$ may be supplied by a point source moving along the $x$ axis with a constant velocity $U$. Therefore, imagine an observer riding along with this heat source and having the material flow by him. If we supply to the observer a moving coordinate system of which he is the

---
[1] *Ibid.*, pp. 73–74.
[2] Rothman, *op. cit.*
[3] D. Rosenthal, *Trans. ASME*, **68**:849 (1946).

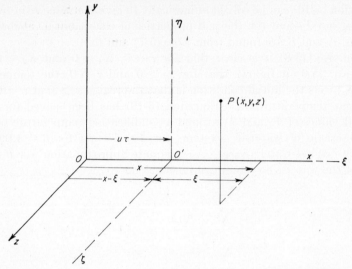

Fig. 5-2. Coordinate system for consideration of moving heat sources.

center, the system will appear in reference to the fixed system as shown in Fig. 5-2.

The point $P(x,y,z)$ in the fixed system becomes $P(\xi,\eta,\zeta)$ in the moving system, and since the translation in the $x$ direction does not change the other coordinates $y = \eta$ and $z = \zeta$, then $P(x,y,z) = P(\xi,y,z)$.

It is necessary to make a transformation from the variables of the stationary coordinate system where $t = f(x,y,z,\tau)$. In the new set of variables $\xi = x - U\tau$; thus $\partial \xi / \partial x = 1$, $\partial \xi / \partial \tau = -U$, and $\partial \tau' / \partial \tau = 1$. Thus the transformation proceeds:

$$\frac{\partial t}{\partial \tau} = \frac{\partial t}{\partial \xi}\frac{\partial \xi}{\partial \tau} + \frac{\partial t}{\partial \tau'}\frac{\partial \tau'}{\partial \tau} = -U\frac{\partial t}{\partial \xi} + \frac{\partial t}{\partial \tau'}$$

$$\frac{\partial^2 t}{\partial x^2} = \frac{\partial^2 t}{\partial \xi^2} \quad \frac{\partial^2 t}{\partial y^2} = \frac{\partial^2 t}{\partial \eta^2} \quad \frac{\partial^2 t}{\partial z^2} = \frac{\partial^2 t}{\partial \zeta^2}$$

If these substitutions are made in Eq. (2-13), we obtain Eq. (5-11).

$$\alpha \left( \frac{\partial^2 t}{\partial \xi^2} + \frac{\partial^2 t}{\partial y^2} + \frac{\partial^2 t}{\partial z^2} \right) = \frac{\partial t}{\partial \tau'} - U \frac{\partial t}{\partial \xi} \qquad (5\text{-}11)$$

The value $\partial t / \partial \tau'$ is zero from the point of view of the observer at the source, and thus Eq. (5-11) becomes, in quasi-steady form,

$$\frac{\partial^2 t}{\partial \xi^2} + \frac{\partial^2 t}{\partial y^2} + \frac{\partial^2 t}{\partial z^2} = -\frac{U}{\alpha}\frac{\partial t}{\partial \xi} \qquad (5\text{-}12)$$

*The Thin Rod.* Consider a rod similar to the one in Sec. 3-4 which has constant cross-sectional area. The temperature distribution at any cross section is to be assumed constant; i.e., the physical condition allowing such an assumption is one where the resistance to heat loss from the rod surface is much greater than the internal resistance to heat flow in the rod. This ratio of resistances allows the equalization of the temperature at each location because of the high thermal conductivity of the conducting material as compared with the low heat-transfer coefficient controlling the convective loss. In such a case there are no temperature gradients $\partial t/\partial y$ and $\partial t/\partial z$. Thus the appropriate differential equation for the problem becomes, in terms of the excess temperature,

$$\frac{\partial \vartheta}{\partial \tau} = \alpha \frac{\partial^2 \vartheta}{\partial x^2} - \frac{Q'}{\rho c}$$

*negative generation or loss by convection*

and if a surface-convection loss term is considered in the manner of Sec. 3-4, the above equation becomes

$$\frac{\partial \vartheta}{\partial \tau} = \alpha \left( \frac{\partial^2 \vartheta}{\partial x^2} - m^2 \vartheta \right) \tag{5-13}$$

where $m = \sqrt{hC/kA}$. Making the transformation to $\xi$ in the quasi-stationary case as above, Eq. (5-13) becomes

$$-\frac{U}{\alpha} \frac{\partial \vartheta}{\partial \xi} = \frac{\partial^2 \vartheta}{\partial \xi^2} - m^2 \vartheta \tag{5-14}$$

which has a solution of the form $\vartheta = e^{-(U/2\alpha)\xi} f(\xi)$, where $f(\xi)$ is to be determined. Thus, differentiating the expression for $\vartheta$ and substituting into Eq. (5-14), we obtain

$$f''(\xi) - \left[\left(\frac{U}{2\alpha}\right)^2 + m^2\right] f(\xi) = 0 \tag{5-15}$$

Equation (5-15) has a solution

$$f(\xi) = A e^{n\xi} + B e^{-n\xi}$$

where

$$n = \sqrt{\left(\frac{U}{2\alpha}\right)^2 + m^2} = \sqrt{\left(\frac{U}{2\alpha}\right)^2 + \frac{hC}{kA}}$$

and since $\vartheta = e^{-(U/2\alpha)\xi} f(\xi)$, the expression for $\vartheta$ becomes

$$\vartheta = A \exp\left\{ +\left[\sqrt{\left(\frac{U}{2\alpha}\right)^2 + \frac{hC}{kA}} - \frac{U}{2\alpha}\right]\xi \right\}$$
$$+ B \exp\left\{ -\left[\sqrt{\left(\frac{U}{2\alpha}\right)^2 + \frac{hC}{kA}} + \frac{U}{2\alpha}\right]\xi \right\} \tag{5-16}$$

The appropriate boundary conditions to be applied to Eq. (5-16) are

$$\vartheta = 0 \qquad \xi \to \pm \infty$$

The solution takes two forms, depending on whether or not $\xi \gtrless 0$. These are

$$\vartheta = A \exp \left\{ + \left[ \sqrt{\left(\frac{U}{2\alpha}\right)^2 + \frac{hC}{kA}} - \frac{U}{2\alpha} \right] \xi \right\} \qquad \xi < 0$$

$$\vartheta = B \exp \left\{ - \left[ \sqrt{\left(\frac{U}{2\alpha}\right)^2 + \frac{hC}{kA}} + \frac{U}{2\alpha} \right] \xi \right\} \qquad \xi > 0$$

At $\xi = 0$, the temperatures of both branches of the solution are equal and represent the maximum temperature. Thus $A = B = \vartheta_{\max}$.

$$\vartheta = \vartheta_{\max} \exp \left\{ + \left[ \sqrt{\left(\frac{U}{2\alpha}\right)^2 + \frac{hC}{kA}} - \frac{U}{2\alpha} \right] \xi \right\} \qquad \xi < 0$$
$$\vartheta = \vartheta_{\max} \exp \left\{ - \left[ \sqrt{\left(\frac{U}{2\alpha}\right)^2 + \frac{hC}{kA}} + \frac{U}{2\alpha} \right] \xi \right\} \qquad \xi > 0 \qquad (5\text{-}17)$$

The heat flows in each direction from $\xi = 0$ can be obtained from Eq. (5-17) by differentiating with respect to $\xi$ and substituting into the expression

$$Q = -kA \frac{dt}{d\xi}.$$

to yield

$$Q_1 = kA\vartheta_{\max} \left[ \sqrt{\left(\frac{U}{2\alpha}\right)^2 + \frac{hC}{kA}} - \frac{U}{2\alpha} \right] \qquad \xi < 0$$
$$Q_2 = kA\vartheta_{\max} \left[ \sqrt{\left(\frac{U}{2\alpha}\right)^2 + \frac{hC}{kA}} + \frac{U}{2\alpha} \right] \qquad \xi > 0 \qquad (5\text{-}18)$$

Since the total output of the heat source must be equal to the sum of the positive and negative heat flow, this relationship enables the calculation of the maximum temperatures $\vartheta_{\max}$, which becomes, in terms of the total heat output of the source,

$$\vartheta_{\max} = \frac{Q_1 + Q_2}{2kA \sqrt{(U/2\alpha)^2 + (hC/kA)}} \qquad (5\text{-}19)$$

The result of Eqs. (5-17) is given in Fig. 5-3, showing that more heat is conducted ahead of the source than is conducted away behind the source.

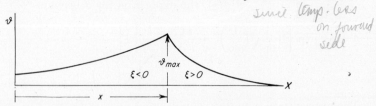

Fig. 5-3. Temperature distribution in a thin rod as a result of a heat source moving along the rod.

The results of Eqs. (5-17) to (5-19) can be modified to express the physical case of a thin rod which is insulated on the surface. For this case $h = 0$ and the referenced equations readily reduce.

*The Point Source.* If the heat flows in the three dimensions from a point source at $0'$ moving at velocity $U$ in the $x$ direction (Fig. 5-2), then the system is treated with more simplicity if the radius from $0'$ is used rather than the coordinates $x$, $y$, and $z$. Equation (5-12) has a solution of the form

$$\vartheta = e^{-(U/2\alpha)\xi} f(\xi,y,z)$$

which when used in Eq. (5-12) results in the differential equation for the unknown function $f(\xi,y,z)$

$$\frac{\partial^2 f}{\partial \xi^2} + \frac{\partial^2 f}{\partial y^2} + \frac{\partial^2 f}{\partial z^2} - \left(\frac{U}{2\alpha}\right)^2 f = 0$$

This auxiliary equation can be written, for $r = \sqrt{\xi^2 + y^2 + z^2}$, in the form

$$\frac{d^2 f}{dr^2} + \frac{2}{r}\frac{df}{dr} - \left(\frac{U}{2\alpha}\right)^2 f = 0 \tag{5-20}$$

when the initial and surface temperatures are such that the isothermal surfaces are concentric spheres, and thus the temperature depends on $r$ alone in this quasi-steady condition. This requires boundary conditions on $\vartheta$ that will satisfy the notion of radial heat flow from the source. These may be

$$\vartheta = 0 \qquad r \to \infty$$
$$\frac{\partial \vartheta}{\partial r} = \frac{q'}{4\pi k r^2} \qquad r = 0 \tag{5-21}$$

where $q'$ is the rate of heat flow from the source.

Solution of Eq. (5-20) can be effected by the transformation[1] $g = fr$ to yield

$$\frac{d^2 g}{dr^2} - \left(\frac{U}{2\alpha}\right)^2 g = 0$$

which has a standard solution

$$g = Ae^{+(U/2\alpha)r} + Be^{-(U/2\alpha)r}$$

We desire the solution for $\vartheta$ which is

$$\vartheta = e^{-(U/2\alpha)\xi} \frac{g}{r}$$

thus
$$\vartheta = \frac{1}{r}[Ae^{(U/2\alpha)(r-\xi)} + Be^{-(U/2\alpha)(r+\xi)}] \tag{5-22}$$

[1] Carslaw and Jaeger, *op. cit.*, p. 198.

The boundary conditions proposed in Eq. (5-21) are used in the solution of Eq. (5-22) to obtain finally

$$\vartheta = \frac{q'}{4\pi k r} e^{-(U/2\alpha)(r+\xi)} \qquad (5\text{-}23)$$

Equation (5-23) gives the temperature distribution about a moving point source in an infinite medium. This result is approximately valid for a source moving on the surface of a semi-infinite medium if the exposed surface is losing heat at a rate which is negligible compared with the source strength. A physical application of this analysis is in the treatment of a welding electrode moving along the surface of a very thick plate where there are minimal losses. In the welding problem only the lower half plane needs to be considered as the body. Equation (5-23) must be modified to account for this to read

$$\vartheta = \frac{q''}{2\pi k r} e^{-(U/2\alpha)(r+\xi)} \qquad (5\text{-}24)$$

where $q''$ is the heat release of the electrode.

The above methods apply to problems in arc welding, extrusion, quenching, annealing, and the passage of projectiles through a gun bore.

## PROBLEMS

**5-1.** Copper wire is being drawn through a thin die at a constant rate. Assuming that the heat of friction is given up to the wire in the plane perpendicular to the die axis, specify the temperature distribution in the wire as a function of the distance away from the die.

**5-2.** A fluid flows unidimensionally in the $x$ direction, and in a plane $x = \xi$ it passes through a fine mesh screen which is electrically heated. Determine the temperature distribution in the fluid.

**5-3.** Ice forms on the surface of a lake at 0 F from water at 32 F. As the layer of ice thickens, the rate of freezing is reduced owing to the thermal resistance of the layer already formed. Obtain an expression giving the thickness of the ice layer as a function of the time. As a first approximation one can neglect the heat capacity of the ice.

**5-4.** A 3,000-lb motorcar traveling 30 mph is stopped in 5 sec by four brakes, with brake bands of 40-in.$^2$ area, each pressing against steel drums of the same surface area. What maximum temperature rise might be expected? Establish carefully the assumptions made in the solution.

**5-5.** Consider the freezing of a spherical volume of water, initially at 60 F, when from the time zero on, the surface of the sphere is maintained at 10 F. In order to simplify the physical problem further, assume that the thermal properties are independent of the temperature, that the density of water and ice is the same, and that the heat conduction is in a radial direction only. For these restrictions the spherical heat-conduction equation can be transformed by $u = tr$ into the heat-conduction equation for the semi-infinite body for which a solution of this kind is known. Make such an equation transformation, and obtain the solution for the spherical case. To what values of time will the obtained solution be valid? Why will it become invalid? What other physical considerations should have been considered for the solution of this problem?

**5-6.** One wishes to weld together two pieces of steel 6 ft long, 2 ft wide, and ⅛ in. thick. The welding electrode produces a temperature of, say, 3000 F locally and moves at a rate of 24 in./min. The steel plate has been treated with a paint coating as a rust inhibitor. Estimate the surface area needing to be repainted at the completion of the welding process. Would convection from the surface appreciably help the situation?

PART B

# HEAT TRANSFER BY CONVECTION

**The Various Types of Heat Transfer.** This chapter deals with heat transfer between a fluid and a solid surface which is in contact with the fluid. A fluid motion can either be set up by some external source like a blower or be caused by temperature differences in the fluid which develop as a result of local heating. An example of the first type of heat transfer is the transport of heat from the walls of a duct to a fluid which is forced by a blower through the duct. An example of the second situation is the heat transfer from a stove, which heats a room, to the air. Heat exchange between a wall and a fluid when the flow is forced along the wall by external means is called *heat transfer by forced convection*. The heat exchange between a wall and a fluid when the fluid is set in motion by temperature differences between the wall surface and the surrounding fluid is called *heat transfer by free or natural convection*. Sometimes situations are encountered where a clear distinction between these modes of heat transfer is difficult to make and where the designation is somewhat arbitrary.

The main portion of the resistance to heat transfer is usually concentrated in a thin layer immediately adjacent to the wall surface. This has already been mentioned in Sec. 1-3. Heat transfer is essentially a problem of the interplay of heat conduction and energy transport by the moving fluid within this layer. As soon as the heat has penetrated this layer, it is carried away readily by the core of the fluid. Hence, the heat-transfer coefficient is determined essentially by the thickness and the characteristics of this boundary layer, which in turn depend on all the parameters determining the flow along the surface.

For forced convection we shall later deduce from the differential equations which control the fluid-flow and the heat-flow processes, the fact that neither the velocity field nor the development of the boundary layer is influenced by heat transfer as long as the fluid property values which enter into the flow equations are independent of temperature. The development of the boundary layer and the flow is then a pure fluid-mechanics problem. However, it can be seen in the property-value tables in the Appendix that all property values are actually dependent on temperature. In such a case an interrelation exists between the heat-flow and the fluid-flow processes. Such an interrelation makes an understanding of the heat-transfer process much more difficult, and in order to avoid this complication, a considerable portion of the discussion in this chapter will be devoted to an idealized fluid with properties which are independent of temperature. An additional advantage exists in this procedure. Only for such an ideal fluid can relationships be developed which are universally applicable, whereas relations for fluids whose properties vary with temperature or pressure hold for a specific fluid or at best for a certain group of fluids. The recognition of this fact and its utilization to develop universal relations were the most significant contributions of W. Nusselt (1916). Finally, many situations arise in applications in which the temperature differences within the flow field are sufficiently small that the variation of properties caused by temperature differences can be neglected. In these cases relationships developed for the ideal constant-property fluid accurately describe the actual heat-transfer process. For such a situation it makes no difference whether the fluid is a liquid or a gas, and the ideal fluid relationships hold for both. In gases, the fluid properties, and especially the density, depend not only on temperature but also on pressure, and therefore the pressure variation has to be small enough that the density variation connected with it can be considered as small. This is the case as long as the flow velocities are small compared with the sound velocity. In air flow, for instance, it is found that constant-property relations describe the actual situation well up to velocities of approximately 300 fps.

Situations are sometimes encountered in engineering applications in which very large variations of the properties arise. For instance, the temperature dependence of the viscosity of oil is such that it influences the heat transfer in oil coolers considerably, even when the temperature differences encountered are very moderate. In nuclear reactors or in many high-temperature processes the temperature differences are so large that they cause very large variations of the properties. In gases at high subsonic and especially at supersonic velocities large pressure and temperature variations are inherently connected with the flow process. It has become a practice in the field of heat transfer to modify relations which have been developed for a constant-property fluid in such a way that they also account for the effects caused by property variations. Some specific relationships have also been developed, and these will be discussed later on in the following chapter.

In free convection, the flow is generated by the temperature differences. Consequently, from the outset the flow and heat-exchange processes are intimately connected.

The present work deals almost exclusively with heat transfer for steady-state processes. In an unsteady process (transient and periodic) the given relationships change as soon as the temperatures in the boundary layer lag noticeably owing to heat accumulation (approximately as in Fig. 4-19). However, this is the case only for very rapid changes, because the boundary-layer thicknesses are generally small. Heat transfer in the cylinders of diesel engines, for instance, can still be computed with steady-state heat-transfer coefficients,[1] and accelerations of missiles or aircraft of the order of 20 to 40 $g$ are necessary to cause a heat transfer to the skin of the craft which is markedly different from steady-state heat transfer.[2]

Since heat transfer, as discussed previously, is determined by hydrodynamic processes, familiarity with these processes is absolutely necessary for an understanding of heat-transfer phenomena. Therefore, the most important factors of the flow theory will be considered in the following chapter.

[1] H. Pfriem, Unsteady Heat Transfer in Gases, Especially in Piston Engines, *VDI-Forschungsheft* 413, 1942.

[2] K. Stewartson, *Quart. J. Mech. Appl. Math.*, **4**(2): 182–198 (1951); E. M. Sparrow, *Jet Propulsion*, **28**: 403 (1958); E. M. Sparrow and J. L. Gregg, *Nat. Advisory Comm. Aeronaut.. Tech. Note* 4311 (1958); R. Siegel, *Trans. ASME*, **80**: 347 (1958).

CHAPTER 6

# FLOW ALONG SURFACES AND IN CHANNELS

**6-1. Boundary Layer and Turbulence.** Primarily, there are two concepts of the flow theory which are essential for an understanding of heat transfer, namely, the concept of the boundary layer and the concept of turbulence. Even though liquids and gases have a measurable viscosity which in flow problems causes frictional forces on solid surfaces, it is still possible in many cases to approximate the actual conditions by the hypothesis of a frictionless fluid. In it, two types of forces occur when body forces are absent, namely, forces of inertia and forces of pressure. The balance of both forces, along each streamline, is given by the Bernoulli equation (D. Bernoulli, 1700–1782)

$$p + \rho \frac{V^2}{2} = \text{const} \tag{6-1}$$

where $p$ = fluid pressure
$\rho$ = density
$V$ = flow velocity

In addition to the forces above, other forces exist in real liquids and gases which are caused by viscosity. These forces manifest themselves as shear stresses between the individual streamlines, where the streamlines are flowing with different velocities. In a flow according to Fig. 6-1, in which the velocity $u$ is directed parallel to the wall $ab$ and velocity differences occur in the $y$ direction normal to the wall, shear stresses $\tau$ occur as a result of the viscosity in planes parallel to the wall, for instance, in plane 1-1. The value of the stress is given by Newton's equation

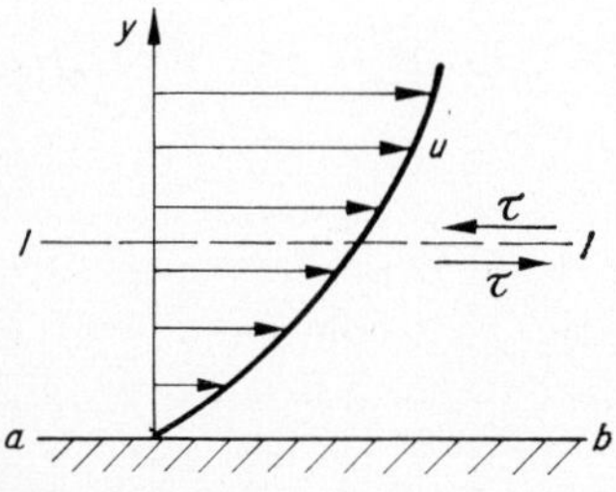

Fig. 6-1. Shear stress in a viscous fluid.

$$\tau = \mu \frac{du}{dy} \tag{6-2}$$

and its direction is indicated by arrows in Fig. 6-1.

For other types of flow, the expression for the shear stress is more complicated. Expressions for the shear stresses in a viscous fluid for the general situation in which velocity components $u$, $v$, $w$ in all three directions in space exist and in which these, in turn, are functions of all three coordinates $x$, $y$, $z$ have been set up by G. G. Stokes (1845). According to Stokes, in a plane normal to the $y$ direction two shear stresses are present of magnitudes

$$\tau_{yx} = \mu \left( \frac{\partial u}{\partial y} + \frac{\partial v}{\partial x} \right) \tag{6-3}$$

$$\tau_{yz} = \mu \left( \frac{\partial v}{\partial z} + \frac{\partial w}{\partial y} \right) \tag{6-4}$$

The first subscript on these stresses $\tau$ indicates that the stresses act in a plane normal to $y$; the second subscript indicates the direction in which the force described by the stress acts. In this more general situation Eq. (6-3) rather than Eq. (6-2) describes the shear stress in the direction indicated by Fig. 6-1. Similar shear stresses exist in the planes normal to $x$ and $z$. In addition to these shear stresses normal stresses exist in all planes. The reader interested in these relations will find more information in texts on fluid mechanics, for instance, in H. Schlichting, "Boundary Layer Theory." In this book we shall deal only with flows for which the simple Eq. (6-2) describes the shear stresses with sufficient accuracy. The shear stress is, according to this relation, proportional to the velocity gradient normal to the direction of flow. The proportionality factor $\mu$ is a property value and is called *dynamic viscosity*. Its dimension is obtained from the equation of definition (6-2) as $lb_f \ sec/ft^2$ or as $lb/sec \ ft$. Besides the dynamic viscosity, the kinematic viscosity is frequently used. Kinematic viscosity $\nu$ is related to dynamic viscosity by the equation

$$\nu = \frac{\mu}{\rho} \tag{6-5}$$

where $\rho$ signifies the density. The dimension of the kinematic viscosity is $ft^2/sec$. In the tables of property values in the Appendix some viscosities are shown. Since the dynamic viscosity in some handbooks is given in the absolute system, namely, in poises (1 poise = 1 g/sec cm), errors frequently occur in applications using the engineering system of units. These errors can be avoided if the calculations are made with the kinematic viscosity which has the dimension length squared per unit time in both systems. For this reason the kinematic viscosity is recorded in the tables of the Appendix.

In liquids as well as in gases, the dynamic viscosity $\mu$ depends primarily upon the temperature and only slightly upon pressure. Only in the vicinity of the critical point does there occur a strong pressure dependency. Figure A-4 shows these relationships for water and steam.

All other liquids for which such information is available are found to behave principally in the same manner. According to Eq. (6-5) the kinematic viscosity $\nu$ in liquids is also practically independent of pressure because of their low compressibility. In gases, according to the equation of state, it is inversely proportional to the pressure. The numerical value of the dynamic viscosity is considerably higher for liquids than for gases. For the kinematic viscosity this relationship is often reversed. For example, the value for water at room temperature is approximately one-tenth the value for air at room temperature.

Since the numerical values of these viscosities are comparatively low, higher shear stresses in a flow, according to Eq. (6-2), occur only where high-velocity gradients $du/dy$ exist. These high gradients are always found near solid walls within a flow field. If the velocity field is measured with a fine pitot tube in the immediate vicinity of such a wall, a velocity profile is obtained as in Fig. 6-2. The velocity begins with the value zero at the wall and increases within a thin layer of thickness $\delta$ to the value of the undisturbed stream which occurs at some distance from the body. This knowledge, which is essential for flow and heat-transfer calculations, was utilized by Ludwig Prandtl to develop his famous boundary-layer theory in 1904. The term *boundary layer* for the thin layer with steep velocity increase was also conceived by Prandtl. Outside the boundary layer, the velocity gradient normal to the direction of flow is usually so low that the viscosity effects can be disregarded. Hence it can be assumed that each flow can be divided into two regions, namely, the boundary layers to which the viscosity effect is confined and the main flow outside the boundary layer which can be considered frictionless and where for this reason Bernoulli's equation applies along each streamline. The fact that the boundary layers are the cause of all flow separations and thus bring about a change in the main flow will be discussed in the following section.

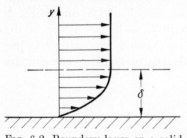

FIG. 6-2. Boundary layer on a solid surface.

In 1883, Osborne Reynolds demonstrated for the first time that two basically different forms of flow exist, namely, a *laminar* and a *turbulent* flow. In laminar flow the individual streamlines run in an orderly manner side by side, while in turbulent flow the streamlines are interwoven with each other in an irregular manner. In the latter case, the individual fluid particles execute fluctuating motions around some mean flow path in an irregular way. In daily life, for example, one can observe both flow processes in the smoke which rises from a cigarette. The smoke, as can be seen in Fig. 6-3, rises first in the form of a straight thread. After a certain distance, however, it becomes wavy and curly and the smoke thread vanishes quickly by mixing with the air. The first part represents the laminar form of flow, and the second part the turbulent form. Turbulence can be generated in an air stream by holding a wire grid in the flow. In every wind tunnel, a certain turbulence is present which is induced in the air stream by the blower and the guide vanes.

A quantitative description of turbulent flow and especially of the intensity of turbulence is customarily obtained in the following way. Imagine that one measures the velocity component $u$ in turbulent flow at a fixed location as a function of time. The result will be a plot like the one presented in Fig. 6-4. The velocity component $u$ at any moment can then be written as

$$u = u' + \bar{u} \qquad (6\text{-}6)$$

where $\bar{u}$ is the time mean value of $u$ and $u'$ is the velocity fluctuation. Turbulent flow is called steady when $\bar{u}$ does not change with time, and $u'$ is defined by the fact that its time mean $\overline{u'}$ is zero when taken over a sufficiently long time interval. Similar relations hold for the velocity components $v$ and $w$. Let us assume that $\bar{v} = \bar{w} = 0$, which means that the mean velocity is in the $x$ direction. Then the turbulence intensity is customarily described by the term

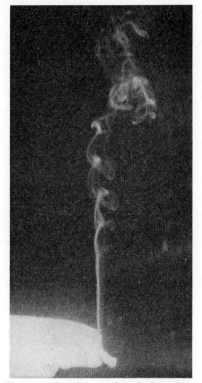

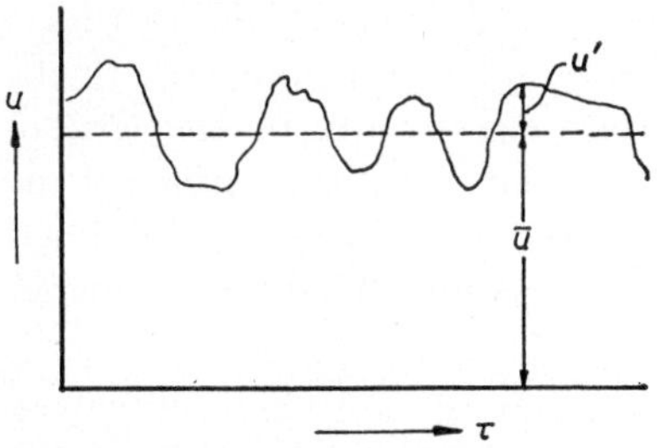

FIG. 6-3. Laminar and turbulent flow of cigarette smoke.

FIG. 6-4. Turbulent velocity fluctuations.

$$J = \frac{\sqrt{\tfrac{1}{3}(\overline{u'^2} + \overline{v'^2} + \overline{w'^2})}}{\bar{u}} \qquad (6\text{-}7)$$

in which the term $\overline{u'^2}$ is the time mean of the square of the velocity fluctuation $u'$. Other terms which are of importance in the description of turbulent flow are the time means of products of velocity fluctuations like $\overline{u'v'}$ because they are connected with turbulent shear stresses. The discussion of these will be resumed in a later section.

Generally, for the discussions in this book we need only time mean velocities. Therefore the symbols $u$, $v$, $w$ will be used for those unless it is specifically stated otherwise.

Heat exchange by convection is promoted by the fluctuating motions in turbulent flow. Consequently, heat transfer in turbulent flow is considerably higher than in laminar flow.

The flow can also be turbulent within the boundary layer. This type of turbulence is of considerable interest in the heat-exchange processes. On a flat plate which is parallel to the flow, a boundary layer is formed as illustrated in Fig. 6-5. Its thickness increases toward the rear, being zero at the plate leading edge. At a certain critical distance $x_c$ from the leading edge, the flow within the boundary layer, which is at first laminar, changes into turbulent motion. If the velocity $u_s$ is increased, the critical length $x_c$ becomes continuously less in such a manner that the product $u_s x_c$ remains unchanged. If finally the tests are made with gases or liquids with various viscosities, it is found that the change from laminar to turbulent flow takes place at a definite value of the dimensionless term $u_s x_c / \nu$. This fact was first discovered by Reynolds. In his honor a dimensionless term, which is obtained by multiplication of a velocity with a length and by division with a kinematic viscosity, was named Reynolds number or Reynolds modulus Re. That value of the Reynolds number at which a flow changes into the turbulent form is called the critical or transition Reynolds number $Re_c$. A more intensive investigation reveals that the value of the critical Reynolds number for a boundary layer can be influenced by external circumstances. If the approaching flow has already been disturbed, e.g., by a turbulence grid or by a special shape of the leading edge of the plate, or if the surface of the plate is rough, the change is attained at a lower critical Reynolds number. If these disturbances are prevented, it is possible to obtain a higher critical Reynolds number.

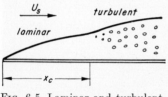

Fig. 6-5. Laminar and turbulent boundary layer on a plate.

It has been established by measurements at the National Bureau of Standards that a definite upper limit exists for laminar boundary-layer flow on a flat plate. In this investigation the turbulence level upstream of the plate was systematically varied. The turbulence intensity, the parameter described by Eq. (6-7), is indicated on the ordinate of Fig. 6-6. The transition Reynolds number $Re_c$ is plotted on the abscissa. It can be seen from the figure that the transition Reynolds number increases with a decrease in the turbulence of the free stream. This, however, holds true only down to a certain turbulence level. Below the value of 0.08 per cent of the intensity of turbulence, the transition Reynolds number cannot be influenced any more by the freestream turbulence.

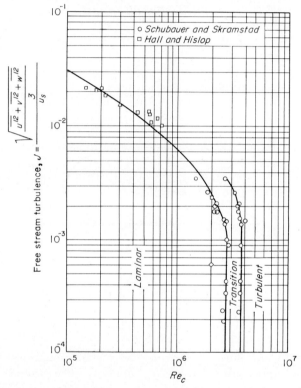

FIG. 6-6. Critical Reynolds number $Re_c$ for transition to turbulence of a boundary layer on a flat plate as function of the stream turbulence J. [*According to G. B. Schubauer and H. K. Skramstad, J. Reasearch. Nat. Bur. Standards,* **38**:281 (1947); *A. A. Hall and G. S. Hislop, Aeronaut. Research Comm., Rep. and Mem., No.* 1843 (1938).]

This Reynolds number, therefore, constitutes an upper limit for laminar boundary-layer flow on a flat plate. The transition Reynolds number depends, in addition to the stream turbulence, on a number of other parameters, for instance, on the way in which the pressure in the flow and the velocity $u_s$ vary along the surface. The transition Reynolds number is higher when the pressure decreases or when the velocity increases along the surface, whereas a flow deceleration along the surface decreases the critical Reynolds number. A prediction of the conditions for which laminar or turbulent flow exists is very important for heat-transfer calculations because laminar and turbulent heat transfer are very different in their magnitude. A large amount of information has been collected through the years on the transition Reynolds number for various flow conditions. Also, a great effort was devoted to research aimed at a better understanding of the nature of turbulent flow and of the transition process. However, our present-day understanding of both situations is still quite limited. Transition seems to be essentially caused by the fact that laminar flow becomes unstable under certain situations and transforms to a turbulent flow under the influence of even very small outside disturbances. The transition process itself has to be regarded in most cases as a continuously fluctuating process. It has been established, for instance, that the transition point often fluctuates continuously in time around a mean location. It also takes a certain time and,

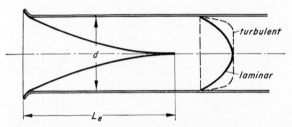

Fig. 6-7. Flow in a tube near the inlet.

therefore, in the flow field a certain distance to develop turbulence fully after the onset of the first disturbances. Thus, observations of the initial disturbances in laminar flow result in a smaller critical Reynolds number than measurements of parameters which indicate the establishment of turbulent flow. Consequently, one distinguishes between a lower and an upper critical Reynolds number. The two curves in Fig. 6-6 indicate both of these values. By special means, for instance by so-called turbulence trips (suitable local roughness), the location of the transition point can be fixed and the establishment of the turbulent flow accelerated. Such means are very useful and are often applied in experimental investigations of heat transfer in turbulent flow.

The conditions in a pipe in the vicinity of the inlet are similar to the conditions on a plate in parallel flow. Here also boundary layers form on the walls, beginning with zero thickness at the pipe inlet and increasing continuously in downstream direction as indicated in Fig. 6-7. It is assumed that by taking a suitable shape for the inlet, the flow into the pipe is smooth and that separation does not occur. At a certain distance $L_e$ from the inlet, the boundary layers become so thick that they make contact with each other. The velocity profile downstream from this point does not change (Fig. 6-7) and has the form of either a parabola or an arched curve, depending on whether the boundary layers are still laminar or already turbulent. This part of the flow is called *developed flow*. The entrance length $L_e$ divided by the diameter $d$ of the tube is a function of the Reynolds number. In the developed-flow region, the flow is turbulent when the Reynolds number exceeds a critical value. If the Reynolds number is defined by the mean velocity $u_m$ over the cross section and the tube diameter $d$, the critical value is found to be

$$\text{Re}_c = \frac{u_m d}{\nu} = 2{,}300$$

By carefully avoiding all disturbances, a critical Reynolds number of 500,000 was reached. Under conditions as they prevail in industrial applications, flow in tubes usually is turbulent when the Reynolds number exceeds the value 3,000.

**6-2. The Momentum Equation of the Boundary Layer.** For the following calculations of heat transfer it is necessary to know the characteristics of the boundary layer, especially its thickness. For laminar boundary layers on bodies of various shapes this problem was solved by integration of the boundary-layer equations derived by L. Prandtl in 1904. These equations and some of their solutions will be discussed in a following chapter. The computations leading to these solutions, however, are difficult and lengthy. Therefore an approximate method originated by T. von Kármán[1] will be developed first. This method is much simpler and has an advantage in that it can be applied to situations for which an exact solution is impossible. In all cases where this approximate method has been applied with sound physical logic, it has given satisfactory results. It starts from the law of momentum, which can be formulated for a steady flow in the following way:

*Imagine in the flow field a certain region enclosed by a stationary control surface of arbitrary form. The fluid particles in flowing through this region in general experience a change of momentum.*

$$Momentum = mass \times velocity$$

*The increase of momentum per unit of time of all particles passing this region can be expressed as the difference between the momentum leaving the region per unit time and the momentum entering it through the surface. This change in momentum per unit time is equivalent to inertia forces and must be in equilibrium with the external forces acting on the surface and within the enclosed region.*

This law can now be applied to steady two-dimensional flow along a

[1] T. von Kármán, *Z. angew. Math. Mech.*, **1**:235 (1921).

plane wall (Fig. 6-8). Use is made of a system of coordinates whose $x$ axis is parallel to the wall and whose $y$ axis is normal to it. The velocities $u$ have essentially a direction parallel to the wall. Only small velocities $v$ normal to the wall will occur. The law of momentum is used for the $x$ direction. The control surface is formed by two planes 1-2 and 3-4, spaced apart from each other by the small distance $dx$; by a plane parallel to the wall at a distance $l$; and by the wall. The distance $l$ is chosen greater than the boundary-layer thickness $\delta$. The velocity profile in the plane 1-2 has the sketched shape. At a distance $y$ from the wall the velocity is $u$. The outside stream velocity $u_s$ is reached at the distance $\delta$ from the wall, therefore within the length $l$. The control surfaces which are parallel to the plane of the paper are a unit distance apart. Through a strip with the height $dy$ at a distance $y$ the mass flow per unit time is $\rho u \, dy$. Its momentum in the $x$ direction is obtained by multiplying it by the velocity component $u$. The momentum flow therefore is $\rho u^2 \, dy$. The momentum flow through the whole plane 1-2 is

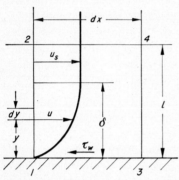

Fig. 6-8. Calculation of the flow boundary layer.

$$\rho \int_0^l u^2 \, dy$$

Progressing in the $x$ direction this momentum flow changes by

$$\rho \frac{d}{dx} \left( \int_0^l u^2 \, dy \right) dx = \rho \, dx \, \frac{d}{dx} \int_0^l u^2 \, dy$$

Through the wall 1-3 there is no flow. Through the plane 2-4, however, a flow usually occurs, the amount of which can be calculated in the following way. The mass flow through plane 1-2 per unit time is $\rho \int_0^l u \, dy$.

In progressing by the distance $dx$ this flow changes by $\rho \, dx \, \dfrac{d}{dx} \int_0^l u \, dy$. This is the difference between the flow through the planes 3-4 and 1-2, and it must have entered the parallelepiped through the plane 2-4. The velocity component in the $x$ direction in this plane is $u_s$. Therefore, the $x$ momentum flow through plane 2-4 is

$$\rho u_s \, dx \, \frac{d}{dx} \int_0^l u \, dy$$

The whole increase in momentum within the parallelepiped is, therefore,

$$-\rho u_s \, dx \, \frac{d}{dx} \int_0^l u \, dy + \rho \, dx \, \frac{d}{dx} \int_0^l u^2 \, dy = -\rho \, dx \, \frac{d}{dx} \int_0^l (u_s - u) u \, dy$$
$$+ \rho \, dx \, \frac{du_s}{dx} \int_0^l u \, dy$$

The following external forces act on the surface of the parallelepiped in the $x$ direction: a shear stress $\tau_w$ along the wall 1-3 and the pressures $p$ and $p + (dp/dx) \, dx$ on the planes 1-2 and 3-4, respectively. In the plane 2-4 no shear stresses occur, since the plane is beyond the boundary layer. By equating these forces with the increase of momentum,

$$\rho \frac{d}{dx} \int_0^l (u_s - u) u \, dy - \rho \frac{du_s}{dx} \int_0^l u \, dy = \tau_w + l \frac{dp}{dx} \qquad (6\text{-}8)$$

Boundary-layer theory shows that the pressure $p$ varies in the $y$ direction throughout the boundary layer by a negligibly small amount only. Bernoulli's equation (6-1) can, therefore, be used to bring the momentum equation (6-8) into a form which is more suitable for numerical calculations. According to Bernoulli's equation, the velocity $u_s$ in the stream outside the boundary layer is related to pressure gradient in the following way:

$$\frac{dp}{dx} = -\rho u_s \frac{du_s}{dx} \qquad (6\text{-}9)$$

This equation can be used to express the last term on the right side of Eq. (6-8), keeping in mind that the stream velocity $u_s$ does not depend on $y$:

$$l \frac{dp}{dx} = -\rho \frac{du_s}{dx} \int_0^l u_s \, dy$$

Substitution into Eq. (6-8) results in

$$\rho \frac{d}{dx} \int_0^\delta (u_s - u) u \, dy + \rho \frac{du_s}{dx} \int_0^\delta (u_s - u) \, dy = \tau_w \qquad (6\text{-}10)$$

The integration limits have now been changed, since in the range $\delta < y < l$ the term $u_s - u$ in the integrands is equal to zero.

The equation can further be written in a simpler form when the following abbreviations are used:

$$\delta^* = \int_0^\delta \left(1 - \frac{u}{u_s}\right) dy \qquad (6\text{-}11)$$

$$\delta_i = \int_0^\delta \left(1 - \frac{u}{u_s}\right) \frac{u}{u_s} \, dy \qquad (6\text{-}12)$$

The first term is called the *displacement thickness* of the boundary layer, since it indicates the extent to which the surface would have to be displaced in the direction toward the flow in order to create the same external flow field in a fluid which is free of viscosity. The second term $\delta_i$ is called the *momentum thickness* of the boundary layer, since it is connected with the flow of momentum or impulse through an area normal to the surface. The boundary-layer equation written with the two boundary-layer thicknesses is

$$\rho \frac{d}{dx}(u_s^2 \delta_i) + \rho u_s \frac{du_s}{dx} \delta^* = \tau_w \qquad (6\text{-}13)$$

Solutions of the boundary-layer momentum equation will be presented in the following paragraphs. At first, however, the exact boundary-layer equations as derived by L. Prandtl will be discussed.

**6-3. The Laminar-flow Boundary-layer Equation.** The boundary-layer equations will be derived in this paragraph from the Navier-Stokes equations. These equations which describe generally the flow of a viscous fluid have been derived by M. Navier (1827) and S. D. Poisson (1831) from a consideration of intermolecular forces and by B. de Saint Venant (1843) and G. G. Stokes (1845) based on the assumption that the normal and shearing stresses in a fluid are proportional to the deformation velocities. The Navier-Stokes equations will not be derived here because such a procedure takes considerable space. Their derivation can be looked up in textbooks on fluid mechanics, for instance, in H. Schlichting's "Boundary Layer Theory." For a fluid with constant properties moving with regard to a stationary coordinate system $x$, $y$, $z$, with velocity components $u$, $v$, $w$, the Navier-Stokes equations, which express the balance of inertia forces, pressure forces, and viscous forces[1] in the three directions, have the form

$$\begin{aligned}
\rho \left( \frac{\partial u}{\partial \tau} + u \frac{\partial u}{\partial x} + v \frac{\partial u}{\partial y} + w \frac{\partial u}{\partial z} \right) &= -\frac{\partial p}{\partial x} + \mu \left( \frac{\partial^2 u}{\partial x^2} + \frac{\partial^2 u}{\partial y^2} + \frac{\partial^2 u}{\partial z^2} \right) \\
\rho \left( \frac{\partial v}{\partial \tau} + u \frac{\partial v}{\partial x} + v \frac{\partial v}{\partial y} + w \frac{\partial v}{\partial z} \right) &= -\frac{\partial p}{\partial y} + \mu \left( \frac{\partial^2 v}{\partial x^2} + \frac{\partial^2 v}{\partial y^2} + \frac{\partial^2 v}{\partial z^2} \right) \\
\rho \left( \frac{\partial w}{\partial \tau} + u \frac{\partial w}{\partial x} + v \frac{\partial w}{\partial y} + w \frac{\partial w}{\partial z} \right) &= -\frac{\partial p}{\partial z} + \mu \left( \frac{\partial^2 w}{\partial x^2} + \frac{\partial^2 w}{\partial y^2} + \frac{\partial^2 w}{\partial z^2} \right)
\end{aligned} \qquad (6\text{-}14)$$

where $\tau$ denotes the time. In addition, the condition of conservation of mass must be fulfilled:

$$\frac{\partial u}{\partial x} + \frac{\partial v}{\partial y} + \frac{\partial w}{\partial z} = 0 \qquad (6\text{-}15)$$

In his famous paper on boundary layers published in 1904, L. Prandtl concluded that in fluids with small viscosity the action of viscosity can be

[1] It is assumed that no body forces are present.

neglected except in a thin layer along solid surfaces. On this basis, he then proceeded to simplify the Navier-Stokes equations by an estimate of the order of magnitude of the different terms in these equations. In following essentially his ideas on the derivation of the boundary-layer equations, we shall restrict ourselves to two-dimensional flow ($w = 0$, $\partial/\partial z = 0$). An extension to rotationally symmetrical flow will be made later on. Three-dimensional boundary-layer flow has recently received attention in connection with secondary flow problems. It is, however, too specialized and complicated to be treated here.

Starting from the consideration that Eqs. (6-14) must describe the flow in the thin boundary layer, we shall assume that the thickness $\delta$ of this layer is very small when compared with any characteristic dimension of a solid object which is immersed in the flow and surrounded by the boundary layer. If the surface of this object is curved, then we shall assume

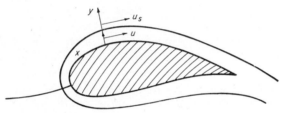

FIG. 6-9. Boundary layer around a streamlined body.

that the boundary layer is also thin as compared with the radius of curvature at any location along the surface. Under this circumstance, we can arrange a coordinate system within the boundary layer as shown in Fig. 6-9 in which the $x$ axis lies along the surface and the $y$ axis is oriented normal to it and points in an outward direction. Within the thin boundary layer we have to expect that the change of parameters like $u$, $v$, $p$ characterizing the flow is much more rapid in the $y$ direction than in the $x$ direction. To take care of this we shall stretch the $y$ direction by measuring it in a smaller scale than a length in $x$ direction. We can express this by stating that $x$ is of order of magnitude 1 whereas $y$ is of order of magnitude $\delta$. For all the other quantities appearing in Eqs. (6-14) and (6-15) we shall a priori assume that they are of order 1 except for the viscosity, which is considered small. For instance, we shall measure the velocity in the main flow $u_s$, in such a scale that it is also of order 1. We have to expect that the velocities in the main-flow direction within the boundary layer are of the same order of magnitude. We can easily determine the order of magnitude of the velocity components $v$ in direction $y$ by inspecting the continuity equation. This equation requires that the two terms in it be of the same order of magnitude. Since $u$ and $x$ are of order 1, the derivative $\partial u/\partial x$ has to be of order 1 also. We have, therefore, to conclude that $\partial v/\partial y$ is of order 1 and that $v$ must be of order $\delta$. The following line shows the continuity equation, and under each term the order of magnitude is indicated:

$$\frac{\partial u}{\partial x} + \frac{\partial v}{\partial y} = 0$$

$$\frac{1}{1} \qquad \frac{\delta}{\delta}$$

Now we proceed to the momentum equation in the $x$ direction. In the following line the momentum equation is written and underneath it the order of magnitude of each term:

$$\rho\left(\frac{\partial u}{\partial \tau} + u\frac{\partial u}{\partial x} + v\frac{\partial u}{\partial y}\right) = -\frac{\partial p}{\partial x} + \mu\left(\frac{\partial^2 u}{\partial x^2} + \frac{\partial^2 u}{\partial y^2}\right)$$

$$1 \qquad\quad 1 \qquad\quad \delta\frac{1}{\delta} \qquad\quad 1 \qquad\quad 1 \qquad\quad \frac{1}{\delta^2}$$

We can immediately indicate that the second term on the left side is of order 1. The first term may be of an order smaller than 1, in which case we consider the actual flow as quasi-steady, or the order of magnitude of this term may be 1 as indicated here, or finally, it could be large compared with 1. This would mean a very rapid change of velocity in time. We shall, however, exclude this possibility here. We can also immediately determine the order of magnitude of the third term of the left-hand side of the equation to be 1. We shall also assume the density $\rho$ to be of order 1. To determine the order of magnitude of the first term on the right-hand side, we remember that Bernoulli's Eq. (6-1) holds in flow outside the boundary layer. This equation can be written for a streamline in the following way:

$$\frac{\partial p}{\partial x} + \rho u_s \frac{\partial u_s}{\partial x} = 0$$

$$1 \quad 1 \quad 1/1$$

Through the thin boundary layer, the pressure gradient cannot change by an order of magnitude. From this we see immediately that the pressure gradient is of the same order of magnitude as the inertia term. The remaining two terms on the right-hand side of the equation have now to be inspected. The order of magnitude of the two terms within the parentheses can be immediately established and is indicated under the corresponding terms. We see that we can neglect the term $\partial^2 u/\partial x^2$ against the term $\partial^2 u/\partial y^2$. The remaining viscosity term must now be of the same order as the other terms in the equation if we want the equation to contain terms which correspond to viscous forces. Therefore, $\mu(\partial^2 u/\partial y^2)$ must be of order 1, which will be fulfilled if we assume that the viscosity is of order $\delta^2$. This means that in order to establish boundary-layer flow the viscosity has to be quite small, namely, of order $\delta^2$.

The same order of magnitude argument for the momentum equation in the $y$ direction leads to the conclusions indicated in the next two lines.

$$\rho\left(\frac{\partial v}{\partial \tau} + u\frac{\partial v}{\partial x} + v\frac{\partial v}{\partial y}\right) = -\frac{\partial p}{\partial y} + \mu\left(\frac{\partial^2 v}{\partial x^2} + \frac{\partial^2 v}{\partial y^2}\right)$$
$$\quad\quad \frac{\delta}{1} \quad\quad 1\frac{\delta}{1} \quad\quad \delta\frac{\delta}{\delta} \quad\quad\quad\quad\quad\quad \delta^2\left(\frac{\delta}{1} + \frac{\delta}{\delta^2}\right)$$

All indicated terms are of order of magnitude $\delta$. Consequently, $\partial p/\partial y$ must also be of order $\delta$, which can be neglected as compared with values of order 1 in the first equation. This indicates that the pressure variation throughout the boundary layer in a direction normal to the surface is negligibly small. In other words, the pressure within the boundary layer is prescribed by the flow of the main stream outside the boundary layer. The system of equations describing two-dimensional boundary-layer flow of a fluid with constant property values is

$$\rho\left(\frac{\partial u}{\partial \tau} + u\frac{\partial u}{\partial x} + v\frac{\partial u}{\partial y}\right) = -\frac{\partial p}{\partial x} + \mu\frac{\partial^2 u}{\partial y^2} \quad\quad (6\text{-}16)$$

$$\frac{\partial p}{\partial y} = 0 \quad\quad (6\text{-}17)$$

$$\frac{\partial u}{\partial x} + \frac{\partial v}{\partial y} = 0 \quad\quad (6\text{-}18)$$

The boundary conditions for these equations are

For $y = 0$: $\quad\quad u = 0 \quad\quad v = 0$
For $y = \infty$: $\quad\quad u = u_s$ $\quad\quad\quad\quad (6\text{-}19)$

assuming that the wall on which the boundary layer develops is at rest. It may be surprising that the second boundary condition is written for $y = \infty$ whereas the system of boundary-layer equations describes the flow in a supposedly thin boundary layer. Solutions obtained from these equations show that actually the stream velocity is approached very rapidly at a short distance from the wall. Solutions for these three equations can be obtained in a considerably simpler way than for the Navier-Stokes equations. These equations are still nonlinear; however, one variable has been eliminated, since the pressure has now to be considered as a quantity prescribed by the main flow. In addition, one of the two viscous terms in the remaining momentum equation was also dropped.

For rotationally symmetrical flow the equations are quite similar. The only change occurs in the continuity equation, which then has to be written in the following way:

$$\frac{\partial(ru)}{\partial x} + \frac{\partial(rv)}{\partial y} = 0 \quad\quad (6\text{-}20)$$

$x$ is again measured along the surface of the rotationally symmetrical object, $y$ normal to the surface, $u$ and $v$ are the corresponding velocity components, and $r$ is the distance of the surface point under consideration from the axis of rotation.

It can be seen that the system of boundary-layer equations is completely independent of temperature, since the properties $\rho$ and $\mu$ appearing in them have been assumed constant. This verifies the conclusion, mentioned on page 121, that for a constant-property fluid the velocity field is completely independent of the temperature field within the fluid. Heat transfer has no influence on the flow characteristics.

On page 123 it was mentioned that almost all flows occurring in engineering applications can be considered steady with sufficient accuracy. A quantitative estimate as to when a boundary-layer flow can be considered as quasi-steady can now be made by a comparison of the first term in Eq. (6-16) with the second term. The velocity $u$ is of order $u_s$; the velocity gradient $\partial u/\partial x$ of order $u_s/L$, with $L$ indicating a characteristic body dimension. If the unit in which we measure time is $\tau_0$, then $\partial u/\partial \tau$ is of order $u_s/\tau_0$. The first term in Eq. (6-16) can be neglected when

$$\frac{u_s}{\tau_0} \ll \frac{u_s^2}{L} \quad \text{or} \quad \tau_0 \gg \frac{L}{u_s}$$

For a body of dimension $L = 1$ ft immersed in a flow with $u_s = 100$ fps, the flow can be considered as quasi-steady when changes in it occur over a period of time greater than $\tau_0 \approx \frac{1}{10}$ sec. Changes in such a short time occur seldom.

Some solutions of the boundary-layer equations will be discussed later on. For the present we shall return to the integrated momentum equation and with its help calculate the boundary-layer thickness. This procedure gives results which are only approximate; however, it has the great advantage, especially for engineering problems, that the calculation procedure is considerably shorter and that the method can be applied very generally, whereas, even with considerable effort, solutions of the boundary-layer differential equations have been obtained for laminar flow only in a limited number of cases. The advantage of the integrated momentum equation consists mainly in the fact that it is a total differential equation in $x$ whereas the full boundary-layer equations are partial differential equations in $x$ and $y$.

**6-4. The Plane Plate in Longitudinal Flow.** Equation (6-8) will now be used to calculate the boundary-layer thickness along a flat plate in steady flow. It has been mentioned that on the surface next to the leading edge a laminar boundary layer develops (Fig. 6-5). The velocity $u_s$ outside the boundary layer may be constant along the plate. The

Bernoulli equation then states that the pressure is also constant. The last term on both sides of Eq. (6-8) therefore vanishes.

The velocity profile within a laminar boundary layer has been found by measurements to have a shape similar to the one shown in Fig. 6-10. The method of solution as suggested by T. von Kármán assumes an arbitrary expression for the velocity profile with a number of constants determined in such a way that they fulfill a number of conditions which the actual velocity profile is known to fulfill or to approximate.

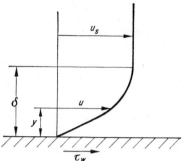

FIG. 6-10. Laminar boundary layer on a flat plate.

Such conditions can be found at the wall surface ($y = 0$) or at the border between boundary layer and stream ($y = \delta$). It is known that

For $y = 0$: $\quad u = 0 \quad$ (6-21)
For $y = \delta$: $\quad u = u_s \quad$ (6-22)

If the boundary-layer equation (6-16) is written for $y = 0$, then for constant pressure the relation

For $y = 0$: $\quad \dfrac{\partial^2 u}{\partial y^2} = 0 \quad$ (6-23)

follows. The approximate velocity profile should join the region of constant velocity $u_s$ smoothly at $y = \delta$. Therefore

For $y = \delta$: $\quad \dfrac{\partial u}{\partial y} = 0 \quad$ (6-24)

is a condition which may be required. One could also go beyond this and ask that the second, third, or any higher derivatives be zero at $y = \delta$. We shall be satisfied here with the four conditions (6-21) to (6-24). Accordingly we can choose for the velocity profile an expression with four undetermined coefficients, for instance,

$$u = a + by + cy^2 + dy^3 \quad (6\text{-}25)$$

The coefficients are determined by applying the four Eqs. (6-21) to (6-24) to Eq. (6-25). They are found to be

$$a = 0 \quad b = \frac{3}{2}\frac{u_s}{\delta} \quad c = 0 \quad d = -\frac{1}{2}\frac{u_s}{\delta^3}$$

and the velocity profile is

$$\frac{u}{u_s} = \frac{3}{2}\frac{y}{\delta} - \frac{1}{2}\left(\frac{y}{\delta}\right)^3 \quad (6\text{-}26)$$

With this equation the momentum integral in Eq. (6-8) becomes

$$I = \rho \int_0^l (u_s - u)u\, dy$$

$$= \rho u_s^2 \int_0^\delta \left[\frac{3}{2}\frac{y}{\delta} - \frac{1}{2}\left(\frac{y}{\delta}\right)^3\right]\left[1 - \frac{3}{2}\frac{y}{\delta} + \frac{1}{2}\left(\frac{y}{\delta}\right)^3\right] dy$$

The upper limit in the integral was changed to $\delta$, since for $y > \delta$, the velocity $u = u_s$ and therefore the integrand is zero, and also, since Eq. (6-26) holds for $y \leq \delta$ only. If the multiplication of the two terms in the brackets is completed and the integration performed for every summand, the integral resolves to

$$I = {}^{39}\!/\!_{280}\, \rho u_s^2 \delta$$

The velocity gradient at the wall is obtained from Eq. (6-26):

$$\left(\frac{\partial u}{\partial y}\right)_w = \frac{3}{2}\frac{u_s}{\delta}$$

Therefore the shear stress at the wall is

$$\tau_w = \mu \left(\frac{\partial u}{\partial y}\right)_w = \frac{3}{2}\mu\frac{u_s}{\delta}$$

By introducing this expression and the value $I$ into the equation of momentum (6-8), the following differential equation results:

$$\frac{39}{280}\rho u_s^2 \frac{d\delta}{dx} = \frac{3}{2}\mu\frac{u_s}{\delta}$$

and by separating the variables,

$$\delta\, d\delta = \frac{140}{13}\frac{\nu}{u_s}\, dx \qquad (6\text{-}27)$$

Integrating gives

$$\delta = 4.64 \sqrt{\frac{\nu x}{u_s}} + \text{const} \qquad (6\text{-}28)$$

The integration constant is zero when $x$ is measured from the leading edge, as, then, for $x = 0$ the thickness $\delta$ is also zero. Equation (6-28) shows that $\delta$ increases with the square root of the distance $x$. It is practicable to write the equation in dimensionless form as

$$\frac{\delta}{x} = \frac{4.64}{\sqrt{u_s x/\nu}}$$

The expression under the radical is the Reynolds number formed with the distance $x$. It may be denoted as $\text{Re}_x$. Therefore

$$\frac{\delta}{x} = \frac{4.64}{\sqrt{\text{Re}_x}} \qquad (6\text{-}29)$$

The boundary-layer thickness $\delta$ is the distance from the wall at which the velocity, as approximated by Eq. (6-26), reaches the value of the outside flow velocity. It is clear that the boundary-layer thickness thus defined is somewhat arbitrary. The exact calculation of laminar boundary layers gives velocity profiles which reach the outside velocity only asymptotically. For such profiles there exists no finite value $\delta$ corresponding to the one above. Other definitions for the boundary-layer thickness were used, e.g., the wall distance where the velocity is 0.99 of the outside flow velocity. At present, however, the value called *displacement thickness* is almost generally accepted. It is obtained by a procedure shown in Fig. 6-11. The area above the velocity profile is transformed into a rectangle $abcd$ of equal area. The width of this rectangle $\delta^*$ is the displacement thickness. Mathematically it is described by the equation

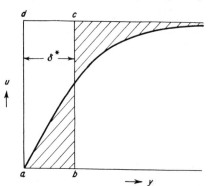

FIG. 6-11. Displacement thickness of the boundary layer.

$$\delta^* = \int_0^\infty \left(1 - \frac{u}{u_s}\right) dy \qquad (6\text{-}30)$$

The name comes from the fact that in a frictionless fluid flow without boundary layers the walls would have to be displaced in outward direction by the amount $\delta^*$ in order to keep the flow field unchanged as compared with the real flow with boundary layers. For the velocity profile given by Eq. (6-26) the above integral gives

$$\delta^* = 0.375\delta \sim \frac{\delta}{3}$$

The momentum thickness which was introduced by Eq. (6-12) becomes

$$\delta_i = \frac{39}{280}\delta = 0.139\delta \sim \frac{\delta}{7}$$

The shearing stress at the wall follows as

$$\tau_w = \frac{0.323\rho u_s^2}{\sqrt{\mathrm{Re}_x}} \qquad (6\text{-}31)$$

Instead of this value, the friction coefficient $f_p$ is generally used. This is the resistance force of the plate divided by the area of the surface and the dynamic pressure in the flow. Two values of this friction coefficient are in use and must be discerned: the local coefficient $f_p$ and the average coefficient $f_{pm}$. The first is defined by the equation

$$\tau_w\, dx = f_p\, dx\, \rho\, \frac{u_s^2}{2}$$

Therefore
$$f_p = \frac{2\tau_w}{\rho u_s^2}$$

The second is given by

$$\int_0^x \tau_w\, dx = f_{pm} x \rho\, \frac{u_s^2}{2}$$

$$f_{pm} = \frac{2}{x\rho u_s^2}\int_0^x \tau_w\, dx = \frac{1}{x}\int_0^x f_p\, dx$$

The result is
$$f_{pm} = \frac{1.292}{\sqrt{\mathrm{Re}_x}}$$

The same formula has the numerator 1.327, which is 3 per cent greater, when it is calculated from an exact solution of the boundary-layer equations. The formula is also in good agreement with the results of experiments.

**Example 6-1.** The thickness of the boundary layer which builds up on a plate 4 in. distant from the leading edge is to be determined for longitudinal air flow. The velocity is 33 fps, the temperature of the air 60 F, and the pressure atmospheric. The Reynolds number is

$$\mathrm{Re}_x = \frac{u_s x}{\nu} = \frac{33\ (\mathrm{fps}) \times \frac{4}{12}\ (\mathrm{ft})}{15.4 \times 10^{-5}\ (\mathrm{ft}^2/\mathrm{sec})} = 71{,}400$$

The kinematic viscosity was taken from the Appendix. The ratio of the boundary-layer thickness to the distance from the leading edge is

$$\frac{\delta}{x} = \frac{4.64}{\sqrt{71{,}400}} = 0.0174$$

The boundary layer is therefore 0.070 in. thick at the 4-in. distance. The displacement thickness is $\delta^* = 0.026$ in. It can be seen that for usual velocities and distances the boundary layers build up to less than 0.1-in. thickness.

In the *turbulent boundary layer* the shape of the velocity profile is much more curved than in the laminar layer. The measured velocity profile agrees satisfactorily with the equation proposed by Prandtl:

$$u = u_s \left(\frac{y}{\delta}\right)^{1/7} \tag{6-32}$$

The equation cannot hold true, however, in the immediate proximity of the wall. This can be shown if the shear stress at the wall is calculated.

FLOW ALONG SURFACES AND IN CHANNELS 143

The velocity gradient is

$$\frac{du}{dy} = \frac{1}{7} \frac{u_s}{(\delta^{1/7})(y^{6/7})}$$

and at the wall ($y = 0$) it is infinite. This would give an infinite value of the shear stress, which is physically impossible. In reality, the turbulence always dies down in the neighborhood of the wall. The real conditions can be simplified, according to Prandtl, through the concept that between the turbulent boundary layer and the wall a laminar sublayer exists within which the velocity increases linearly with the distance $y$. Outside this sublayer Eq. (6-32) holds true. At the border between both layers the assumed velocity profiles join at a small angle as is shown in Fig. 6-12.

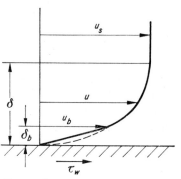

FIG. 6-12. Laminar sublayer within a turbulent boundary layer.

For this calculation the shearing stress at the wall must be determined by measurement. For not too large Reynolds numbers and smooth surfaces, the equation of Blasius

$$\tau_w = 0.0228 \rho u_s^2 \left(\frac{\nu}{u_s \delta}\right)^{1/4} \tag{6-33}$$

holds true. This equation was determined by experiments on flow through tubes. It was confirmed in experiments by Schultz-Grunow[1] up to a Reynolds number $10^7$ for flat plates. For higher Reynolds numbers a more complicated equation exists which will be discussed later in this section.

With Eq. (6-32), the momentum integral in Eq. (6-8) becomes

$$I = \rho \int_0^l u(u_s - u) \, dy = \rho u_s^2 \int_0^\delta \left(\frac{y}{\delta}\right)^{1/7} \left[1 - \left(\frac{y}{\delta}\right)^{1/7}\right] dy = \frac{7}{72} \rho u_s^2 \delta$$

Here again the upper limit of the integral must be changed to $\delta$, since Eq. (6-32) holds true only for $y < \delta$. For $y > \delta$, $u = u_s$; therefore the integrand equals zero.

Introducing the integral $I$ and Eq. (6-33) into the equation of momentum gives the differential equation

$$\frac{7}{72} \rho u_s^2 \frac{d\delta}{dx} = 0.0228 \rho u_s^2 \left(\frac{\nu}{u_s \delta}\right)^{1/4}$$

[1] F. Schultz-Grunow, A New Resistance Law for Smooth Plates, *Luftfahrt-Forsch.*, **17**:239–246 (1940).

Separation of the variables results in

$$\delta^{1/4}\, d\delta = 0.235 \left(\frac{\nu}{u_s}\right)^{1/4} dx$$

and by integration

$$\delta = 0.376 \left(\frac{\nu}{u_s}\right)^{1/5} x^{4/5} + \text{const} \tag{6-34}$$

Some difficulty arises in determining the integration constant. According to Fig. 6-5 the turbulent boundary layer begins at the critical distance $x_c$ from the leading edge. There it already has a certain thickness, as it originates from the laminar layer. Both layers have to be joined at this point. Some make the assumption that the momentum thickness of the laminar and the turbulent layer are equal here. According to L.

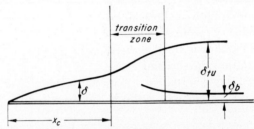

Fig. 6-13. Laminar and turbulent boundary layers on a flat plate: $\delta$ laminar boundary layer, $\delta_{tu}$ turbulent boundary layer, $\delta_b$ laminar sublayer.

Prandtl[1] the expression (6-34) agrees satisfactorily with measurements if the thickness of the turbulent boundary layer is determined as if it began at the leading edge with the thickness zero. Newer measurements indicate that this is not quite correct. We shall, however, retain this simple assumption. The constant in Eq. (6-34) is then zero, where $x$ denotes the distance from the leading edge. In dimensionless form the equation is written

$$\frac{\delta}{x} = \frac{0.376}{(u_s x/\nu)^{1/5}} = \frac{0.376}{(\text{Re}_x)^{1/5}} \tag{6-35}$$

On the right side again the Reynolds number calculated with the distance $x$ appears. The displacement thickness of the turbulent profile according to Eq. (6-32) is $\delta^* = \delta/8$, and the momentum thickness is $\delta_i = 7/72\,\delta$. If the laminar and the turbulent boundary layers are calculated for the critical distance $x_c$, it can be seen that the turbulent layer is thicker. In reality an instantaneous increase in boundary-layer thickness is not possible. The transition from the laminar to the turbulent boundary layer takes place in a transition zone as is indicated in Fig. 6-13.

[1] L. Prandtl, *Ergeb. aerodynamischen Versuchsanstalt Göttingen*, **3**:4 (1927).

The transition from laminar to turbulent flow within the boundary layer starts with oscillations of comparatively long wavelength. They are basically of the same nature as the waves, which can be clearly observed in the thermal free-convection boundary layer on a vertical plate in Fig. 11-11. It was shown mathematically by W. Tollmien[1] and H. Schlichting[2] that the boundary-layer flow becomes unstable for such oscillations with a certain wavelength as soon as the boundary-layer thickness reaches a certain value. Such oscillations are always present from time to time because of small disturbances coming from the outer flow. In the unstable region of the boundary layer they are not damped but increase in amplitude as they travel downstream. On their way they become irregular in shape and transform into vortices, which finally dissolve into the smaller scaled turbulence. Because of the random nature of the disturbances, the location of the transition zone is subject to irregular fluctuations with time. This region is extremely difficult to study, and therefore our knowledge of this phase is still quite limited. The calculations given in this section deal with the zone of fully established turbulence.

For the following heat-transfer calculations the thickness $\delta_b$ of the laminar sublayer is needed. To determine this, we must first calculate the velocity $u_b$ at the border between the turbulent layer and the laminar sublayer. The linear-velocity increase in the sublayer is derived from the shear stress on the wall:

$$\tau_w = \mu \frac{du}{dy} = \mu \frac{u}{y}$$

Introducing the expression for $\tau_w$ gives

$$u = 0.0228 \rho \frac{u_s^2}{\mu} \left(\frac{\nu}{u_s \delta}\right)^{1/4} y$$

Solving this equation for $y$ and taking into account that for $y = \delta_b$, $u = u_b$, the sublayer thickness is obtained as

$$\frac{\delta_b}{\delta} = \frac{u_b}{u_s} \frac{1}{0.0228} \left(\frac{\nu}{u_s \delta}\right)^{3/4}$$

On the other hand at this place Eq. (6-32) and therefore

$$\frac{\delta_b}{\delta} = \left(\frac{u_b}{u_s}\right)^7$$

---

[1] W. Tollmien, *Nachr. Ges. Wiss. Göttingen, Math. physik. Kl.*, 1929, p. 21.

[2] H. Schlichting, *Nachr. Ges. Wiss. Göttingen, Math. physik. Kl.*, 1933, p. 181; 1935, p. 48.

also hold true.  Equating both expressions gives

$$\frac{u_b}{u_s} = 1.878 \left(\frac{u_s \delta}{\nu}\right)^{-1/8} = \frac{1.878}{(\mathrm{Re}_\delta)^{1/8}} \tag{6-36}$$

The Reynolds number is calculated here with the boundary-layer thickness. By Eq. (6-35) the distance $x$ can be introduced:

$$\frac{u_b}{u_s} = 2.12 \left(\frac{\nu}{u_s x}\right)^{0.1} = \frac{2.12}{(\mathrm{Re}_x)^{0.1}} \tag{6-37}$$

The sublayer thickness is then

$$\frac{\delta_b}{\delta} = \left(\frac{u_b}{u_s}\right)^7 = \frac{194}{(\mathrm{Re}_x)^{0.7}} \tag{6-38}$$

The shear stress at the wall is given by

$$\tau_w = \mu \frac{u_b}{\delta_b}$$

$$\tau_w = \frac{0.0296}{(\mathrm{Re}_x)^{0.2}} \rho u_s^2$$

The resistance force of the plate can be calculated from the shear stresses on the first part with the laminar boundary layer and the second part with the turbulent boundary layer. For a critical Reynolds number of 485,000 this calculation gives

$$f_{pm} = \frac{0.074}{(\mathrm{Re}_x)^{1/5}} - \frac{1{,}700}{\mathrm{Re}_x}$$

There corresponds to a critical Reynolds number of 85,000

$$f_{pm} = \frac{0.074}{(\mathrm{Re}_x)^{1/5}} - \frac{300}{\mathrm{Re}_x}$$

Figure 6-14 shows the velocity field in the turbulent boundary layer as measured by van der Hegge-Zijnen.[1] Here also, instead of a sharp boundary, a transition curve arises in which the turbulence dies down. Figure 6-15 shows a comparison of the above two formulas for the friction coefficient with experimental results.

At large Reynolds numbers (above $10^7$) Blasius's equation does not describe results of measurements with sufficient accuracy. Various interpolation formulas have been developed which are accurate up to $\mathrm{Re} = 10^9$. Well known is the Kármán-Schoenherr equation

$$\frac{0.242}{\sqrt{f_{pm}}} = \log_{10}(\mathrm{Re}_x f_{pm})$$

[1] B. G. van der Hegge-Zijnen, "Measurements of the Velocity Distribution in the Boundary Layer along a Plane Surface," I. Waltmann, Delft, 1924.

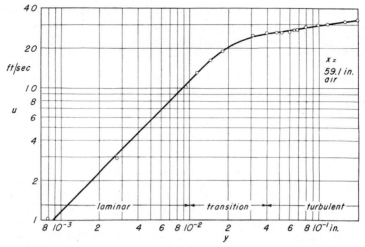

FIG. 6-14. Velocity profile in the turbulent boundary layer on a flat plate as measured by van der Hegge-Zijnen. ("*Measurements of the Velocity Distribution in the Boundary Layer along a Plane Surface,*" I. Waltmann, Delft, 1924.)

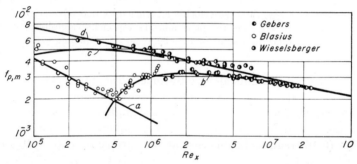

FIG. 6-15. Friction coefficient for laminar and turbulent boundary-layer flow on a flat plate. (a) $f_{pm} = 1.327/\sqrt{\text{Re}_x}$. (b) $f_{pm} = 0.074/\sqrt[5]{\text{Re}_x} - 1700/\text{Re}_x$. (c) $f_{pm} = 0.074/\sqrt[5]{\text{Re}_x} - 300/\text{Re}_x$. (d) $f_{pm} = 0.074/\sqrt[5]{\text{Re}_x}$. [*From L. Prandtl, Ergeb. aerodyn. Versuchsanstalt Göttingen,* **3**:1 (1927).]

A relation proposed by Prandtl and Schlichting is simpler to use:

$$f_{pm} = \frac{0.455}{(\log_{10} \text{Re}_x)^{2.58}}$$

For the local friction factor Schultz-Grunow gives the relation

$$f_p = \frac{0.370}{(\log_{10} \text{Re}_x)^{2.584}}$$

The contribution of the laminar boundary layer can be neglected at higher Reynolds numbers. This is evident from Fig. 6-15.

The above relations hold for a perfectly smooth surface. The friction on rough surfaces is generally greater than the friction on smooth surfaces. More will be discussed about surface roughness in Sec. 6-7.

**Example 6-2.** The thickness is to be calculated of the turbulent boundary layer which builds up in a 12-in. distance from the leading edge of a plane plate in an air stream flowing with 33 fps velocity, at 60 F, and 14.2 lb$_f$/in.$^2$ pressure. The Reynolds number is

$$\text{Re}_x = \frac{33 \times 1}{15.4 \times 10^{-5}} = 214{,}000$$

Equation (6-35) gives the ratio of the boundary-layer thickness to the distance from the leading edge: $\delta/x = 0.0326$. The boundary-layer thickness itself is 0.4 in., and the displacement thickness is 0.05 in. The ratio of the laminar sublayer to the turbulent layer is, according to Eq. (6-38), $\delta_b/\delta = 0.036$. The sublayer thickness is 0.014 in. Because of this small thickness, the shear stress at the wall for turbulent flow depends on the roughness of the wall, whereas in laminar flow the roughness has only a small influence. This may be explained by the fact that the shear stresses and the resistance of the wall increase when the roughnesses are not filled out by the laminar sublayer. Exact measurements show that an increase in resistance arises as soon as the height of the roughnesses is greater than approximately one-third the thickness of the sublayer. For the example calculated above, a plate with roughnesses smaller than 0.004 in. can be called hydraulically smooth. The laminar sublayer thickness decreases with the velocity $u_s$, whereas it depends very little on the distance $x$. For high velocities the plate must be especially carefully finished if an increase of the resistance by roughness is to be avoided. Equation (6-33) holds true for the smooth plate.

**6-5. Pressure Gradients along a Surface.** In a flow field in which the pressure varies along the surface of the plate, the calculation procedure has to be varied to account for this pressure gradient. Such a calculation will be performed for a laminar boundary layer. The boundary-layer equation (6-16) reduces in this case at the surface itself ($y = 0$) to

$$\frac{\partial p}{\partial x} = \mu \left(\frac{\partial^2 u}{\partial y^2}\right)_w$$

This expression replaces Eq. (6-23) in the previous paragraph. With Bernoulli's equation and remembering that $\partial p/\partial x$ is constant through the boundary layer, it can be written in the following form:

$$\left(\frac{d^2 u}{dy^2}\right)_w = -\frac{1}{\nu} u_s \frac{du_s}{dx} \tag{6-39}$$

This equation together with Eqs. (6-21), (6-22), and (6-24) determines the coefficients in the velocity profile

$$u = a + by + cy^2 + dy^3$$

Differentiating this equation twice and setting $y = 0$ results in the following equation for the coefficient $c$:

$$c = -\frac{1}{2\nu} u_s \frac{du_s}{dx}$$

The other boundary conditions determine the coefficients $a$, $b$, and $d$. Thus the profile becomes

$$\frac{u}{u_s} = \left(\frac{3}{2} + \frac{\lambda}{2}\right)\frac{y}{\delta} - \lambda\left(\frac{y}{\delta}\right)^2 - \left(\frac{1}{2} - \frac{\lambda}{2}\right)\left(\frac{y}{\delta}\right)^3 \tag{6-40}$$

with
$$\lambda = \frac{\delta^2}{2\nu}\frac{du_s}{dx} \tag{6-41}$$

The shear stress $\tau_w$ is

$$\tau_w = \mu\left(\frac{3}{2} + \frac{\lambda}{2}\right)\frac{u_s}{\delta}$$

Introducing the velocity profile and the shear stress into the boundary-layer momentum equation and integrating result in

$$\rho \frac{d}{dx}\left[\left(\frac{39}{280} - \frac{1}{280}\lambda - \frac{1}{420}\lambda^2\right)u_s^2\delta\right] + \rho\left(\frac{3}{8} - \frac{\lambda}{24}\right)u_s\frac{du_s}{dx}\delta$$
$$= \left(\frac{3}{2} + \frac{\lambda}{2}\right)\frac{u_s\mu}{\delta}$$

or
$$\frac{d}{dx}\left[\left(\frac{39}{280} - \frac{1}{280}\lambda - \frac{1}{420}\lambda^2\right)u_s^2\delta\right] = \frac{\nu u_s}{\delta}\left(\frac{3}{2} - \frac{\lambda}{4} + \frac{\lambda^2}{12}\right) \tag{6-42}$$

At this point the specific variation of the stream velocity along the surface has to be known. Equation (6-42) constitutes a first-order differential equation for the unknown value $\delta$, since the parameter $\lambda$ is a function of the boundary-layer thickness $\delta$ [Eq. (6-41)] and of known parameters. The integration is best carried out numerically or graphically. As a result of such a calculation the boundary-layer thickness $\delta$ along the surface is obtained. This value at any position $x$ along the surface also determines the form parameter $\lambda$ [Eq. (6-41)] and with it the shape of the boundary-layer profile. It can be seen that a negative $\lambda$ indicates a positive curvature of the velocity profile at the surface ($y = 0$) and accordingly an S-shaped profile. The profile with $\lambda = -3$ has additionally a velocity gradient $du/dy = 0$ at the wall. Larger negative $\lambda$ values would result in a velocity profile with negative velocities near the wall indicating backflow. Therefore, when the boundary-layer calculation indicates a value $\lambda = -3$ anywhere, then this fact is interpreted as indicating flow separation. The first calculations of this form have been carried out by K. Pohlhausen.[1] He used a fourth-order polynomial instead of Eq.

[1] K. Pohlhausen, Z. angew. Math. Mech., **1**:252 (1921).

(6-40) to describe the velocity profile throughout the boundary layer and calculated the development of a boundary layer around a circular cylinder.

Similar calculation procedures have also been developed for turbulent boundary layers. They will not be discussed here because turbulent boundary layers in flows with large pressure gradients occur less frequently than laminar ones.

**6-6. Exact Solutions of the Laminar Boundary-layer Equations for a Flat Plate.** The laminar boundary-layer development along a flat plate with constant pressure and for steady state was studied quite early through integration of the flow boundary-layer equations. This calculation has been carried out by H. Blasius after L. Prandtl had indicated that, for the flow situation considered, a transformation of the partial differential equations (6-16) and (6-18) into total differential ones is possible. The two differential equations describing the flow process can be united into one equation by the introduction of a stream function $\psi$ defined in the following way:

$$u = \frac{\partial \psi}{\partial y} \qquad v = -\frac{\partial \psi}{\partial x}$$

It is seen that introduction of the stream function into the continuity equation (6-18) satisfies this equation. The momentum equation (6-16) becomes

$$\frac{\partial \psi}{\partial y}\frac{\partial^2 \psi}{\partial x \partial y} - \frac{\partial \psi}{\partial x}\frac{\partial^2 \psi}{\partial y^2} = \nu \frac{\partial^3 \psi}{\partial y^3} \qquad (6\text{-}43)$$

L. Prandtl pointed out that this equation can be transformed into a total differential equation by introducing a new independent variable

$$\eta = \frac{1}{2} y \sqrt{\frac{u_s}{\nu x}}$$

and by assuming that the stream function can be written in the following way:

$$\psi = \sqrt{\nu u_s x}\, f(\eta)$$

in which $f$ is a function of $\eta$ only. Introducing these two new parameters $f$ and $\eta$ into the momentum equation results in

$$\frac{d^3 f}{d\eta^3} + f\frac{d^2 f}{d\eta^2} = 0 \qquad (6\text{-}44)$$

The boundary conditions (6-19) in the new variables are

For $\eta = 0$: $\qquad f = 0 \qquad \frac{df}{d\eta} = 0$

For $\eta = \infty$: $\qquad \frac{df}{d\eta} = 2$

It is seen that the function $f$ immediately describes the two velocity components according to the relations

$$u = \frac{u_s}{2}\frac{df}{d\eta} \qquad v = \frac{1}{2}\sqrt{\frac{\nu u_s}{x}}\left(\eta\frac{df}{d\eta} - f\right)$$

Blasius solved this equation by developing the function $f$ into a series. Later on Piercy and Preston[1] pointed out another method which results in a simple solution by successive approximations. For this purpose we shall write the second derivative of $f$ as $z$. The differential equation then assumes the form

$$\frac{dz}{d\eta} = -fz$$

If for the moment we treat $f$ like a given function of $\eta$, we can separate the variables

$$\frac{dz}{z} = -f\,d\eta$$

and integrate

$$\ln z = -\int f\,d\eta + \ln C_1 \qquad z = C_1 e^{-\int f\,d\eta}$$

One additional integration results in the following expression for the velocity $u$ within the boundary layer:

$$\frac{u}{u_s} = \frac{1}{2}\frac{df}{d\eta} = \frac{1}{2}\int z\,d\eta = \frac{C_1}{2}\int e^{-\int f\,d\eta}\,d\eta + C_2$$

The two constants $C_1$ and $C_2$ are determined from the boundary conditions and give

$$\frac{u}{u_s} = \frac{\int_0^\eta e^{\int_0^\eta f\,d\eta}\,d\eta}{\int_0^\infty e^{\int_0^\eta f\,d\eta}\,d\eta} \tag{6-45}$$

This equation cannot be considered as a solution of Eq. (6-44), since the function $f$ appearing in the exponent on the right-hand side is unknown. However, it can be utilized to obtain a solution by successive approximations in the following way: A first estimate is made of the function $f$. This estimate is introduced into Eq. (6-45), and the equation is solved for $u$. One additional integration yields $f$, which can now be inserted again into Eq. (6-45) as a second approximation of the function $f$. By integration a third approximation is obtained, and this procedure can be continued until successive approximations agree with one another satisfactorily. Piercy and Preston started their calculation with the rough

[1] N. A. V. Piercy and G. H. Preston, *Phil. Mag.*, **21**:995 (1936).

assumption $f = 2\eta$ and obtained a very good solution for the velocity profile after only three successive iterations. The velocity profile obtained in this way is shown in Fig. 6-16. Also indicated is the profile described by Eq. (6-26) with equal displacement thickness. It can be seen that the agreement is quite satisfactory. Exact solutions of the

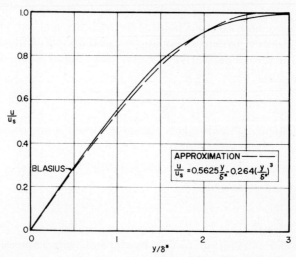

Fig. 6-16. Velocity profile in the laminar boundary layer along a flat plate as calculated by H. Blasius and as approximated by a third-power polynomial.

laminar boundary-layer equations, which proved very useful, have also been obtained for two-dimensional flow over a surface when the stream velocity $u_s$ varies as

$$u_s = Cx^m$$

This velocity distribution is established along the surface of an infinite wedge with the opening angle $\alpha = 2m\pi/(m+1) = \beta\pi$ in an incompressible flow directed symmetrically toward the apex. Accordingly these solutions are refered to as "wedge-type flow solutions." Again a transformation of the boundary-layer equations to total differential ones is possible by the same substitutions as used before. The resulting total differential equation is[1]

$$\frac{d^3f}{d\eta^3} + f\frac{d^2f}{d\eta^2} - \frac{2m}{m+1}\left[\left(\frac{df}{d\eta}\right)^2 - 1\right] = 0 \qquad (6\text{-}46)$$

The velocity profiles for this type of flow are shown in Fig. 6-17. The profile with the parameter $\beta = 0$ ($m = 0$) is the Blasius profile. Positive $\beta$ or $m$ values indicate a velocity increase along the surface. The value

[1] V. M. Falkner and S. W. Skan, *Phil. Mag.*, **12**: 865 (1931).

$m = 1$ ($\beta = 1$) corresponds to a wedge angle $\alpha = 180°$, or to two-dimensional flow directed normally toward a flat plate. This flow also occurs near the stagnation line of a blunt cylinder and is referred to as stagnation flow. Negative $m$ values correspond to a flow whose velocity decreases along the surface. The value $m = -0.1104$ ($\beta = -0.1988$) is characterized by the fact that the velocity gradient of this profile at the surface is zero (flow separation profile).

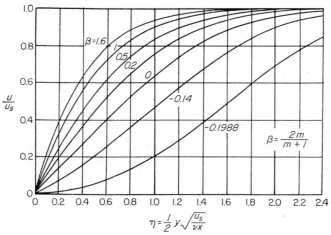

Fig. 6-17. Laminar velocity profiles for boundary-layer flow over wedges with various pressure parameters $\beta$. [*According to D. R. Hartree, Proc. Cambridge Phil. Soc.*, **33**(II): 223 (1937).]

**6-7. The Flow in a Tube.** The flow through a pipe near the entrance can also be calculated with the momentum equation (6-8). For a tube with circular cross section the momentum equation has to be derived and solved for rotationally symmetric flow. This calculation was made by L. Schiller.[1] He approximated the velocity profile by a curve built up by two parabolas and a straight line. The vertices of the parabolas lie on the border of the boundary layer, as shown in Fig. 6-18. The velocity in the core of the flow outside the boundary layer increases with increasing distance from the entrance, since through any cross section the same amount of fluid flows and the boundary-layer thickness increases. The

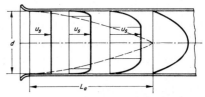

Fig. 6-18. Development of the laminar-velocity profile in the intake region of a tube.

[1] L. Schiller, Strömung in Rohren in Wien-Harms, "Handbuch Experimental Physik," vol. IV/4, p. 48, Akademie-Verlag G.m.b.H., Berlin, 1931.

pressure therefore decreases according to Bernoulli's equation, which is valid for the core. After this entrance region (at $L_e$), the velocity profile is a parabola. The difference in pressures between the cross section where the flow is already fully developed and a place outside the tube where the flow has still a negligible velocity can be calculated according to L. Schiller by adding the pressure drop

$$\Delta p = 2.16 \rho \frac{u_m{}^2}{2} \tag{6-47}$$

to the pressure loss which the flow would have if it were fully developed along the whole tube length. The results of this calculation agree well with test results. The ratio of the entrance length $L_e$ to the tube diameter $d$ is obtained by the same calculation as a function of the Reynolds number:

$$\frac{L_e}{d} = 0.0288 \frac{u_m d}{\nu} = 0.0288 \, \text{Re}_d \tag{6-48}$$

The Reynolds number is calculated with the mean velocity in a cross section and the tube diameter $d$. This is the way the Reynolds number is generally used for tube flow in technical literature. It must be kept in mind that a sharp limit for the entrance-flow region does not exist in reality. The flow approaches the fully developed condition asymptotically. Therefore, for distances greater than the entrance length given in Eq. (6-48), small deviations from the developed condition can still be measured.

If the Reynolds number exceeds the critical value, the boundary layer changes into turbulent flow somewhere in the entrance-flow region. The transition place moves toward the entrance when the Reynolds number increases. Since the turbulent boundary layer increases its thickness faster than the laminar layer [Eq. (6-35)], the entrance length $L_e$ decreases in this range. For a Reynolds number $\text{Re}_d = 3,000$, the whole entrance flow is laminar. The entrance length $L_e$ is, according to Eq. (6-48), approximately $100d$. When the Reynolds number increases over this value, the entrance length first decreases to about $40d$ and then increases again.

Developed laminar flow through a circular tube is one of the few cases for which a simple, exact solution of the Navier-Stokes equations can be obtained. This solution indicates that the velocity profile is described by a parabola and gives for the friction coefficient as defined by Eq. (6-53) the relation

$$f = \frac{64}{\text{Re}_d} \tag{6-49}$$

This type of flow is customarily referred to as *Poiseuille flow*.

FLOW ALONG SURFACES AND IN CHANNELS    155

The velocity profile in the fully developed turbulent flow is well represented by Eq. (6-32) up to Re = 100,000 if the boundary-layer thickness is replaced by the radius $r$. This corresponds to the concept that the two boundary layers have touched each other in the tube axis. The velocity $u_s$ is now the velocity at the axis. Also Eq. (6-33) for the shear stress on the wall holds true, as does Eq. (6-36) for the velocity at the border between the turbulent flow and the laminar sublayer. This sublayer builds up here in the same way as on the plate. If in these equations the radius $r$ is replaced by the diameter $d$ and the velocity $u_s$ by the mean velocity $u_m$ [$u_m = 0.82 u_s$ is found by integration of Eq. (6-32)], the following equations arise which will be used later:

$$\tau_w = 0.0384 \rho u_m^2 \left(\frac{\nu}{u_m d}\right)^{1/4} = \frac{0.0384}{(\text{Re}_d)^{1/4}} \rho u_m^2 \tag{6-50}$$

$$\frac{u_b}{u_m} = 2.44 \left(\frac{\nu}{u_m d}\right)^{1/8} = \frac{2.44}{(\text{Re}_d)^{1/8}} \tag{6-51}$$

$$\frac{\delta_b}{d} = 63.5 \left(\frac{\nu}{u_m d}\right)^{7/8} = \frac{63.5}{(\text{Re}_d)^{7/8}} \tag{6-52}$$

In reality here as on the plate, no strictly laminar layer exists, but only a gradual decrease of turbulence as the wall is approached. Equations taking into account this transition zone are shown in Fig. 6-20. Instead of the shearing stress the friction coefficient $f$ is generally used as defined by the equation

$$\Delta p = f \frac{L}{d} \rho \frac{u_m^2}{2} \tag{6-53}$$

where $\Delta p$ is the pressure drop along the tube length $L$. In a piece of the tube as shown in Fig. 6-19 equilibrium exists between the pressures in the cross sections 1-1 and 2-2 and the shear stresses on the wall of the tube length $L$ in steady flow according to the momentum law. This gives for developed flow

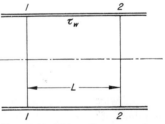

FIG. 6-19. Using the momentum law for fully developed flow in a tube.

$$\Delta p \frac{\pi d^2}{4} = \tau_w \pi \, dL \tag{6-54}$$

With Eqs. (6-50), (6-53), and (6-54) the friction coefficient can be calculated from the shear stress:

$$f = \frac{8\tau_w}{\rho u_m^2} = \frac{0.316}{(\text{Re}_d)^{1/4}} \tag{6-55}$$

This expression is usually called Blasius's law. Temperature differences exist where the flow is associated with heat transfer. Then the property values are evaluated at the temperature $(t_w + t_m)/2$ (where $t_w$ is the wall temperature and $t_m$ is the mean temperature in flow) for gases according to McAdams,[1] or the friction coefficient calculated with the property values at the temperature $t_m$ is multiplied by $(\mu_m/\mu_w)^{0.14}$ for oils where $\mu_m$ is the viscosity at $t_m$ and $\mu_w$ is the viscosity at $t_w$ according to Sieder and Tate.[2] Experiments of Rohonczy[3] with water agreed best with Eq. (6-55) when the property values were introduced at the wall temperature $t_w$.

For Reynolds numbers greater than $10^5$, Eq. (6-55) must be replaced by the general resistance law developed by L. Prandtl, T. von Kármán, and coworkers:[4]

$$\frac{1}{\sqrt{f}} = 2.0 \log_{10} [(\text{Re}_d) \sqrt{f}\,] - 0.8 \tag{6-56}$$

This law is more complicated than Eq. (6-55), since here the friction coefficient appears on both sides of the equation.

Recently it has become customary to present the turbulent-velocity profile in a semilogarithmic plot as shown in Fig. 6-20.[5] The velocities are made dimensionless by the value $\sqrt{\tau_w/\rho}$ which has the dimension of a velocity and is called *shear-stress velocity:* $u^+ = u/\sqrt{\tau_w/\rho}$. They are plotted in this way on the ordinate. The distance from the tube wall is transformed into a Reynolds number by multiplying it by the shear-stress velocity and dividing it by the kinematic viscosity:

$$y^+ = y \frac{\sqrt{\tau_w/\rho}}{\nu}$$

In this way of plotting, the velocity profile becomes approximately independent of the Reynolds number $\text{Re}_d$. It is therefore called the *universal-velocity profile*. Measurements by Nikuradse, Reichardt, and coworkers are shown in Fig. 6-20. In the semilogarithmic plot the linear relationship between the velocity and the wall distance expected in the laminar sublayer is represented by the curve on the left-hand side of the figure. It can be seen that the measured values follow this curve up to approxi-

---

[1] W. H. McAdams, "Heat Transmission," 3d ed., McGraw-Hill Book Company, Inc., New York, 1954.

[2] E. N. Sieder and G. E. Tate, *Ind. Eng. Chem.*, **28**:1429–1435 (1936).

[3] G. Rohonczy, *Schweiz. Arch. angew. Wiss. Tech.*, **5**:121–140, 167–175, 239–275, 349–366 (1939).

[4] L. Prandtl, "Führer durch die Strömungslehre," Vieweg-Verlag, Brunswick, Germany, 1942.

[5] Taken from R. C. Martinelli, Heat Transfer to Molten Metals, *Trans.* ASME, **69**:947–959 (1947).

mately $y^+ = 5$. The straight line on the right-hand side gives the velocities in the turbulent core. The figure shows that a buffer layer exists between both regions within which the turbulence gradually dies down in the direction toward the wall. T. von Kármán[1] proposed another straight line shown in the figure as approximation for the velocities in this range and fixed the limits of the buffer layer at $y_1^+ = 5$ and $y_2^+ = 30$. The equation for this line as well as the equations for the laminar sublayer and the turbulent zone are included in Fig. 6-20.

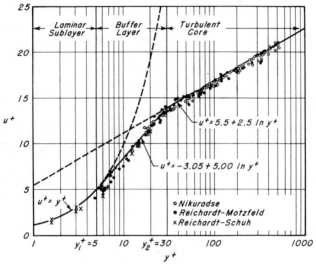

FIG. 6-20. Universal-velocity profile. [*From R. C. Martinelli, Trans. ASME,* **69**:947–959 (1947).]

For rough tubes similar equations have been developed in which the degree of roughness enters as a new parameter.[2]

An inherent difficulty in investigations of the influence of surface roughness is caused by the fact that no satisfactory geometric description of a rough surface by a limited number of parameters has been found as yet. It is generally assumed that the most important parameter is the ratio of the average height of the roughness elements to the tube diameter. In his extensive experiments on friction in tubes with rough surfaces, Nikuradse produced a defined roughness pattern by gluing sand of fairly uniform size to the tube surface to form a cover which was made as dense as possible. Friction factors determined in this way are plotted in Fig. 6-21 with the Reynolds number on the abscissa and with $R/k_s$ as

---

[1] T. von Kármán, *Trans. ASME*, **61**:705–710 (1939).
[2] H. Schlichting, "Boundary Layer Theory," McGraw-Hill Book Company, Inc., New York, 1955.

parameter ($k_s$ is the average roughness height; $R$ is the tube radius). It can be seen that in laminar flow and in turbulent flow with small Reynolds numbers, the roughness has no influence on friction. The tube wall in this range is said to be "hydraulically smooth." This fact can be visualized as being caused by a situation in which the roughness elements are completely embedded in the laminar sublayer. For sufficiently large Reynolds numbers, the friction-factor curves become horizontal, indicating that the pressure drop in this range increases in proportion to the square of the mean velocity. In this range, local separation on the

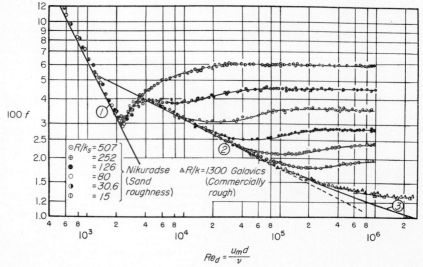

FIG. 6-21. Friction factors for flow through tubes with rough walls. (1) Eq. (6-49). (2) Eq. (6-55). (3) Eq. (6-56). (*From H. Schlichting, "Boundary Layer Theory," McGraw-Hill Book Company, Inc., New York, 1955.*)

roughness elements probably constitutes the main contribution to the pressure drop. Investigations of other roughness geometries have revealed that the shape of the friction curves is often different from the ones in Fig. 6-21. Measurements on commercially rough tubes which have been summarized by Moody[1] indicate that the friction factor on such surfaces decreases monotonically toward a constant value with increasing Reynolds number and that the dip apparent in the curves in Fig. 6-21 is missing. Such a curve is shown in Fig. 6-21 as the line representing the measurements by Galavics. Semiempirical calculations based on mixing length concepts have succeeded in correlating the measurements of friction factors and velocity profiles in rough tubes. An excellent presentation of these is contained in the book by H. Schlichting.[2]

[1] L. F. Moody, *Trans. ASME*, **66**:671 (1944).
[2] Schlichting, *op. cit.*

As a first approximation, Fig. 6-21 can also be used to obtain friction coefficients for flat plates with rough surfaces when the boundary-layer thickness is set equal to the tube radius.

Tubes which are used in engineering applications often have cross sections with shapes other than circular. The friction coefficient for laminar developed flow through tubes with noncircular cross section is given by equations of the same form as Eq. (6-49), with the values for the numerical constant, however, depending on the shape of the cross section. It has become customary to base the Reynolds number for such cross sections on a length which is called "hydraulic diameter" and which is defined by the equation

$$d_h = \frac{4A}{C} \tag{6-57}$$

with $A$ indicating the area and $C$ the perimeter of the cross section. For a slot between two parallel plane walls the hydraulic diameter equals twice the distance between the walls. The numerical constant in Eq. (6-49) has for this cross section the value 96 when the hydraulic diameter is used instead of $d$ in Eqs. (6-49) and (6-53).

Experiments on developed turbulent flow through noncircular passages have indicated that Eqs. (6-55) and (6-56) describe their friction factors with sufficient accuracy when the diameter in these equations and in Eq. (6-53), which defines the friction factor, is replaced by the hydraulic diameter. This statement holds as long as the cross section does not have sharp corner angles. The numerical constants remain unchanged.

Interesting observations have been made on the transition process from laminar to turbulent flow in passages with polygonal cross section. The transition occurs in this case not at a certain Reynolds number but in a Reynolds-number range in such a way that with increasing Reynolds number the flow at first becomes turbulent in the core of the fluid while remaining laminar in the corner regions. The turbulence penetrates gradually into the corners as the Reynolds number increases. This behavior is especially pronounced in corners of a small opening angle. It became evident by flow visualization and was substantiated by means of measured velocity profiles.[1] Figure 6-22 presents such profiles as measured along the height of the triangular cross section of a duct. The duct cross section was an isosceles triangle with a ratio of height $L$ to base equal to 5. At a Reynolds number $Re_{d_h} = 500$, the flow is completely laminar and the velocity profile has a peaked shape with a parabolic velocity increase near the small angle corner. At $Re_{d_h} = 3,000$ turbulence has flattened the peak of the profile but the parabolic velocity increase still persists over approximately half of the height, indicating

[1] E. R. G. Eckert and T. F. Irvine, Jr., *J. Aeronaut. Sci.*, **22**:65–66 (1955).

that the flow in this range near the small-angle corner is still essentially laminar. Only at $\text{Re}_{dh} = 20{,}580$ has the turbulence spread over the whole cross section.

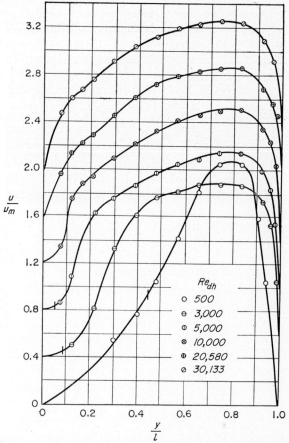

Fig. 6-22. Velocity profiles along the center line of ducts with triangular cross section. $y$ is the distance from the corner, $l$ is the height of the cross section. The profiles for $\text{Re} > 500$ are shifted in vertical direction. [*From E. R. G. Eckert and T. F. Irvine, Trans. ASME,* **77**:709 (1956).]

**6-8. The Cylinder in Crossflow.** The boundary layer has a great influence on the flow around a cylinder. Figure 6-23 shows a photograph of such a flow. It can be seen that the flow separates on both sides of the cylinder surface and builds up a dead water filled with vortices on the downstream side. If the pressure distribution along the surface is calculated for a frictionless flow without separation, the dash-dotted line in Fig. 6-24 results. The pressure decreases first from the stagnation point ($\alpha = 0$) and reaches its minimum value at $\alpha = 90°$. On the rearward side it increases again and attains for $\alpha = 180°$ the same value as at the forward stagnation point. In a fluid with friction, a boundary layer builds up along the surface. The fluid particles outside the boundary

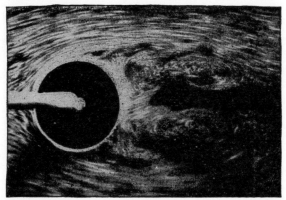

FIG. 6-23. Flow around a cylinder. (*Photograph by H. L. Rubach, Mitt. Forschungsarb.*, 1916, p. 185.)

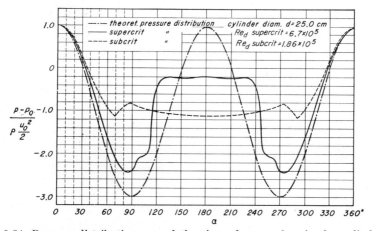

FIG. 6-24. Pressure distribution around the circumference of a circular cylinder $p$, local pressure; $p_0$ pressure at large distance from cylinder, $\rho(V_0^2/2)$, freestream impact pressure; $\alpha$, angle measured from stagnation point. [*From L. Flachsbart, "Handbuch Experimental Physik," vol. IV, part 2, (1932), p. 316, Fig. 37.*]

layer are able to move against the pressure increase on the back side of the cylinder by changing their kinetic energy into pressure energy. The fluid particles in the boundary layer do not possess so much kinetic energy. They can therefore move only a certain distance into the region of increasing pressure before their kinetic energy is consumed. Then they reverse their flow direction. In this way the flow separates from the surface. Because of this separation, the pressure distribution along the back side of the cylinder is changed also. How far the fluid particles penetrate the region of increasing pressure depends on their kinetic energy. This is on an average greater in the turbulent boundary layer (Fig. 6-12) than in the laminar one. As a consequence, the turbulent boundary layer separates at approximately $\alpha = 110°$ and the laminar boundary layer at $\alpha = 82°$. The corresponding pressure distributions are shown in Fig. 6-24. The pressures on the rear side are smaller than on the front. This gives a force in the flow direction which is called *form resistance*. To it is added a frictional resistance caused by the shear stresses along the cylinder walls. Both parts are included in the drag coefficient $f_c$ defined by the equation

$$D = f_c L d\, \rho\, \frac{u_0^2}{2} \qquad (6\text{-}58)$$

where $d$ = diameter
$L$ = length
$D$ = drag of cylinder
$\rho$ = density
$u_0$ = freestream velocity
$Ld$ = cross section normal to air flow

Figure 6-25 shows the drag coefficient $f_c$ for various Reynolds numbers. The Reynolds number is calculated with the freestream velocity $u_0$ and

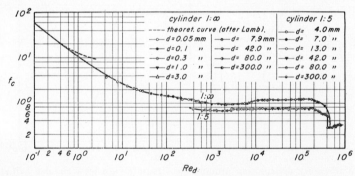

FIG. 6-25. Drag coefficient of a circular cylinder in a flow normal to its axis. (*From C. Wieselsberger, Ergeb. aerodyn. Versuchsanstalt Göttingen*, **2**, 24, 1932)

the diameter $d$. At very small freestream velocities no separation occurs and the drag is caused only by shear stresses. From $\mathrm{Re}_d \approx 1$ a dead water builds up behind the cylinder. With further increasing Reynolds numbers the dead water becomes greater. From $\mathrm{Re}_d \approx 100$ vortices separate alternately from the left and right sides (von Kármán's vortex street) and are carried away by the flow in a regular pattern. The sudden decrease in drag at $\mathrm{Re}_d \approx 4 \times 10^5$ is originated by the fact that the boundary layer has become turbulent ahead of the separation point. At $Re_d > 4 \times 10^5$ no vortex wake forms but only small irregular vortices in the dead water. At small velocities the drag of the cylinder is mostly frictional resistance and at $Re_d > 1{,}000$ mostly form resistance. Each of these flow regimes influences the heat transfer in a distinct way.

The development of the laminar boundary layer along the front portion of the cylinder can be calculated by the method presented in Sec. 6-5 when the pressure distribution in Fig. 6-24 is introduced into Bernoulli's equation to determine the local stream velocity $u_s$. Such a calculation also determines the form parameter $\lambda$ and it is found that this form parameter changes from positive values near the forward stagnation point to the value zero which is obtained at the location where the pressure gradient is zero and to negative values for that part of the surface along which the pressure increases in the flow direction. The point where the laminar boundary layer separates from the surface is determined as that location along the cylinder where the form parameter assumes the value $\lambda = -3$. Similar calculation methods have also been developed for turbulent boundary layers. One of the main purposes of such calculations is a prediction of the point at which the flow separates from a cylinder with arbitrary cross section. It has been found, however, that for such calculations the shape of the velocity profile must be well described. A large number of methods have been presented in the literature which either utilize some integral relationship or attempt to solve the boundary-layer equations themselves. Any discussion of these methods would exceed the scope of this book.

**6-9. Flow around Rotationally Symmetric Objects.** Rotationally symmetric flow around any immersed object exhibits essentially the same characteristic behavior as flow around cylindrical objects. Flow around a sphere, for instance, shows again a separation on the sides of the sphere and a stagnant region filled with vortices in the back. This separation occurs at larger Reynolds numbers, whereas for very low Reynolds numbers (below 1) laminar flow closes around the rear portion of the sphere. The drag connected with this type flow is described by Stokes's law:

$$D = 3\pi d\, \mu u_0 \tag{6-59}$$

The drag coefficient of a sphere for a large Reynolds-number range is indicated in Fig. 6-26. This coefficient is defined in the same way as the one for a cylinder. Again a characteristic drop is observed at a Reynolds number around $3 \times 10^5$. It was found that the Reynolds number at which the drop in drag occurs for a sphere with smooth surface depends on the turbulence in the freestream, because the turbulence level determines whether the boundary layer ahead of the separation point is laminar or turbulent. This connection between the turbulence level in the free-stream and the critical Reynolds number indicated by the drop in the drag of a sphere was interpreted correctly by L. Prandtl. It makes a sphere useful for measuring turbulence in an air stream, for instance, in wind tunnels. Today this method has been almost universally replaced by a direct measurement of the turbulent-velocity fluctuations $(u',v',w')$ with the hot-wire anemometer.

The development of a boundary layer around the front portion of a rotationally symmetric object can be handled, with small modifications, by the methods described for two-dimensional boundary layers in the previous paragraph. Very useful, also, is a transformation which has been pointed out by Mangler.[1]

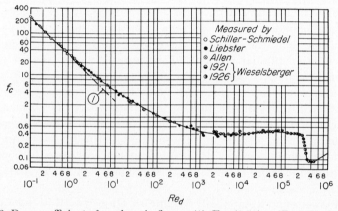

FIG. 6-26. Drag coefficient of a sphere in flow. (1) Eq. (6-59). (*From H. Schlichting, "Boundary Layer Theory," McGraw-Hill Book Company, Inc., New York, 1955.*)

The Mangler transformation offers the possibility of determining the field for a related rotationally symmetrical boundary layer from a known velocity field within a steady, two-dimensional, laminar boundary layer. The coordinates $x$ and $y$ of the rotationally symmetric boundary layer are related to the coordinates $\bar{x}$ and $\bar{y}$ for the two-dimensional flow by the following equations:

$$d\bar{x} = \frac{r^2}{C^2} dx \qquad d\bar{y} = \frac{r}{C} dy \qquad (6\text{-}60)$$

or integrated,

$$\bar{x} = \frac{1}{C^2} \int r^2 \, dx \qquad \bar{y} = \frac{r}{C} y \qquad (6\text{-}61)$$

$r$ is the distance of the surface point of the rotationally symmetric object under consideration from the axis, and $C$ an arbitrary constant with the

[1] W. Mangler, *Z. angew. Math. Mech.*, **28**:97 (1948).

dimension of length. That the above relationships transform a rotationally symmetric boundary layer into a two-dimensional one can be checked by writing down the rotationally symmetric boundary-layer flow equations and by introducing into these equations the above transformation. The continuity equation (6-20) of a rotationally symmetrical boundary layer is satisfied by introduction of the stream function $\psi(\partial\psi/\partial y = ru/C,\ \partial\psi/\partial x = -rv/C)$. With it the momentum equation (6-16) assumes the following form:

$$\frac{C}{r}\frac{\partial\psi}{\partial y}\frac{\partial}{\partial x}\left(\frac{C}{r}\frac{\partial\psi}{\partial y}\right) - \frac{C}{r}\frac{\partial\psi}{\partial x}\frac{\partial}{\partial y}\left(\frac{C}{r}\frac{\partial\psi}{\partial y}\right) = \frac{\partial}{\partial y}\left[\nu\frac{\partial}{\partial y}\left(\frac{C}{r}\frac{\partial\psi}{\partial y}\right)\right] - \frac{1}{\rho}\frac{\partial p}{\partial x}$$

Introduction of the new independent variables $\bar{x}$ and $\bar{y}$ leads to the expression

$$\frac{r^2}{C^2}\frac{\partial\psi}{\partial\bar{y}}\frac{\partial}{\partial\bar{x}}\left(\frac{\partial\psi}{\partial\bar{y}}\right) - \frac{r^2}{C^2}\frac{\partial\psi}{\partial\bar{x}}\frac{\partial}{\partial\bar{y}}\left(\frac{\partial\psi}{\partial\bar{y}}\right) = \frac{r^2}{C^2}\frac{\partial}{\partial\bar{y}}\left[\nu\frac{\partial}{\partial\bar{y}}\left(\frac{\partial\psi}{\partial\bar{y}}\right)\right] - \frac{r^2}{C^2}\frac{1}{\rho}\frac{\partial p}{\partial\bar{x}}$$

It can be seen that both sides of the equation can be divided by $r^2/C^2$ and that in this way the equation becomes identical with Eq. (6-43) (with the additional pressure term) for a two-dimensional boundary layer. This means that the stream function of a two-dimensional boundary layer expressed in the coordinates $\bar{x}$ and $\bar{y}$ is at the same time a solution of a rotationally symmetrical boundary layer in $x$ and $y$ provided the boundary conditions are the same for both situations. For a solid surface the similarity in the boundary conditions at the surface ($u = 0$) is a priori fulfilled. The similarity in the boundary conditions at the outer edge of the boundary layer requires that the pressure gradient $\partial p/\partial x$ for the rotationally symmetric flow is identical with the pressure gradient $\partial p/\partial \bar{x}$ for the two-dimensional flow. An example of such a transformation will be given in a later chapter. From a knowledge of the stream function the velocity $u$ can be easily calculated.

## PROBLEMS

**6-1.** Calculate the laminar boundary-layer thickness and the friction coefficient for flow over a flat plate with a linear approximation or a parabolic approximation of the velocity profile. Compare these results with the ones obtained in Sec. 6-4.

**6-2.** Calculate the boundary-layer development near the stagnation line of a cylinder with circular cross section, assuming that the stream velocity in this range can be approximated by $u_s = 4\ u_0 x/d$ ($u_o$ is the freestream velocity, $d$ is the diameter; see also page 184). The boundary-layer thickness for $x = 0$ can be determined from the condition that at this location $d\delta/dx = 0$. (Otherwise the boundary-layer thickness plotted over $x$ for positive and negative $x$ would have a cusp at $x = 0$.)

**6-3.** Compare the shape of the velocity profiles described by Eq. (6-40) with the exact boundary-layer solutions presented in Fig. 6-17 by plotting both versus the ratio of wall distance to boundary-layer displacement thickness as abscissa. (The parameters $\lambda$ and $\beta$ can be compared by expressing each one as function of the boundary-layer momentum thickness.)

**6-4.** Carry out the iteration procedure for the solution of Eq. (6-45) with the velocity $u = u_s$ as a first approximation. Compare the results of the third iteration with the solution obtained by Blasius (Fig. 6-16).

**6-5.** Calculate the velocity profile, the wall shear stress, and the friction factor for developed laminar flow through a channel built with two parallel flat plates. Introduce the hydraulic diameter into the relation for the friction factor [note that the constant in this relation is different from the one in Eq. (6-49)].

**6-6.** Calculate the development of the velocity profile in the entrance region of a duct built with two parallel walls. Assume the same velocity profile shape as in Fig. 6-18. Use the continuity equation and Bernoulli's equation to calculate the velocities $u_s$ in the central part and the integrated momentum equation for the whole velocity profile.

**6-7.** Repeat the calculation as in the preceding problem, assuming, however, that the flow is turbulent from the duct entrance and using a seventh-power law for the velocity profile near the walls. Compare entrance lengths for laminar and turbulent flow.

**6-8.** Discuss why the drag coefficient of a cylinder with finite length is smaller than for an infinitely long cylinder. This fact can be observed in Fig. 6-25. Consider for this purpose the pressure distribution around the cylinder, including the end regions for the cylinder with finite length.

**6-9.** Calculate the size which a sphere must have if it is to be used to determine the turbulence level in an airstream of 100 or of 600 fps. Assume standard atmospheric conditions.

**6-10.** Calculate the size of a droplet which can remain suspended in the atmosphere without falling to the ground when an upward velocity of 1 fps exists in the air. Assume standard atmospheric conditions.

CHAPTER 7

# FORCED CONVECTION IN LAMINAR FLOW

**7-1. The Heat-flow Equation of the Boundary Layer.** When a body submerged in a flow is heated or cooled, a temperature field builds up in the surrounding medium. This field comprises only a region within a small distance from the body surface as long as very small velocities and the wake within which the heat is carried away are excluded. This region along the body is called the *thermal boundary layer*. Within this layer the temperature changes from the value $t_w$ on the surface of the body to the value $t_s$ in the undisturbed flow. The temperature profile in the proximity of the wall is shown in Fig. 7-1. At the distance $\delta_t$ from the wall, the same temperature $t_s$ is measured in the flowing fluid as in the case when the body is neither heated nor cooled. The length $\delta_t$ is the thickness of the thermal boundary layer. Figure 7-1 shows also the velocity profile and the hydrodynamic boundary layer with the thickness $\delta$. The two boundary layers generally have different thicknesses. The energy transfer in the thermal boundary layer can be described by a boundary-layer equation which is derived from an energy equation by an order of magnitude argument in the same way as the flow boundary-layer equations are obtained from the Navier-Stokes equations. An approximate but faster and simpler determination of the thermal boundary layer and the heat transfer can be made with a *heat-flow equation of the thermal boundary layer*. It is derived from a heat balance on a volume element at the surface, built up by two planes 1-2 and 3-4 in Fig. 7-2, separated from each other by the length $dx$, by a plane 2-4 parallel to the wall at the distance $l$, and by the wall 1-3. The length $l$ is assumed greater than both boundary-layer thicknesses $\delta$ and $\delta_t$. The calculation may be again confined to steady two-dimensional problems, to a constant-property fluid, and to velocities which are sufficiently small so that temperature increases caused in the boundary layer by internal friction can be neglected. In Fig. 7-2 the volume element has unit length normal to the plane of the paper. Through the plane 1-2 a quantity of heat, which is determined by the following integral, is transported per unit time:

$$\rho c_p \int_0^l tu \, dy$$

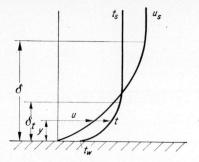

FIG. 7-1. Temperature and velocity profiles on a wall.

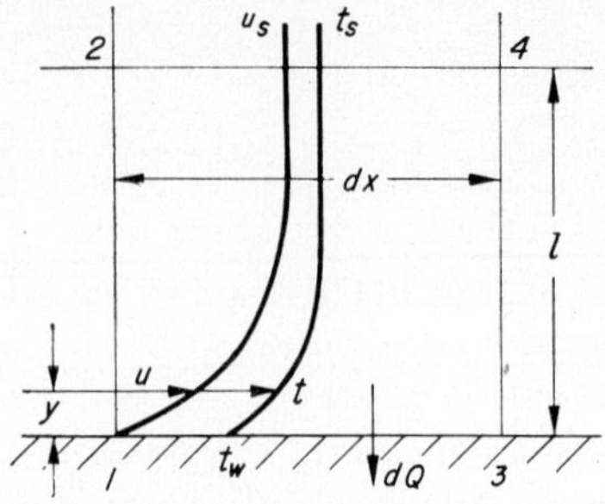

FIG. 7-2. Deriving the heat-flow equation for the boundary layer.

where $c_p$ is the specific heat at constant pressure per unit mass.[1] This heat flow changes with the length $dx$ by

$$\rho c_p \, dx \, \frac{d}{dx} \int_0^l tu \, dy$$

The heat flow leaving the volume element through plane 3-4 is greater by this amount than the heat flow entering the element through plane 1-2. Through the plane 2-4 there flows the fluid mass $\rho \, dx \, \frac{d}{dx} \int_0^l u \, dy$ as was derived in the preceding chapter. This transports an amount of heat given by $\rho c_p t_s \, dx \, \frac{d}{dx} \int_0^l u \, dy$ into the volume element. Through the plane 1-3 a certain amount of heat flows from the volume element to the wall or in the opposite direction. Since the fluid particles adjacent to the wall have the velocity zero, this heat must be transferred by conduction through this fluid layer. Therefore Eq. (2-2) is applicable. With the nomenclature used here the heat flow per unit time is

$$dQ = -k \, dx \left(\frac{dt}{dy}\right)_w$$

where $k$ is the thermal conductivity of the fluid. The sum of the heat carried into the volume element by convection with the flow and the heat lost from the element by conduction to the wall must be zero for steady state:

$$\rho c_p \frac{d}{dx} \int_0^l t_s u \, dy - \rho c_p \frac{d}{dx} \int_0^l tu \, dy - k \left(\frac{dt}{dy}\right)_w = 0 \qquad (7\text{-}1)$$

[1] It will be explained on page 170 why the specific heat at constant pressure is to be used.

Introducing the thermal diffusivity of the fluid $\alpha = k/\rho c_p$ gives

$$\frac{d}{dx}\int_0^l (t_s - t)u\, dy = \alpha \left(\frac{dt}{dy}\right)_w \qquad (7\text{-}2)$$

This is the *heat-flow equation of the boundary layer* by which the heat transfer can be calculated.[1] It holds for turbulent as well as for laminar flow. The property values used in Eq. (7-2) are tabulated in the Appendix. The specific heat $c_p$ and the thermal conductivity $k$ as well as the viscosity $\mu$ depend upon pressure only in the proximity of the critical point. Figures A-3 to A-7 show this for water vapor. The specific heat at the critical point theoretically approaches infinity. This is indicated by the very steep increase of $c_p$ in Fig. A-3 near the critical point. The density $\rho$ for fluids depends very little on the pressure. For gases it can be calculated from the equation of state. In the proximity of the critical point the dependency on the pressure is more complicated. As an example, Fig. A-1 shows this relationship for water and steam. The thermal diffusivity $\alpha$ and the kinematic viscosity $\nu$ for liquids depend only in a very small degree on the pressure outside the critical pressure range. For gases both values are inversely proportional to the pressure.

The dependence of density, specific heat, and heat conductivity on temperature can also be observed in Figs. A-1, A-3, A-6, and A-7. It is especially pronounced near the critical state. Equation (7-2) has been derived under the assumption of constant properties. It can, therefore, be used for heat-transfer processes in which the temperature differences involved are such that the actual variation of properties is small. By introduction of appropriately chosen mean values this range of validity can be extended.

**7-2. Laminar Boundary-layer Energy Equation.** The heat-flow equation as derived in the preceding paragraph offers the possibility to calculate heat transfer by forced convection for various specific situations if appropriate assumptions regarding the shape of the temperature profile are made. Before such a calculation is discussed, the differential equation describing the energy conditions within a flowing medium will be developed. This equation is derived from an energy balance on a stationary volume element located within the flow field. Heat may be transported by conduction into the volume element, or it may be carried with the moving fluid through the boundaries of the element. Heat may be stored in the interior of the element. Heat may also be generated by internal heat sources. Such heat sources are actually always present in a moving viscous fluid, since the shear stresses cause internal friction and

---

[1] G. Kroujiline [*Tech. Phys. U.S.S.R.*, **3**:183, 311 (1936)] was apparently the first one to use Eq. (7-2) for heat-transfer calculations.

convert kinetic energy into heat. For small velocities the temperature changes caused by this internal friction are small and are usually neglected. For large flow velocities frictional effects become important. In the development of high-speed aircraft they are of major concern (aerodynamic heating). In the preceding section internal heat generation was neglected. It will be included here, however. In addition a separate paragraph will be devoted later on to high-velocity heat transfer.

The energy equation for a viscous fluid with constant properties can be obtained easily by an extension of Eq. (2-13). This equation gives a balance among the heat stored in the interior of a small volume element with sides $dx, dy, dz$; the heat which is conducted into the volume element through its surfaces; and the heat which is generated internally. If a stationary volume element is considered through which a fluid flows with velocity components $(u,v,w)$, then additional heat will be transported into the volume element by convection. This convective heat transport can be calculated in the same way as in the previous paragraph.† Its inclusion into the energy balance results in

$$\rho c_p \left( \frac{\partial t}{\partial \tau} + u \frac{\partial t}{\partial x} + v \frac{\partial t}{\partial y} + w \frac{\partial t}{\partial z} \right) = k \left( \frac{\partial^2 t}{\partial x^2} + \frac{\partial^2 t}{\partial y^2} + \frac{\partial^2 t}{\partial z^2} \right) + \Phi \quad (7\text{-}3)$$

$$\quad\quad 1 \quad\; 1 \quad\; \tfrac{1}{1} \quad\; \delta \quad\; \tfrac{1}{\delta} \quad\;\; 0 \quad\quad\; \delta^2 \quad\quad 1 \quad\quad \tfrac{1}{\delta^2} \quad\quad 0$$

The heat generated per unit time and unit volume by internal friction is indicated by the letter $\Phi$ in the above equation. It is called *heat of dissipation*. The derivation of this term from the velocity field is a lengthy procedure and will not be discussed here. For its derivation see, for instance, Schlichting's "Boundary Layer Theory." The following equation describes the heat dissipation for a fluid with constant properties:

$$\Phi = 2\mu \left[ \left(\frac{\partial u}{\partial x}\right)^2 + \left(\frac{\partial v}{\partial y}\right)^2 + \left(\frac{\partial w}{\partial z}\right)^2 \right] + \mu \left[ \left(\frac{\partial v}{\partial x}\right) + \left(\frac{\partial u}{\partial y}\right) \right]^2$$

$$\quad\;\; \delta^2 \quad\quad\quad 1 \quad\quad\quad 1 \quad\quad\quad 0 \quad\quad\;\; \delta^2 \quad\quad\; \delta^2 \quad\quad\;\; \tfrac{1}{\delta^2}$$

$$+ \mu \left[ \left(\frac{\partial w}{\partial y}\right) + \left(\frac{\partial v}{\partial z}\right) \right]^2 + \mu \left[ \left(\frac{\partial u}{\partial z}\right) + \left(\frac{\partial w}{\partial x}\right) \right]^2 \quad (7\text{-}4)$$

$$\quad\quad\;\; \delta^2 \quad\quad 0 \quad\quad\quad 0 \quad\quad\quad 0 \quad\quad\quad 0$$

Equation (7-3) together with Navier-Stokes equations describes the temperature field in a viscous fluid. For ordinary fluids, the numerical values of the heat conductivity are so small that the heat transport by

† There is actually no difference between the specific heat at constant volume and that at constant pressure for a fluid with constant properties (including the density). The value $c_p$ is used here and in the following equations because many of them apply to a compressible fluid also.

conduction becomes important only in the region where the convective heat transport is small because of small velocities. We know that such a region always exists near the surface of walls of solid objects because there the velocity decreases toward zero. As a consequence, it must be expected that heat conduction for such fluids warrants consideration only near solid surfaces. In other words, it is expected that a thin layer along solid surfaces will exist within which heat conduction is equal in importance to heat convection whereas outside this layer the conduction is relatively so small that it can be neglected. This layer will be called the *thermal boundary layer*. We shall now simplify by an order-of-magnitude estimate the differential equation describing the energy flow in this temperature boundary layer. The derivation will be essentially the same as for the velocity boundary layer. It will consider two-dimensional flow. Accordingly the terms in Eqs. (7-3) and (7-4) which are indicated by a zero will vanish.

Since we are concerned with the effect of a small value of the boundary-layer thickness and of a small heat-conductivity value on the energy equation, all other quantities in Eq. (7-3) will be measured in such units that they are of order 1. The thermal boundary-layer thickness is assumed of order $\delta$. The order of the heat-conductivity value will be left open. The velocity component $v$ was found to be of order $\delta$. The order of magnitude of the terms on the left-hand side of Eq. (7-3) can now be readily established. The orders of magnitude are again indicated underneath the equation. On the right-hand side the first term in the bracket can be neglected as compared with the second one. The whole bracketed term is of the same order of magnitude as the terms of the left-hand side of the equation when $k$ is of order $\delta^2$. The only term remaining in the dissipation function $\Phi$ is the value $\mu(du/dy)^2$, which is of order 1. The boundary-layer energy equation for an incompressible constant property fluid is therefore

$$\rho c_p \left( \frac{\partial t}{\partial \tau} + u \frac{\partial t}{\partial x} + v \frac{\partial t}{\partial y} \right) = k \frac{\partial^2 t}{\partial y^2} + \mu \left( \frac{\partial u}{\partial y} \right)^2 \qquad (7\text{-}5)$$

The following boundary conditions belong to this differential equation: The temperature for the fluid is prescribed outside the boundary layer ($t = t_s$). For constant properties this temperature is constant. At the wall a larger variety of boundary conditions exists for the temperature boundary layer than for the velocity boundary layer. The most commonly prescribed condition is either a fixed value of the surface temperature $t_{y=0} = t_w(x)$ or a prescribed value for the heat flow through the solid surface:

$$q_{y=0} = k \left( \frac{\partial t}{\partial y} \right)_{y=0} = q_w(x)$$

The boundary-layer energy equation looks in form quite similar to the momentum equation of the boundary layer. However, two important differences exist. In the energy equation (7-5) the values $u$ and $v$ have to be considered as known parameters, determined from a solution of the flow equations. Correspondingly, the boundary-layer energy equation is a linear equation in the temperature $t$, which from a mathematical standpoint makes the problem of obtaining solutions to this equation considerably simpler because the "law of superposition" holds. That means that as soon as a number of solutions of this equation are known, new solutions can readily be obtained by addition or subtraction of any of the known solutions. Another difference in the two equations is connected with the fact that a term corresponding to the pressure gradient is not contained in the energy equation. From this it can be suspected, and this will be confirmed later on, that the influence of pressure variations along the surface is smaller on heat transfer than on flow parameters like the drag.

For flow with small velocity over a flat plate, the momentum equation (without the term containing $\partial p/\partial x$) and the energy equation (without the dissipation term) look very similar in form. Additionally, when the numerical value for the thermal diffusivity is equal to the value of the kinematic viscosity, then the equations are identical and can be readily transformed into each other. As a consequence, if the boundary conditions are the same in these cases also, the solution of the momentum equation (the velocity profile within the boundary layer) and the solution of the energy equation (the temperature profile within the boundary layer) are completely similar in form and the flow boundary-layer thickness is equal to the thermal boundary-layer thickness. This will be discussed in more detail later on when actual solutions of the boundary-layer energy equation are presented.

The preceding derivation assumed that both boundary-layer thicknesses are of equal order. This means that the combination of properties, known as the Prandtl number

$$\text{Pr} = \frac{\mu c_p}{k} = \frac{\nu}{\alpha}$$

is of order 1. This is the case for most fluids.

In order to arrive at a criterion for the condition when the dissipation term in Eq. (7-3) can be neglected, let us consider that we measure temperature with a unit which is of order of a temperature difference imposed on the problem (for instance, the difference between wall temperature and stream temperature) whereas the unit with which velocities are measured is of order of the stream velocity $u_s$. Then the second term on the right-hand side of Eq. (7-3) is of order $\delta^2(\Delta t_0/\delta^2)$, the dissipation term of order $\delta^2(u_s^2/\delta^2)$, and both are of the same order when $u_s^2/\Delta t_0$ is of order 1.

This ratio can be divided by $c_p$ (of order 1) to make it dimensionless. The dissipation is, therefore, of the same order as the other terms when $u_s^2/c_p \Delta t_0$ is of order 1. When the group $u_s^2/c_p \Delta t_0$ is of an order which is small as compared with 1, then the dissipation can be neglected. By introduction of numerical values, it is found that in air for a temperature difference of 10 F the velocity has to be of order 500 fps to make the group $u^2/c_p \Delta t_0$ of order 1. Usually considerably smaller velocities or larger temperature differences are employed in engineering applications. The dissipation term will, therefore, be neglected in general in the following calculations.

**7-3. The Plane Plate in Longitudinal Flow.** The heat transfer from a plane wall with a constant temperature $t_w$ in a flow of constant velocity will be calculated. The first part of the plate with the length $x_0$ may

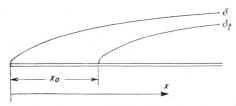

FIG. 7-3. Flow boundary layer and thermal boundary layer on a flat plate.

not be heated and has the same temperature as the fluid, as shown in Fig. 7-3. Thus the hydrodynamic boundary layer begins at the edge of the unheated plate, while the temperature boundary layer starts at the edge of the heated part. Both increase their thicknesses $\delta$ and $\delta_t$ in flow direction. A calculation of the thermal boundary-layer thickness $\delta_t$ and with it of the heat transfer will be made with the heat-flow equation (7-2). For this purpose a suitable statement must be made as to the shape of the temperature profile in the boundary layer.

The result of our calculation will be the better the closer the assumed profile agrees with the real one. Therefore, an expression with a number of free functions is chosen. The functions are determined such that the assumed profile satisfies conditions which are known to hold for the real one.

For $y = 0$: $\quad t = t_w$

For $y = \infty$: $\quad t = t_s \quad \dfrac{\partial t}{\partial y} = 0$

Equation (7-5), written for steady, low-velocity flow and for $y = 0$, results in

For $y = 0$: $\quad \dfrac{\partial^2 t}{\partial y^2} = 0$

According to these four conditions, a polynomial with four functions will be used.

$$t = a + by + cy^2 + dy^3$$

If, in addition, the temperature difference $\theta = t - t_w$ is introduced, and if the functions are determined from the above conditions, then the temperature profile becomes

$$\theta' = \frac{\theta}{\theta_s} = \frac{3}{2}\frac{y}{\delta_t} - \frac{1}{2}\left(\frac{y}{\delta_t}\right)^3$$

with
$$\theta = t - t_w \qquad \theta_s = t_s - t_w$$

Now the integral in the heat-flow equation can be calculated:

$$\int_0^l (t_s - t)u\, dy = \int_0^l (\theta_s - \theta)u\, dy$$
$$= \theta_s u_s \int_0^{\delta_t}\left[1 - \frac{3}{2}\frac{y}{\delta_t} + \frac{1}{2}\left(\frac{y}{\delta_t}\right)^3\right]\left[\frac{3}{2}\frac{y}{\delta} - \frac{1}{2}\left(\frac{y}{\delta}\right)^3\right] dy$$

It may be presumed that the thermal boundary layer is smaller than the hydrodynamic boundary layer. Then the second integral must be extended only to $y = \delta_t$, since for $y > \delta_t$, $\theta = \theta_s$ and therefore the integrand is zero. By introducing the ratio $\zeta = \delta_t/\delta$ of the thermal to the hydrodynamic boundary layer and evaluating the integral, there follows

$$\int_0^l (\theta_s - \theta)u\, dy = \theta_s u_s \delta(\tfrac{3}{20}\zeta^2 - \tfrac{3}{280}\zeta^4)$$

As we assumed $\delta_t$ to be smaller than $\delta$ and therefore $\zeta < 1$, the second term in the right-hand expression is small compared with the first and can be neglected. According to Fig. 7-3 the ratio $\zeta$ is a function of $x$.

If the above integral is introduced into Eq. (7-2), there is obtained

$$\frac{3}{20}\theta_s u_s \frac{d}{dx}(\zeta^2 \delta) = \frac{3}{2}\alpha \frac{\theta_s}{\zeta \delta}$$

or
$$\frac{1}{10} u_s \left(\zeta^3 \delta \frac{d\delta}{dx} + 2\zeta^2 \delta^2 \frac{d\zeta}{dx}\right) = \alpha$$

Introducing Eq. (6-27) for $\delta\, d\delta/dx$ and (6-28) for $\delta^2$ gives

$$\frac{14}{13}\frac{\nu}{\alpha}\left(\zeta^3 + 4x\zeta^2 \frac{d\zeta}{dx}\right) = 1$$

The ratio $\nu/\alpha$ in the equation is a dimensionless value very common in heat-transfer calculations. It is called the Prandtl number or Prandtl modulus and is written as Pr.

$$\text{Pr} = \frac{\nu}{\alpha} = \frac{c_p \mu}{k} \qquad (7\text{-}6)$$

FORCED CONVECTION IN LAMINAR FLOW

The Prandtl number is composed of property values and is therefore itself a property value. It has the advantage that it is dimensionless. The tables in the Appendix give this value for liquids and gases. The Prandtl number for liquids and gases depends on temperature. Only near the critical point does a marked dependence on pressure occur. The variation with temperature is small for gases.

Introducing Pr in the above equation gives

$$\zeta^3 + \frac{4}{3} x \frac{d(\zeta^3)}{dx} = \frac{13}{14\,\mathrm{Pr}}$$

or with the substitution $\zeta^3 = y$,

$$y + \frac{4}{3} x \frac{dy}{dx} = \frac{13}{14\,\mathrm{Pr}}$$

A particular integral is $y = 13/(14\,\mathrm{Pr})$. The general solution of the homogeneous equation can be found by trying the expression $y = x^n$; $n$ becomes $-3/4$. Therefore the complete solution of the above equation is

$$y = \frac{13}{14\,\mathrm{Pr}} + Cx^{-3/4}$$

From the boundary condition for $x = x_0$, $\zeta = 0$, or $y = 0$, there follows

$$\zeta = \frac{1}{1.026 \sqrt[3]{\mathrm{Pr}}} \sqrt[3]{1 - \left(\frac{x_0}{x}\right)^{3/4}} \qquad (7\text{-}7)$$

If the plate is heated over the whole length ($x_0 = 0$),

$$\zeta = \frac{1}{1.026 \sqrt[3]{\mathrm{Pr}}} \qquad (7\text{-}8)$$

Viscous oils have a Prandtl number $\mathrm{Pr} = 1{,}000$ or more. For these fluids the thickness of the thermal boundary layer is only one-tenth the thickness of the hydrodynamic boundary layer. Gases have Prandtl numbers smaller than 1. For these, $\zeta$ is greater than 1, and therefore an assumption in the preceding calculation does not hold true. Since the smallest value for gases is $\mathrm{Pr} = 0.6$ and therefore $\zeta = 1.16$, the error introduced by this fact is quite small. The only materials with a very small Prandtl number are liquid or molten metals. For these, the above results are not valid.

The heat flow from the plate per unit area is given by

$$q = -k \left(\frac{d\theta}{dy}\right)_w$$

On the other hand a heat-transfer coefficient $h$ can be introduced:

$$q = h(t_w - t_s) = -h\theta_s$$

Equating these two expressions gives

$$h = \frac{k}{\theta_s}\left(\frac{d\theta}{dy}\right)_w = \frac{3}{2}\frac{k}{\delta_t} = \frac{3}{2}\frac{k}{\zeta\delta} \qquad (7\text{-}9)$$

The heat-transfer coefficient is therefore inversely proportional to the thermal boundary-layer thickness. With the ratio $\zeta$ and Eq. (6-28),

$$h = 0.332k \frac{\sqrt[3]{\text{Pr}}}{\sqrt[3]{1 - (x_0/x)^{3/4}}} \sqrt{\frac{u_s}{\nu x}} \qquad (7\text{-}10)$$

and for the plate heated over its entire length,

$$h = 0.332k \sqrt[3]{\text{Pr}} \sqrt{\frac{u_s}{\nu x}} \qquad (7\text{-}11)$$

As shown in Fig. 7-4, the heat-transfer coefficient begins with an infinite value at the starting point of the heated section of the plate and decreases with increasing $x$. It is advantageous to write Eq. (7-10) in dimensionless form:

$$\frac{hx}{k} = 0.332 \sqrt[3]{\text{Pr}} \sqrt{\frac{u_s x}{\nu}} \frac{1}{\sqrt[3]{1 - (x_0/x)^{3/4}}}$$

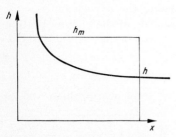

FIG. 7-4. Local film heat-transfer coefficient on a flat plate as a function of the distance from the leading edge.

On the right there appears again the Reynolds number calculated with the freestream velocity $u_s$ and the length $x$. The dimensionless expression on the left side is called *Nusselt number* or Nusselt modulus and is indicated as Nu. $\text{Nu}_x$ is written to indicate that the Nusselt number is calculated with the length $x$.

$$\text{Nu}_x = \frac{hx}{k} \qquad (7\text{-}12)$$

With this notation the above equation can be written

$$\text{Nu}_x = 0.332 \sqrt[3]{\text{Pr}} \sqrt{\text{Re}_x} \frac{1}{\sqrt[3]{1 - (x_0/x)^{3/4}}} \qquad (7\text{-}13)$$

and for the plate heated over its entire length,

$$\text{Nu}_x = 0.332 \sqrt[3]{\text{Pr}} \sqrt{\text{Re}_x} \qquad (7\text{-}14)$$

The Nusselt number can be expressed as the ratio of two lengths. In Sec. 1-3 there was introduced a boundary-layer thickness $\delta'_t$ (Fig. 1-3) which really is the length of the subtangent to the temperature profile within the boundary layer at the wall. Figure 7-5 shows this in detail.

For the temperature profile represented by the cubic parabola, the relationship $\delta'_t = \tfrac{2}{3}\delta_t$ holds. The film heat-transfer coefficient is, according to Sec. 1-3,

$$h = \frac{k}{\delta'_t}$$

and therefore the Nusselt number

$$\mathrm{Nu}_x = \frac{hx}{k} = \frac{x}{\delta'_t}$$

is the ratio of the length $x$ to the thermal boundary-layer thickness $\delta'_t$. Since the boundary-layer thickness is always small as compared with the length $x$, the Nusselt number will be large. The thermal boundary-layer thickness decreases with increasing Reynolds and Prandtl numbers. Both values therefore increase the Nusselt number.

For calculations on industrial heat exchangers, not the local but the average value of the heat-transfer coefficient is important. For the plate heated over its entire length this value is

$$\bar{h} = \frac{1}{x}\int_0^x h\,dx = \frac{C}{x}\int_0^x \frac{dx}{\sqrt{x}} = \frac{C}{x}2\sqrt{x} = 2\frac{C}{\sqrt{x}} = 2h \quad (7\text{-}15)$$

In $C$ are combined all values which do not depend on $x$. The average value of the heat-transfer coefficient is therefore twice the local value at the end of the plate. The heat transfer from a plate heated over its entire length was also calculated by an exact solution of the differential equations of the boundary layer. This calculation gives the expression in (7-14) with a numerical value 0.332. Such close agreement of the results with our approximate procedure is, of course, coincidental.

If the property values in the equation depend on temperature, then they must be introduced at a mean temperature. From calculations which solved the laminar boundary-layer equations for a fluid with variable properties (air), it can be shown that Eq. (7-14) gives the right answer for air when the property values are introduced at a reference temperature determined by the equation

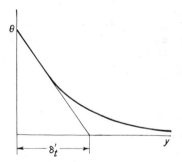

FIG. 7-5. Determination of the Nusselt number.

$$t^* - t_s = 0.50(t_w - t_s)\dagger \quad (7\text{-}16)$$

† E. R. G. Eckert, *Trans. ASME*, **56**:1273–1283 (1956).

This reference temperature can be used for other gases. For liquids our knowledge of the reference temperature is less certain. Experiments have verified Eq. (7-14) well for gases.

Equations (7-13) to (7-15) have been obtained on the assumption that the ratio $\zeta$ of the thermal to the flow boundary-layer thickness is smaller than 1. It has been found that this is true for fluids with a Prandtl number larger than 1. An extension of the calculation procedure presented in this section to a ratio $\zeta$ larger than 1 is quite straightforward. One only has to carry out the integration in two steps, since the velocity $u$ is given by Eq. (6-26) for $0 < y < \delta$ and by $u = u_s$ for $\delta < y < \delta_t$. The following equation results for the plate heated to a constant temperature on its entire length:

$$\mathrm{Nu}_x = \frac{\sqrt{\mathrm{Re}\,\mathrm{Pr}}}{1.55\sqrt{\mathrm{Pr}} + 3.09\sqrt{0.372 - 0.15\,\mathrm{Pr}}}$$

This relation is useful for calculating heat transfer in liquid metals which have Prandtl numbers between 0.005 and 0.05. In this range, the denominator is only a weak function of Pr, so that the Nusselt number depends essentially on the product Re Pr which is also referred to as the *Peclet number*. An exact solution of the laminar-boundary-layer equations results in a relation which has, in place of the denominator in the above equation, a weak function of Pr which varies by $\pm 5$ per cent around the value 1.98 for the above Prandtl-number range.[1] It will be found that the agreement of the above equation with this result is very good.

**Example 7-1.** The heat-transfer coefficient is to be calculated for a plate in an air stream for a distance of 4 in. from the leading edge. The plate is heated over its entire length; the air velocity $u_s = 33$ fps; its temperature $t_s = 125$ F; the surface temperature of the plate $t_w = 255$ F.

The kinematic viscosity of air at $t^* = 190$ F and ambient pressure is $23.6 \times 10^{-5}$ ft$^2$/sec (Appendix). Therefore the Reynolds number is $Re_x = 46{,}600$. The Prandtl number is found in the Appendix to be 0.695. Equation (7-14) gives $\mathrm{Nu}_x = 0.332 \times \sqrt[3]{0.695} \times \sqrt{46{,}600} = 63.3$. The local heat-transfer coefficient at a distance 4 in. from leading edge is, from Eq. (7-12), $h = (k/x)63.3$. The thermal conductivity $k$ for air at 190 F is $k = 0.0178$ Btu/hr ft F. Therefore

$$h = \frac{0.0178 \times 12}{4} \times 63.3 = 3.38 \text{ Btu/hr ft}^2 \text{ F}$$

The average value over a 4-in. length is $\bar{h} = 6.76$ Btu/hr ft$^2$ F. The heat flow from a plate $b = 1$ ft wide and 4 in. long is

$$Q = \bar{h}bx(t_w - t_s) = 6.76 \times \tfrac{4}{12} \times (255 - 125) = 293 \text{ Btu/hr}$$

**7-4. The Plane Plate with Arbitrarily Varying Wall Temperature.** In the preceding section the case was investigated in which a plate exposed

[1] E. M. Sparrow and J. L. Gregg, *J. Aeronaut. Sci.*, **24**:852 (1957).

to steady flow has a temperature $t_s$ up to $x = x_0$ and a temperature $t_w$ downstream from $x_0$. The wall temperature therefore varies along the plate in a stepwise fashion.

In engineering applications other wall-temperature variations are often of interest. For certain wall-temperature variations the boundary-layer differential equations were integrated by D. Chapman and M. W. Rubesin.[1] Such cases can also be calculated with the method used in the previous paragraph. A variation of the surface temperature along $x$ has two effects on the temperature boundary layer. It influences the shape of the temperature profile and the boundary-layer thickness. The first effect becomes evident when Eq. (7-5) without the dissipation term is differentiated with respect to $y$ and then written for $y = 0$. The result of this calculation is

$$\alpha \left(\frac{\partial^3 t}{\partial y^3}\right)_{y=0} = \left(\frac{\partial u}{\partial y}\right)_{y=0} \frac{\partial t_w}{\partial x}$$

This equation expresses a relation between the third derivative of the temperature profile and the wall-temperature gradient. It can be used again to determine a constant in the assumed expression by which the temperature profile is approximated. In addition, the temperature difference $\theta$ is now a function of $x$ and has to be kept within the differential operator $d/dx$ in the equation on page 174.

Another calculation method makes use of the fact that the energy equation (7-5) is a linear differential equation and develops a solution of the temperature field for an arbitrary wall-temperature variation by superposition of a large number of stepwise variations. This method, usually referred to as Duhamel's theorem (1833), is used extensively in electrical-circuit analysis. It was applied for the first time to the present problem by M. W. Rubesin in a thesis in 1945. It proved very convenient and has the advantage that it can be applied not only for a flat plate with constant velocity $u_s$ but also for bodies of different shapes and for turbulent as well as for laminar flow conditions.

Equation (7-5) is a linear differential equation describing the temperature field $t$ within the boundary layer. Let us assume that we have a number of particular solutions $t_i$ of this equation for specific boundary conditions. Then it can be easily seen by substitution into Eq. (7-5) that

$$t = \sum_{1}^{n} C_i t_i$$

is also a solution. The constants $C_i$ can be used to adjust this new solu-

[1] D. Chapman and M. W. Rubesin, *J. Aeronaut. Sci.*, **16**:547–565 (1949).

tion to the required boundary conditions. The heat flow at the surface is

$$q = -k\left(\frac{\partial t}{\partial y}\right)_w = -k\sum_1^n C_i \left(\frac{\partial t_i}{\partial y}\right)_w$$

We shall now assume that each particular solution corresponds to a condition in which the wall temperature is equal to the stream temperature up to a certain location and then suddenly changes by an amount $\Delta t_{wi}$. For each particular solution a film heat-transfer coefficient can be defined according to the equation

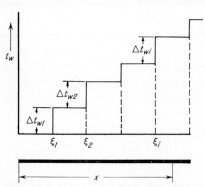

FIG. 7-6. Stepwise variation of wall temperature.

$$-k\left(\frac{\partial t_i}{\partial y}\right)_w = h_i \, \Delta t_{wi}$$

with $\Delta t_{wi}$ denoting the jump in the wall temperature or, in other words, the difference between stream and wall temperature for the particular situation. Then for the temperature field $t$,

$$q = \sum_1^n C_i h_i \, \Delta t_{wi}$$

From the wall temperature variation it is immediately seen that in this case all constants $C_i$ have the value 1.

The method outlined above will now be used to calculate heat transfer to a plate with arbitrarily varying wall temperature. When the wall temperature varies in a stepwise fashion with steps $\Delta t_{w1}, \Delta t_{w2}, \ldots,$ $\Delta t_{wi}, \ldots$ occurring at locations $\xi_1, \xi_2, \ldots, \xi_i \ldots$ (see Fig. 7-6) then the heat flow leaving the wall at location $x$ is, according to the last equation above,

$$q = \sum_1^n h(x, \xi_i) \, \Delta t_{wi} \qquad (7\text{-}17)$$

In this equation $h$ denotes the heat-transfer coefficient which describes the heat flow at location $x$ when only one step $\Delta t_{wi}$ in the wall temperature occurs at location $\xi_i$. This heat-transfer coefficient for laminar-flow conditions is given by Eq. (7-13) when $x_0$ is replaced by $\xi_i$. In Fig. 7-6 it is assumed that the plate has an unheated starting section up to $x = \xi_1$. Otherwise an additional temperature step $\Delta t_{w0} = t_{w0} - t_s$ at $\xi = 0$ has to be added, where $t_s$ indicates the fluid temperature outside the boundary layer.

A continuously varying wall temperature can be interpreted as a succession of infinitesimally small temperature steps $dt_w$ occurring at infinitesimally closely spaced locations $d\xi$. From this consideration the heat flow at the location $x$ for a wall temperature which varies as indicated in Fig. 7-7 can be obtained by replacing the series in the above equation by an integral. This results in the following relation:

$$q = \int_0^x h(x,\xi) \, dt_w(\xi)$$

In order to express this integral in the independent variable $\xi$, the equation is transformed into

$$q = \int_0^x h(x,\xi) \frac{dt_w(\xi)}{d\xi} d\xi \tag{7-18}$$

If finite steps occur in the wall temperature simultaneously with the continuous variation as shown in Fig. 7-8, then the following equation has to be used:

$$q = \int_0^x h(x,\xi) \frac{dt_w(\xi)}{d\xi} d\xi + \sum_0^n h(x,\xi_i) \, \Delta t_w(\xi_i) \tag{7-19}$$

It must be kept in mind that a finite temperature difference between wall and stream temperature at the leading edge of the plate (for $x = 0$) must also be considered as a finite step $\Delta t_w$ and must be taken care of by a term in the series in Eq. (7-17) or (7-19).

By a summation and integration in the above equations the heat flow along a flat plate can be calculated for any prescribed wall-temperature variation. In some engineering applications the heat flow rather than the wall temperature is prescribed at the surface of the plate. Tribus and Klein have considered this problem in a paper in which they improved and generalized the method.[1] The method can be used for a turbulent boundary-layer flow also when the turbulent heat-transfer coefficients for a plate with stepwise varying wall temperature are inserted for $h(x,\xi)$.

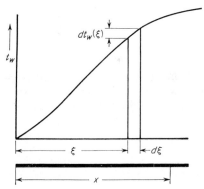

FIG. 7-7. Continuous variation of wall temperature.

---

[1] H. Tribus and I. Klein, Forced Convection from Nonisothermal Surfaces, "Heat Transfer" (a symposium), University of Michigan, Engineering Research Institute, 1953.

For engineering calculations the evaluation of the integral in Eq. (7-18) or (7-19) is somewhat tedious. For this reason an approximate procedure has been developed,[1] which can be briefly outlined. The way in which the temperature $t_w$ varies along the surface of a flat plate with length $L$ in the flow direction is prescribed. The task is to find the local heat flux $q(x)$ from the surface into a fluid stream of temperature $t_s$ at an arbitrary location $x$ and the total heat flux $Q(x)$ from a surface of length $x$ in the flow direction and of unit width. The length $L$ is subdivided into an arbitrary number of equal steps of length $\Delta L$. Figure 7-9 shows a plot of the local temperature difference $\Delta t = t_w - t_s$. The temperature differences $\Delta t_0, \Delta t_1, \Delta t_2$ occur at locations 0, 1, 2. The following equation then gives the local or total heat flux for laminar- or turbulent-

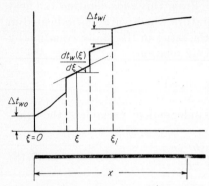

FIG. 7-8. Stepwise and continuous variation of wall temperature.

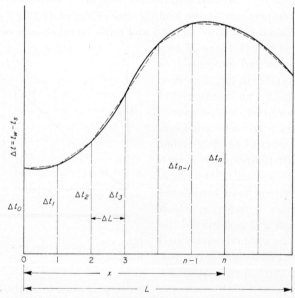

FIG. 7-9. Approximation of continuously varying wall temperature by straight-line segments.

[1] E. R. G. Eckert, J. P. Hartnett, and R. Birkebak, *J. Aeronaut. Sci.*, **24**: 549–550 (1957).

flow conditions when the proper constants are inserted from Table 7-1:

$$q(x) \text{ or } \frac{Q(x)}{x} = h_0(x) \left\{ \Delta t_0 + a(\Delta t_n - \Delta t_0) \right. $$
$$\left. + b \frac{\Delta L}{x} [(2n - 1) \Delta t_n - \Delta t_0 - 2(\Delta t_1 + \Delta t_2 + \Delta t_3 + \cdots + \Delta t_{n-1})] \right\}$$
(7-20)

$h_0(x)$ is the local or average heat-transfer coefficient, respectively, for a plate with constant wall temperature [Eqs. (7-10), (7-15), (8-17), (8-18)]. The equation gives the heat flux for any location $x$ on the surface which coincides with one of the stations (1,2,3, . . .). $n$ indicates the number of the station coinciding with location $x$.

TABLE 7-1

|  | Local heat flow | | Total heat flow | |
| --- | --- | --- | --- | --- |
|  | a | b | a | b |
| Laminar............ | 0.995 | 0.446 | 0.969 | −0.432 |
| Turbulent.......... | 0.991 | 0.117 | 0.982 | −0.478 |

For the development of Eq. (7-20), the actual difference between wall temperature and stream temperature was replaced by a broken line (Fig. 7-9). The integral appearing in Eq. (7-18) was solved for a linear wall-temperature variation, and the result was approximated by a second-order equation. From these elements the heat flux for the broken line could be derived. For turbulent flow the equation appearing on page 217 which describes the Nusselt number for a stepwise temperature variation was used in the integration of Eq. (7-18). In a comparison with available exact solutions Eq. (7-20) was found to be in error by only a few per cent.

The equation $q_s(x) = h_0(x) \Delta t_n$ describes the local heat flux at location $n$ when the temperature difference $\Delta t$ is constant above the surface and equal to $\Delta t_n$. Since the constant $a$ in Eq. (7-20) is almost equal to 1, the difference between the local heat flux $q(x)$ and $q_c(x)$ depends on the value for the constant $b$. It is evident from Table 7-1 that the difference between $q$ and $q_c$ is much larger for laminar than for turbulent flow. In other words, the upstream temperature history makes itself felt to a much larger degree in laminar than in turbulent flow. This statement holds also for flow through tubes and ducts. For turbulent flow it is necessary to take the previous history into account only when the temperature upstream from the position $n$ varies very rapidly.

## 7-5. Cylindrical Bodies in Flow Normal to Their Axes.

In Sec. 6-8 it was mentioned that on the forward side of a body in a flow, a boundary layer builds up. This boundary layer is always laminar near the stagnation point. If the body is heated, a thermal boundary layer exists in the same way. In the immediate proximity of the stagnation point the velocity outside the boundary layer always increases linearly with the distance from the stagnation point as measured along the body surface. This is expressed by the equation $u_s = Cx$. The heat transfer in this region on cylindrical bodies in a flow normal to their axes was calculated by Squire[1] by an exact solution of the differential equations when the temperature of the body is constant over its surface. This calculation gives the heat-transfer coefficient

$$h = Bk \sqrt{\frac{u_s}{\nu x}} = Bk \sqrt{\frac{C}{\nu}} \tag{7-21}$$

The heat-transfer coefficient is therefore independent of the distance from the stagnation point in its neighborhood. The dimensionless value $B$ is given in Table 7-2 for some Prandtl numbers. Equation (7-21) as written in dimensionless form is

$$\mathrm{Nu}_x = \frac{hx}{k} = B \sqrt{\frac{u_s x}{\nu}}$$

TABLE 7-2. CONSTANT $B$ FOR CALCULATION OF HEAT TRANSFER AT THE STAGNATION POINT BY EQ. (7-21)*

| Pr | 0.7 | 0.8 | 1.0 | 5 | 10 |
|---|---|---|---|---|---|
| $B$ | 0.496 | 0.523 | 0.570 | 1.043 | 1.344 |

* E. Eckert, *VDI-Forschungsheft* 416, 1942.

On the right-hand side the Reynolds number again appears. For potential flow around a cylinder with circular cross section the velocity $u_s$ outside the boundary layer is, according to potential theory,

$$u_s = 2u_0 \sin\left(\frac{2x}{d}\right)$$

where $u_0$ is the velocity, $x$ the distance along the surface from the stagnation point, and $d$ the diameter of the cylinder. In the neighborhood of the stagnation point the sine can be replaced by the angle. This gives $u_s = 4u_0(x/d)$ and

$$h = 2Bk \sqrt{\frac{u_0}{\nu d}}$$

[1] S. Goldstein, "Modern Developments in Fluid Dynamics," vol. 2, Oxford Unverisity Press, New York, 1938.

or in the dimensionless form,

$$\text{Nu}_d = \frac{hd}{k} = 2B\sqrt{\frac{u_0 d}{\nu}} = 2B\sqrt{\text{Re}_d} \tag{7-22}$$

$\text{Nu}_d$ and $\text{Re}_d$ are based on freestream velocity and diameter. The local heat-transfer coefficients along the surface of a cylindrical body at a greater distance from the stagnation point can be calculated with Eq.

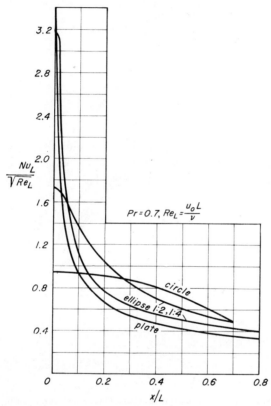

FIG. 7-10. Local laminar film heat-transfer coefficients along the circumference of cylinders with circular and elliptical cross section and along a flat plate. Nu and Re are based on freestream velocity and major axis $L$. (*From E. Eckert, VDI-Forschungsheft* 416, 1942.)

(7-2). Methods for such calculations were developed by Kroujiline, Frössling, Eckert, Schuh, Squire, and others. A summary of several of them is found in a paper by E. Eckert.[1] Figure 7-10 shows the heat-transfer coefficients along the circumference of cylinders with circular and elliptical cross sections and for the flat plate. The distance $x$ from

[1] E. Eckert, Die Berechnung der Wärmeüberganges in der laminaren Grenzschicht umströmter Körper, *VDI-Forschungsheft* 416, 1942.

the stagnation point is divided by the greatest diameter of the cylinder (its major axis $L$). In all laminar boundary layers the Nusselt number increases with the square root of the Reynolds number. Therefore the values $\mathrm{Nu}_L/\sqrt{\mathrm{Re}_L}$ are plotted in Fig. 7-10 against the ratio $x/L$. It can be seen that the distribution of the heat transfer becomes more and more similar to the plate the more slender the cross section of the cylinder.

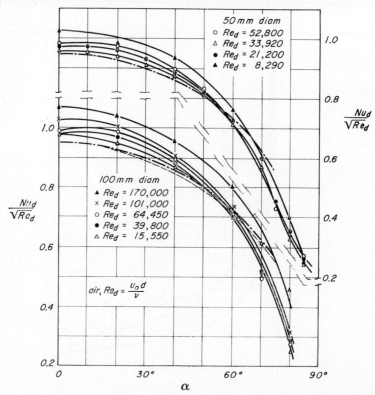

Fig. 7-11. Comparison between calculated and measured film heat-transfer coefficients on circular cylinders. $\alpha$ is the angular distance from the stagnation line. (*Measured values according to E. Schmidt and K. Wenner, from E. Eckert, VDI-Forschungsheft 416, 1942. Dashed line calculated from Kroujiline, dash-dotted line calculated from E. Eckert.*)

The smaller the radius of curvature at the stagnation point, the greater the heat-transfer coefficient at this place. A comparison between calculated values and results of experiments by E. Schmidt and K. Wenner[1] is shown in Fig. 7-11. The calculation gives the heat-transfer coefficients only for the laminar boundary layer. The distance of the place of transition to the turbulent boundary layer from the stagnation point is deter-

[1] E. Schmidt and K. Wenner, *Forsch. Gebiete Ingenieurw.*, **12**:65–73 (1941).

mined for the flat plate to be at a Reynolds number 80,000 to 500,000. For a flow with pressure decrease along the surface, the transition takes place at a greater Reynolds number; for a flow with pressure increase, at a smaller Reynolds number. Experiments on an inclined flat plate which are interesting in this connection were made by R. M. Drake.[1]

**7-6. Exact Solutions of the Laminar-boundary-layer Energy Equation.** The temperature field near a flat plate and the associated heat transfer have been calculated also by an exact solution of the boundary-layer equations for steady two-dimensional flow. The solution for a plate with constant surface temperature was obtained by E. Pohlhausen in 1921.[2] He assumed that the flow velocities are sufficiently small so that the viscous dissipation term can be neglected in the boundary-layer energy equation. This equation has then the following form:

$$u \frac{\partial t}{\partial x} + v \frac{\partial t}{\partial y} = \alpha \frac{\partial^2 t}{\partial y^2}$$

The corresponding boundary conditions are

For $y = 0$: $\quad t = t_w$
For $y = \infty$: $\quad t = t_s$

In Sec. 6-6 it was shown that for the particular flow situation the flow boundary-layer equations can be transformed into total differential equations by introducing the following new variables:

$$\eta = \frac{1}{2} y \sqrt{\frac{u_s}{\nu x}}, \qquad f = \frac{\psi}{\sqrt{\nu x u_s}}$$

If the same variables are introduced into the above energy equation, and if in addition the following dimensionless parameter is used to describe the temperature within the boundary layer

$$\theta' = \frac{t - t_w}{t_s - t_w}$$

then the total differential equation

$$\frac{d^2\theta'}{d\eta^2} + \Pr f \frac{d\theta'}{d\eta} = 0 \qquad (7\text{-}23)$$

with the boundary conditions for $\eta = 0$, $\theta' = 0$; $\eta = \infty$, $\theta' = 1$ result. This equation can be integrated in the same way as Eq. (6-44). The result is

$$\theta' = \frac{\int_0^\eta e^{-\int_0^\eta \Pr f\, d\eta}\, d\eta}{\int_0^\infty e^{-\int_0^\eta \Pr f\, d\eta}\, d\eta} \qquad (7\text{-}24)$$

---

[1] R. M. Drake, Jr., *J. Appl. Mech.*, **16**:1–8 (1949).
[2] E. Pohlhausen, *Z. angew. Math. Mech.*, **1**:115 (1921).

In this case $f$ has to be considered as a known function and Eq. (7-24) is an explicit solution of Eq. (7-23). Temperature profiles obtained in this way and expressed in the conventional manner as $1-\theta'$ are shown in Fig. 7-12 for various Pr numbers. Heat-transfer coefficients are given by

$$\mathrm{Nu}_x = f(\mathrm{Pr}) \sqrt{\mathrm{Re}_x}$$

The parameter $f(\mathrm{Pr})$ as calculated by E. Pohlhausen can be approximated in a Prandtl-number range from 0.6 to 15 very closely by

$$f(\mathrm{Pr}) = 0.332 \sqrt[3]{\mathrm{Pr}}$$

In the section on exact solutions of the flow boundary-layer equations, wedge-type flow was also discussed. For this type of flow the stream

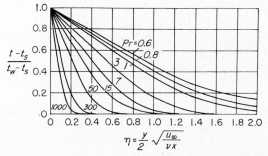

FIG. 7-12. Temperature profiles for laminar boundary-layer flow over a flat plate. [*From E. Eckert and O. Drewitz, Forsch. Gebiete Ingenieurw.*, **11**:116 (1940).]

velocity outside the boundary layer is given by the equation

$$u_s = Cx^m$$

The corresponding energy equation will be discussed now. The discussion will also be extended to the case of a variable wall temperature. It has been shown by Fage and Falkner[1] that the energy equation can be transformed to a total differential equation for wedge-type flow when the difference between wall and stream temperature varies according to the following law:

$$t_w - t_s = C_t x^\gamma$$

The transformation gives the following energy boundary-layer equation:

$$\frac{d^2\theta'}{d\eta^2} + \mathrm{Pr}\, f \frac{d\theta'}{d\eta} - 2\,\mathrm{Pr}\,\gamma \frac{df}{d\eta}(\theta' - 1) = 0 \qquad (7\text{-}25)$$

This equation has been investigated by several authors. The resultant

---

[1] A. Fage and V. M. Falkner, *Brit. Aeronaut. Research Comm. Repts. Mem.* 1408, 1931.

temperature profiles (again as 1-$\theta'$) are indicated in Figs. 7-13 and 7-14. Figure 7-13 holds for the specific case in which the value $\gamma$ is equal to zero. This means that a constant temperature difference and correspondingly a constant wall temperature are prescribed. It can be seen

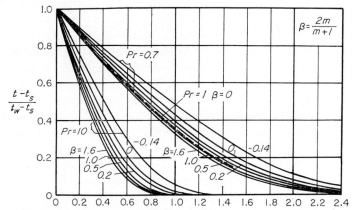

FIG. 7-13. Temperature profiles for laminar wedge-type flow. (*From E. Eckert, Die Berechnung des Wärmeübergangs in der laminaren Grenzschicht, VDI-Forschungsheft 416, 1942.*)

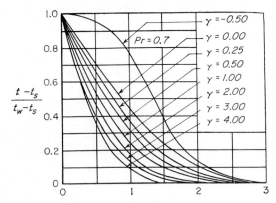

FIG. 7-14. Temperature profiles for laminar flow over a flat plate with variable wall temperature. [*From S. Levy, J. Aeronaut. Sci.* **19**:341 (1952).]

that the influence of the parameter $\beta = 2m/(m + 1)$, which describes the particular flow condition, on the temperature profile is not too great, especially not so great as the influence on the velocity profile. Figure 7-14 on the other hand indicates the temperature profiles for flow over a flat plate ($\beta = 0$) for the situation wherein the wall temperature varies (according to different values of $\gamma$). The variation of the wall temperature has a stronger effect on the temperature profile than the variation of the stream velocity. A remarkable situation exists for negative $\gamma$

values. The profiles assume an S shape, and the temperature gradient at the wall (and correspondingly the heat flow at the wall) may become zero or even negative (for $\gamma < -0.5$). This means that, even if the local wall temperature is greater than the stream temperature, this particular situation may lead to a heat flow which is directed from the stream into the wall. This peculiar fact can be explained in the following way. A negative $\gamma$ value means a wall temperature which decreases in flow direction. As a consequence, the fluid layers in the immediate neighborhood of the wall at any location come from upstream regions where they were in contact with a hotter wall. They carry this temperature downstream and cause the fluid near the wall to be hotter than the wall itself. This situation finally causes a heat flow from the fluid into the wall. The particular wall-temperature variation ($\gamma = -0.5$) for which the temperature profile in Fig. 7-14 was calculated is difficult to reproduce experimentally in so far as it requires a wall temperature which is infinite at the leading edge. Qualitatively, however, it is found that inversions of the temperature profile and of the heat flow occur for other wall-temperature variations also. Such cases are, for instance, calculated and discussed by Chapman and Rubesin[1] and by Schlichting.[2] A situation similar to the one shown in Fig. 7-14 for $\gamma = -0.5$ occurs in film cooling when the film of coolant is ejected into the flow at $x = 0$. The cold boundary layer created by the coolant protects the plate surface from the hotter fluid in the stream. This influence decreases in the downstream direction as the boundary layer heats up by conduction from the stream.

**7-7. Flow through a Tube.** If the wall of a tube through which a liquid or gas flows is heated or cooled, a thermal boundary layer builds up along the walls. The thermal boundary layers meet on the axis at a certain distance from the starting cross section just as the hydrodynamic boundary layers did (Fig. 6-18). At this cross section by convention the entrance region ends. First, an approximate calculation of the heat transfer in the region of fully developed thermal conditions may be made. As for the plate, the temperature profile in the tube is assumed to be a cubic parabola. Again, a condition can be derived for the curvature of the temperature profile at the wall. The heat flow through the layer immediately adjacent to the wall must be constant, as this layer has a negligibly small velocity and therefore no heat is carried away by convection. The heat flows through this layer only by conduction. For a cylindrical area of length $L$ and radius $r$ (Fig. 7-15) the equation

$$Q = -k2\pi rL \frac{dt}{dy}$$

[1] D. Chapman and N. W. Rubesin, *J. Aeronaut. Sci.*, **16**:547–65 (1949).
[2] H. Schlichting, *Forsch. Gebiete Ingenieurw.*, **17**:1 (1951).

holds, where $y$ is again the distance from the wall and $r$ the radius of the tube. Introducing $r = R - y$ into the above equation, solving for the temperature gradient, and differentiating with respect to $y$ gives

$$\frac{d^2t}{dy^2} = -\frac{1}{(R-y)^2}\frac{Q}{2\pi Lk} = \frac{1}{R-y}\frac{dt}{dy}$$

at the wall ($y = 0$); therefore,

$$\left(\frac{d^2t}{dy^2}\right)_w = \frac{1}{R}\left(\frac{dt}{dy}\right)_w \tag{7-26}$$

The rate of change of the temperature gradient is not zero as for the flat plate but is connected by the above relationship with the temperature gradient. The reason is that the cross section through which the heat flows decreases with increasing distance from the wall whereas it remains constant on the flat plate. To fulfill the above condition we must write for the temperature difference between the wall and the stream

$$\theta = ay + by^2 + cy^3$$

By differentiation and introduction into Eq. (7-26) there is obtained

$$b = \frac{a}{2R}$$

FIG. 7-15. Determination of the temperature profile in a tube with laminar flow distant from the inlet.

The temperature difference between the tube axis and the wall may be denoted by $\theta_s$. The temperature gradient must be zero at this location:

For $y = R$: $\qquad \theta = \theta_s, \dfrac{d\theta}{dy} = 0$

This gives

$$\theta' = \frac{\theta}{\theta_s} = \frac{6}{5}\frac{y}{R} + \frac{3}{5}\left(\frac{y}{R}\right)^2 - \frac{4}{5}\left(\frac{y}{R}\right)^3 \tag{7-27}$$

The heat flow per unit area and time at the wall is

$$q = -k\left(\frac{d\theta}{dy}\right)_w = \frac{6}{5}k\frac{\theta_s}{R} \tag{7-28}$$

The heat-transfer coefficient for the flow in a tube is commonly calculated with the difference between the average temperature of the fluid and the wall temperature. The average temperature most commonly used is one which would result if the fluid were mixed after the cross section considered. This value is called the *bulk temperature* $t_B$. It is defined by the equation

$$t_B = \frac{\int tu\, dA}{\int u\, dA} = \frac{\int_0^R tur\, dr}{\int_0^R ur\, dr} \qquad (7\text{-}29)$$

Introducing the temperature profile [Eq. (7-27)] and the velocity profile

$$\frac{u}{u_s} = 2\frac{y}{r} - \left(\frac{y}{r}\right)^2$$

where $u_s$ is the velocity on the axis, we obtain for the difference between the bulk temperature and the wall temperature

$$\theta_B = 0.583\theta_s$$

The heat-transfer coefficient is defined by

$$q = h(t_B - t_w) = h\theta_B$$

With Eq. (7-28) there results

$$h = k\frac{6}{5}\frac{1}{R}\frac{\theta_s}{\theta_B} = 4.12\frac{k}{d}$$

where $d$ is the diameter of the tube. In dimensionless form

$$Nu_d = \frac{hd}{k} = 4.12 \qquad (7\text{-}30)$$

The heat transfer to the walls of a tube was also calculated by Grätz,[1] Callendar,[2] and Nusselt[3] by a solution of the differential equations. In order to derive these solutions, one can choose the procedure to transform the Navier-Stokes equations, the continuity equation, and the energy equation into cylindrical coordinates. Then a certain number of terms in this equation can be dropped because of the special conditions prevailing in a cylindrical tube with fully developed flow. The solution of the flow equation is quite simple and indicates that in developed flow the velocity profile has the shape of a parabola. This type of flow is usually referred to as *Poiseuille flow*. The energy equation can be derived directly from an energy balance on a ring-shaped volume element of length $dx$, radius $r$, and width $dr$ located in the flow concentrically with respect to the tube axis. Heat is transported into this volume element by conduction and

---

[1] L. Grätz, *Ann. Physik*, **18**:79–94 (1889); **25**: 337–357 (1895). Additional eigenfunctions of the Grätz solution which permit the investigation of general boundary conditions, are contained in a paper by J. R. Sellars, M. Tribus, and J. S. Klein, *Trans. ASME*, **78**:441–448 (1956).

[2] H. L. Callendar, *Trans. Roy. Soc. London*, **199A**:55 (1902).

[3] W. Nusselt, *Z. Ver. deut. Ingr.*, **54**:1154–1158 (1910).

by convection. Let us first consider heat conduction in a radial direction. The heat flow through the ring-shaped area $2\pi r\, dx$ at the distance $r$ from the axis is

$$Q = -k 2\pi r\, dx\, \frac{\partial t}{\partial r}$$

In proceeding to a ring-shaped area at a distance $r + dr$ from the axis the heat flow changes by

$$\frac{\partial Q}{\partial r} dr = -k 2\pi\, dx\, \frac{\partial}{\partial r}\left(r\, \frac{\partial t}{\partial r}\right) dr$$

This term, therefore, gives the difference between the heat leaving the volume element through the area at $r + dr$ and entering the volume element through the area at a distance $r$. There may also be heat flow by conduction in the axial direction. However, it is to be expected that ordinarily this heat flow will be considerably smaller than that in the radial direction because the temperature gradients are expected to be larger in the radial direction. Accordingly, heat conduction in the axial direction has been neglected in the calculations by Grätz and Nusselt. In liquid metals, however, longitudinal conduction may contribute essentially to the development of the temperature field. Accordingly, newer calculations take this condition into account.[1] Heat will be transported into the volume element additionally by convection. The transport occurs only in the axial direction, and the amount of heat stored in the volume element by the flow into and out of the ring-shaped element is

$$2\pi r\, dr\, \rho c_p u\, \frac{\partial t}{\partial x}\, dx$$

For steady state the heat transport by conduction and convection must be equal. Therefore,

$$\frac{1}{ur}\frac{\partial}{\partial r}\left(r\,\frac{\partial t}{\partial r}\right) = \frac{\rho c_p}{k}\frac{\partial t}{\partial x}$$

This is the equation describing the energy flow and determining the temperature field. The boundary conditions belonging to this equation are

At $r = 0$: $\quad\quad\quad\quad\dfrac{\partial t}{\partial r} = 0$

At $r = R$: $\quad\quad t = t_w \quad$ or $\quad -k\left(\dfrac{\partial t}{\partial r}\right)_R = q_w$

At the wall either the wall temperature or the heat flow is prescribed. If heat flow at a constant rate occurs ($q_w$ = const) at the tube surface, then

---

[1] K. Millsaps and K. Pohlhausen, "Proceedings of the Conference on Differential Equations," University Bookstore, University of Maryland, College Park, Md., 1955.

an energy balance of the fluid flowing through the tube immediately results in the statement that for constant fluid properties the fluid bulk temperature increases linearly in the flow direction. For thermally developed flow this must also hold for the temperature at any distance $r$ from the tube axis. Therefore, constant heat rate results in

$$\frac{\partial t}{\partial x} = C$$

The energy equation thus reduces in this case to a total differential equation in $r$ which can be solved by simple integration and results in the following expression for the temperature profile:

$$\theta = t - t_w = \frac{2\rho c_p u_m R^2}{k} \frac{\partial t}{\partial x} \left[ \frac{1}{4}\left(\frac{r}{R}\right)^2 - \frac{1}{16}\left(\frac{r}{R}\right)^4 - \frac{3}{16} \right]$$

The temperature profile is the same at any station $x$. The heat-transfer coefficient is determined by the temperature gradient at the wall. For the Nusselt number based on the local difference between wall temperature and fluid bulk temperature, a short calculation gives

$$\text{Nu}_d = 4.36 \qquad (7\text{-}31)$$

For a constant wall temperature the above equation can be solved by separation of the variables, assuming that the temperature can be expressed as a product of a function depending only on the radius and another one depending on the axial location. It is found that the temperature difference $\theta$ decreases in the axial direction like an $e$ function. The temperature variation in the radial direction is described by Bessel functions. The temperature gradient at the tube surface again determines the Nusselt number. The calculations by Grätz and Nusselt gave for the local Nusselt number (based on the local difference between wall and bulk temperature)

$$\text{Nu}_d = 3.65 \qquad (7\text{-}32)$$

For constant wall temperature, profiles change in the $x$ direction continuously. In the thermally developed region, however, this change occurs in such a way that the profiles at all stations along the tube are similar to one another and that only the scale changes. Figure 7-18 shows this condition beyond a value $(1/\text{Re}_d \, \text{Pr})(x/d) = 0.05$. The result of the approximate calculation at the beginning of this section [Eq. (7-30)] is 6 per cent smaller than Eq. (7-31) and 13 per cent larger than Eq. (7-32). The temperature profile [Eq. (7-27)] agrees quite well with the exact calculation (Fig. 7-16). The heat transfer in a flat duct formed by two plane walls at a distance $b$ from each other was calculated by W. Nusselt and L. Ehret and H. Hahnemann.[1] They found

$$\mathrm{Nu}_b = \frac{hb}{k} = 3.75 \tag{7-33}$$

Most of the existing experiments on heat transfer in laminar flow are not very well suited for a comparison with the above equations for three reasons:

1. Experiments were conducted principally with viscous liquids, because in heat exchangers laminar flow is usually encountered with such

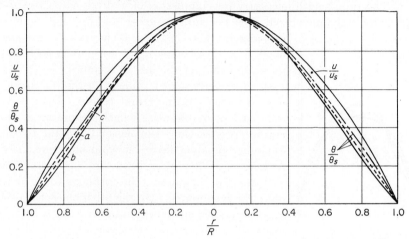

FIG. 7-16. Laminar velocity and temperature profiles in a tube at a great distance from the entrance. (a) Approximate solution by E. Eckert. (b) Exact solution by L. Grätz and W. Nusselt for constant wall temperature. (c) Exact solution for constant heat flux.

fluids. Liquids with high viscosities (oils) have a peculiarity in that their viscosity changes greatly with temperature. The assumption of constant property values in the above calculations is therefore fulfilled only to a small degree of accuracy in these experiments. The velocity profile, which is a parabola in isothermal flow, changes as a consequence of the variable viscosity. Velocity profiles according to Keevil and McAdams[2] are reproduced in Fig. 7-17. If heat flows from the tube wall to the liquid, the velocity profile (curve b) is flatter than a parabola (curve a) because the fluid layers in the neighborhood of the wall are warmer and consequently the viscosity is less than in the fluid core. For heat flow from the liquid to the wall, the fluid in the core is warmer and therefore less viscous than the layer near the wall; thus, the velocity profile has the form of curve c. Because of this change in the shape of the velocity profile, both the temperature profile and the heat transfer are influenced.

[1] L. Ehret and H. Hahnemann, *Wärme- u Kältetech.*, **44**:176 (1942).
[2] C. S. Keevil and W. M. McAdams, *Chem. Met. Eng.*, **36**:464 (1929).

The heat-transfer coefficient now depends on the direction of the heat flow and on its amount. A calculation of heat transfer in viscous liquids was made by K. Yamagata.[1]

2. A second difficulty in comparing the calculated and measured values is caused by the fact that frequently, for the low velocities necessary for laminar flow, eddy currents by free convection change the flow pattern. A combination of forced and free convection results.[2] This situation will be discussed in Sec. 11-5.

3. The tube length necessary for completion of hydrodynamically and thermally fully developed flow becomes so great for oils that in heat exchangers there usually is no fully developed flow. Thus, the experiments are concerned with the exploration of the intake region.

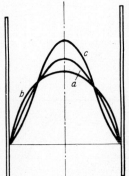

Fig. 7-17. Distortion of the velocity profile in a heated or cooled tube when the viscosity of the fluid depends on temperature.

It is necessary to distinguish two different modes of heat transfer in the intake region. If the tube is heated over its entire length beginning at the intake cross section, the thermal and the hydrodynamic boundary layers develop simultaneously. As long as the boundary layers are small compared with the tube diameter, the formulas for a plane plate can be applied. On the other hand, the tube may be heated only beginning at a cross section where the velocity profile is already fully developed. This case was calculated in 1889 by Grätz and later by Nusselt (1910). They assumed that the tube wall temperature changes suddenly at the cross section $x = 0$ from a temperature equal to the temperature of the entering fluid to another temperature which is constant downstream from $x = 0$. From the results of these calculations Figs. 7-18 and 7-19 are reproduced here. Figure 7-19 contains also some results obtained by W. H. Kays for simultaneous thermal and hydrodynamic development. Figure 7-18 shows the temperature profile, and Fig. 7-19 the local Nusselt number. The temperature profile, which has a rectangular shape in the section where the heat transfer starts, changes in the direction of the flow by the boundary layers forming along the walls. The temperature profile is fully developed as soon as the boundary layers meet at the tube axis. From here on the temperature profile does not change its shape; it only decreases gradually in the direction of flow. In Fig. 7-20 the Nusselt

---

[1] K. Yamagata, A Contribution to the Theory of Nonisothermal Laminar Flow of Fluids inside a Straight Tube of Circular Cross Section, *Mem. Fac. Eng. Kyushu Imp. Univ.*, **8**:365–445 (1940).

[2] E. R. G. Eckert and A. Y. Diagula, *Trans. ASME.*, **76**:497–504 (1954).

number resulting from the calculations of Grätz and Nusselt is compared with formulas derived by Kraussold[1] and by Sieder and Tate[2] from their experiments. Notwithstanding the fact that the experiments were carried out principally with oil, the agreement with the calculated results is fairly reasonable. Some results of experiments with gases by Nusselt[3]

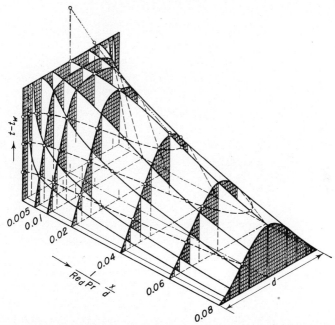

FIG. 7-18. Temperature profile in laminar flow through the entrance region of a tube. (*From L. Prandtl, "Stroemungslehre," 3d ed., p. 377, Fig. 305, Vieweg–Verlag, Brunswick, Germany, 1949.*)

are inserted and shown by points. New experiments by J. Boehm[4] give similar results. The logarithmic mean temperature difference is to be used in connection with the Nusselt numbers in Figs. 7-19 and 7-20. H. Hausen[5] developed a formula for the average Nusselt number which represents the results of Grätz and Nusselt very well:

$$\overline{\mathrm{Nu}}_d = 3.65 + \frac{0.0668(d/x)\,\mathrm{Re}_d\,\mathrm{Pr}}{1 + 0.04[(d/x)\,\mathrm{Re}_d\,\mathrm{Pr}]^{2/3}} \qquad (7\text{-}34)$$

[1] H. Kraussold, *VDI-Forschungsheft* 351, 1931.
[2] E. N. Sieder and G. E. Tate, *Ind. Eng. Chem.*, **28**:1429–1435 (1936).
[3] W. Nusselt, "Habilitationsschrift," Technische Hochschule, Dresden, 1909.
[4] J. Boehm, *Wärme*, **66**:143–152 (1943).
[5] H. Hausen, *Z. Ver. deutsch. Ing., Beih. Verfahrenstech.*, no. 4, 1943, pp. 91–98.

For greater temperature differences and oils the influence of the variable viscosity can be taken into account by multiplying the right-hand side of Eq. (7-34) by the ratio $(\mu_B/\mu_w)^{0.14}$, where $\mu_B$ is the viscosity at the fluid bulk temperature and $\mu_w$ the viscosity at wall temperature. The properties in the dimensionless parameters of the equation are then inserted at the bulk temperature. For gases it is recommended that Eq. (7-34) be used without the addition of the viscosity ratio and that the properties be introduced at the reference temperature as given by Eq. (7-16).

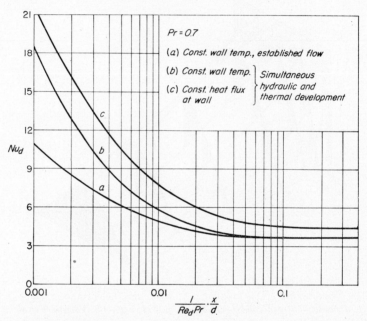

FIG. 7-19. Local Nusselt number for the entrance region of a tube. Curve $a$ is valid for all Prandtl numbers; curves $b$ and $c$ for Pr = 0.7 only. [$a$, according to W. Nusselt, $b$ and $c$ according to Kays. *Trans. ASME*, **77**:1265 (1955).]

In the literature, information on heat transfer related to laminar flow through tubes in the inlet region is often presented as a function of a dimensionless parameter which is called Grätz number and which is the reciprocal of the parameter used as abscissa in Fig. 7-19 and 7-20. A summarizing paper considering various temperature conditions along the tube wall was published by Norris and Streid.[1]

Some peculiar features are exhibited by heat transfer connected with laminar flow through ducts with noncircular cross sections. This situation was studied analytically[2] for hydrodynamically and thermally developed flow through a duct whose cross section has the shape of the sector of

---

[1] R. H. Norris and D. D. Streid, *Trans. ASME*, **62**:525–533 (1940).
[2] E. R. G. Eckert, T. F. Irvine, Jr., and J. T. Yen, *ASME Paper* 57-A-133.

a circle and for the condition that the heat flux from the duct wall into the fluid is constant in the direction of the duct axis. It was found that the local heat-transfer coefficient changes considerably around the duct periphery, approaching the value zero in the corners, and that the average heat-transfer coefficient depends very much on the boundary conditions. Two boundary conditions around the circumference of the duct were

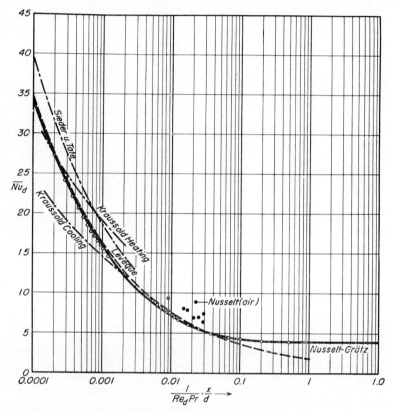

FIG. 7-20. Comparison of measured and calculated average film heat-transfer coefficients in laminar flow through the entrance region of a tube. The open circles indicate values obtained with Eq. (7-34). [*From H. Hausen, Z. Ver. deut. Ingr., Beih. Verfahrenstech.*, **4**:91–98 (1943).]

investigated: a wall temperature which is constant around the periphery and a locally constant heat flux. The Nusselt number averaged around the circumference for a constant wall temperature was found to be seven times the value for a constant heat flux when the opening angle of the sector was 20°. For a 60° opening angle, the ratio of the two Nusselt numbers is 2.5. The heat-transfer coefficient in the Nusselt numbers was defined as the average heat flux $q_w$ at the wall divided by the difference between the fluid bulk temperature and the average wall temperature (averaged around the duct periphery). The above result indicates that in using published information on heat-transfer coefficients, one has to establish the pertinent boundary conditions carefully.

## PROBLEMS

**7-1.** Derive the integrated boundary-layer energy equation for steady rotationally symmetric flow over a cylinder with its axis parallel to the flow direction. Assume that the properties vary with pressure and temperature. Define the convection thickness for this situation.

**7-2.** Generalize the integrated boundary-layer energy equation to high-velocity flow by including a term describing heat generation by internal friction.

**7-3.** Calculate the thermal boundary-layer thickness for steady two-dimensional laminar flow over a flat plate for the situation in which the surface temperature at the plate leading edge is equal to the fluid stream temperature and in which the surface temperature increases linearly in the flow direction. Use the integrated boundary-layer equations. Be careful to repeat the calculation in Sec. 7-3 and to check which steps in this procedure have to be changed to account for the varying temperature difference along $x$.

**7-4.** Calculate the local Nusselt number for the situation described in the preceding problem, but use the method of Sec. 7-4. The integrals appearing in the calculation can be transformed to tabulated gamma functions.[1]

**7-5.** Prove that the velocity profile is identical in shape with the temperature profile for the laminar flow of a fluid with $Pr = 1$ with constant velocity over a flat plate.

**7-6.** Derive Eq. (7-25) for heat transfer in flow over an infinite wedge by introducing the variables $f$, $\vartheta'$, and $\eta$ into the boundary-layer energy equation.

**7-7.** Engine oil with properties as listed in the Appendix, Table A-3, flowing through a 10-ft-long brass tube ($d_i = \frac{1}{2}$ in.) at the rate of 45 lb/min is cooled from 200 to 100 F. The inside surface temperature of the tube is 80 F. Find the inside-surface heat-transfer coefficient.

**7-8.** Calculate the temperature profiles for hydraulically and thermally developed steady laminar flow through a tube with circular cross section, prescribing constant heat flow through the tube wall and constant wall temperature. Assume for the calculation that the velocity is constant over the tube cross section. Calculate the Nusselt numbers for both boundary conditions. It will be pointed out later on that the results obtained are good approximations to heat transfer in developed turbulent tube flow for a fluid with very small Prandtl number.

**7-9.** Make the same calculation as in the preceding problem, however, for a channel built with two parallel flat plates.

**7-10.** Calculate the heat-transfer coefficients and Nusselt numbers for thermally and hydraulically developed laminar flow through a duct built by two parallel plane walls under the following boundary conditions:

  *a.* One wall is kept at a prescribed temperature different from the fluid bulk temperature; the other wall is insulated.

  *b.* The two walls are kept at two different constant temperatures.

  *c.* A prescribed heat flow is transferred from one wall into the fluid, the other wall being insulated.

[1] See, for instance, E. Jahnke and F. Emde, "Tables of Functions," Teubner, Verlagsgesellschaft, Leipzig, 1933.

CHAPTER 8

# FORCED CONVECTION IN TURBULENT FLOW

**8-1. Analogy between Momentum and Heat Transfer.** In the preceding chapters, heat transport from a solid surface into a moving fluid by the combined effects of conduction and convection was studied. In the immediate neighborhood of the surface the fluid is practically at rest and conduction is the only mode by which heat flows away from the surface. As the velocities increase with increasing distance from the wall, heat is picked up and carried away in an increasing amount by the fluid (convection). In regions farther away from the wall convection is by far the dominant mode by which heat is transported. In turbulent flow a continuous mixing of fluid particles is related to the turbulent-velocity fluctuations. This mixing causes a transport of heat when temperature differences are present in the fluid. In this way, in turbulent flow a third mode of heat transfer occurs in addition to the heat conduction and the convection associated with the bulk movement of the fluid. The turbulent mixing process is so little understood at present that to this date no one has succeeded in predicting from calculations alone the heat transfer in turbulent flow. On the other hand it is possible in a way as shown by Reynolds, Prandtl, Taylor, von Kármán, and others to derive formulas for heat transfer from hydrodynamic measurements and to gain a good insight into the mechanism of turbulent heat exchange.

In this section the basic ideas will be developed in a simple way following essentially Reynolds' analogy and Prandtl's concept of a laminar sublayer. In a later section more recent improvements in the calculation of turbulent heat transfer will be discussed.

Only the fluid layers in the proximity of a wall influence the heat transfer essentially. The velocities in these layers are practically parallel to the wall. The heat flow is normal to the wall. Therefore, in the following considerations the laws of heat transfer in a flow parallel to the wall (the $x$ direction) are investigated. It is assumed that the velocity changes materially in the $y$ direction only and that the heat flows in the $y$ direction. Therefore the temperature changes materially only in the $y$ direction. Following Prandtl, we simplify the actual conditions by assuming that a laminar sublayer in which no turbulent mixing occurs exists in the immediate neighborhood of the wall and that in the rest of the flow, laminar conduction and shear are small compared with the turbulent exchange and can be neglected. In any arbitrary plane parallel to the wall, i.e., parallel to the $x$ axis, within the laminar sublayer there exists a shear stress

$$\tau = \mu \frac{du}{dy}$$

The heat flow per unit area of the plane is, according to Eq. (2-2),

$$q = -k \frac{dt}{dy}$$

Now let us consider a plane parallel to the wall in turbulent flow. By the turbulent mixing motions fluid particles are continuously carried through this plane. This is presented in Fig. 8-1, simplified in such a way that the fluid mass $m'$ penetrating the plane $aa$ per unit time and unit area in an upward direction comes from the plane 1-1, where the velocity is $u$ and the temperature is $t$, and is transported to the plane 2-2. In steady flow the same amount $m'$ must be transported from plane 2-2, where the velocity is $u'$ and the temperature is $t'$, to the plane 1-1. The fluid flowing upward transports an amount of heat equal to $m'c_p t$, and the fluid flowing downward carries the heat $m'c_p t'$. If the temperature $t$ is greater than $t'$, a heat flow

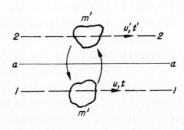

FIG. 8-1. Simplified picture of turbulent exchange.

$$q_t = m'c_p(t - t') \tag{8-1}$$

is transported as a result of the mixing movements per unit area and time. The fluid particles also carry with them their kinetic energy. If the velocity $u$ is greater than $u'$, the fluid above the plane $aa$ is accelerated by the particles moving upward whereas the fluid below the plane is decelerated by the particles coming down. By such a turbulent exchange the difference in the velocities $u$ and $u'$ is therefore decreased. The turbulent mixing acts in the same way as a shear stress in plane $aa$. This is the basis for introducing the concept of a "virtual turbulent shear stress." According to the law of momentum as formulated in Sec. 6-2, this virtual turbulent shear stress equals the increase or decrease of momentum generated by the action of the fluid mass $m'$ exchanged per unit time. Therefore the equation

$$\tau_t = m'(u' - u) \tag{8-2}$$

holds for the turbulent shear stress $\tau_t$. If the unknown value $m'$ is eliminated from Eqs. (8-1) and (8-2), the following relation between heat flow and shear stress in turbulent flow is obtained:

$$q_t = \tau_t c_p \frac{t - t'}{u' - u}$$

It can also be written in differential form

$$q_t = -\tau_t c_p \frac{dt}{du} \tag{8-3}$$

since the location of the planes 1-1 and 2-2 is arbitrary. This relation was first derived by Reynolds in 1874 and is therefore called *Reynolds' analogy*. If a corresponding relation is constructed for a plane in the laminar sublayer by dividing the two equations at the beginning of this section, there follows

$$q = -\tau \frac{k}{\mu} \frac{dt}{du} \tag{8-4}$$

For the ratio of the heat flow to the shear stress the same law holds true in turbulent or laminar flow, when

$$\frac{k}{\mu} = c_p \tag{8-5}$$

or when

$$\frac{c_p \mu}{k} = \frac{\nu}{\alpha} = \Pr = 1 \tag{8-6}$$

In a fluid or gas with a Prandtl number equal to 1, the heat transfer in laminar and turbulent flow is connected with the shear stress by the same equation.

For boundary-layer flow, it is desired to obtain a relation between the values $(t_w, u = 0)$ at the wall surface and those $(t_s, u_s)$ in the stream at the outer edge of the boundary layer. Such a relation is obtained for a fluid with $\Pr = 1$ by integration of Eq. (8-3) through the boundary layer:

$$\int_\delta dt = \frac{1}{c_p} \int_\delta \frac{q}{\tau} du$$

Since the boundary-layer thickness $\delta$ is small, it is argued that the ratio $q/\tau$ can be considered as constant and equal to the ratio $q_w/\tau_w$ of the values at the surface. Then the integration results in

$$t_s - t_w = \frac{q_w}{\tau_w} \frac{u_s}{c_p}$$

For fully developed flow in a tube, the stream values are usually replaced by the mean velocity $u_m$ and the bulk temperature $t_B$. This gives for the tube

$$q_w = \tau_w c_p \frac{t_B - t_w}{u_m}$$

By multiplying both sides of this equation by the tube wall surface $A$, there is obtained

$$Q_w = R \frac{c_p}{u_m} (t_B - t_w) \tag{8-7}$$

where $R$ is the resistance ($R = \tau_w A$). This simple relation between the heat flow $Q_w$ and the resistance $R$ holds true only for a medium with $\Pr = 1$. Inasmuch as all gases have Prandtl numbers which deviate only little from 1, Eq. (8-7) is very useful in obtaining a first approximation for the heat transfer where the resistance is known. The resistance can be replaced by the power which is necessary to drive the gas through the tube. The resistance of a tube with a pressure loss $\Delta p$ is $R = \Delta p(\pi d^2/4)$, and the volume flowing per unit time through the tube is $V = u_m(\pi d^2/4)$; therefore the power necessary for the flow is

$$W = V \Delta p = R u_m$$

The ratio of the heat flow to this power is therefore

$$\frac{Q}{W} = \frac{c_p}{u_m^2} (t_B - t_w) \tag{8-8}$$

This important relation shows that the power necessary for a certain heat transfer becomes smaller when the velocity $u_m$ is decreased. This fact is utilized in coolers or heat exchangers for airplanes by locating the cooler in a duct. The air stream is thus decelerated before it flows through the cooler, and afterward the pressure increase gained by deceleration is utilized to accelerate the air again. Such a ducted cooler is shown schematically in Fig. 8-2. By this ducting, which is today in general use in airplanes but which is useful as well for fast locomotives and motorcars, the power necessary to drag the cooler through the air can be very much decreased. Something must be paid, however, for this gain, for the lower the velocity of the air through the cooler, the greater the surface area necessary for a given heat flow. The same balancing between the amount of surface on one hand and the power necessary on the other hand is necessary in the design of a steam boiler. If it is desired to decrease the weight and dimensions of such a boiler by increasing the gas velocity, use must be made of a cheap source of energy which drives the gases through the boiler. This problem was solved in the Velox boiler by using a gas turbine to drive the blower for the combustion air.

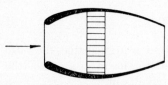

Fig. 8-2. Ducted cooler for aircraft.

**Example 8-1.** Air flows with 100-fps velocity through a tube 0.8 in. in diameter and 3 ft long. The air intake temperature is 68 F; its pressure is 14.2 $lb_f/in.^2$. The pressure loss in the tube is 3.2 in. of water column = 0.115 $lb_f/in.^2$. How much heat is transferred from the tube wall to the air when the wall is heated to a temperature $t_w = 200$ F?

For a first approximation Eq. (8-7) can be used, although air has a Prandtl number equal to 0.72. The resistance of the tube is

$$R = \Delta p \frac{\pi d^2}{4} = 0.115 \frac{lb_f}{in.^2} \times \frac{\pi \times 0.64}{4} in.^2 = 0.0578 \; lb_f$$

The heat transfer to the air is

$$Q_w = 0.0578 \frac{32.2 \times 0.24}{100} (t_w - t_B) \quad \text{Btu/sec F}$$

$$= 0.00447 \times 3{,}600(t_w - t_B) \frac{\text{Btu}}{\text{hr F}} = 16.1(t_w - t_B) \quad \text{Btu/hr F}$$

The tube-wall temperature is $t_w = 200$ F. The bulk temperature $t_B$ of the air is not yet known. The heat $Q_w$ taken up by the air can be written as $Q_w = \rho c_p u_m (\pi d^2/4)(t_e - t_i)$, where $\rho$ is the density of the air ($\rho = 0.075 \; lb/ft^3$ for 68 F and 14.2 $lb_f/in.^2$), $t_i$ is the intake temperature into the tube, and $t_e$ is the bulk exit temperature from the tube. Inserting the numerical values into this equation gives

$$Q_w = 0.075 \times 0.24 \times 100 \times \frac{\pi \times 0.64}{144 \times 4}(t_e - t_i) \frac{\text{Btu}}{\text{sec F}}$$

$$= 0.00628(t_e - t_i) \quad \text{Btu/sec F}$$

The mean temperature difference $\Delta t_m = t_w - t_B$ is, according to Sec. 1-4, $\Delta t_m = a(\Delta t_i + \Delta t_e)/2$. The wall temperature is assumed constant. Therefore $\Delta t_m = t_w - t_B = a[t_w - (t_i + t_e)/2]$. By equating the two equations for $a$ there is obtained

$$0.00447a \left( t_w - \frac{t_i + t_e}{2} \right) = 0.00628(t_e - t_i)$$

The factor $a$ cannot be determined from Table 1-2, as the ratio $\Delta t_i/\Delta t_e$ is unknown; therefore, estimate $a = 1$. Now the equation gives $t_e = 137$ F. To establish $a$, use the ratio

$$\frac{\Delta t_i}{\Delta t_e} = \frac{200 - 68}{200 - 137} = 2.1$$

Therefore $a = 0.96$ from Table 1-2. As the whole calculation gives only a first approximation, it is not necessary to repeat it with the new value for $a$. The heat flow is $Q = 0.00628(137 - 68) \; 3{,}600 = 1560$ Btu/hr. An exact calculation with the formulas in the next paragraph gives a heat flow greater by 15 per cent.

For a medium whose Prandtl number deviates greatly from 1, the heat resistance of the laminar sublayer and the turbulent flow must be calculated separately. Let the temperature of the wall be $t_w$ (Fig. 8-3), the temperature at the border between the laminar sublayer and the turbulent boundary layer be $t_b$, and the temperature in the stream outside the boundary layer be $t_s$. The corresponding velocities are $u_w = 0$, $u_b$, and

$u_s$. An integration of Eq. (8-4) through the sublayer, again with the assumption $q/\tau = \text{const} = q_w/\tau_w$, gives

$$q_w = \tau_w \frac{k}{\mu} \frac{1}{u_b} (t_w - t_b) \qquad (8\text{-}9)$$

The heat flow in the turbulent boundary layer is obtained by integration of Eq. (8-3) between the values $(t_b, u_b)$ and $(t_s, u_s)$. This results in

$$q_w = \tau_w c_p \frac{t_b - t_s}{u_s - u_b} \qquad (8\text{-}10)$$

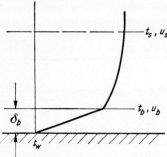

FIG. 8-3. Turbulent temperature and velocity profile with laminar sublayer.

The film heat-transfer coefficient $h$ at the wall is defined by the equation

$$q_w = h(t_w - t_s) \qquad (8\text{-}11)$$

If Eqs. (8-9) to (8-11) are solved for the temperature differences and the first and second are subtracted from the third, we get

$$\frac{1}{h} = \frac{1}{\tau_w c_p}(u_s - u_b) + \frac{\mu u_b}{\tau_w k} = \frac{u_s}{\tau_w c_p}\left(1 - \frac{u_b}{u_s} + \frac{\mu c_p}{k}\frac{u_b}{u_s}\right)$$

$$= \frac{u_s}{\tau_w c_p}\left[1 + \frac{u_b}{u_s}(\text{Pr} - 1)\right]$$

The heat-transfer coefficient is

$$h = \frac{\tau_w c_p/u_s}{1 + (u_b/u_s)(\text{Pr} - 1)} \qquad (8\text{-}12)$$

This equation was derived by L. Prandtl (1910 and 1928) and G. I. Taylor (1916).[1]

**8-2. Flow in a Tube.** In fully developed flow the boundary layers have met at the tube axis. For the velocity $u_s$ and temperature $t_s$, the values at the axis must therefore be used. It is usual, however, to establish the heat transfer with the bulk temperature $t_B$ and to use the mean velocity $u_m$ in Eq. (8-12). This is not quite correct, but the resultant error is small. If the shear stress is taken from Eq. (6-50) and the velocity ratio from Eq. (6-51), the above equation becomes

$$\frac{h}{\rho c_p u_m} = \frac{0.0384(u_m d/\nu)^{-\frac{1}{4}}}{1 + A(u_m d/\nu)^{-\frac{1}{8}}(\text{Pr} - 1)} \qquad (8\text{-}13)$$

The constant $A$ has the value 2.00. A comparison with experimental results on viscous fluids shows, however, that this value is too great and furthermore that $A$ is a function of the Prandtl number. In Fig. 8-4 this

---
[1] L. Prandtl, *Z. Physik*, **11**:1072 (1910); **29**:487 (1928). G. I. Taylor, *Rept. Mem. Brit. Aeronaut. Comm.* 272, 1916, p. 423.

function is given according to W. Bühne.[1] The expression on the left-hand side of Eq. (8-13) is equivalent to the following product of dimensionless values: $\mathrm{Nu}_d/(\mathrm{Re}_d\,\mathrm{Pr})$. It is called Stanton number and denoted by St. The final formula for heat transfer in a tube with turbulent flow is therefore

$$\mathrm{St} = \frac{\mathrm{Nu}_d}{\mathrm{Re}_d\,\mathrm{Pr}} = \frac{0.0384(\mathrm{Re}_d)^{-1/4}}{1 + A(\mathrm{Re}_d)^{-1/8}(\mathrm{Pr}-1)} \qquad (8\text{-}14)$$

The value $A$ can be taken from Fig. 8-4. E. Hoffman[2] gives the formula $A = 1.5\,\mathrm{Pr}^{-1/6}$.

The shape of the velocity profile for different Prandtl numbers can also be derived from the above equations. By equating (8-9) and (8-10), the ratio of the temperature drop in the laminar sublayer to the temperature drop in the turbulent zone is found to be

$$\frac{t_b - t_w}{t_s - t_b} = \frac{\mu c_p}{k}\frac{u_b}{u_s - u_b} = \mathrm{Pr}\,\frac{u_b}{u_s - u_b}$$

For a medium with $\mathrm{Pr} = 1$, the ratio of the temperature differences equals the ratio of the corresponding velocity differences. Both profiles are similar. For two Prandtl numbers greater than 1, the temperature profiles are shown in Fig. 8-5. For large Prandtl numbers, the temperature drop in the laminar sublayer is the essential one.

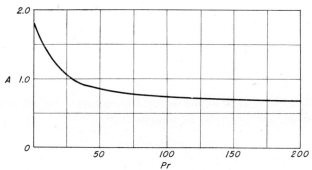

Fig. 8-4. Factor $A$ in Eqs. (8-14) and (8-17). [*From W. Bühne, Wärme*, **61**:162 (1938).]

The temperature profiles for low Prandtl numbers, when determined in the same way as the profiles in Fig. 8-5, would look S-shaped, with a very small temperature rise in the laminar sublayer and most of the temperature difference concentrated in the turbulent core. This behavior is determined by a very low heat resistance in the laminar sublayer as compared with the turbulent zone. This, of course, does not correspond to

---

[1] W. Bühne, *Wärme*, **61**:162 (1938).
[2] E. Hoffman, *Z. ges. Kälte-Ind.*, **44**:99–107 (1937).

reality. In any turbulent flow, the turbulent heat exchange is superimposed on a heat flow by conduction. For Prandtl numbers near or greater than 1, the heat conduction is small compared with the turbulent exchange and accordingly was neglected in the previous calculations. At low Prandtl numbers, however, it is of equal or greater importance, and in the limiting case the turbulent exchange can be neglected and the heat conduction is the dominating factor present.

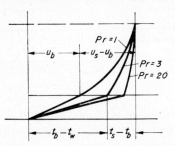

FIG. 8-5. Turbulent temperature profiles for various Prandtl numbers.

Calculations as performed in Sec. 7-7 should, therefore, give good information on heat transfer in a fluid with $Pr \to 0$ flowing turbulently through a tube when the parabolic velocity profile used for laminar flow in Sec. 7-7 is replaced by the turbulent profile. A fairly good answer is already obtained when the full turbulent profile is approximated by one of constant velocity (slug flow). For thermally developed conditions and constant heat flow at the wall this procedure results in the equation $Nu_d = 8$; for constant wall temperature, in the relation $Nu_d = 5.8$. More information on liquid-metal heat transfer will be presented in Sec. 10-4.

The combination of the values $Nu_d$, $Re_d$, and $Pr$ on the left-hand side of Eq. (8-14) is very useful for heat-transfer calculations. The heat flow to the wall of a tube of length $L$ is

$$Q = h\pi \, dL(t_B - t_w)$$

The heat flow can also be expressed by the cooling of the liquid from the intake temperature $t_i$ to the exit temperature $t_e$:

$$Q = \rho c_p u_m \frac{\pi d^2}{4} (t_i - t_e)$$

Both equations give

$$\frac{h}{\rho c_p u_m} = \frac{Nu_d}{Re_d \, Pr} = \frac{d}{4L} \frac{t_i - t_e}{t_B - t_w} \qquad (8\text{-}15)$$

From this equation the temperature $t_e$ with which the liquid leaves the tube can be calculated without using any property values as soon as the tube dimensions, the intake temperature $t_i$, and the wall temperature $t_w$ are given and the value $Nu_d/(Re_d \, Pr)$ is known. For the bulk temperature $t_B$, the logarithmic mean of intake and exit temperatures must be used.

For the derivation of Eq. (8-12) the property values were considered as constant. E. Hoffman[1] investigated the influence of the variability of the property values on heat transfer. He found that Eq. (8-14) holds true when the property values are inserted at the reference temperature

$$t^* = t_B - \frac{0.1 \, \text{Pr} + 40}{\text{Pr} + 72} (t_B - t_w) \tag{8-16}$$

For gases (Prandtl number near 1) this equation gives approximately the arithmetic mean between liquid and wall temperature for $t^*$. The greater the Prandtl number, the more $t^*$ approaches the mean liquid temperature $t_B$. The temperature $t_b$ between the laminar sublayer and turbulent zone (Fig. 8-5) has the same trend. Such a reference temperature, which is a function of Pr only, has to be considered as a first approximation, since it has to be expected that the heat transfer in a fluid with variable properties depends on how these properties vary with temperature and pressure.

Water at supercritical pressure is used as coolant in nuclear reactors. Heat transfer to water near the critical pressure is characterized by the fact that the properties, especially the specific heat, vary strongly with temperature. R. G. Deissler[2] and K. Goldmann[3] have calculated friction and heat transfer to water flowing turbulently through a tube at a pressure of 5,000 atm/in.². Established velocity and temperature distribution were assumed. The results of Deissler's calculation will be discussed here. Figure 8-6 shows how the properties vary at the assumed pressure. The results of his calculations can be well approximated by a constant-property relation [for instance, Eq. (8-14)] when the properties are introduced at a properly selected reference temperature. Eckert[4] obtained the relation presented in Fig. 8-7 for the reference temperature $t^*$. The upper figure gives the reference temperature for the calculation of the friction factor, and the lower figure contains the reference temperature for the heat transfer. It can be observed that the magnitude of the bulk temperature $t_B$ and of the wall temperature $t_w$ relative to the temperature $t_m$ at which the specific heat goes through a maximum has the greatest influence on the reference temperature. The Prandtl number has to be introduced into the equation for the heat-transfer coefficient at the wall temperature. Results of experiments on heat transfer in fluids near their critical condition do not present a uniform picture. Measurements on carbon dioxide[5] gave good agreement with Deissler's calculations, whereas measurements[6] with supercritical steam at 4,500 lb$_f$/in.²

[1] E. Hoffmann, *Z. Ver. deut. Ingr.*, **82**:741–782 (1938).
[2] R. G. Deissler, *Trans. ASME.*, **76**:73–85 (1954).
[3] K. Goldmann, *Chem. Eng. Progr. Symposium Ser.* 50, no. 11, 1954, pp. 105–113.
[4] E. R. G. Eckert, discussion on paper by Deissler.
[5] R. B. Bringer and J. M. Smith, *J. Am. Inst. Chem. Engrs.*, **3**:49–55 (1951).
[6] N. L. Dickinson and C. P. Welch, *Trans. ASME.*, no. 57-HT-7.

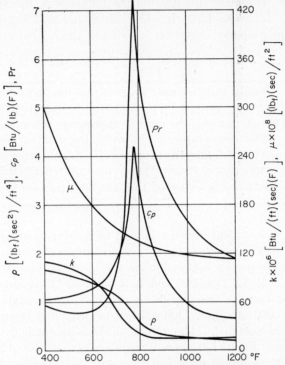

FIG. 8-6. Properties of water at a pressure of 5,000 atm/in.² as a function of temperature. [*From R. G. Deissler, Trans. ASME,* **76**:73–85 (1954).]

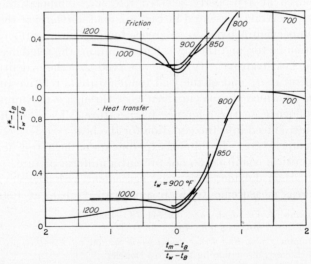

FIG. 8-7. Reference temperature $t^*$ for calculation of heat transfer of water at 5,000 atm/in.² pressure flowing turbulently through a pipe. $t_w$ is wall temperature; $t_b$ bulk temperature. [*From R. G. Deissler, Trans. ASME,* **76**:73–85 (1954).]

pressure resulted in heat-transfer coefficients which were generally higher than the calculated values by 70 per cent and near the critical temperature by 100 per cent. The deviations may be caused by our poor knowledge of properties in the high-temperature, high-pressure range or by free-convection effects which have been found very large near the critical state.[1]

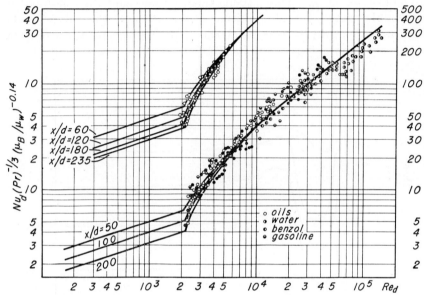

FIG. 8-8. Film heat-transfer coefficients in a tube within the transition zone from laminar to turbulent flow. [*According to* E. N. *Sieder and* G. E. *Tate, Ind. Eng. Chem.*, **28**:1429–1435 (1936).]

Besides the theoretically founded Eq. (8-14), empirical formulas are often used. For example, Dittus's and Boelter's[2] equations are well known:

$$\mathrm{Nu}_d = 0.0243(\mathrm{Re}_d)^{0.8}(\mathrm{Pr})^{0.4}$$

for heat flow from the wall to the liquid and

$$\mathrm{Nu}_d = 0.0265(\mathrm{Re}_d)^{0.8}(\mathrm{Pr})^{0.3}$$

for heat flow from the liquid to the wall. Use of two different expressions for heating and cooling[3] is somewhat unsatisfactory.

The equations up till now are good for fully developed turbulent flow. In the intake region greater heat-transfer coefficients may be expected.

---

[1] E. Schmidt, E. Eckert, and U. Grigull, "Jahrbuch 1939 der deutschen Luftfahrtforschung," vol. II, pp. 53–58.

[2] F. W. Dittus and L. M. K. Boelter, *Univ. Calif. Publs. Eng.*, **2**:443 (1930).

[3] E. Eckert, *Z. Ver. deut. Ingr.*, **80**:137–138 (1936).

H. Hausen[1] gave the expression for the average Nusselt number:

$$\overline{\mathrm{Nu}}_d = 0.116[(\mathrm{Re}_d)^{2/3} - 125](\mathrm{Pr})^{1/3}\left[1 + \left(\frac{d}{x}\right)^{2/3}\right]\left(\frac{\mu_B}{\mu_w}\right)^{0.14}$$

where $\mu_B$ is the viscosity at bulk liquid temperature and $\mu_w$ the viscosity at tube-wall temperature. Apart from the latter, the property values are to be inserted at $t_B$. This formula takes into account the conditions in the intake region. It also satisfactorily reproduces the values in the transition zone $\mathrm{Re}_d = 2{,}300$ to $6{,}000$. This relation is expected to be especially applicable to fluids for which the variation of viscosity is the

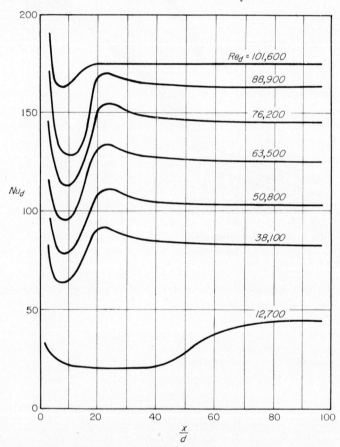

Fig. 8-9. Local Nusselt numbers for flow through a tube near the entrance with simultaneous development of the flow and temperature field. [*From W. Linke and H. Kunze, Allgem. Wärmetech.*, **4**:73–79 (1953).]

[1] H. Hausen, *Z. Ver. deut. Ingr., Beih. Verfahrenstech.*, no. 4, 1943, pp. 91–98.

dominating one (oils). Figure 8-8 shows the transition region according to a summary of available test results by Sieder and Tate.[1]

Hausen's equation gives the average heat-transfer coefficient between the beginning of the heated tube section and the position $x$. The local heat-transfer coefficient can be obtained from this relation by differentiation. In carrying out this operation one finds that the same relation describes the local Nusselt number when the term within the brackets is replaced by $1 + (1/3)(d/x)^{2/3}$. Experiments by J. Hartnett[2] and calculations by R. Deissler[3] indicate that the heat-transfer coefficients for fluids with Prandtl numbers of order 1 or larger drop even faster with increasing length $x$ than indicated by this equation. A length of 15 diameters brings the local coefficient within approximately 1 per cent of the value which is reached asymptotically in a long tube. This statement holds when the flow is fully developed and turbulent at the beginning of the heated tube section.

Conditions are more involved when the tube is heated over its entire length, that is, when the hydrodynamic and thermal starting sections coincide. Figure 8-9 presents the local Nusselt number at position $x$ as measured by W. Linke and H. Kunze.[4] The dip in the curves is explainable by the fact that the boundary layers along the tube walls are at first laminar and change through a transition region to turbulent flow.

With the equations in this section the heat transfer in a tube with noncircular cross section can also be calculated. Instead of the diameter $d$, the hydraulic diameter $d_h = 4A/C$ (where $A$ is the cross-sectional area and $C$ is the wetted perimeter) must be inserted. The whole perimeter must be used also in the case when only part is heated or cooled.[5] Only in calculating the heat flow with Eq. (1-13) is the heated area alone to be used.

Heat transfer for flow of air through tubes with rough surfaces has been studied thoroughly in a paper by W. Nunner.[6] The roughnesses were rings of various cross sections attached to the inner surface of the pipe. Some naturally rough tubes were also investigated. It was found that regardless of the shape of the roughness elements the measured Nusselt numbers correlated as functions of Reynolds number and friction-factor ratio $f/f_0$ ($f$ is the friction factor of a rough surface; $f_0$ is the friction factor of a tube with a smooth surface at the same Reynolds number). Figure

[1] E. N. Sieder and G. E. Tate, *Ind. Eng. Chem.*, **28**:1429–1435 (1936).
[2] J. P. Hartnett, *Trans. ASME*, **77**:1211–1220 (1955).
[3] R. G. Deissler, *Natl. Advisory Comm. Aeronaut. Tech. Note* 3016, 1953.
[4] W. Linke and H. Kunze, *Allgem. Wärmetech.*, **4**:73–79 (1953).
[5] M. ten Bosch, "Die Wärmeübertragung," 3d ed., Springer-Verlag, Berlin, 1936.
[6] W. Nunner, Wärmeübergang und Druckabfall in rauhen Rohren, *VDI-Forschungsheft* 455, 1956.

8-10 presents the results of the investigations. Nunner also found that the measured Nusselt numbers agreed closely with the ones calculated from Eq. (8-12) or (8-14) when the term $Pr - 1$ in these equations was replaced by $Pr\, f/f_0 - 1$. Values calculated in this way also give fair agreement with previous measurements.

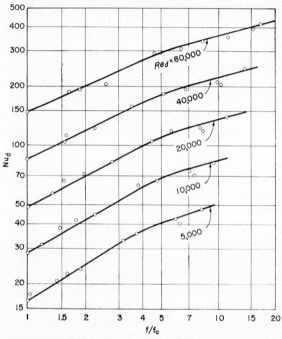

FIG. 8-10. Nusselt number for turbulent flow through a tube with rough surfaces. (*From W. Nunner, Wärmeübergang und Druckabfall in rauhen Rohren, VDI-Forschungsheft* 455, 1956.)

**Example 8-2.** The radiator of a motorcar is constructed of finned tubes. The tubes are 0.24 in. in diameter and 24 in. long. Cooling water flows through them with a 3-fps velocity and 140 F temperature. The film heat-transfer coefficient on the tube wall is to be calculated. The kinematic viscosity of water at 140 F can be found in the Appendix as $\nu = 0.514 \times 10^{-5}$ ft$^2$/sec and the Prandtl number $Pr = 3.02$. The Reynolds number is $(3 \times 0.24/12)/(0.514 \times 10^{-5}) = 11{,}700$. Therefore the flow is turbulent. The intake length for turbulent flow is $L_e/d = 15$ for the development of the temperature field. Here the ratio of the tube length to the diameter is 100. Therefore the greater part of the flow is fully developed, and calculations can be made with Eq. (8-13). From Fig. 8-4 the value $A$ is found to be 1.5. With it $Nu_d/(Re_d\, Pr) = \dfrac{0.0396/11{,}700^{1/4}}{1 + 1.5 \times 2.02/11{,}700^{1/8}} = 0.00196$. The Nusselt number is $Nu_d = 0.00196 \times 11{,}700 \times 3.02 = 69$, and the heat-transfer coefficient, with $k = 0.376$ Btu/ft hr F, is

$$h = Nu_d \frac{k}{d} = 69 \frac{0.376}{0.24/12} = 1297 \text{ Btu/hr ft}^2 \text{ F}$$

Within the first eighth of the tube length the heat-transfer coefficient drops to this value. The heating of the water can be calculated immediately from $\mathrm{Nu}_d/(\mathrm{Re}_d\,\mathrm{Pr})$. We get

$$\frac{t_e - t_i}{t_B - t_W} = \frac{4L}{d} 0.00196 = 0.784$$

**8-3. The Plane Plate in Longitudinal Flow.** In order to calculate the heat transfer from a plane plate with Eq. (8-12), the freestream velocity must be used for the value of $u_s$ and the temperature in the freestream accordingly used for $t_s$. Introducing the shear stress from Eq. (6-33) and the velocity ratio from Eq. (6-37) gives

$$\mathrm{St} = \frac{\mathrm{Nu}_x}{\mathrm{Re}_x\,\mathrm{Pr}} = \frac{0.0296(\mathrm{Re}_x)^{-1/5}}{1 + 0.87A(\mathrm{Re}_x)^{-1/10}(\mathrm{Pr} - 1)} \qquad (8\text{-}17)$$

The value $A$ can be taken from Fig. 8-4 or from the equation

$$A = 1.5\,\mathrm{Pr}^{-1/6}$$

The factor 0.87 arises by the change from the mean velocity used for the tube to the velocity $u_s$ used here, which corresponds to the velocity on the tube axis. Equation (8-17) gives the local heat-transfer coefficient. It was pointed out in Sec. 6-1 that a laminar boundary layer exists in the neighborhood of the leading edge. Only where the Reynolds number based on the distance from the leading edge has reached the critical value (approximately $5 \times 10^5$) does the flow in the boundary layer become turbulent. Now Eq. (8-17) gives the heat-transfer coefficient in the turbulent zone, whereas for the laminar part Eq. (7-13) is to be used.[1] For gases the above equation can be simplified, since the Prandtl numbers are close to 1 in this case and the denominator can be regarded as constant. By integrating this simplified equation over the plate length, the average heat-transfer coefficient is obtained. Assuming the boundary layer to be turbulent over the whole length, the Nusselt number based on the average heat-transfer coefficient is[2]

$$\overline{\mathrm{Nu}}_x = 0.037(\mathrm{Re}_x)^{0.8}(\mathrm{Pr})^{1/3} \qquad (8\text{-}18)$$

In reality a certain part of the boundary layer near the leading edge is always laminar. Integration must then be carried out in two steps, over the laminar part and over the turbulent part separately. Assuming that the heat-transfer coefficients in the turbulent zone are the same as if the turbulent boundary layer started right on the leading edge, such an integration gives

$$\overline{\mathrm{Nu}}_x = 0.037(\mathrm{Pr})^{1/3}[(\mathrm{Re}_x)^{0.8} - 23{,}100]$$

[1] M. Jakob and W. M. Dow, *Trans. ASME*, **68**:124–134 (1946).
[2] H. A. Johnson and M. W. Rubesin, *Trans. ASME*, **71**:447–456 (1949).

for $\mathrm{Re}_{cr} = 5 \times 10^5$ and

$$\overline{\mathrm{Nu}}_x = 0.037(\mathrm{Pr})^{1/3}[(\mathrm{Re}_x)^{0.8} - 4{,}200]$$

for $\mathrm{Re}_{cr} = 10^5$.

For boundary-layer flow, the Stanton number in Eq. (8-17) can be interpreted as the ratio of two lengths. To show this, we use the fact that all the heat $Q$ transferred from a hot plate to the flow past its surface within the length $x$ (Fig. 8-11) must in steady state be carried with the flow through the plane $I$-$I$. In mathematical form this gives

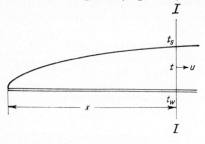

$$Q = \rho c_p \int_0^\infty u(t - t_s)\, dy$$

Actually the integration has to be carried out only within the thermal boundary layer. The heat flow $Q$ can be expressed by a heat-transfer coefficient averaged over the length $x$:

$$Q = \bar{h}x(t_w - t_s)$$

From both equations the average Stanton number can be determined:

$$\overline{\mathrm{St}} = \frac{\bar{h}}{\rho c_p u_s} = \frac{1}{x}\int_0^{\delta_t} \frac{u(t - t_s)}{u_s(t_w - t_s)}\, dy$$

The integral has the dimension of length and can be regarded as a kind of boundary-layer thickness. Since it characterizes the heat carried by convection within the boundary layer we shall call it *convection thickness* ($\delta_c$)

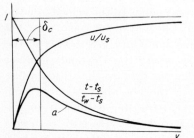

FIG. 8-11. Determination of the convection thickness of the thermal boundary layer.

It corresponds closely to the *impulse or momentum thickness* of the boundary layer which is used in hydrodynamics. The lower diagram of Fig. 8-11 shows the determination of the convection thickness. Multiplying for all values $y$ the ordinates of the nondimensional velocity profile $u/u_s$ by the nondimensional temperature profile $(t - t_s)/(t_w - t_s)$ leads to curve $a$, the area under which is the above integral. It can be transformed into a rectangle with unit height and the length $\delta_c$. It is easily understood that the convection thickness is smaller than the thermal boundary-layer thickness $\delta_t$ (page 177) and also smaller than the *displacement thickness* as defined on page 141. Since the boundary-layer thickness is always small as compared with the length $x$, it can be seen that the Stanton number always has a small value.

In the derivation of Eq. (8-12) it has been assumed that the temperature changes only in a direction normal to the wall surface. The relations

FORCED CONVECTION IN TURBULENT FLOW

in this section should, therefore, hold best for a plate with constant surface temperature. For a wall temperature which varies along the surface, heat transfer can be calculated by the method presented in Sec. 7-4 provided a relation is known which describes heat transfer for a stepwise variation of the surface temperatures. Seban, in an M.S. thesis by Scesa,[1] gives such a relation for heat transfer at location $x$ on a plate with an unheated starting section of length $\xi$:

$$\mathrm{Nu}_x = \frac{hx}{k} = 0.0289(\mathrm{Pr})^{1/9}(\mathrm{Re}_x)^{0.8}\left[1 - \left(\frac{\xi}{x}\right)^{9/10}\right]^{-1/9}$$

The power $1/9$ of the Prandtl number appears low and should probably be replaced by $1/3$. Otherwise the equation appears to be in agreement with existing experiments.

**Example 8-3.** A flat plate heated to 176 F is cooled by an air stream at 68 F, 14.2 $\mathrm{lb}_f/\mathrm{in.}^2$, and a velocity $u_s = 30$ fps. The local heat-transfer coefficient at a distance $x = 12$ in. is to be calculated.

According to page 177 the property values for gases are to be inserted at the arithmetic mean between wall and gas temperature. This gives $t^* = 122$ F. From the Appendix we get $\nu = 19.45 \times 10^{-5}$ ft²/sec, $k = 0.0162$ Btu/hr ft F, Pr = 0.703,

$$\mathrm{Re}_x = \frac{u_s x}{\nu} = \frac{30 \times 12/12}{19.45 \times 10^{-5}} = 154{,}000$$

At this Reynolds number the boundary layer may be laminar or turbulent. If it is turbulent, there is obtained from Eq. (8-17)

$$\mathrm{St} = \frac{\mathrm{Nu}_x}{\mathrm{Re}_x\,\mathrm{Pr}} = \frac{0.0296(154{,}000)^{-1/5}}{1 - 0.87 \times 1.75(154{,}000)^{-1/10} \times 0.297} = 0.00315$$

It is questionable as to which length must be inserted as $x$, since the boundary layer is laminar in the vicinity of the leading edge. Following the suggestion of Prandtl (page 144) the distance from the leading edge was used as $x$. The Nusselt number becomes

$$\mathrm{Nu}_x = 0.00315 \times 0.703 \times 154{,}000 = 341$$

and the heat-transfer coefficient

$$h = \mathrm{Nu}_x \frac{k}{x} = 341\,\frac{0.0162}{1} = 5.52 \text{ Btu/hr ft}^2 \text{ F}$$

If the flow is laminar at the distance $x = 12$ in., there is obtained from Eq. (7-14)

$$\mathrm{Nu}_x = 0.332\,\sqrt[3]{\mathrm{Pr}}\,\sqrt{\mathrm{Re}_x} = 0.332 \times \sqrt[3]{0.703}\,\sqrt{154{,}000} = 116$$

and the heat-transfer coefficient

$$h = 116 \times 0.0162 = 1.88 \text{ Btu/hr ft}^2 \text{ F}$$

Therefore, the heat transfer increases very much when the boundary layer becomes turbulent.

[1] S. Scesa, "Experimental Investigation of Convective Heat Transfer to Air from a Flat Plate with a Stepwise Discontinuous Surface Temperature," M.S. thesis, University of California, Berkeley, Calif., 1949.

## 8-4. Recent Developments in the Theory of Turbulent Heat Transfer.

The possibility of improving the calculation of turbulent heat transfer rests on a better knowledge of the details of turbulent flow. A complete description of such a flow with its continuously fluctuating nature would require information on the flow parameters—velocity and pressure—at each location in the flow field and at each instant. At present, such a description goes far beyond our capabilities, and we have to be satisfied with a knowledge of time averaged values. The procedure by which this transformation of the Navier-Stokes equations is accomplished was outlined in 1883 by O. Reynolds. The instantaneous-flow parameters are written as the sum of the time-averaged value (indicated by a bar) and the instantaneous deviation from this value (fluctuation, indicated by a prime):

$$u = \bar{u} + u' \qquad v = \bar{v} + v' \qquad p = \bar{p} + p' \tag{8-19}$$

We then introduce these expressions into the Navier-Stokes equations (6-14); average all fluctuating quantities involved over the time, taking into account that according to our definition $\overline{u'} = 0$, $\overline{v'} = 0$, $\overline{p'} = 0$ (but products like $\overline{u'v'}$ are not necessarily zero); and obtain for steady flow

$$\rho \left( \bar{u} \frac{\partial \bar{u}}{\partial x} + \bar{v} \frac{\partial \bar{u}}{\partial y} + \bar{w} \frac{\partial \bar{u}}{\partial z} \right) = -\frac{\partial \bar{p}}{\partial x} + \mu \left( \frac{\partial^2 \bar{u}}{\partial x^2} + \frac{\partial^2 \bar{u}}{\partial y^2} + \frac{\partial^2 \bar{u}}{\partial z^2} \right) - \rho \left( \frac{\partial \overline{u'^2}}{\partial x} + \frac{\partial \overline{u'v'}}{\partial y} + \frac{\partial \overline{u'w'}}{\partial z} \right) \tag{8-20}$$

and two corresponding equations in $\bar{v}$ and $\bar{w}$. It is also found that the continuity equation does not change its form when it is written in time-averaged quantities:

$$\frac{\partial \bar{u}}{\partial x} + \frac{\partial \bar{v}}{\partial y} + \frac{\partial \bar{w}}{\partial z} = 0$$

The boundary-layer equation in time-averaged values becomes, according to Prandtl's boundary-layer simplifications,

$$\rho \left( \bar{u} \frac{\partial \bar{u}}{\partial x} + \bar{v} \frac{\partial \bar{u}}{\partial y} \right) = -\frac{\partial \bar{p}}{\partial x} + \mu \frac{\partial^2 \bar{u}}{\partial y^2} - \rho \frac{\partial \overline{u'v'}}{\partial y} \tag{8-21}$$

The last term in the boundary-layer equation describes essentially the same turbulent exchange process as the one investigated in Sec. 8-1. The instantaneous mass flow normal to the main flow direction per unit time and area is $\rho v'$, and the $x$-momentum transport connected with this flow is $\rho v' u'$. This averaged term is, therefore, nothing else than another expression for the shearing stress as given by Eq. (8-2). In the complete Navier-Stokes equations several such terms exist corresponding to the $x$-momentum transport in the three coordinate directions. Equations (8-20) and (8-21) give an improved understanding of the process by which turbulent shear stresses are set up. For an actual calculation, which means integration of the equations, additional suitable expressions have to be available which connect the terms in which fluctuating values appear ($\overline{u'^2}$, $\overline{u'v'}$, and so on) with the time-averaged quantities. T. V. Boussinesq was the first to propose such an expression by writing for the turbulent shear stress

$$\tau_t = -\rho(\overline{u'v'}) = \rho\epsilon_m \frac{\partial \bar{u}}{\partial y} \qquad (8\text{-}22)$$

The term $\epsilon_m$ is called *turbulent diffusivity for momentum*. We shall see that the above expression has become very useful in connection with heat-transfer calculations. For the calculation of flow fields it did not prove so useful because $\epsilon_m$ was found by experiments to depend in a complicated manner on location and velocity. L. Prandtl, from certain concepts of the turbulent exchange, arrived at the relation

$$\tau = \rho l^2 \left| \frac{\partial \bar{u}}{\partial y} \right| \frac{\partial \bar{u}}{\partial y} \qquad (8\text{-}23)$$

(where by the bars in the equation it is indicated that the shear stress has the same sign as the velocity gradient) and found that relatively simple relations for the "mixing length" $l$ (for instance, $l$ proportional to the wall distance near a wall) are in fairly good agreement with experiments. Equation (8-23) has been used as the basis for many flow calculations. T. von Kármán arrived at a similar expression by similarity considerations. Values like $\overline{u'^2}$, $\overline{u'v'}$ have been measured by hot-wire anemometry for various types of flow. Statistical theory of turbulence is also concerned with their evaluation.

O. Reynolds' procedure applied to the energy equation (7-3) and L. Prandtl's boundary-layer simplification result, for small velocities, in the following turbulent boundary-layer energy equation in time-averaged values:

$$\rho c_p \left( \bar{u} \frac{\partial \bar{t}}{\partial x} + \bar{v} \frac{\partial \bar{t}}{\partial y} \right) = k \frac{\partial^2 \bar{t}}{\partial y^2} - \rho c_p \frac{\partial(\overline{v't'})}{\partial y} \qquad (8\text{-}24)$$

This equation again differs from the corresponding laminar equation for steady flow by the last term on the right-hand side, which is an expression for the turbulent exchange of heat equivalent to Eq. (8-1). Boussinesq introduced for this turbulent heat flow the expression

$$q_t = -\rho c_p \overline{v't'} = -\rho c_p \epsilon_q \frac{\partial \bar{t}}{\partial y} \qquad (8\text{-}25)$$

with $\epsilon_q$ called the *turbulent diffusivity for heat*. This expression, together with Eq. (8-22), has become very useful for a calculation of heat transfer from flow information, since even the simplest assumption for the ratio

$\epsilon_m/\epsilon_q$, namely, that it is a constant, leads to results which are in satisfactory agreement with measurements. The still simpler assumption $\epsilon_m/\epsilon_q = 1$ is equivalent to Reynolds' analogy, as will be seen shortly.

A large number of papers have been published in recent years which calculate heat transfer for various turbulent-flow situations on the basis of Reynolds' analogy or from a more general formulation for the ratio of the two diffusivities. In this section the idea underlying all these calculations will be sketched in a general way at first, and afterward a specific procedure published by von Kármán will be discussed in more detail.

In a laminar flow in which the changes of velocity and temperature normal to the flow direction are considerably larger than the ones in a direction parallel to the flow, the equations on page 202 describe the shear stress and heat flow. They can be written in the form

$$\tau_l = \nu\rho \frac{du}{dy} \tag{8-26}$$

$$q_l = -k \frac{dt}{dy} = -\frac{\nu}{\Pr} \rho c_p \frac{dt}{dy} \tag{8-27}$$

The shear stress and heat flow connected with the turbulent exchange can be written according to Eqs. (8-22) and (8-25):

$$\tau_t = \epsilon_m \rho \frac{du}{dy} \tag{8-28}$$

$$q_t = -\epsilon_q \rho c_p \frac{dt}{dy} \tag{8-29}$$

From here on, $u$ and $t$ are interpreted as time mean values. The two parameters $\epsilon_m$ and $\epsilon_q$ have the same dimensions as the kinematic viscosity $\nu$ and are called *eddy diffusivity* or *turbulent diffusivity*. It has to be kept in mind that these parameters are complicated functions of wall distance, Reynolds number, and other variables. Reynolds' analogy demands that the turbulent diffusivities for momentum ($\epsilon_m$) and for heat ($\epsilon_q$) be equal. This can easily be seen by dividing the equations for the turbulent heat flow and the turbulent shear stress. The result is

$$\frac{q_t}{\tau_t} = -\frac{\epsilon_q}{\epsilon_m} c_p \frac{dt}{du} \tag{8-30}$$

For $\epsilon_q = \epsilon_m$, this expression is identical with Eq. (8-3). Recent experiments have indicated, however, that the two diffusivities are not exactly identical. It appears that the ratio of $\epsilon_m/\epsilon_q$, which is called turbulent Prandtl number ($\Pr_t$) for reasons which will be evident soon, has a value of approximately 0.7 in boundary-layer flow and approximately 0.5 for wake flow behind blunt objects and for vortex flow.

$$\text{Pr}_t = \frac{\epsilon_m}{\epsilon_q}† \tag{8-31}$$

Measurements by Sage[1] and coworkers and by Ludwieg[2] have even indicated that the turbulent Prandtl number is not a constant; for instance, in a boundary-layer type of flow it depends on the distance from the wall. Nevertheless, many calculations are still carried out on the basis of a turbulent Prandtl number equal to 1, and they usually agree quite well with reality. It causes only minor complications to use a numerical value different from 1 for the turbulent Prandtl number as long as this value is considered a constant for a particular flow situation. If, however, one attempts to make the Prandtl number a variable depending on wall distance and other parameters, then the whole calculation procedure based on Reynolds' analogy loses much of its effectiveness.

In Sec. 8-1 it was assumed that the flow is either completely laminar in the sublayer or completely turbulent in the core of the flow. Actually, newer measurements have indicated that in a turbulent flow a certain amount of turbulence exists right up to the surface. Therefore, laminar and turbulent shear exist simultaneously, and the total shear stress and heat flow have to be written according to Eqs. (8-21) and (8-24) in the following way:

$$\tau = \tau_l + \tau_t = (\nu + \epsilon_m)\rho \frac{du}{dy} \tag{8-32}$$

$$q = q_l + q_t = -\left(\frac{\nu}{\text{Pr}} + \frac{\epsilon_m}{\text{Pr}_t}\right)\rho c_p \frac{dt}{dy} \tag{8-33}$$

The first one of these equations can be used to calculate the eddy diffusivity $\epsilon_m$ as soon as the flow field is known. This value, then, can be introduced into Eq. (8-33), and the temperature field and heat flow can be calculated.

The momentum and energy equations (8-21) and (8-24) for turbulent boundary-layer flow, for instance, can be written in the following form:

$$\rho\left(u\frac{\partial u}{\partial x} + v\frac{\partial u}{\partial y}\right) = \rho\frac{\partial}{\partial y}\left[(\nu + \epsilon_m)\frac{\partial u}{\partial y}\right] - \frac{\partial p}{\partial x} \tag{8-34}$$

$$\rho c_p\left(u\frac{\partial t}{\partial x} + v\frac{\partial t}{\partial y}\right) = \rho c_p \frac{\partial}{\partial y}\left[\left(\frac{\nu}{\text{Pr}} + \frac{\epsilon_m}{\text{Pr}_t}\right)\frac{\partial t}{\partial y}\right] \tag{8-35}$$

When the velocity and pressure fields are known by measurements, then the first of these equations can be solved for the unknown $\epsilon_m$ which will

† In the German literature $\text{Pr}_t$ is sometimes differently defined as $(\epsilon_q/\epsilon_m)\text{Pr}$.
[1] W. H. Corcoran, F. Page, Jr., W. A. Schlinger, and B. H. Sage, *Ind. Eng. Chem.*, **44**:401 (1952).
[2] H. Ludwieg, *Z. Flugwissenschaften*, **4**:73 (1956).

be obtained as a function of the coordinates and possibly of Reynolds number. This value, then, can be introduced into the second equation, and this equation can be solved for the temperature field as soon as $\mathrm{Pr}_t$ is known. A very exact knowledge of the velocity field is required to get the turbulent diffusivity with sufficient accuracy, since gradients of the velocity components must be calculated and inserted into the momentum equation. Therefore, Eqs. (8-32) and (8-33) are generally used in conjunction with assumptions on the variation of $\tau$ and $q$ along the $y$ coordinate. In some cases the shear stress is known. For developed flow through a tube, for instance, a force balance immediately indicates that the shear stress increases linearly with the radius $r$, and (8-32) can be used to calculate the eddy diffusivity $\epsilon_m$ from a knowledge of the wall shear stress $\tau_w$ and the velocity profile. In boundary-layer type of flow the main variation of the velocity occurs near the wall and it is argued that the shear stress cannot vary appreciably along this small dimension. Accordingly, for boundary layers the assumption is often made that the shear stress is constant on normals to the surface. For any case, solving Eq. (8-32) for $\epsilon_m$ results in the expression

$$\epsilon_m = \frac{\tau}{\rho} \frac{1}{du/dy} - \nu \tag{8-36}$$

which can now be used to calculate the eddy diffusivity as soon as the shear stress and the velocity gradient are known from measurements. The temperature field then can be determined when a statement on the value of $q$ as a function of wall distance is possible. In boundary-layer type of flow, the assumption of a constant $q$ is expected to give a reasonable approximation to reality. In channel or tube flow, $q$ varies greatly over the cross section whereas the ratio $q/\tau$ exhibits a smaller variation as long as the Prandtl number is not very small (say above 0.1). To utilize this fact, Eq. (8-33) is divided by Eq. (8-32):

$$\frac{q}{\tau} = c_p \frac{(\nu/\mathrm{Pr}) + (\epsilon_m/\mathrm{Pr}_t)}{\nu + \epsilon_m} \frac{dt}{du} \tag{8-37}$$

Integration of this equation with the assumption that $q/\tau$ is constant and equals $q_w/\tau_w$ gives the following relation for the temperature profile:

$$t - t_w = \frac{1}{c_p} \frac{q_w}{\tau_w} \int_0^u \frac{\nu + \epsilon_m}{(\nu/\mathrm{Pr}) + (\epsilon_m/\mathrm{Pr}_t)} \, du \tag{8-38}$$

The calculation of the eddy diffusivity $\epsilon_m$ and the integration of Eq. (8-38) is carried out either numerically on the basis of measured velocity profiles or analytically if an analytic expression for the velocity profile is known.

Von Kármán subdivided the whole velocity profile into three layers—a laminar sublayer, a buffer layer, and a turbulent core region—in order to have simple expressions for an analytical calculation. The extent of these three regions and the equations by which the velocity field in each one is described can be seen in Fig. 6-20. In addition von Kármán assumed that the flow in the laminar sublayer is completely laminar, so that the term with the turbulent diffusivities vanishes in Eqs. (8-32) and (8-33). On the other hand, in the turbulent core he assumed that the turbulent diffusivity is very much larger than the laminar viscosity or heat conductivity. Accordingly, he neglected the laminar terms in Eqs. (8-32) and (8-33) in the turbulent region. In the buffer layer the flow is expected to change gradually from laminar to turbulent, and all terms are maintained in the equations describing this region. In addition von Kármán assumed that $\epsilon_m = \epsilon_q$ or $\mathrm{Pr}_t = 1$.

With this and with the assumption that $q$ and $\tau$ can be considered constant, Eqs. (8-32) and (8-33) become for the buffer layer

$$\tau_b = (\nu + \epsilon)\rho \frac{du}{dy}$$
$$q_b = \left(\frac{\nu}{\mathrm{Pr}} + \epsilon\right)\rho c_p \frac{dt}{dy} \qquad (8\text{-}39)$$

We shall now calculate the drop in temperature in the buffer layer with the help of these equations. The second equation gives

$$\Delta t_b = \frac{q_b}{\rho c_p} \int_{y_1}^{y_2} \frac{dy}{(\nu/\mathrm{Pr}) + \epsilon}$$

The eddy diffusivity is determined from the first equation:

$$\epsilon = \frac{\tau_b}{\rho} \frac{dy}{du} - \nu$$

The laminar sublayer and the buffer layer are very thin compared with the tube radius; therefore, the shear stress and the specific heat flow can change by only a small amount within these layers. They were therefore assumed constant throughout the layer. Accordingly we replace the values $\tau_b$ and $q_b$ by $\tau_w$ and $q_w$. In addition we introduce the dimensionless values $y^+$ and $u^+$. From the equations which define these values (page 156) it follows that

$$\frac{\rho}{\tau_w} \frac{du}{dy} = \frac{1}{\nu} \frac{du^+}{dy^+}$$

von Kármán's equation for the velocity in the buffer layer

$$u^+ = 5\left(1 + \ln \frac{y^+}{5}\right)$$

gives $du^+/dy^+ = 5/y^+$. Therefore

$$\epsilon = \nu \frac{y^+}{5} - \nu$$

and

$$\Delta t_b = \frac{q_w}{\rho c_p \nu} \int_{y_1}^{y_2} \frac{dy}{(1/\text{Pr}) + (y^+/5) - 1} = \frac{q_w}{\rho c_p} \sqrt{\frac{\rho}{\tau_w}} \int_{5}^{30} \frac{dy^+}{(1/\text{Pr}) + (y^+/5) - 1}$$

Integrated,

$$\Delta t_b = \frac{q_w}{\rho c_p} \sqrt{\frac{\rho}{\tau_w}} \, 5 \ln (5 \, \text{Pr} + 1)$$

The linear temperature drop in the laminar sublayer is represented by

$$\Delta t_l = \frac{q_w}{k} y_1 = \frac{q_w}{\rho c_p} \frac{\text{Pr}}{\nu} y_1$$

Introducing the dimensionless value $y^+(y_1^+ = 5)$ gives

$$\Delta t_l = \frac{q_w}{\rho c_p} \sqrt{\frac{\rho}{\tau_w}} \, \text{Pr} \, y_1^+ = \frac{q_w}{\rho c_p} \sqrt{\frac{\rho}{\tau_w}} \, 5 \, \text{Pr}$$

Within the turbulent layer one obtains from Reynolds' analogy [Eq. (8-10)] or from an integration of Eq. (8-37), assuming $q/\tau = \text{const}$ and $\text{Pr}_t = 1$,

$$\Delta t_t = \frac{q_w}{\rho c_p} \frac{\rho}{\tau_w} (u_s - u_b)$$

Introducing again the dimensionless value $u_b^+$ gives

$$\Delta t_t = \frac{q_w}{\rho c_p} \sqrt{\frac{\rho}{\tau_w}} \left( \frac{u_s}{\sqrt{\tau_w/\rho}} - u_b^+ \right)$$

The border of the turbulent zone, where the velocity is $u_b$, is fixed by the wall distance $y_2^+ = 30$. Therefore with von Kármán's equation $u_b^+ = 5(1 + \ln y^+/5)$ and $y^+ = 30$,

$$\Delta t_t = \frac{q_w}{\rho c_p} \sqrt{\frac{\rho}{\tau_w}} \left[ \frac{u_s}{\sqrt{\tau_w/\rho}} - 5(1 + \ln 6) \right]$$

By summing up all the individual temperature drops, the whole temperature difference between the tube wall and the end of the turbulent zone (the tube axis) is found:

$$t_w - t_s = \frac{q_w}{\rho c_p} \sqrt{\frac{\rho}{\tau_w}} \left[ \frac{u_s}{\sqrt{\tau_w/\rho}} + 5(\text{Pr} - 1) + 5 \ln \frac{5 \, \text{Pr} + 1}{6} \right]$$

The dimensionless heat-transfer coefficient is

$$\mathrm{St} = \frac{\mathrm{Nu}}{\mathrm{Re}\,\mathrm{Pr}} = \frac{q_w}{\rho c_p u_s (t_w - t_s)}$$

$$= \frac{\tau_w/\rho u_s^2}{1 + \sqrt{\tau_w/\rho u_s^2}\,\{5(\mathrm{Pr} - 1) + 5\ln[(5\,\mathrm{Pr} + 1)/6]\}}$$

This equation supersedes Eq. (8-12) derived by Prandtl and Taylor. It is valid for the flat plate as well as for the tube.

To adapt it to the *flat plate* one can introduce the equation

$$\frac{\tau_w}{\rho u_s^2} = \frac{f_P}{2}$$

from page 142. This gives

$$\mathrm{St} = \frac{\mathrm{Nu}_x}{\mathrm{Re}_x\,\mathrm{Pr}} = \frac{f_P/2}{1 + \sqrt{f_P/2}\,\{5(\mathrm{Pr} - 1) + 5\ln[(5\,\mathrm{Pr} + 1)/6]\}}$$

For a smooth wall $f_P/2 = 0.0296/(\mathrm{Re}_x)^{0.2}$.

In adapting it to the *tube* one must keep in mind that the heat-transfer coefficient is usually built with the bulk temperature $t_B$ of the flowing substance, whereas in the above equation the temperature $t_s$ of the tube axis appears. The equations for the shear stress at the tube wall also contain the average velocity $u_m$, not the velocity $u_s$ in the tube axis as used here. Denoting the ratio of the two velocities $\varphi_m = u_m/u_s$ and the temperature ratio

$$\theta'_B = \frac{t_B - t_w}{t_s - t_w}$$

and using the expression $\tau_w/\rho u_m^2 = f/8$ (page 155), we get for the *tube*

$$\mathrm{St} = \frac{\mathrm{Nu}_d}{\mathrm{Re}_d\,\mathrm{Pr}} = \frac{(\varphi_m/\theta'_B)\,f/8}{1 + \varphi_m\sqrt{f/8}\,\{5(\mathrm{Pr} - 1) + 5\ln[(5\,\mathrm{Pr} + 1)/6]\}}$$

The velocity ratio can be derived from Eq. (6-32) for the turbulent-velocity profile $\varphi_m = 0.82$. In reality, it changes somewhat with Reynolds number. The friction coefficient $f$ is obtained from Eq. (6-55) or (6-56).

With the use of especially adapted coordinates, temperature profiles can be constructed which depend only slightly on Reynolds number and which can therefore be called universal temperature profiles in analogy to the universal velocity profile. In the formulas for the temperature drop in the different layers there appeared the factor $q_w/(\rho c_p \sqrt{\tau_w/\rho})$ which has the dimension of temperature. Dividing the difference in the temperature at an arbitrary point within the boundary layer and the wall $(t - t_w)$ by the above factor gives a dimensionless $\Delta t^+$ which serves as ordinate for the universal temperature profile. The abscissa is the same ($y^+$) as in Fig. 6-20. Figure 8-12 shows these temperature profiles for different Prandtl numbers. The profile for Pr = 1 is identical with the universal velocity profile when the eddy diffusivity of momentum and heat are equal. Plotting of measured temperature profiles in this form is useful for a check of the assumptions of the theory presented in this section.

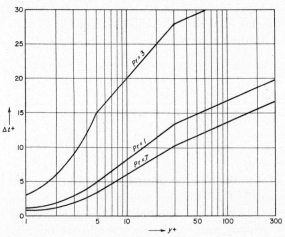

FIG. 8-12. Universal temperature profiles for turbulent heat transfer in a tube.

Many researchers have modified and improved von Kármán's approach along the lines indicated at the beginning of this section. Especially extensive and successful was the work of R. G. Deissler.[1] He uses the relation

$$\epsilon_m = k^2 \frac{(du/dy)^3}{(d^2u/dy^2)^2} \qquad (8\text{-}40)$$

to describe the turbulent diffusivity of momentum in the interior of a turbulent flow. This relation had been developed by T. von Kármán from similarity considerations. For a region close to the wall, which corresponds roughly to the laminar sublayer and the buffer layer in Fig. 6-20, he developed the relation

$$\epsilon_m = n^2 uy \qquad (8\text{-}41)$$

A universal velocity profile calculated with these equations and Eq. (8-32) agrees very well with the measured profile (Fig. 6-20) when the values $k = 0.36$ and $n = 0.109$ are introduced, when Eq. (8-41) is used between $y^+ = 0$ and $y^+ = 26$ and Eq. (8-40) for values $y^+$ greater than 26.

The universal temperature profiles (analogous to Fig. 8-12) are obtained by an integration of Eq. (8-33) with the assumption $Pr_t = 1$. The tem-

[1] R. G. Deissler, *Trans. ASME.*, **73**:101–105 (1951); **76**:73–85 (1954).

perature gradient at the wall then determines the heat-transfer coefficient.

Deissler was in this way able to predict heat transfer for turbulent pipe and boundary-layer flow of air in very good agreement with experimental results. In this calculation he assumed the Prandtl number and the specific heat to be constant and the viscosity and heat conductivity to vary proportionally to the power 0.68 of the absolute temperature. Heat transfer and flow then depend on an additional parameter

$$\beta = \frac{q_w}{\rho_w c_{pw} \sqrt{\tau_w/\rho_w}\, t_w}$$

which can be recognized as the reciprocal of a dimensionless temperature as used in Fig. 8-12. Deissler also extended his analysis to entrance effects in pipe flow and to other fluids, including those near the critical state. Analogous work on turbulent heat transfer is reported in the German literature by H. Reichardt.[1]

An extension of the derivations in this section is necessary for substances with low Prandtl numbers, e.g., for molten metals. In these substances, the molecular exchange of momentum and heat cannot be neglected in the turbulent zone as it has been done in the preceding calculations. Such an extension was made by R. C. Martinelli.[2]

## PROBLEMS

**8-1.** Calculate and compare the power required to overcome the pressure drop of air flowing through a cooler on an airplane flying at 500 mph when the cooler is arranged in a duct so that (a) the approach velocity of the air relative to the airplane is reduced to one-tenth before the air enters the cooler, (b) the approach velocity is reduced only by a negligible amount. The atmospheric air temperature is $-50$ F, and it increases by 40 F in flowing through the cooler. The cooler is used to cool 6 lb/sec of oil from 250 to 150 F in counterflow. Compare also the surface areas required for the heat exchange, assuming that the heat-transfer coefficient increases proportional to the 0.8th power of the mass velocity $\rho u$. Use the approximation Pr = 1.

**8-2.** Calculate the development of the temperature field and the thermal entrance length for turbulent flow through a tube, approximating the actual velocity field by a velocity constant over the tube cross section. Approximate the temperature profile near the wall by a seventh-power law, and use Eq. (11-14) to describe the local heat flow through the tube surface. Compare the results of this calculation with the information in Sec. 8-2, and discuss in what points actual conditions are expected to be different from assumptions made in this calculation.

**8-3.** Calculate the development of the thermal boundary layer along a flat plate on the basis of the following assumptions: The flow is laminar to a critical Reynolds number $Re_c$. It then suddenly changes to turbulent flow in such a way that at the critical location the convection thickness of the turbulent boundary layer is equal

[1] H. Reichardt, *Z. angew. Math. Mech.*, **20**:297–328 (1940).
[2] R. C. Martinelli, *Trans. ASME*, **68**:947–959 (1947).

to the convection thickness of the laminar layer. The fluid has a Prandtl number equal to 1. Derive a relation for the average Nusselt number, and compare with the relation on page 215.

**8-4.** Calculate the local heat-transfer coefficients for a turbulent boundary layer along a flat plate, assuming that the temperature difference between stream and surface increases linearly with distance from the leading edge. Use the procedure outlined in Sec. 7-4 and the heat-transfer relation in Sec. 8-3 for a stepwise-varying wall temperature.

**8-5.** Derive an equation for heat transfer caused by turbulent flow over a flat plate for the condition that the plate is unheated over a distance $x_0$ so that on this part it has a temperature equal to the stream temperature and that downstream from $x_0$ it is heated to a constant temperature. Use a seventh-power law to describe the temperature profile and Eq. (11-14) for the local heat flux. Base the calculation on the concept that a thermal boundary layer starts to develop at $x = x_0$ and that this boundary layer grows into the velocity boundary layer. Fulfill the integrated boundary-layer energy equation. Compare the result with the equation on page 217.

**8-6.** Plot the variation of the turbulent and the laminar shear stress as determined by the equations in Fig. 6-20 and in Sec. 8-4 over a normal to the surface. Discuss how far you expect actual conditions to be different from the ones in the plot.

CHAPTER 9

# FORCED CONVECTION IN SEPARATED FLOW

**9-1. Dimensional Analysis of Heat Transfer.** Information on the various types of heat transfer which have been discussed up to this point could be obtained by analytical methods. For other types, however, our basic understanding of the underlying processes is still too poor to permit any attempt of a calculation. This is especially true for all bodies on which the flow separates at some location on the surface. A very common and important example is the tube with circular cross section in a stream flowing normal to its axis. It has been shown in the preceding sections that heat transfer to such a cylinder can be calculated along the upstream portion over which a boundary layer exists. The downstream part of the surface, however, borders on the separated flow region which is filled with irregular vortices, and no calculation of convective heat transfer appears feasible at present for this portion. The only remaining possibility is to rely on experiments and to generalize the results by dimensional analysis. This method postulates that all physical processes can be expressed as a relationship among dimensionless parameters and specifies how to find those parameters. The application of dimensional analysis to heat transfer enabled W. Nusselt[1] in a fundamental paper published in 1915 systematically to coordinate for the first time earlier experimental results and to plan new experiments. This year can, therefore, be considered as the birth year of a science of heat transfer. There are today several methods in existence by which the dimensionless parameters can be determined on which a certain physical process depends. The most important and difficult part in such an analysis is to find not the form of the parameters but the number which fully describes the process under consideration. According to the author's experience, the most effective and reliable way to obtain this answer is the method which starts from the differential equations and makes them dimensionless. This method will, therefore, be applied here. The objection may be raised that in some cases the equations governing the process may not be known. However, it will be seen later that only an approximate knowledge of the equations is required, and if this knowledge is missing, then a

[1] W. Nusselt, *Gesundh. Ingr.*, **38**:477–490 (1915).

satisfactory dimensional analysis cannot be carried out by any other method either.

In all flow processes which were discussed in Chap. 6, it was found that the flow parameters of interest (boundary-layer thickness, friction factor), when presented in a dimensionless form, were functions of the Reynolds number. In the same way, the heat-transfer parameters like Nusselt or Stanton numbers and thermal boundary-layer thickness, presented in dimensionless form, turned out to be functions of the dimensionless Reynolds and Prandtl numbers. We shall now try to obtain this information directly from the differential equations without solving them.

Let us first look at the flow process. Since we want our results to include separated flow regions, we have to use the Navier-Stokes equations to describe this process. We shall restrict our considerations to steady flow. In Sec. 6-3 these equations were given in Cartesian coordinates for a fluid with constant properties and in the absence of body forces. The equation for the $x$ direction and the continuity equation will be repeated here:

$$\rho \left( u \frac{\partial u}{\partial x} + v \frac{\partial u}{\partial y} + w \frac{\partial u}{\partial z} \right) = -\frac{\partial p}{\partial x} + \mu \left( \frac{\partial^2 u}{\partial x^2} + \frac{\partial^2 u}{\partial y^2} + \frac{\partial^2 u}{\partial z^2} \right)$$

$$\frac{\partial u}{\partial x} + \frac{\partial v}{\partial y} + \frac{\partial w}{\partial z} = 0$$

These equations together with the momentum equations for the $y$ and $z$ directions are four equations for the four unknowns $u, v, w, p$. In addition to the differential equations, boundary conditions must be given describing the specific flow situation. To fix our ideas, we shall be concerned with flow of a fluid normal to the axis of a cylinder of circular cross section and infinite length. Figure 9-1 is a sketch of this situation and of the necessary boundary conditions. The diameter of the cylinder is $d$, and the cylinder is assumed at rest. Accordingly, the velocity on its surface has to be zero. The velocity upstream and on both sides at a sufficient distance from the cylinder is $u_0$ and is assumed constant along the stream boundaries. The pressure is also constant on the outer boundary of the flow field. Its absolute value is immaterial, since only the pressure differential appears in the above equation.

As a first step in the dimensional analysis, we shall make the differential equations and the boundary conditions dimensionless by dividing all parameters which have the dimension length by the prescribed length $d$ and all parameters with the dimension velocity by the prescribed velocity $u_0$. The pressure can be made dimensionless with the term $\rho u_0^2$ based on the constant density and the prescribed velocity $u_0$ on the

# FORCED CONVECTION IN SEPARATED FLOW

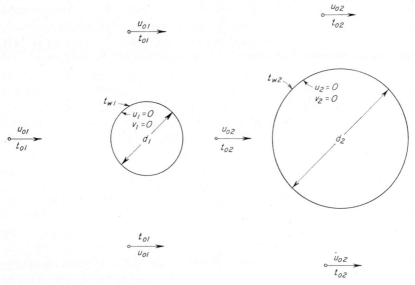

FIG. 9-1. Dimensional analysis on two circular cylinders.

boundary. Indicating dimensionless quantities by a prime, we write

$$x' = \frac{x}{d} \quad y' = \frac{y}{d} \quad z' = \frac{z}{d} \quad u' = \frac{u}{u_0}$$
$$v' = \frac{v}{u_0} \quad w' = \frac{w}{u_0} \quad p' = \frac{p}{\rho u_0^2} \quad (9\text{-}1)$$

Introducing the dimensionless quantities into the momentum equation results in

$$\rho \frac{u_0^2}{d}\left(u'\frac{\partial u'}{\partial x'} + v'\frac{\partial u'}{\partial y'} + w'\frac{\partial u'}{\partial z'}\right) = -\rho \frac{u_0^2}{d}\frac{\partial p'}{\partial x'} + \mu \frac{u_0}{d^2}\left(\frac{\partial^2 u'}{\partial x'^2} + \frac{\partial^2 u'}{\partial y'^2} + \frac{\partial^2 u'}{\partial z'^2}\right)$$

This equation is still not dimensionless but can be made so by dividing it with $\rho(u_0^2/d)$:

$$u'\frac{\partial u'}{\partial x'} + v'\frac{\partial u'}{\partial y'} + w'\frac{\partial u'}{\partial z'} = -\frac{\partial p'}{\partial x'} + \frac{1}{\text{Re}}\left(\frac{\partial^2 u'}{\partial x'^2} + \frac{\partial^2 u'}{\partial y'^2} + \frac{\partial^2 u'}{\partial z'^2}\right) \quad (9\text{-}2)$$

The dimensionless continuity equation is

$$\frac{\partial u'}{\partial x'} + \frac{\partial v'}{\partial y'} + \frac{\partial w'}{\partial z'} = 0 \quad (9\text{-}3)$$

and the boundary conditions are

In the freestream, upstream of the cylinder: $u' = \dfrac{u_0}{u_0} = 1$ (9-4)

At the cylinder surface: $u' = v' = w' = 0$

The notation $Re_d$ has been introduced for the dimensionless group $\rho u_0 d/\mu$. Now it is obvious that the solution of the differential equations will present the dimensionless dependent variables $u'$, $v'$, $w'$, $p'$ as functions of the independent variables $x'$, $y'$, $z'$ and of all constant parameters contained in the differential equations and in the boundary conditions. In our case the only constant parameter is Re. Therefore, the solution must have the form

$$u' = f_u(x',y',z',\text{Re}) \qquad (9\text{-}5)$$
$$v' = f_v(x',y',z',\text{Re}) \qquad (9\text{-}6)$$
$$w' = f_w(x',y',z',\text{Re}) \qquad (9\text{-}7)$$
$$p' = f_p(x',y',z',\text{Re}) \qquad (9\text{-}8)$$

The significance of this result may need some further discussion. Consider for this purpose two cylinders with diameters $d_1$ and $d_2$ exposed to two fluid streams with upstream velocities $u_{01}$ and $u_{02}$, respectively (Fig. 9-1). The above equations indicate that the dimensionless velocity components and the dimensionless pressures are for both cases the same functions of the dimensionless coordinates and of the Reynolds number. When the Reynolds number has the same value for both cases, then at similarly located points (with the same dimensionless coordinates) the dimensionless velocities and pressures will have the same values in the field around cylinders 1 and 2. We call physical quantities, which in a dimensionless presentation—such as dimensionless values $(u',v',w',p')$ plotted versus the dimensionless location $(x',y',z')$—become identical, "physically similar," and we can now state: The velocity fields and pressure fields for flow around cylinders are similar for all situations regardless of cylinder diameter and upstream velocity provided that the Reynolds number has the same value for all cases considered.

When we want to generalize our results to flow around or through other body geometries, it is immediately evident that physical similarity in the sense defined above can exist only for geometrically similar bodies. In addition, other conditions, which have to be fulfilled to produce physical similarity, may come from the velocities at the boundaries of the problem. When, for instance, the cylinder rotates with a circumferential velocity $v_c$ at its surface, then the solution of the differential equations must fulfill the condition that the velocity vector **V** described by the components $u$, $v$, $w$ is, at the cylinder surface, identical in amount and direction with the velocity $v_c$. In dimensionless form this boundary condition prescribes that at the cylinder surface $\mathbf{V}' = \mathbf{V}_c/u_0$ and the parameter will appear in the functional relations (9-5) to (9-8).

Therefore, it can be stated generally that the velocity and pressure field for steady flow of fluids with constant properties are similar when geometrical similarity of the boundaries of the fields exists, when the velocities along the boundaries are similar, and when the Reynolds number has a constant value. The condition of similar velocities on the boundaries sometimes imposes restrictions. It implies, for instance, that velocity variations in time, when they occur, are similar. Such fluctuations occur in turbulent flow, and it is a well-known fact that flow parameters often depend on the degree of turbulence in the stream.

Let us now assume that the cylinder in Fig. 9-1 is kept at a temperature $t_w$ whereas the upstream fluid has a temperature $t_0$. Accordingly, a temperature field will establish itself in the fluid around the cylinder which can, in principle, be calculated from the energy equation (7-3). For steady state and constant properties this equation reads

$$\rho c_p \left( u \frac{\partial t}{\partial x} + v \frac{\partial t}{\partial y} + w \frac{\partial t}{\partial z} \right) = k \left( \frac{\partial^2 t}{\partial x^2} + \frac{\partial^2 t}{\partial y^2} + \frac{\partial^2 t}{\partial z^2} \right) + \mu \left[ 2 \left( \frac{\partial u}{\partial x} \right)^2 + \cdots \right]$$

The last term on the right-hand side indicates the dissipation function. It is sufficient for dimensional analysis to consider only the first term in this function, since the other terms are dimensionally identical with the first one. It is immediately evident that the problem under consideration is independent of the temperature level, since only temperature differentials appear in the above equation. We can, therefore, measure the temperatures from an arbitrary reference value, and we shall choose as such the temperature $t_0$. Introducing $\vartheta = t - t_0$, building a dimensionless excess temperature $\vartheta' = \vartheta/\vartheta_w = (t - t_0)/(t_w - t_0)$ and introducing this parameter, together with the others defined before, into the energy equation results in

$$\frac{\rho c_p u_0 \vartheta_w}{d} \left( u' \frac{\partial \vartheta'}{\partial x'} + v' \frac{\partial \vartheta'}{\partial y'} + w' \frac{\partial \vartheta'}{\partial z'} \right) = \frac{k \vartheta_w}{d^2} \left( \frac{\partial^2 \vartheta'}{\partial x'^2} + \cdots \right) \\ + \mu \frac{u_0^2}{d^2} \left[ 2 \left( \frac{\partial u'}{\partial x'} \right)^2 + \cdots \right]$$

The equation is made dimensionless by multiplying it by $d/\rho c_p u_0 \vartheta_w$:

$$u' \frac{\partial \vartheta'}{\partial x'} + v' \frac{\partial \vartheta'}{\partial y'} + w' \frac{\partial \vartheta'}{\partial z'} = \frac{1}{\text{Re Pr}} \left( \frac{\partial^2 \vartheta'}{\partial x'^2} + \cdots \right) \\ + \frac{\text{E}}{\text{Re}} \left[ 2 \left( \frac{\partial u'}{\partial x'} \right)^2 + \cdots \right] \quad (9\text{-}9)$$

The dimensionless parameter $k/\rho c_p u_0 d$ appearing in the first right-hand term can be identified as the reciprocal of the product Re Pr, and the parameter $\mu u_0 / \rho c_p \vartheta_w d$ is the quotient of a parameter $u_0^2 / c_p \vartheta_w$, to which H. Schlichting[1] gave the symbol E and the name Eckert number, and

---

[1] H. Schlichting, "Boundary Layer Theory," p. 250, McGraw-Hill Book Company, Inc., New York, 1955.

of the Reynolds number. The boundary conditions are

In the freestream, upstream of the cylinder: $\vartheta' = 0$
At the cylinder surface: $\vartheta' = 1$

In order to obtain a solution, the velocity components have to be introduced into the energy equation from Eqs. (9-5) to (9-7). The solution, if it could be obtained, would be of the form

$$\vartheta' = f\left(x',y',z',\text{Re},\text{Re Pr},\frac{\text{E}}{\text{Re}}\right)$$

A function of the parameters Re, Re Pr, and E/Re must also be expressible as a function of Re, Pr, and E. Therefore,

$$\vartheta' = f(x',y',z',\text{Re},\text{Pr},\text{E}) \tag{9-10}$$

It has been discussed in Sec. 6-3 that for velocities as they usually occur in engineering heat transfer, the dissipation term in the energy equation can be neglected. Correspondingly the last term in Eq. (9-10) drops out and the dimensionless temperature field for low-velocity flow is

$$\vartheta' = f(x',y',z',\text{Re},\text{Pr}) \tag{9-11}$$

With the generalizations made before, the situation is equivalent to the following statement:

The temperature fields around geometrically similar objects are similar when the temperatures and velocities around the boundaries are similar and when the Reynolds and Prandtl numbers as well as the parameter E have a constant value.

For engineering purposes, the heat exchange between the fluid and the cylinder wall is a very important quantity. It can be calculated as soon as the heat-transfer coefficient is known, which is defined by the equation

$$h(t_w - t_0) = -k\left(\frac{\partial t}{\partial n}\right)_w$$

in which $n$ indicates the direction normal to the wall surface. Introducing the dimensionless temperature $\vartheta'$ one obtains

$$h\vartheta_w = \frac{k\vartheta_w}{d}\left(\frac{\partial \vartheta'}{\partial n'}\right)_w$$

or
$$\text{Nu} = \frac{hd}{k} = \left(\frac{\partial \vartheta'}{\partial n'}\right)_w$$

The Nusselt number is, therefore, nothing else than the dimensionless temperature gradient at the surface. By differentiating Eq. (9-11) and by introducing the result into the above expression, one obtains for low-velocity flow

$$\text{Nu} = f(x',y',z',\text{Re},\text{Pr}) \tag{9-12}$$

where $x'$, $y'$, $z'$ now indicate the coordinates of an arbitrary surface point and where $\mathrm{Nu}_d$ is the local Nusselt number at that point. The average Nusselt number of the whole surface is given by a relation

$$\overline{\mathrm{Nu}} = f(\mathrm{Re},\mathrm{Pr}) \tag{9-13}$$

Summarizing the results obtained up to now, it can be stated that in steady flow of a fluid with constant properties around geometrically similar objects or through geometrically similar ducts, all dimensionless flow parameters, e.g., friction factors, are functions of location and Reynolds number provided the velocities on the boundaries are similar. The dimensionless heat-transfer parameters like Nusselt or Stanton numbers are functions of location, of Reynolds number, of Prandtl number, and, for high velocities, of the E parameter provided the velocities and the temperatures along the boundaries are similar.

As an extension of the similarity considerations we shall now assume that the density of our fluid changes with temperature. As a consequence of this, buoyancy forces arise as soon as temperature differences exist within the fluid, and these forces will produce free-convection currents which influence the heat transfer. The buoyancy force per unit volume of a fluid element is $g(\rho - \rho_u)$, where $g$ indicates the gravitational acceleration, $\rho$ the actual fluid density at the location of the fluid element, and $\rho_u$ the density which the fluid would have at that location if it were not heated by the heat-transfer process. It will be assumed that the density difference $\Delta\rho = \rho - \rho_u$ is small as compared with the density $\rho$ itself, so that the fluid properties are still practically constant. The term $\Delta\rho$ can then be expressed by the temperature difference, when the thermal expansion coefficient $\beta$ is introduced as defined by the equation

$$\beta = \frac{1}{v}\left(\frac{\partial v}{\partial t}\right)_p$$

From $v = 1/\rho$, $dv = -(1/\rho^2)\,d\rho$, this expression becomes

$$\beta = -\frac{1}{\rho}\left(\frac{\partial \rho}{\partial t}\right)_p$$

For a small density difference caused by temperature variation we can therefore write with good accuracy

$$\Delta\rho = -\rho\beta\,\Delta t$$

and the buoyancy force per unit volume becomes $-g\rho\beta\,\Delta t$. When this force acts in the $x$ direction, the momentum equation for this direction reads

$$\rho\left(u\frac{\partial u}{\partial x} + v\frac{\partial u}{\partial y} + w\frac{\partial u}{\partial z}\right) = -\frac{\partial p}{\partial x} + \mu\left(\frac{\partial^2 u}{\partial x^2} + \frac{\partial^2 u}{\partial y^2} + \frac{\partial^2 u}{\partial z^2}\right) - g\rho\beta\,\Delta t$$

In a constant-property fluid the temperature changes only in the neighborhood of surfaces with temperatures different from the fluid temperature. In this case $\Delta t = \vartheta$. When the equation is made dimensionless by introducing the primed values, then the last term on the right-hand side changes to $(g\beta d\vartheta_w/u_0{}^2)\vartheta'$. The constant term ahead of $\vartheta'$ then constitutes a new parameter on which the solution of the dimensionless equation depends. This parameter is inconvenient because it contains two values prescribed at the boundaries: $u_0$ and $\vartheta_w$. The velocity $u_0$ can be replaced by the Reynolds number: $(g\beta\, d^3\vartheta_w/\nu^2)(1/\mathrm{Re}^2)$. The parameter $g\beta\, d^3\vartheta_w/\nu^2$ is called Grashof number (Gr). The dimensionless momentum equation then reads

$$u'\frac{\partial u'}{\partial x'} + v'\frac{\partial u'}{\partial y'} + w'\frac{\partial u'}{\partial z'} = -\frac{\partial p'}{\partial x'} + \frac{1}{\mathrm{Re}}\left(\frac{\partial^2 u'}{\partial x'^2} + \frac{\partial^2 u'}{\partial y'^2} + \frac{\partial^2 u'}{\partial z'^2}\right) \\ -\frac{\mathrm{Gr}}{\mathrm{Re}^2}\vartheta' \quad (9\text{-}14)$$

Now the dimensionless temperature $\vartheta'$ appears in this equation. Therefore, the system of Eqs. (9-3), (9-9), and (9-14) and the two momentum equations in the $y$ and $z$ directions have to be solved simultaneously. The result will be

$$u' = f_u(x',y',z',\mathrm{Re},\mathrm{Pr},\mathrm{Gr}) \quad (9\text{-}15)$$

and two corresponding equations for $v'$ and $w'$

$$p' = f_p(x',y',z',\mathrm{Re},\mathrm{Pr},\mathrm{Gr}) \quad (9\text{-}16)$$
$$\vartheta' = f_\vartheta(x',y',z',\mathrm{Re},\mathrm{Pr},\mathrm{Gr}) \quad (9\text{-}17)$$

Physically this means that the velocity and pressure fields depend now on the temperature difference between cylinder and fluid whereas it was not influenced at all by heating or cooling of the cylinder when body forces were not present.

Buoyancy forces generate a flow in a fluid even when it is otherwise at rest ($u_0 = 0$). The heat transport caused by this movement is called free or natural convection. The Reynolds number is now zero and drops out of Eqs. (9-15) to (9-17) and out of the equations describing dimensionless heat-transfer parameters. The average Nusselt number for heat transfer to the cylinder is, for instance,

$$\overline{\mathrm{Nu}} = f(\mathrm{Gr},\mathrm{Pr}) \quad (9\text{-}18)$$

In many cases of free-convection flow the velocities are so small that inertia forces and pressure forces can be neglected relative to viscous forces and buoyancy forces. This means that the left-hand terms in Eq. (9-14) and in the other two momentum equations can be set equal to zero. The velocity $u_0$ is zero and, therefore, not available to make the velocities $u$ dimensionless. The parameter $\sqrt{g\beta d\vartheta_w}$, however, has the dimension of velocity and can be used for this purpose. Repeating otherwise the procedure which has been applied before, one ends up with dimensionless equations which contain as constant parameter only the term Gr Pr. As a consequence, all dimensionless flow and heat-transfer parameters will be functions of Gr Pr. A relation of the form

$$\overline{\mathrm{Nu}} = f(\mathrm{Gr}\,\mathrm{Pr}) \tag{9-19}$$

will, for instance, describe the average Nusselt number.

Now we shall turn to a discussion of dimensional analysis for a fluid with variable properties. It will be seen later on that a general treatment for such a fluid is not possible. We shall therefore start by considering flow of a gas for which the properties vary in the following way:

$$\rho = \frac{p}{RT} \qquad \mu = C_\mu T^\alpha \qquad k = C_k T^\alpha \qquad c_p = \text{const} \qquad \mathrm{Pr} = \text{const}$$

$R$ denotes the gas constant, and $T$ the absolute temperature. It is easily verified that the exponent in the expressions for viscosity and heat conductivity must be the same when Pr and $c_p$ are constants. Considering again the flow around a circular cylinder, we have now to deal not only with velocity, pressure, and temperature fields but also with density, viscosity, and heat-conductivity fields. The latter can again be expressed conveniently in a dimensionless form as

$$\rho' = \frac{\rho}{\rho_0} \qquad \mu' = \frac{\mu}{\mu_0} \qquad k' = \frac{k}{k_0}$$

with the subscript 0 indicating the conditions at a sufficient distance from the cylinder. These fields can easily be determined from the following relations when the temperature and pressure fields are known:

$$\rho' = \frac{p'}{T'} \qquad \mu' = T'^\alpha \qquad k' = T'^\alpha \tag{9-20}$$

For a further analysis, the momentum, continuity, and energy equations for a fluid with variable properties have to be known. A knowledge of the dimensional structure is, however, sufficient, and the equations are, therefore, written below in an abbreviated way:

$$\rho u \frac{\partial u}{\partial x} + \cdots = -\frac{\partial p}{\partial x} + \left[\frac{\partial}{\partial x}\left(\mu \frac{\partial u}{\partial x}\right) + \cdots\right] \tag{9-21}$$

$$\frac{\partial(\rho u)}{\partial x} + \cdots = 0 \tag{9-22}$$

$$\rho c_p u \frac{\partial T}{\partial x} + \cdots = \left[u \frac{\partial p}{\partial x} + \cdots\right] + \left[\frac{\partial}{\partial x}\left(k \frac{\partial T}{\partial x}\right) \cdots\right]$$
$$+ \left[2\mu \left(\frac{\partial u}{\partial x}\right)^2 + \cdots\right] \tag{9-23}$$

It can be seen that the equations are quite similar to the ones for a constant-property fluid. An additional term $u(\partial p/\partial x)$ appears in the energy equation, expressing the fact that in a compressible fluid the temperature changes when compression or expansion is caused by a change in pressure.

The equation can be made dimensionless by introduction of the primed values:

$$\frac{p'}{T'}\left(u'\frac{\partial u'}{\partial x'} + \cdots\right) = -\frac{\partial p'}{\partial x'} + \frac{1}{\text{Re}}\left[\frac{\partial}{\partial x'}\left(T'^\alpha \frac{\partial u'}{\partial x'}\right) + \cdots\right] \quad (9\text{-}24)$$

$$\frac{\partial}{\partial x'}\left(\frac{p'}{T'}u'\right) + \cdots = 0 \quad (9\text{-}25)$$

$$\frac{p'}{T'}\left(u'\frac{\partial T'}{\partial x'} + \cdots\right) = \text{E}\left(u'\frac{\partial p'}{\partial x'}\right) + \frac{1}{\text{Re Pr}}\left[\frac{\partial}{\partial x'}\left(T'^\alpha \frac{\partial T'}{\partial x'}\right) + \cdots\right]$$
$$+ \frac{\text{E}}{\text{Re}}\left[2T'^\alpha\left(\frac{\partial u'}{\partial x'}\right)^2 + \cdots\right] \quad (9\text{-}26)$$

The boundary conditions have to prescribe velocity, pressure, and temperature on the boundaries of our problem. The temperatures cannot, however, be expressed as measured from an arbitrary reference condition, since the absolute temperature appears in the previous equations. In the dimensionless parameters the boundary conditions are

In the freestream, upstream of the cylinder: $u' = 1 \quad T' = 1 \quad p' = 1$

At the cylinder surface: $u' = v' = w' = 0 \quad T' = \dfrac{T_w}{T_0}$

(No pressure can be arbitrarily prescribed at the cylinder surface.) The solution of the system of equations with the boundary conditions will be

$$u' = f_u\left(x',y',z',\text{Re},\text{Pr},\text{E},\alpha,\frac{T_w}{T_0}\right)$$

$$T' = f_T\left(x',y',z',\text{Re},\text{Pr},\text{E},\alpha,\frac{T_w}{T_0}\right)$$

The parameter E in this case is proportional to the square of the Mach number in the upstream, since the velocity of sound for a gas with constant $c_p$ is given by $a = \sqrt{\gamma RT} = \sqrt{(\gamma - 1)c_p T}$.

$$\text{E} = \frac{u_0^2}{c_p(T_w - T_0)} = (\gamma - 1)\frac{u_0^2}{a_0^2}\frac{T_0}{T_w - T_0} = (\gamma - 1)(\text{Ma})^2\frac{1}{(T_w/T_0) - 1}$$

The average heat transfer to the cylinder is then described by the equation

$$\overline{\text{Nu}} = f\left(\text{Re},\text{Pr},\text{Ma},\frac{T_w}{T_0},\alpha,\gamma\right) \quad (9\text{-}27)$$

The number of parameters in the function $f$ is already quite large, and the number of situations with physical similarity is correspondingly reduced. Actually the equations used above to describe the properties of a gas are only approximations to the actual situation. The more accurately an equation describes the actual variation of the properties, the more constants it will contain and the more parameters appear in the dimensionless equations describing flow and heat transfer. This restricts the similarity to such a degree that in a practical sense it soon fails to exist. Only by describing the properties of a class of fluids in a dimensionless form by the same equations can similarity be maintained within this class and common dimensionless equations be expected which describe flow and heat transfer.

**9-2. Tubes and Tube Bundles in Crossflow.** Among the differently shaped bodies from whose surface there is flow separation, circular tubes

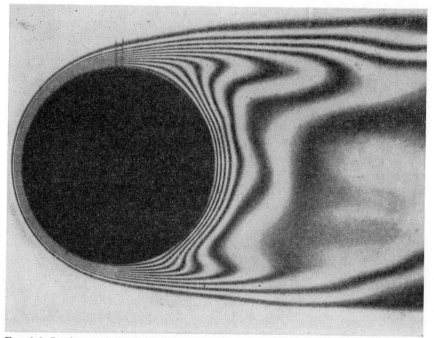

Fig. 9-2. Isotherms around a cylinder cooled by a fluid flowing normal to its axis, as revealed by an interference photograph. Re = 1,260. (*Photograph by E. Eckert and E. Soehngen.*)

are those of the greatest engineering importance; therefore, a great number of experimental investigations of heat transfer from tubes have been made. The flow around a circular tube was discussed in Sec. 6-8.

The temperature field around a heated cylinder is revealed by the interference photograph in Fig. 9-2. The dark lines around the shadow of the cylinder are lines of constant temperature. This figure thus shows the thermal boundary layer on the front half of the cylinder. A kind of boundary layer exists also along the back half. Within it the temperature drops from the surface value to the temperature within the wake. The flow shown in this photograph has a small Reynolds number. At higher Reynolds numbers the boundary layers are thinner and the temperatures in the wake more fluctuating and irregular.

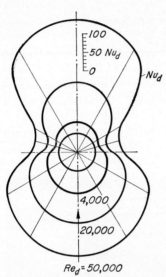

FIG. 9-3. Local film heat-transfer coefficients around a cylinder in flow of air normal to its axis. (*According to W. Lorisch from M. ten Bosch, "Die Wärmeübertragung," 3d ed., Springer-Verlag, Berlin, 1936.*)

The distribution of the heat-transfer coefficient around the tube surface is shown in Figs. 9-3 and 9-4. Heat transfer on the front side of the tubes was discussed in Sec. 7-5 and is shown by Fig. 7-11. The heat transfer on the back side occurs in the wake. Our knowledge of this kind of heat transfer is very limited. At low Reynolds numbers, the heat transfer on the front side is much greater than that on the back side. With increasing Reynolds number, the heat flow from the back side increases at a greater rate than on the front, and for $Re_d = 50,000$, it is as high as on the front side. The reason for this fact is that the vortices, while separating alternately from the right and left flank of the cylinder, wash the surface of the rear half of the cylinder with an intensity that increases with the Reynolds number. If the boundary layer changes to turbulent flow before it separates from the surface, the distribution of the heat-transfer coefficient along the surface changes. This is the case when the Reynolds number exceeds the value $Re_d = 4 \times 10^5$. Figure 9-4 shows that for a high Reynolds number the heat-transfer coefficient increases very much at an angle of 100°. This indicates transition in the flow within the boundary layer. It was seen also on the plate in longitudinal flow that the heat transfer increases appreciably at the place where the laminar boundary layer changes to a turbulent one. The place on the cylinder surface where the flow separates probably lies downstream of the second maximum for the heat-transfer coefficient.

For calculations on heat exchangers, the interest is mostly in the total heat flow from or to a tube, or, what is equivalent, the average heat-transfer coefficient around the circumference. R. Hilpert[1] made accurate

---

[1] R. Hilpert, *Forsch. Gebiete Ingenieurw.*, **4**:215 (1933).

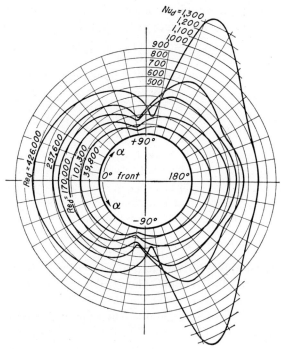

Fig. 9-4. Local film heat-transfer coefficients around a cylinder in flow of air normal to its axis. [*From E. Schmidt and K. Wenner, Forsch. Gebiete Ingenieurw.*, **12**:65–73 (1933).]

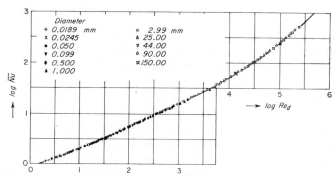

Fig. 9-5. Average film heat-transfer coefficient on a cylinder in flow of air normal to its axis. [*From R. Hilpert, Forsch. Gebiete Ingenieurw.*, **4**:215 (1933).]

measurements of this average value for air flow. Figure 9-5 shows the result. The Nusselt and the Reynolds numbers are calculated with the tube diameter as the reference length and with the freestream velocity as the reference velocity. From Fig. 9-5 and from the results of other experiments, especially at low Reynolds numbers, it can be seen that in various ranges of the Reynolds number, the Nusselt number can be presented in the form[1]

$$\text{Nu}_d = 0.43 + C(\text{Re}_d)^m \tag{9-28}$$

The numerical values for the constant $C$ and the exponents $m$ are tabulated in Table 9-1.

TABLE 9-1. COEFFICIENTS FOR CALCULATION OF HEAT TRANSFER ON A CIRCULAR CYLINDER TO AIR FLOW NORMAL TO ITS AXIS, BY EQ. (9-28)*

| $\text{Re}_d$ | $C$ | $m$ |
|---|---|---|
| 1–4,000 | 0.48 | 0.5 |
| 4,000–40,000 | 0.174 | 0.618 |
| 40,000–400,000 | 0.0239 | 0.805 |

* R. Hilpert, *Forsch. Gebiete Ingenieurw.*, **4**:215 (1933).

H. Y. Douglas and S. W. Churchill[2] in a recent paper came to the conclusion that all available, reliable information, including experiments at gas temperatures from 60 to 1800 F and wall temperatures from 70 to 1915 F, correlated best on a single line when the heat conductivity in the Nusselt number and the kinematic viscosity in the Reynolds number are introduced at a reference temperature equal to the arithmetic mean of upstream fluid temperature and wall temperature.

A high turbulence level in the approaching stream increases not only the average heat-transfer coefficient but also the local heat transfer on the upstream part of the cylinder circumference which is covered by a laminar boundary layer.[3] Increases in $\text{Nu}_d$ up to 25 per cent have been measured for turbulence levels up to 7 per cent.

Heat transfer from the front side of the tube can be calculated with the heat-flow equation of the boundary layer. Figure 7-11 shows the comparison between the calculated and the measured values. The ratio of heat flow to loss of momentum (friction loss) in the boundary layer on the front side of the tube corresponds approximately to Eq. (8-7), although in the accelerated flow along the surface the assumptions necessary for this equation are not strictly fulfilled. For small Reynolds numbers the

---

[1] E. R. G. Eckert and E. Soehngen, *Trans. ASME*, **74**: 343–347 (1952). See also F. M. Sauer and R. M. Drake, *J. Aeronaut. Sci.*, **20**:175–180 (1953).

[2] W. Y. M. Douglas and S. W. Churchill, "Recorrelation of data for convective heat transfer between gases and single cylinders with large temperature differences," *Preprint* 16, American Institute of Chemical Engineers, Heat Transfer Symposium, National Meeting, Louisville, Ky., 1955.

[3] W. H. Giedt, *J. Aeronaut. Sci.*, **18**:725–731 (1951). E. W. Comings, I. T. Clapp, and T. F. Taylor, *Ind. Eng. Chem.*, **40**:1076 (1948).

whole resistance of the tube is essentially friction resistance. Equation (8-7) then gives approximately the ratio of the total heat flow $Q$ to the total resistance $R$. For Reynolds numbers higher than 1,000, a wake arises in the flow and the form resistance caused by this wake exceeds the friction loss, as was mentioned in Sec. 6-8. The heat transfer on the rear half of the tube, however, is only approximately the same as on the front half. Therefore it follows that bodies which cause wakes in a flow are inferior to those bodies which do not with respect to the ratio of the heat flow to the resistance. For a quantitative comparison the dimensionless value $\Omega = (W/Q)[c_p(t_m - t_w)/u_m^2]$ is very suitable (see page 204). For boundary-layer flow of a fluid with $Pr = 1$, this value is 1 according to Eq. (8-8). For a tube in crossflow at Reynolds numbers from 2,000 to 40,000, the value of $\Omega$ is 10 to 40. With respect to heat transfer *per unit area*, surfaces exposed to wakes are approximately as effective as those covered by a boundary layer (Figs. 9-3 and 9-4). In this respect a tube with streamline-shaped cross section has no advantage over a tube with circular cross section.

By experiments with different liquids (water, oils) the dependence of the heat transfer on the Prandtl number was determined. It is given by the relation

$$\mathrm{Nu}_d = 0.43 + K(\mathrm{Re}_d)^m (\mathrm{Pr})^{0.31} \qquad (9\text{-}29)$$

With the values in Table 9-1, the heat transfer in any liquid can be calculated from Eq. (9-29) using the relation $K = 1.11C$. In Table 9-2 the constant $C$ and the exponent $m$ of Eq. (9-28) are presented for cylinders with different cross sections, according to R. Hilpert. The Nusselt and Reynolds numbers are based on the diameter of a circular tube with equal surface.

TABLE 9-2. COEFFICIENTS FOR CALCULATION OF HEAT TRANSFER FROM CYLINDERS WITH DIFFERENT CROSS SECTIONS TO AN AIR FLOW NORMAL TO THEIR AXES BY EQ. (9-28)*

| Cross section | $\mathrm{Re}_d$ | $C$ | $m$ |
|---|---|---|---|
| → □ | 5,000–100,000 | 0.0921 | 0.675 |
| → ◇ | 5,000–100,000 | 0.222 | 0.588 |
| → ○ | 5,000–100,000 | 0.138 | 0.638 |
| → ○ | 5,000– 19,500 | 0.144 | 0.638 |
| → ⬠ | 19,500–100,000 | 0.0347 | 0.782 |

* R. Hilpert, *Forsch. Gebiete Ingenieurw.*, **4**:215 (1933).

Heat transfer of tube bundles with the flow normal to the tube axis depends on the configuration of the tubes. The available test results in a Reynolds-number range from 2,000 to 40,000 were summarized by E. D.

Grimison.[1] There are two possible arrangements: The tubes may be either aligned or staggered in flow direction (Fig. 9-6). The heat-transfer coefficients can be presented again in the form of Eq. (9-28), in which the first term can be neglected for the Reynolds-number range considered. As a reference velocity, the average velocity in the narrowest cross section between the tubes is used. The values of the constant $C$ and the exponent $m$ are tabulated in Table 9-3. For ratios $a = s_1/d$ and $b = s_2/d$ greater than the values in the table, the equation

$$\mathrm{Nu}_d = 0.297(\mathrm{Re}_d)^{0.602} \quad (9\text{-}30)$$

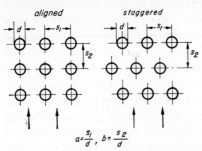

FIG. 9-6. Tube bundles with tubes aligned and staggered.

given by R. Benke can be used. This equation and Table 9-3 are for

TABLE 9-3. COEFFICIENTS FOR CALCULATION OF HEAT TRANSFER FROM A TUBE BUNDLE TO AN AIR FLOW NORMAL TO ITS AXIS BY EQ. (9-28)*
$\mathrm{Re}_d = 2{,}000\text{--}40{,}000 \qquad a = s_1/d \qquad b = s_2/d$ (Fig. 9-6)

| $b$ \ $a$ | 1.25 | | 1.5 | | 2 | | 3 | |
|---|---|---|---|---|---|---|---|---|
| | $C$ | $m$ | $C$ | $m$ | $C$ | $m$ | $C$ | $m$ |
| Aligned tubes | | | | | | | | |
| 1.25 | 0.348 | 0.592 | 0.275 | 0.608 | 0.100 | 0.704 | 0.0633 | 0.752 |
| 1.5  | 0.367 | 0.586 | 0.250 | 0.620 | 0.101 | 0.702 | 0.0678 | 0.744 |
| 2    | 0.418 | 0.570 | 0.299 | 0.602 | 0.229 | 0.632 | 0.198  | 0.648 |
| 3    | 0.290 | 0.601 | 0.357 | 0.584 | 0.374 | 0.581 | 0.286  | 0.608 |
| Staggered tubes | | | | | | | | |
| 0.6   |       |       |       |       |       |       | 0.213 | 0.636 |
| 0.9   |       |       |       |       | 0.446 | 0.571 | 0.401 | 0.581 |
| 1     |       |       | 0.497 | 0.558 |       |       |       |       |
| 1.125 |       |       |       |       | 0.478 | 0.565 | 0.518 | 0.560 |
| 1.25  | 0.518 | 0.556 | 0.505 | 0.554 | 0.519 | 0.556 | 0.522 | 0.562 |
| 1.5   | 0.451 | 0.568 | 0.460 | 0.562 | 0.452 | 0.568 | 0.488 | 0.568 |
| 2     | 0.404 | 0.572 | 0.416 | 0.568 | 0.482 | 0.556 | 0.449 | 0.570 |
| 3     | 0.310 | 0.592 | 0.356 | 0.580 | 0.440 | 0.562 | 0.421 | 0.574 |

* E. D. Grimison, *Trans. ASME*, **59**:583–594 (1937).

[1] E. D. Grimison, *Trans. ASME*, **59**:583–594 (1937).

heat transfer in air. For other liquids, the constants are to be determined in the same way as for single tubes. It is recommended that property values be introduced at the temperature given by Eq. (8-16). The equations hold true for tube banks with 10 or more rows. For a smaller number of rows the average heat-transfer coefficient of the bundle is smaller; as an example, for four rows the heat-transfer coefficient is smaller by 12 per cent. The first row has approximately the same heat transfer as a single tube.

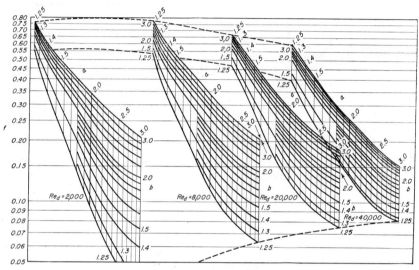

FIG. 9-7. Resistance coefficient for tube bundles—tubes aligned. [*According to E. D Grimison, Trans. ASME,* **59**: 583–594 (1937).]

The pressure loss $\Delta p$ in the flow through the bank is given by the equation

$$\Delta p = nf\rho \frac{u^2}{2} \qquad (9\text{-}31)$$

where $n$ = number of rows in flow direction
$\rho$ = density of the fluid
$u$ = average velocity between tubes

The resistance coefficient $f$ can be taken from Figs. 9-7 and 9-8 according to Grimison. The property values should again be introduced at the temperature given by Eq. (8-16). The values in Figs. 9-7 and 9-8 hold true for 10 and more rows. Banks with a smaller number of rows have greater resistance coefficients. For example, a bank with four rows has a value greater by 8 per cent. For rough surfaces the heat transfer and the resistance are enlarged. With very rough surfaces increases up to 20 per cent have been measured.

Much experimental information has recently been collected on compact heat-exchanger surfaces which have large heat-transfer areas per unit volume, attaining this by use of passages with small dimensions and by liberal application of fins and other extended surfaces. The available information can be found in the book by W. M. Kays and A. L. London.[1] The book also contains valuable information on friction and heat transfer connected with flow through single tubes.

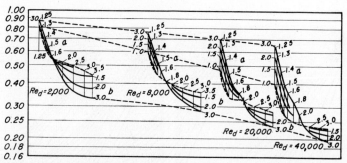

Fig. 9-8. Resistance coefficient for tube bundles—tubes staggered. [*According to E. D. Grimison, Trans. ASME*, **59**:583–594 (1937).]

Little information exists on tube bundles with flow in the direction of the tube axes. Tentatively, they can be calculated with Eq. (8-14). The hydraulic diameter which should be introduced here is

$$d_h = d\left(\frac{4ab}{\pi} - 1\right)$$

If different configurations of tube banks are to be compared with respect to the power necessary to cause the air flow through the tube banks, the dimensionless value $\Omega$ can be used as a basis for coolers on vehicles (airplanes, motorcars, locomotives, etc.). For such coolers a small frontal area is most important, as the drag of the cooler is determined primarily by this area. If such coolers with different configurations of the tubes are to have the same frontal area, the average velocity $u_m$ between the tubes used in the expression $\Omega = (W/Q)[c_p(t_m - t_w)/u_m^2]$ is fixed. Besides, a certain heat flow $Q$ is to be transferred at a given temperature difference $(t_m - t_w)$. Therefore, the power necessary per unit heat flow is proportional to the value $\Omega$. For banks of circular tubes in a gas flow normal to the tube axes this value lies between 1.5 and 10. Smaller tube spacings result in smaller values. Tubes in a gas flow parallel to their axes and tubes with streamlined cross section in crossflow have a value $\Omega \approx 1$. They are therefore better for such coolers.

[1] "Compact Heat Exchangers," National Press, Palo Alto, Calif., 1956.

For heat exchangers in stationary plants the magnitude of the heating surface is most important, since the weight and price depend primarily upon it. Now the question is which configuration of the tubes transfers a certain heat flow with a given pressure loss and with the smallest heating surface. In this problem each configuration demands a different velocity, depending on the heat-transfer coefficient. A calculation[1] shows that tubes in crossflow are better than in flow parallel to the axis. The latter have a pressure loss 3 to 15 times greater when the heating surface area is the same. By spacing a given number of tubes closer together in the

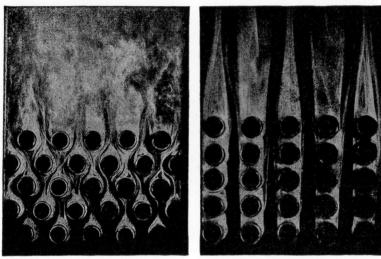

FIG. 9-9      FIG. 9-10

FIGS. 9-9 and 9-10. Visualization of the flow through tube banks. [*According to H. Thoma, photograph from W. Lorisch, Mitt. Forschungsarb., 322 (1929).*]

rows, thus decreasing the number of tube rows required, a given quantity of heat can be transferred with a reduction in the over-all pressure loss through the tube bundle. Therefore, it is advantageous to make the distance between the tubes in such a heat exchanger as small as manufacturing and maintenance allow. There is almost no difference in pressure loss whether the tubes are aligned or staggered. Only at small Reynolds numbers are staggered tubes somewhat better.

Figures 9-9 and 9-10 show photographs of the flow through two tube banks; they were made according to a proposal of H. Thoma.[2] The porous surfaces of the tubes were wetted with hydrochloric acid, and ammonia vapor was mixed with the air. Where this vapor mixes with the acid

---

[1] E. Eckert, *Forsch. Gebiete Ingenieurw.*, **16**:133–140 (1949–1950).

[2] H. Thoma, "Hochleistungskessel," Springer-Verlag, Berlin, 1921; also W. Lorisch, *Mitt. Forschungsarb.*, 1929, p. 322.

vapor, a white fog of sal ammoniac arises. In this way the boundary layers and the wakes can be seen in the photographs. These are also the regions within which the air stream is cooled or heated from the tubes.

**Example 9-1.** A tube tank of four rows in a steam boiler is traversed by combustion gases with a temperature of 1100 F and a velocity of 24 fps. The steam pressure is 1,420 $lb_f/in.^2$. The tubes have a diameter of 2.4 in. and are arranged on the corners of squares with sides 4.8 in. long. The heat-transfer coefficient and the pressure loss are to be calculated.

The saturation temperature of water vapor at 1,420 $lb_f/in.^2$ is 589 F. Because of the large heat-transfer coefficient on the water side and the high thermal conductivity of the tube wall, the temperature on the outer surface of the tube is only slightly greater than the steam temperature. It can be assumed to be 600 F. The gas property values are introduced at the reference temperature $t^*$ given by Eq. (8-16); that is,

$$t^* = \frac{600 + 1{,}100}{2} = 850 \text{ F}$$

where the factor $(0.1 \text{ Pr} + 40)/(\text{Pr} + 72)$ has been taken as $\tfrac{1}{2}$ for simplicity. Since the property values of combustion gases deviate very little from the values for air, the calculations will be made with the latter. From the Appendix,

$$\nu = 75.9 \times 10^{-5} \text{ ft}^2/\text{sec} \qquad k = 0.0311 \text{ Btu/hr ft F}$$

This gives $\text{Re}_d = 24 \times (2.4/12)/75.9 \times 10^{-5} = 6{,}324$. The ratios $a$ and $b$ are $a = b = 2$. Therefore from Table 9-3, $C = 0.229$, $m = 0.632$, and Eq. (9-28) gives $\text{Nu}_d = 0.43 + 0.229 \times 6{,}324^{0.632} = 58.2$. The heat-transfer coefficient becomes

$$h = \frac{k}{d} \text{Nu}_d = \frac{0.0311 \times 12}{2.4} \, 58.2 = 9.05 \text{ Btu/hr ft}^2 \text{ F}$$

This value holds true for 10 and more tube rows. For four rows the heat-transfer coefficient is smaller by 12 per cent. Therefore

$$h = 7.96 \text{ Btu/hr ft}^2 \text{ F}$$

The pressure loss follows from Eq. (9-31). The density is best introduced at the gas temperature: $\rho = 0.0254 \text{ lb/ft}^3$. The resistance coefficient is, from Fig. 9-7, $f = 0.225$. Therefore

$$\Delta p = 4 \times 0.225 \times 0.0254 \, \frac{24^2}{2 \times 32.2} = 0.204 \text{ lb}_f/\text{ft}^2$$

**9-3. Spheres and Packed Beds.** Heat transfer at the surface of spheres is determined by the flow conditions which have been discussed in Sec. 6-9. A laminar boundary layer covers the upstream portion of the surface, and this boundary layer separates from the side of the sphere, creating an irregularly fluctuating flow condition along the downstream portion. On the side of the sphere the boundary layer may become turbulent at large Reynolds numbers. This will influence the location of the flow separation. Near the forward stagnation point of a blunt object the stream velocity varies with the distance from the stagnation point according to the equation

$$u = Cx \qquad (9\text{-}32)$$

For potential flow around a sphere with radius $r$, the constant $C$ is $3u_0/2r$ ($u_0$ is the freestream velocity). The heat-transfer coefficient in this region can be calculated from the solution of the boundary-layer equations for rotationally symmetric flow. The result is an equation which holds for any rotationally symmetric blunt object

$$\mathrm{Nu}_x = f(\mathrm{Pr}) \sqrt{\mathrm{Re}_x} \tag{9-33}$$

with $\mathrm{Nu}_x = hx/k$ and $\mathrm{Re}_x = ux/\nu$ ($x$ distance from stagnation point measured along surface, $u$ = local velocity).[1] The function of Pr can be approximated for $0.5 < \mathrm{Pr} < 5$ by $f(\mathrm{Pr}) = 0.763\,\mathrm{Pr}^{0.4}$. Near a stagnation point, according to Eqs. (9-32) and (9-33), the heat transfer is independent of location. For larger distances from the stagnation point, an exact solution of the differential equations describing flow and heat transfer in the boundary layer requires a tedious process. Approximate solutions of the laminar boundary-layer equations can be obtained by methods very similar to the ones discussed for two-dimensional flow around cylindrical objects.[2] The integrated momentum and energy equations for two-dimensional flow apply also to rotationally symmetric flow, with the exception that the first term on the left-hand side of Eq. (6-13) describing the momentum flow through the boundary layer changes to

$$\rho \frac{1}{r} \frac{d}{dx} (u_s{}^2 r \delta_i)$$

and the first term in the energy equation (7-2) changes to $(1/r)(d/dx) r\int u(t_s - t)\,dy$. The development of the flow boundary layer and the temperature boundary layer can be calculated with these equations for any surface for which the distance $r$ of the surface from the axis of rotation is known as a function of $x$ ($x$ is the distance from the stagnation point measured along the surface). Information on rotationally symmetric flow can also be obtained by Mangler's transformation. This transformation of the momentum equation was discussed in a previous paragraph. The same transformation can also be carried out on the energy equation of the boundary layer and indicates that a rotationally symmetric flow can be found for each two-dimensional boundary layer such that the temperature fields for both situations are identical provided the boundary conditions on this field are the same for both flow configurations.

**Example 9-2.** Mangler's transformation will be applied to a calculation of heat transfer for laminar supersonic flow around a cone. The theory of inviscid flow around a cone shows that for rotationally symmetric flow with supersonic velocities the pressure is constant along the cone surface when the Mach number is sufficient

---

[1] M. Sibulkin, *J. Aeronaut. Sci.*, **19**:570–71 (1952).
[2] R. M. Drake, Jr., *J. Aeronaut. Sci.*, **20**:309–316 (1953).

to create an attached shock. Therefore, if this rotationally symmetric flow is compared with a two-dimensional flow along a flat plate for which the pressure is also constant along the surface, then the requirement of the same pressure distribution along both surfaces is immediately satisfied. The radius $r$ for the cone with an opening angle $\alpha$ is

$$r = x \sin \alpha$$

Introduction of this value into Eq. (6-61) results in

$$\bar{x} = \frac{1}{C^2} \int_0^x r^2 \, dx = \frac{\sin^2 \alpha}{C^2} \frac{x^3}{3} = \frac{r^2}{C^2} \frac{x}{3} \qquad \bar{y} = \frac{r}{C} y$$

For flow of a fluid with constant properties in a laminar boundary layer along a flat plate, the heat-transfer coefficient can be expressed by the following relationship [Eq. (7-14)]:

$$\frac{\mathrm{Nu}_{\bar{x}}}{\sqrt{\mathrm{Re}_{\bar{x}}}} = 0.332 \sqrt[3]{\mathrm{Pr}}$$

In the dimensioned parameters this equation reads

$$\frac{h\bar{x}}{k} \Big/ \sqrt{\frac{u_s \bar{x}}{\nu}} = 0.332 \sqrt[3]{\mathrm{Pr}}$$

The heat-transfer coefficient is connected with the temperature field within the boundary layer by the following equation:

$$h = \frac{q}{t_w - t_s} = \frac{k}{t_w - t_s} \left(\frac{\partial t}{\partial y}\right)_w$$

The last two equations can now be combined, and the resulting expression be transformed to the situation on a cone by replacing the variables $\bar{x}$ and $\bar{y}$ by the values $x$ and $y$.

$$\frac{\bar{x}}{t_w - t_s} \left(\frac{\partial t}{\partial \bar{y}}\right)_w \Big/ \sqrt{\frac{u_s \bar{x}}{\nu}} = 0.332 \sqrt[3]{\mathrm{Pr}} = \frac{x}{t_w - t_s} \frac{1}{\sqrt{3}} \left(\frac{\partial t}{\partial y}\right)_w \Big/ \sqrt{\frac{u_s x}{\nu}}$$

Returning again to the dimensionless parameters, Nusselt and Reynolds numbers give, therefore, the following solution describing heat transfer on the surface of a cone in laminar, rotationally symmetric flow:

$$\frac{\mathrm{Nu}_x}{\sqrt{\mathrm{Re}_x}} = \sqrt{3} \times 0.332 \sqrt[3]{\mathrm{Pr}} = 0.576 \sqrt[3]{\mathrm{Pr}}$$

The average heat-transfer coefficient for a sphere cannot be obtained by calculation, since we are not yet able to calculate the heat transfer in the separated flow region on the downstream part of the sphere. The following relation is proposed by U. Grigull[1] from experiments:

$$\mathrm{Nu}_d = 0.37 (\mathrm{Re}_d)^{0.6} (\mathrm{Pr})^{1/3} \tag{9-34}$$

for a range $20 < \mathrm{Re} < 150{,}000$. For smaller Reynolds numbers the relation

$$\mathrm{Nu}_d = 2 + 0.37 (\mathrm{Re}_d)^{0.6} (\mathrm{Pr})^{1/3} \tag{9-35}$$

[1] H. Gröber, S. Erk, and U. Grigull, "Grundgesetze der Wärmeübertragung," Springer-Verlag, Berlin, 1955.

should describe actual conditions well. For Re → 0 this equation goes toward $Nu_d = 2$, which is the value for heat transfer by pure conduction [see Eq. (3-10) for $r_0 = \infty$].

The flow and heat transfer through packed beds are of considerable importance in engineering applications. Storage-type heat exchangers, air-conditioning equipment, and devices in chemical industry use such arrangements. Flow through sintered and porous materials belongs in the same category. One basic difficulty for a description of pressure drop and heat transfer in such beds is the large variety and often the poor definition of the geometries which are involved. In this paragraph only one well-defined arrangement will be discussed, namely, a bed consisting of spherical particles with uniform size.[1] It was found that the relationships describing conditions in such a bed can be utilized with sufficient accuracy for beds built of more irregularly shaped particles as long as their shape does not deviate too much from that of a sphere. In this case it is recommended to use, instead of the particle diameter, the expression

$$d_p = \frac{6}{S_p} \tag{9-36}$$

in which $S_p$ denotes the surface area per unit volume. An important parameter in the description of a packed bed is the porosity $\epsilon$, defined as the ratio of the volume of the voids to the over-all volume. When a fluid is directed upward through a packed bed, it is found that at a certain velocity the fluid begins to carry the particles so that they start moving apart. The bed is now in a "fluidized state." In such a state the average porosity is considerably greater than under the condition of low-flow velocities. Fluidization often does not produce a uniform density. For instance, when the particles are small, then the air tends to create diverse channels. In beds consisting of particles with nonuniform size, often a separation of the particles takes place, with the smaller particles gradually locating themselves near the upper end of the layer. The particles in a fluidized bed are in continuous movement, and a vigorous mixing of the particles takes place. Relationships for pressure drop and heat transfer in a fluidized bed depend on the particular state it is in and are therefore extremely difficult to describe. Much research has still to be done in this connection.

Ergun[2] made a recent survey of the investigations on the pressure drop $\Delta p$ in a nonfluidized packed bed of height $L$ consisting of spherical par-

---

[1] Measurements on wire screen matrices have been reported by J. E. Coppage and A. L. London, *Chem. Eng. Progr.*, **52**:57–63 (1956).

[2] Sabri Ergun, *Chem. Eng. Progr.*, **48**:227 (1952).

ticles with diameter $d_p$ and arrived at the following relation:

$$\frac{\Delta p}{L} \frac{d_p}{\rho V_0^2} \frac{\epsilon^3}{1-\epsilon} = 150 \frac{1-\epsilon}{\mathrm{Re}_p} + 1.75 \qquad (9\text{-}37)$$

from experiments which covered a range of the porosity $\epsilon$ between 0.40 and 0.65. In the above equation, $\rho$ is the density of the fluid and $V_0$ is the velocity which the fluid would have if no particles would be present

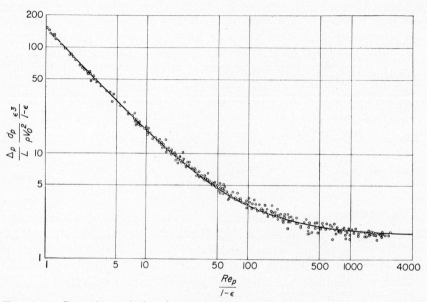

Fig. 9-11. General correlation for pressure drop in packed beds. [*According to Sabri Ergun, Chem. Eng. Progr.*, **48**: 227 (1952).]

(the approach velocity). The Reynolds number is based on this approach velocity $V_0$ and the particle diameter

$$\mathrm{Re}_p = \frac{V_0 d_p}{\nu}$$

Equation (9-37) can be transformed into the form

$$\Delta p = a\mu V_0 + b\rho V_0^2$$

in which the parameters $a$ and $b$ depend only on the geometry of the bed. This equation is often interpreted as indicating that the total resistance consists of a viscous part (first term) and a form resistance (second term). Figure 9-11 indicates the correlation which was obtained by this relationship.

The arrangement of the particles and the porosity of the bed are usually different near the side walls from in the interior of the bed. Accordingly, the above relationship holds only as long as the diameter of the whole bed is larger than at least ten times the particle diameter. For smaller beds the resistance is found to be larger because of friction of the fluid at the wall.

In the fluidized state the pressure drop must be equal to the weight of the bed; therefore, the beginning of fluidization can be calculated in the following way:

$$\frac{\Delta p}{L} = g\rho_p$$

The particle density $\rho_p$ is defined as the density of the bed in the packed state before fluidization occurs. Equation (9-37) for the pressure drop can then be used to calculate the approach velocity $V_0$ for beginning fluidization. The expansion of the bed by fluidization ranges between 20 to 50 per cent.

Heat transfer in randomly packed beds of spheres is quite difficult to measure and therefore not very well known. A relationship which appears to describe heat-transfer conditions reasonably well has been obtained by W. H. Denton.[1]

$$\mathrm{Nu}_p = 0.80(\mathrm{Re}_p)^{0.7}(\mathrm{Pr})^{1/3}$$

This relationship holds in a range of Reynolds numbers between approximately 500 and 50,000 and for a porosity $\epsilon = 0.370$. The parameters in the equation are based on sphere diameter. A considerable spread of data obtained by different investigators must be attributed to unknown effects and especially to complicated geometry. The situation is still more uncertain in a bed in the fluidized state. Generally lower heat-transfer coefficients have been found in this state.

In applications another type of heat transfer is often important, namely, the heat exchanged from the fluid and bed to the walls of the vessel containing the bed. This type will not be treated here because of space limitation.

## PROBLEMS

**9-1.** Carry out the dimensional analysis for flow and heat transfer of oil through a tube or duct assuming that all properties can be considered as constants except the viscosity. It has been found that the dependence of the viscosity of most oils on temperature can be well approximated by the relation $\mu = C/(T - T_0)^4$, where $T_0$ is a characteristic temperature which has to be established for each oil experimentally. Assume this relation for the viscosity. Establish the dimensionless parameters on which Nusselt number depends.

[1] W. H. Denton, "General Discussion on Heat Transfer," p. 370, Institution of Mechanical Engineers and American Society of Mechanical Engineers, London, 1951. The Prandtl-number relation was incorporated by the first author of this book.

**9-2.** Heat-transfer coefficients to single tubes in a bundle are often measured in the following way: The bundle is fabricated out of unheated tubes, only one tube in the arrangement is heated, and the heat flow through its surface as well as the surface temperature and the gas temperature are measured. From that information, the mean heat transfer-coefficient at the surface of this tube is calculated. Discuss whether or not you expect this heat-transfer coefficient to be the same as for the situation where the tube is surrounded in the bundle by tubes which are heated to the same surface temperature. If you expect differences, explain why and under which conditions the differences should vanish.

CHAPTER 10

# SPECIAL HEAT-TRANSFER PROCESSES

**10-1. Heat Transfer at High Velocities.** In a fluid which moves with sufficiently high velocity, the energy which is converted by internal friction into heat causes considerable temperature increases. Applications in which this temperature increase is important are, for instance, bearings in which the temperature in the oil film is dictated by the internal heat generation or high-speed aircraft where the heat transfer to the surfaces occurs at high subsonic or supersonic velocities. Two flow geometries in which heat transfer is influenced by internal friction will be considered in this section. The first one is chosen because the calculation is especially simple and illustrative; the second one because of its importance for engineering applications. In both cases, it will be assumed that the properties of the fluid can be considered constant.

*Couette Flow.* Consider two parallel flat plates, one of which is at rest whereas the other one moves with a velocity $u_1$. A coordinate system is chosen as indicated in Fig. 10-1. The space between the two plates is taken up by a fluid, and we consider flow in sufficient distance from the entrance so that it has become fully developed. This means that no flow parameter depends upon the $x$ direction parallel to the plate surfaces. The moving plate may be cooled to a constant temperature $t_1$, whereas plate 0 may be adiabatic to heat flow. If the fluid has viscosity, then a force is required to move plate 1 and correspondingly, a shear stress $\tau = \mu\, du/dy$ is set up within the fluid. From a simple force balance it follows that the shear stress $\tau$ in any plane parallel to the plate has the same value. According to the equation $du/dy = \tau/\mu =$ const, this means that the velocity profile $u$ is linear as indicated in the figure.

Fig. 10-1. Sketch of Couette flow.

Considering now the energy within the fluid, we see that work is transferred continuously down through the fluid from plate 1 toward plate 0. The amount of work transferred through a plane at distance $y$ per unit area and time is $u\tau = u\mu(du/dy)$. If we consider the space between two planes distant from each other by the amount $dy$, then the energy stored in this layer becomes

$$\frac{d}{dy}\left(u\mu\frac{du}{dy}\right)dy = \mu\left(\frac{du}{dy}\right)^2 dy$$

The right-hand expression is obtained, since $d^2u/dy^2 = 0$. This energy is converted into heat. From a thermal standpoint the internal friction is therefore equivalent to heat sources distributed throughout the fluid. In the present situation these heat sources are locally constant and have the strength

$$\phi = \mu\left(\frac{du}{dy}\right)^2 = \mu\frac{u_1^2}{b^2}$$

per unit volume. Under steady-state conditions this heat must be removed from the layer by conduction. No heat transport by convection will occur because the temperature is assumed to be constant in flow direction. We have, therefore, exactly the same situation as in a plane wall with constant internal heat sources. Correspondingly, the solution for this case, which has been treated in Sec. 3-6, can be immediately transferred to our situation, and the temperature throughout the fluid is given by Eq. (3-54).

$$t = -\frac{\mu}{2k}\left(\frac{du}{dy}\right)^2 y^2 + C_1 y + C_2 = -\frac{\mu u_1^2}{2kb^2} y^2 + C_1 y + C_2 \quad (10\text{-}1)$$

The boundary conditions for our temperature profile are

$$t = t_1 \quad \text{at } y = b$$
$$\frac{dt}{dy} = 0 \quad \text{at } y = 0$$

With these, the temperature profile can be expressed as

$$t - t_1 = \frac{\mu}{k}\frac{u_1^2}{2}\frac{b^2 - y^2}{b^2} = (\text{Pr})\frac{u_1^2}{2c}\frac{(b^2 - y^2)}{b^2}$$

with $c$ indicating the specific heat of the fluid. We see that the temperature profile has the shape of a parabola with its vertex in the plane 0. Therefore this plane assumes by the internal friction process a temperature which is higher than the temperature of the cooled plate 1. The temperature which a surface assumes under the influence of internal friction is called *recovery temperature*. The difference between the recovery temperature $t_{r0}$ of plate 0 and the temperature of plate 1 becomes

$$t_{r0} - t_1 = (\text{Pr})\frac{u_1^2}{2c} \quad (10\text{-}2)$$

Often, this temperature difference is made dimensionless by dividing it by the parameter $u_1^2/2c$, and the resulting value is called *temperature*

SPECIAL HEAT-TRANSFER PROCESSES 257

*recovery factor r.* For Couette flow the recovery factor is therefore

$$r = \text{Pr} \tag{10-3}$$

A calculation for the situation in which surface 0 is kept at a temperature $t_1$ by cooling and the moving surface 1 is adiabatic to heat flow results in a temperature difference $t_{r0} - t_1$, which is now the recovery temperature of surface 1 minus $t_1$, of exactly the same magnitude as the one found in the preceding calculation. This geometry resembles quite closely the condition in a bearing in which the shaft constituting the moving surface is not cooled whereas the outside shell of the bearing which resembles surface 0 is kept at a certain temperature by cooling. Equation (10-2) can be used to estimate the temperature difference between the shaft and the outside shell of the bearing. Actually, conditions in a bearing are more complicated, especially because of the finite length of the bearing which causes heat conduction in the axial direction.

As a next step, we want to consider the situation in which both plates are kept at prescribed temperatures by some outside means. The boundary conditions for this situation are

$$t = t_0 \quad \text{at } y = 0$$
$$t = t_1 \quad \text{at } y = b$$

Calculating the constants in Eq. (10-1) from these boundary conditions results in

$$t - t_0 = -\frac{\mu}{k}\frac{u_1^2}{2}\frac{y^2}{b^2} + \frac{\mu}{k}\frac{u_1^2}{2}\frac{y}{b} + (t_1 - t_0)\frac{y}{b}$$

The heat flow into surface 0 is given by

$$q_0 = k\left(\frac{dt}{dy}\right)_0 = \mu\frac{u_1^2}{2}\frac{1}{b} + k\frac{t_1 - t_0}{b} = \frac{k}{b}\left[(\text{Pr})\frac{u_1^2}{2c} + t_1 - t_0\right]$$

This expression can be simplified by introduction of the recovery temperature. Combining the last equation with Eq. (10-2) results in

$$q_0 = k\frac{t_{r0} - t_0}{b} \tag{10-4}$$

This equation indicates a law which we shall find to apply quite generally in high-velocity flow. The heat flow at a surface bounding a high-velocity flow can be calculated with the equation valid for low velocities when the temperature difference determining the heat flow is introduced properly, namely, as the difference between the recovery temperature of the surface and its actual temperature. For Couette flow, the heat transfer at any surface is given by the normal heat-conduction equation through a slab [Eq. (1-6)] with the provision that instead of the difference between the two surface temperatures $t_{w1}$ and $t_{w2}$, the mentioned difference between the recovery temperature and actual surface temperature of the plate under consideration is used.

*Laminar and Turbulent Boundary Layer on a Flat Plate.* The energy equation which describes heat transfer in a laminar, steady boundary layer, including the effect of internal friction, is according to Eq. (7-5),

$$u\frac{\partial t}{\partial x} + v\frac{\partial t}{\partial y} = \alpha\frac{\partial^2 t}{\partial y^2} + \frac{\mu}{\rho c}\left(\frac{\partial u}{\partial y}\right)^2 \qquad (10\text{-}5)$$

Again the situation will be considered first in which the surface is adiabatic. The boundary condition, therefore, is

$$\frac{dt}{dy} = 0 \quad \text{at } y = 0$$
$$t = t_s \quad \text{at } y = \infty$$

The velocity field is identical with the one for low flow velocities as long as the properties are considered constant. A solution to the problem to determine the temperature field in the boundary layer was obtained for the first time by E. Pohlhausen.[1] Again, a transformation to a total differential equation is possible by introduction of the parameters $f$ and $\eta$ as used in Sec. 6-6 and by use of the following parameter:

$$\vartheta_r = \frac{t - t_s}{u_s^2/2c}$$

which expresses the temperature field within the boundary layer in a dimensionless form. Introduction of these variables into the above equation results in

$$\frac{d^2\vartheta_r}{d\eta^2} + (\text{Pr})f\frac{d\vartheta_r}{d\eta} + \frac{(\text{Pr})}{2}\left(\frac{d^2f}{d\eta^2}\right)^2 = 0 \qquad (10\text{-}5a)$$

This is a second-order inhomogeneous linear differential equation for the temperature parameter $\vartheta_r$. The independent variable is $\eta$, and the parameter $f$ has to be considered as a known function of $\eta$ determined by Blasius's solution of the flow equation. A solution of this equation can be obtained, for instance, by the method of variation of coefficients and results in

$$\vartheta_r = \frac{\text{Pr}}{2}\int_0^\infty \varphi\, d\eta - \frac{\text{Pr}}{2}\int_0^\eta \varphi\, d\eta$$

with $\quad \varphi = \exp\left(-\text{Pr}\int_0^\eta f\, d\eta\right)\left[\int_0^\eta \left(\frac{d^2f}{d\eta^2}\right)^2 \exp\left(\text{Pr}\int_0^\eta f\, d\eta\right) d\eta\right] \qquad (10\text{-}6)$

The temperature which the plate surface assumes is again its recovery temperature, and the value of the parameter $\vartheta_r$ at the wall is equal to

---

[1] E. Pohlhausen, *Z. angew. Math. Mech.*, **1**:115–121 (1921).

the recovery factor. The recovery factor, therefore, is

$$r = \frac{\Pr}{2} \int_0^\infty \exp\left(-\Pr \int_0^\eta f\, d\eta\right) \left[\int_0^\eta \left(\frac{d^2f}{d\eta^2}\right)^2 \exp\left(\Pr \int_0^\eta f\, d\eta\right) d\eta\right] d\eta$$

Values for the recovery factor have been obtained by numerical integration of this equation. A. Busemann has shown that these values can be approximated in the Prandtl-number range from 0.5 to 5 by the simple expression

$$r = \sqrt{\Pr} \tag{10-7}$$

Next, we want to consider heat transfer to the surface of the plate when its temperature is kept by cooling or heating at a value $t_w$ which is different from the recovery temperature. In order to find the solution for the temperature field, we can make use of the following consideration: The energy equation describing the temperature field within the boundary layer as given by Eq. (10-5a) is a linear equation in the temperature parameter $\vartheta_r$. Correspondingly, the general solution of this linear inhomogeneous equation can be composed of a particular solution of the inhomogeneous equation plus a general solution of the corresponding homogeneous equation. A particular solution of the inhomogeneous equation has been found in Eq. (10-6). The solution for the homogeneous equation has been found in Sec. 7-6. Therefore, the temperature field for the high-velocity boundary layer on a cooled plate can be written as

$$t = \vartheta_r \frac{u_s^2}{2c} + (t_w - t_r)(1 - \theta') + t_s$$

in which $\vartheta_r$ is given by Eq. (10-6) and $\theta'$ by Eq. (7-24). It is easily shown that this equation also satisfies the following boundary conditions:

$$t = t_w \quad \text{at } y = 0$$
$$t = t_s \quad \text{at } y = \infty$$

Of special importance for engineering calculations is the heat flow at the plate surface. This heat flow per unit area and time is given by

$$q = k \left(\frac{\partial \theta'}{\partial y}\right)_w (t_w - t_r)$$

since

$$\left(\frac{\partial \vartheta_r}{\partial y}\right)_w = 0$$

For low-velocity flow the corresponding heat flow can be expressed according to Sec. 7-6 as

$$q = k \left(\frac{\partial \theta'}{\partial y}\right)_w (t_w - t_s)$$

A comparison of the last two equations again indicates the following rule: The heat flow in a high-velocity boundary layer is given by the same relation as the heat flow in a low-velocity boundary layer except that the temperature potential determining the heat flow for high velocity is the difference between the actual wall temperature and its recovery temperature. In order to derive this conclusion, we had only to make use of the fact that the energy equation for the boundary layer is linear in the temperature. Therefore, the rule must apply quite generally for all fluids with constant properties. It holds also for turbulent flow, and as a result, all relations for heat transfer found for low-velocity flow can be immediately used to calculate the heat transfer under high-velocity conditions.[1] The only additional knowledge required is a knowledge of the recovery factor for the particular situation from which the recovery temperature can be determined. For laminar boundary-layer flow over a flat plate Eq. (10-7) describes the recovery factor. For turbulent flow the following relation has been derived theoretically and verified for Prandtl values near 1:

$$r = \sqrt[3]{\text{Pr}} \qquad (10\text{-}8)$$

It has been established that heat-transfer relations found for a constant-property fluid approximate very well conditions in high-velocity flow of gases as long as the pressure in the flow field is constant, provided the property values are introduced at an appropriately chosen reference temperature. This will be discussed in more detail in the next section. When the pressure varies, differences exist between a constant property fluid and a gas. One of the main differences is caused by the fact that gases expand as a consequence of a pressure drop and are compressed by a pressure increase and that temperature variations are connected with these processes. The effect of this process can be well illustrated by one-dimensional treatment of the following example:

Consider the flow through a tube with adiabatic walls. Thermodynamics shows that the energy transported with the flow through a cross section of the tube is the sum of internal ($u$) plus pressure $p/\rho$ plus kinetic $v^2/2$ energy and that in a tube with adiabatic walls this energy flux must be the same through any cross section:

$$u + \frac{p}{\rho} + \frac{v^2}{2} = \text{const}$$

For an incompressible fluid flowing in a tube of constant cross section, the velocity has to be constant along the tube and correspondingly, the sum of internal energy and pressure energy has to be constant:

$$u + \frac{p}{\rho} = \text{const}$$

[1] E. Eckert and O. Drewitz, *Forsch. Gebiete Ingenieurw.*, **11**:116–124 (1940).

SPECIAL HEAT-TRANSFER PROCESSES

The internal friction causes the pressure to decrease in the flow direction. Since the density is considered constant, the term $p/\rho$ in the above equation decreases in flow direction and the internal energy $u$ has to increase correspondingly. The internal energy can be written as the product of specific heat times temperature, and therefore, the temperature will increase in such a flow in the flow direction, indicating the fact that by internal friction, pressure energy has been converted into internal energy. For the flow of a gas, the situation is different. We can introduce the enthalpy $i$ for the sum of internal and pressure energy, and the above equation transforms into

$$i + \frac{v^2}{2} = \text{const} \tag{10-9}$$

Although the tube cross section is constant, the velocity will increase in flow direction because, connected with the drop in pressure, a decrease in density occurs. Therefore, in this type of flow, the enthalpy decreases in flow direction, and according to the relation $i = c_p t$, the temperature also decreases.

**10-2. Heat Transfer in Gases at High Velocities.** Heat transfer under conditions where the temperature increase due to internal friction is important is today of special interest in connection with the "aerodynamic heating problem" of high-velocity aircraft and missiles. At high supersonic speeds, the heat dissipation in the boundary layers which surround the skin of the vehicle creates extremely high temperatures. Values of order 10,000 F are mentioned in this connection. As a consequence, the proper choice of material and of a method for cooling the skin is one of the vital problems in the development of such vehicles. As a basis for such considerations, the heat transfer from the boundary layer into the surface of the vehicle must be known. Consequently, a large effort has been made in recent years to obtain this needed information by analysis and experiments.[1] For laminar boundary layers, present knowledge in this field rests almost completely on analytical results, whereas for turbulent boundary layers one has to rely heavily on experiments.[2]

An analysis of laminar heat transfer under the conditions expected in aerodynamic heating is made difficult by the fact that at the high flight velocities under consideration (Mach numbers up to 10 and more), very large temperature variations occur throughout the boundary layer. Accordingly, all properties, including the specific heat and Prandtl number, have to be considered variable. This not only increases the

[1] The first one to measure this effect was E. Brun, *Comptes rendus des séances de l'Academie des Sciences*, **194**:594 (1932).

[2] An analysis of turbulent heat transfer at high Mach numbers is contained in R. A. Deissler and A. L. Loeffler, *Natl. Advisory Comm. Aeronaut., Tech. Notes*, no. 4262 (1958).

number of parameters on which the problem depends but also makes the solution of the boundary-layer equations with the many unknowns extremely difficult. In addition, the air in the boundary layer will be dissociated in many cases and even ionized, and the question of how fast such processes approach equilibrium conditions arises. We do not have a satisfactory answer to this question as yet. Finally, when the vehicle flies at high altitude, the air density is often so small that studies on heat transfer have to be made on a molecular basis rather than by considering the air as a continuum. Some material on the problems mentioned above will be found in Sec. 9-1 on dimensional analysis and in Sec. 10-3 on heat transfer in gases at low densities. A more detailed discussion exceeds the scope of this book.[1]

In the following paragraphs, some basic considerations on high-velocity gas flows will be taken up, such as the distinction between static and total states and measurement of state parameters. Information on the recovery temperature which an unheated flat plate and cylinder assume in a high-velocity gas flow will be presented, as well as relations which allow the calculation of heat transfer for a flat plate in a flow of uniform velocity for laminar and turbulent boundary layers with an accuracy which is sufficient for design purposes.

The state at any point in a gas stream is fixed by the velocity and by two state parameters. The easiest to measure are the pressure and the temperature. It is useful to distinguish between two types of state parameters. The first is given by the pressure and temperature, which are measured with instruments moving along with the flowing gas. We call this state the *static state*, fixed by the static pressure $p_{st}$ and the static temperature. The importance of the static state stems from the fact that, except in extreme cases, to an observer that moves along with the flow, the gas in a small region behaves in the same way as a gas at rest and in equilibrium. This means that, for instance, the density in the flow can be calculated from the state equation using the static pressure and temperature and that viscosity, specific heat, and heat conductivity are also functions of the static pressure and temperature. The static pressure can be measured by a small hole in a wall parallel to the flow direction. The measurement of the static temperature is much more difficult. At present, no simple instrument for measuring this value is known.[2]

---

[1] More information on the subject and a compilation of literature can be found, for instance, in E. R. G. Eckert, Survey on Heat Transfer at High Speeds, *WADC Tech. Rept.* 54-70, Wright Air Development Center, Wright Patterson Air Force Base, Ohio, 1954; also E. R. G. Eckert, *Trans. ASME*, **78**:1273–1283 (1956).

[2] As an introduction to the general laws of high-velocity flow see, for example, H. W. Liepmann and A. Roshko, "Elements of Gasdynamics," John Wiley & Sons, Inc., New York, 1957.

SPECIAL HEAT-TRANSFER PROCESSES

The second state of a gas stream is the one which arises by isentropically decreasing the velocity to zero. We call this the *total state*. It is fixed by the total pressure $p_T$ and the total temperature $t_T$. The total pressure is indicated by a pitot tube at subsonic velocities. At supersonic velocities the pitot tube indicates a lower pressure, since a shock wave arises before the instrument and since the transformation of the kinetic energy to pressure in this shock wave is not isentropic. It is, however, possible to calculate the total pressure when the static pressure and the pressure indicated by the pitot tube are measured.

Instruments with which the total temperature can be measured are available and will be discussed soon. The importance of the total state lies partially in the fact that total pressure and total temperature can be measured readily. These measurements together with a measurement of the static pressure also indirectly determine the static temperature and fix the static state. For an ideal gas with constant specific heat the absolute static temperature $T_{st}$ can be calculated from the relation valid for an isentropic process $T_{st}/T_T = (p_{st}/p_T)^{(\gamma-1)/\gamma}$. For a gas with variable specific heat the determination of $T_{st}$ can readily be made with the help of available gas tables.[1]

A measurement of $p_T$, $p_{st}$, $T_T$ at the same time fixes the velocity $V$ at the point where the measurement has been made: The values $p_T$ and $T_T$ determine an enthalpy $i_T$, and the pair $p_{st}$ and $T_{st}$ determine an enthalpy $i_{st}$, which, for instance, can be obtained from gas tables. For an adiabatic deceleration to zero velocity, therefore also for an isentropic one, Eq. (10-9) can be written

$$i_T = i_{st} + \frac{V^2}{2} \tag{10-10}$$

From this equation the velocity $V$ can be calculated. When the difference between $T_T$ and $T_{st}$ is sufficiently small so that the variation of the specific heat can be neglected, then Eq. (10-10) can be written as

$$T_T - T_{st} = \frac{V^2}{2c_p} \tag{10-11}$$

Replacing the static temperature in this equation by the static pressure and solving for the velocity lead to the well-known formula by Barre de Saint-Venant and Wantzel:

$$V = \sqrt{2\frac{\gamma}{\gamma-1}RT_T\left[1 - \left(\frac{p_{st}}{p_T}\right)^{(\gamma-1)/\gamma}\right]}$$

with $\gamma$ denoting the ratio of the specific heat at constant pressure to that at constant volume and $R$ indicating the gas constant.

[1] J. H. Keenan and T. Kaye, "Gas Tables," John Wiley & Sons, Inc., New York, 1948.

The difference between total pressure and static pressure is called *dynamic pressure*. Analogously, the difference between total temperature and static temperature may be called *dynamic temperature*.

$$\vartheta_{dy} = T_T - T_{st}$$

For air at moderate temperatures the specific heat is

$$c_p = 0.24 \text{ Btu/lb F} = 187 \text{ ft lb}_f/\text{lb F}$$

and the dynamic temperature in degrees Fahrenheit becomes

$$\vartheta_{dy} = \frac{V^2}{12{,}000} \qquad (10\text{-}12)$$

if the velocity $V$ is introduced in feet per second. As Eq. (10-10) is derived from the first law of thermodynamics, it holds true not only for isentropic changes of the velocity but always when the velocity is diminished to zero without addition or subtraction of heat and without external work. In the boundary layer on a surface the velocity is diminished by friction forces and its kinetic energy is transformed to heat. This process is connected, however, with exchange of heat and work among the different gas layers even when the solid surface does not exchange heat with the gas. Therefore, the temperature of the gas layer immediately adjacent to the surface whose velocity is zero may be greater or smaller than the total temperature in the gas stream. The wall assumes the same temperature if it is not heated or cooled either by radiation or by heat flow through the surface into the interior of the wall. We call this temperature assumed by an unheated wall in a gas stream the *recovery temperature* $T_r$ and denote by $\vartheta_r$ the difference between this temperature and the static temperature $T_{st}$ in the gas stream. The recovery temperature is expressed nondimensionally by a recovery factor $r$ defined by the equation

$$T_r - T_{st} = r \frac{V^2}{2c_p} \qquad (10\text{-}13)$$

The factor $r$ was calculated by E. Pohlhausen[1] for a plate in longitudinal laminar flow of a fluid with constant properties. His results can be expressed by the equation $r = \sqrt{\text{Pr}}$ for Prandtl numbers from 0.5 to 10.

The calculations have been extended by a large number of papers to gases with variable properties. It was found that the relation $r = \sqrt{\text{Pr}}$ describes the results of these calculations with good accuracy as long as the temperature difference $T_r - T_{st}$ is such that the variation of specific heat in this temperature range can be neglected. For very large supersonic velocities where the variation of the specific heat becomes important, the following procedure[2] gives results which agree very well with

[1] Pohlhausen, *loc. cit.*
[2] E. R. G. Eckert, *loc. cit.*

the results of boundary-layer solutions. An enthalpy-recovery factor $r_i$ is defined by the equation

$$i_r - i_{st} = r_i \frac{V^2}{2} \qquad (10\text{-}14)$$

This enthalpy recovery is calculated from the equation $r_i = \sqrt{\text{Pr}}$. The Prandtl number is introduced into this relation at a reference enthalpy

$$i^* = i_s + 0.72(i_r - i_s)$$

No difference exists between the temperature and the enthalpy-recovery factor as long as the specific heat is considered constant. Recovery factors calculated in this way are also in good agreement with measured values. For laminar air flow at moderate temperatures the recovery factor is 0.84. In turbulent boundary-layer flow of air over a flat plate a value of 0.88 was measured. In the transition region between laminar and turbulent boundary layers the recovery factor rises from the value 0.84 to a peak and then decreases to the turbulent value 0.88 (Fig.

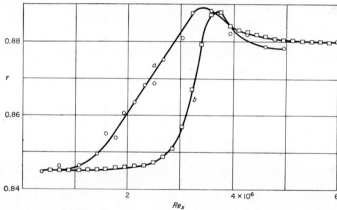

FIG. 10-2. Laminar, transitional, and turbulent temperature-recovery factors $r$ for air (measured on a cone at Ma = 3.12). (a) High stream turbulence. (b) Low stream turbulence (~1 per cent). (*From J. C. Evvard, H. Tucker, and W. C. Burgess, Inst. Aeronaut. Sci. Preprint* 438, 1954.)

10-2). Interesting are the results of measurements on a cylinder in a subsonic air flow normal to its axis.[1] The cylinder was manufactured of rubber in order to eliminate temperature equalization by conduction in the solid material. The measured surface temperatures were used to calculate local recovery factors using Eq. (10-13) and introducing the upstream static temperature and the upstream velocity. An example of the results is shown in Fig. 10-3. It is seen that the recovery value at the

[1] E. Eckert and W. Weise, *Forsch. Gebiete Ingenieurw.* **13**:246–254 (1942).

stagnation point is 1, which means that the surface temperature is equal to the total temperature in the stream. The low recovery values along the downstream part of the surface are remarkable. Over some range the recovery factor is even negative, which means that the surface temperature is lower than the upstream static temperature. Ryan[1] confirmed this observation, and Ackeret showed that it is intimately

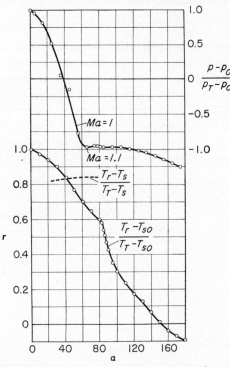

Fig. 10-3. Pressure distribution and temperature-recovery factor $r$ for high-velocity subsonic air flow normal to a circular cylinder. $\alpha$ is the angular distance from the stagnation line; $T_T$, $p_T$ are total temperature and pressure upstream; $T_{s0}$, $p_0$ are static temperature and pressure upstream; $p$, $T_s$ are static pressure and temperature outside the boundary layer; $T_r$ is the recovery temperature. [*From E. Eckert and W. Weise, Forsch. Ingenieurw.*, **13**:246 (1942).]

connected with the periodic shedding of vortices accompanying the separation of the flow. No low temperatures in separated regions were observed in supersonic flows.

Another type of flow which leads to regions with low total temperatures is vortex flow. Ranque utilized this effect to separate a gas stream into a portion with high energy content (high total temperature) and another

[1] L. F. Ryan, *Mitt. Inst. Aerodynamik Eidgen. Tech. Hochschule*, Zurich, no. 18, 1951, pp. 7–50.

one with low energy content. Figure 10-4 shows such a vortex tube as investigated by Hilsch. It is, therefore, sometimes called a Hilsch tube. It consists of a tube into which air (or another gas) is blown tangentially through one or more nozzles $a$. In this way a strong vortex flow is created in the tube. Part of the flow leaves the tube through the orifice $b$; the rest leaves at the far end of the tube $c$. A valve at this location serves to adjust the strength of the two air streams. It is found that the total temperature in the air leaving the tube through the orifice is considerably lower (as much as 100 F at 40 $lb_f/in^2$ inlet pressure) than the temperature of the air upstream of the inlet nozzle. The air leaving the

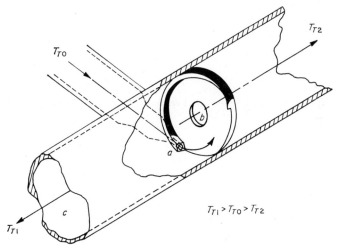

FIG. 10-4. Sketch of a vortex tube (Hilsch tube).

tube at $c$ is correspondingly higher. Sprenger[1] found that various flow types exist which lead to strong energy separation effects, and Vonnegut[2] developed a thermometer which measures the static temperature in a high-velocity gas stream. It consists essentially of a vortex tube with suitable opening and a thermometer arranged in the low-energy air at the tube axis.

For temperature measurements in high-velocity gas streams two instruments have proved suitable: the cylinder thermometer[3] (Fig. 10-5) and the diffuser thermometer[4] (Fig. 10-6). The cylinder thermometer measures the surface temperature on a cylinder in longitudinal flow. Since the boundary-layer thickness is small compared with the cylinder diam-

[1] H. Sprenger, *Z. angew. Math. Physik*, **2**:293 (1951); *Mitt. Inst. Aerodynamik, Eidgen, Tech. Hochschule, Zurich.*, no. 21, 1954.
[2] B. Vonnegut, *Rev. Sci. Instr.*, 1950, pp. 21–22, 136.
[3] E. Eckert, *Z. Ver. deut. Ingr.*, **84**:813–817 (1940).
[4] A. Franz, *Jahrb.* 1938 *deut. Luftfahrt-Forsch.*, **II**:215–218.

eter, the $r$ values for the flat plate (Fig. 10-2) can be used and the static temperature calculated from Eq. (10-13) when the velocity is known. The diffuser thermometer (Fig. 10-6) measures the total temperature.

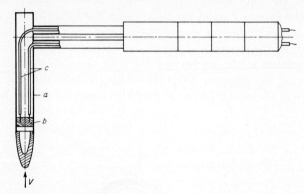

Fig. 10-5. Cylinder thermometer. (*a*) Thin-walled tube. (*b*) Copper bar. (*c*) Thermocouples. [*From E. Eckert, Z. Ver. deut. Ingr.*, **84**:813–817 (1940).]

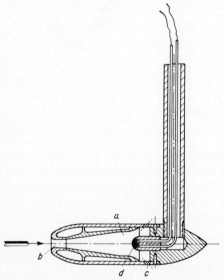

Fig. 10-6. Diffuser thermometer. (*From A. Franz, Jahrb. 1938 deut. Luftfahrt-Forsch.*, II, *pp.* 215–218.)

The gas enters the diffuser through the opening $b$ and is decelerated in the tube $a$. The temperature of the gas is measured by the thermocouple $d$. Afterward the gas escapes through the small holes $c$. If the holes were not present, the gas in the interior would cool off and would indicate an erroneous temperature. Both thermometers have their field of application. The advantage of the diffuser thermometer is that it immediately measures the total temperature. On the other hand, the cylinder thermometer can be made in very small dimensions and follows changes in gas temperature very rapidly.

As soon as the surface temperature of a body deviates from its recovery temperature, heat transfer arises between the body and the flowing gas. The static-temperature profiles in the boundary layer for heat transfer to and from the wall are shown in Fig. 10-7. The horizontal tangent to curve $a$ at the wall shows that no heat exchange occurs between the gas and the wall. The static temperature, which has the value $T_s$ in the outside flow, rises within the boundary layer and on this curve reaches the recovery temperature $T_r$ at the wall. As soon as the wall temperature is lower, heat flows from the gas into the wall. This holds for the temperature profiles below curve $a$ in Fig. 10-7.

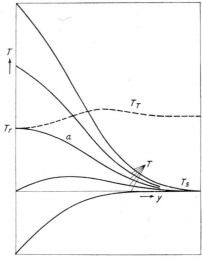

Fig. 10-7. Temperature profiles in the laminar boundary layer of a heated or cooled plate in a high-velocity gas stream. $T_T$ is total temperature, $T$ is static temperature, $y$ is wall distance.

At wall temperatures higher than the adiabatic, the heat flows in the opposite direction, from the wall to the gas.

The dashed curve indicates the total-temperature profile in the boundary layer for the case in which no heat flows into the wall. It is seen that the total temperature for some distance away from the wall is lower than in the stream outside the boundary layer whereas a layer farther away from the surface has a total temperature higher than the freestream value. Therefore, a high-velocity boundary layer separates air into low-energy and high-energy portions in the same way as vortex flow in the Hilsch tube. At the same velocity, however, the temperature differences are considerably larger in vortex flow.

Heat transfer in a high-velocity gas flow is best calculated[1] with a heat-transfer coefficient which is based on enthalpies according to the equation

$$q_w = h_i(i_r - i_w) \tag{10-15}$$

where $q_w$ = heat flow into wall per unit time and area
$i_r$ = recovery enthalpy
$i_w$ = enthalpy of gas at wall temperature

[1] E. R. G. Eckert, *Trans. ASME.*, **78**:1273–1283 (1956).

The recovery enthalpy is calculated with Eq. (10-14). However, the properties are introduced into Eq. (10-7) or (10-8) at a reference enthalpy given by the equation

$$i^* = i_s + 0.5(i_w - i_s) + 0.22(i_r - i_s) \tag{10-16}$$

For two-dimensional laminar flow along a surface with constant pressure and temperature, for instance, the local heat-transfer coefficient is obtained from a Stanton number

$$\text{St}_i = \frac{h_i}{\rho u_s} \tag{10-17}$$

and from Eq. (7-14) valid for low-velocity flow

$$\text{St}_i = \frac{0.332}{(\text{Re}_x)^{1/2}(\text{Pr})^{2/3}} \tag{10-18}$$

All properties are introduced into Eqs. (10-17) and (10-18) at the reference enthalpy given by Eq. (10-16). The local friction factor is obtained for laminar flow from the relation which was determined for low-velocity flow:

$$f_p = \frac{0.664}{\sqrt{\text{Re}_x}} \tag{10-19}$$

For turbulent flow over a flat plate the local friction factor is obtained from Blasius's equation:

$$f_p = \frac{0.0296}{(\text{Re}_x)^{0.2}} \tag{10-20}$$

or for Reynolds number above $10^7$ from a relation by Schultz-Grunow:

$$f_p = \frac{0.370}{(\log \text{Re}_x)^{2.584}} \tag{10-21}$$

The local heat-transfer coefficient can then be derived from the relation

$$\text{St}_i = \frac{f_p}{2(\text{Pr})^{2/3}} \tag{10-22}$$

All properties are again introduced into all these equations at the reference enthalpy as defined by Eq. (10-16).

It was found that the procedure outlined above accounts not only for the variation of properties with temperature in a wide range with an accuracy better than 4 per cent but also for dissociation of the air at high temperatures, where dissociation becomes appreciable.

As long as the temperatures are so low that the specific heat can be considered constant, the heat transfer calculated with temperatures leads to the same results as calculated with enthalpies.

It has been shown[1] that Eqs. (10-15) and (10-16), together with the

[1] M. R. Romig, *Jet Propulsion*, **26**:1098–1101 (1956).

incompressible relation (9-33) in which $h = c_p h_i$, establish heat transfer at the stagnation point of a blunt body in hypersonic flow, when the constant $C$ in Eq. (9-32) is replaced by $0.8(\text{Ma})^{0.232}$ (Ma is the upstream Mach number). Figure 10-8 shows a comparison between the values calculated with the above formulas and the results of experiments on a flying V-2 rocket by Fischer and Norris.[1] The measurements were made at different positions on the conical head of the missile. The different symbols distinguish these positions. The lower line represents the heat-transfer values as expected from the formulas for low velocities with laminar boundary layer. On the cone they are greater by the factor

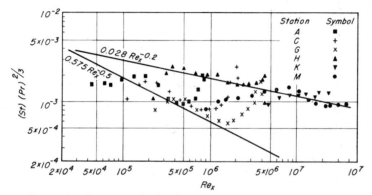

FIG. 10-8. Comparison between calculated and measured film heat-transfer coefficients for high-velocity flow. [*From W. Fischer and R. H. Norris, Trans. ASME*, **71**:457–469 (1949).]

$\sqrt{3}$ than on the plane plate because the surface increases in downstream direction and the boundary layer therefore does not grow so fast (see Sec. 9-3). The higher line holds for heat transfer with turbulent boundary layer as calculated with the formulas for small velocities. It can be seen that the measured points follow the laminar line for small Reynolds numbers and with increasing Reynolds number change over to the turbulent line. The transition depends obviously on some characteristic value other than the Reynolds number.

**Example 10-1.** A plate $x = 4$ in. long is suspended in an air flow with $u_s = 400$ mph $= 587$ fps. The temperature of the plate is $t_w = 196$ F; the air temperature $t_s = 79$ F; its pressure 14.2 lb$_f$/in.². The heat transfer from the plate to the air is to be calculated.

First we must determine the adiabatic wall temperature of the plate in the air stream. The property values can be introduced at a temperature given by $(t_w + t_s)/2 = 137.5$ F. The Reynolds number is

$$\text{Re}_x = \frac{u_s x}{\nu} = \frac{587 \times \frac{4}{12}}{20.4 \times 10^{-5}} = 9.59 \times 10^5$$

[1] W. W. Fischer and R. H. Norris, *Trans. ASME*, **71**:457–469 (1949).

From Fig. 10-2 take $r = 0.845$. Equation (10-13) gives

$$t_r = 79 + 0.845 \times \frac{587^2}{12{,}000} = 79 + 24.3 = 103.3 \text{ F}$$

This is the temperature the plate assumes if it is not heated. The heat-transfer coefficient can be calculated by introducing property values at a reference enthalpy given by Eq. (10-16), with values of $i$ for air taken from Keenan and Kaye.[1]

$$i^* = 128.7 + 0.5(156.9 - 128.7) + 0.22(134.6 - 128.7)$$
$$= 128.7 + 14.1 + 1.3 = 144.1$$

which corresponds to a temperature $t^* = 143$ F. Then by Eqs. (10-17) and (10-18) the local heat-transfer coefficient is

$$h_i = \frac{0.0657 \times 587 \times 0.0296(9.59 \times 10^5)^{-1/5} \times 3600}{1 + 0.87 \times 1.75(9.59 \times 10^5)^{-1/10}(0.700 - 1)} = 295.4 \text{ lb/hr ft}^2$$

The average value is

$$\bar{h}_i = 1.25 h_i = 369.3 \text{ lb/hr ft}^2$$

The average heat flow per unit area is, by Eq. (10-15),

$$q_w = 369.3(134.6 - 156.9) = -8{,}235 \text{ Btu/ft}^2 \text{ hr}$$

## 10-3. Heat Transfer in Rarefied Gases.

In the conventional treatment of heat transfer in gases, the structure of the gas is considered to be continuous; that is to say, no consideration needs to be given to the molecular structure of the gas. Flow and heat-transfer phenomena under these continuum conditions can be adequately described in terms of the Reynolds, Mach, Nusselt, and Prandtl numbers. However, at small absolute pressures a gas partly loses its continuum characteristics and important phenomena occur which are understandable only if account is taken of the coarseness of the molecular structure of the fluid. The study in aerodynamic terms of the flow and heat transfer of rarefied gases is a relatively recent development, and many fundamental questions remain to be answered by analysis and experiment.

The tool in dealing with these low-density phenomena is the kinetic theory of gases developed first in a quantitative way by Daniel Bernoulli in 1738. This theory considers the gases as consisting of molecules which travel about on straight paths, deviating only as a result of collision with other molecules. The average kinetic energy of the molecular aggregate determines the temperature of the gas. The collisions are considered to occur according to the laws of elastic impact. The statistical average length of the rectilinear paths between molecular collisions is called the *molecular mean free path* $\lambda$. The velocity of each molecule changes direction upon collision; thus for any specific instance molecules of many different velocities are present. Most have velocities near a mean velocity $\bar{v}$;

[1] J. H. Keenan and T. Kaye, "Gas Tables," John Wiley & Sons, Inc., New York' 1948.

that is, there are few molecules of exceedingly high or exceedingly low velocities. The velocity distribution in a gas at equilibrium was calculated by J. C. Maxwell in 1859 and is called the *Maxwellian velocity distribution*.

One of the early successes of the kinetic theory of gases was that it predicted the astonishing fact that the dynamic viscosity and the thermal conductivity of ideal gases are independent of pressure. This means, for example, that a certain amount of heat is conducted through a stagnant layer of gas under given temperature conditions independent of the pressure of the gas. By means of simplified concepts we shall derive expressions for the viscosity and thermal conductivity. Kinetic theory explains the shear stresses in a flowing gas by the fact that molecules move back and forth between layers of gas flowing at different velocities. Thus a molecule from a low-velocity gas layer may jump into a layer of

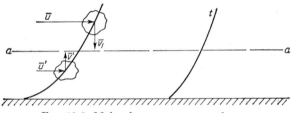

Fig. 10-9. Molecular momentum exchange.

gas moving at a higher velocity where after some collisions its velocity is increased and the velocities of the molecules collided with are slowed down. In this way momentum is exchanged between the fluid layers. The same motion of the molecules causes an exchange of energy when there are temperature differences in the gas.

Consider the laminar flow of a gas parallel to a wall with a velocity and temperature gradient normal to the wall (Fig. 10-9). Molecules pass back and forth across the arbitrary plane $aa$ parallel to the wall. During the time interval $d\tau$ molecules with velocity components $\bar{v}_1$ normal to plane $aa$ will pass that plane when they are in a proximity $\bar{v}_1 \, d\tau$ of the plane. The number of molecules per unit volume is $n$. The average velocity component toward plane $aa$ of all molecules will be some fraction $i'$ of the mean molecular velocity. Therefore all the molecules within the volume $i'\bar{v} \, d\tau$ will pass a unit area of the plane. The number of such molecules is $i'n\bar{v} \, d\tau$, and the number crossing a unit area per unit time of plane $aa$ from one side is $i'n\bar{v}$. The molecules are characteristic in the flow velocity $u$ from their last until their next collision. On the average, therefore, they transport the flow velocity from plane 1-1 to plane 2-2 separated by a distance related to the molecular mean-free-path length $\lambda$. This distance we shall call $j\lambda$, where $j$ cannot be very much different from unity. A molecular stream of equal magnitude passes plane $aa$ in the opposite direction and carries the flow velocity $\bar{u}'$ from plane 2-2 to plane 1-1. The momentum exchange connected with this process is

$$i'n\bar{v}m(\bar{u}-\bar{u}')$$

where $m$ is the mass of one molecule. When the velocity gradient at plane $aa$ in a direction normal to that plane is $d\bar{u}/dy$, then $(\bar{u}-\bar{u}')$ can be expressed as $j\lambda(d\bar{u}/dy)$. Therefore the exchange of momentum per unit area, or the shear stress, is

$$\tau = i'jnm\bar{v}\lambda \frac{d\bar{u}}{dy} = inm\bar{v}\lambda \frac{d\bar{u}}{dy}$$

The product $i'j$ can be denoted by $i$. The viscosity $\mu$ is defined by the equation $\tau = \mu(d\bar{u}/dy)$, from which by comparison with the above expression for $\tau$ we deduce

$$\mu = inm\bar{v}\lambda$$

The product $nm$ is the mass per unit volume $\rho$. A thorough calculation for spherical molecules by S. Chapman and D. Enskog, which accounted for the molecular velocity distribution, yielded for the numerical factor $i$ the value 0.499 and therefore for the dynamic viscosity

$$\mu = 0.499\rho\bar{v}\lambda \tag{10-23}$$

Kinetic theory shows that $\rho\lambda$ is independent of the pressure and further yields for the mean molecular velocity

$$\bar{v} = \sqrt{\frac{8RT}{\pi}}$$

$R$ is the gas constant for the gas in question; $T$ is the absolute temperature. Therefore $\mu$ as shown is independent of the pressure.[1]

When a thermal gradient $dt/dy$ is present in the gas, thermal energy is exchanged in a similar way as momentum by the molecules. If $c_m$ is the specific heat per molecule, the heat exchange per unit area of the plane $aa$ is

$$q = in\bar{v}c_m\lambda \frac{dt}{dy}$$

with the specific heat for constant volume $c_v$ per unit mass $nc_m = \rho c_v$. The thermal conductivity is defined by the equation $k = q/(dt/dy)$. Therefore we can substitute these values into the expression for $q$ to obtain

$$k = i\bar{v}\lambda\rho c_v = \mu c_v \tag{10-24}$$

---

[1] This result is not true for gases at very high pressures where ultimately the viscosity depends on pressure.

Thus this simple theory relates mechanical and thermal properties which might have been suspected of being quite unrelated. Equation (10-24) yields acceptable order-of-magnitude values but does not agree with experimental data; however, the possibility remains open for refinements in calculation to establish a numerical factor that will cause good agreement. The refinements in the calculation of the thermal conductivity are more difficult than the refinements afforded the viscosity, since the thermal conductivity is a much more complicated property. Chapman[1] introduced such refinements, which resulted in an expression for thermal conductivity that agrees with experimental data for monatomic molecules but not for more complex molecules. That is,

$$k = \tfrac{5}{2}\mu c_v \tag{10-25}$$

Eucken[2] developed an expression for the thermal conductivity for complex molecules as

$$k = \tfrac{1}{4}(9\gamma - 5)\mu c_v \tag{10-26}$$

which agrees surprisingly well with the experimental data for simple and complex molecules. $\gamma$ is the ratio of specific heats of constant pressure and constant volume.

From Eq. (10-26) one can write an expression for the *Prandtl number* on the basis of kinetic theory.

$$\Pr = \frac{4\gamma}{9\gamma - 5} \tag{10-27}$$

Equation (10-27) agrees reasonably well with measured values of the Prandtl number.

The *Reynolds number* can be interpreted from the kinetic theory as well.

$$\text{Re} = \frac{\rho V l}{\mu} = \frac{1}{0.499} \frac{V}{\bar{v}} \frac{l}{\lambda} \tag{10-28}$$

Thus Reynolds number is the product of a velocity ratio and a length ratio. The velocity ratio is between the macroscopic flow velocity $V$ and the mean molecular velocity $\bar{v}$; the length ratio is between the significant body dimension $l$ and the mean free molecular path $\lambda$.

In a similar manner an interpretation of the *Mach number* is also possible. The expression for the velocity of sound is

$$V_s = \sqrt{\gamma R T}$$

---

[1] E. H. Kennard, "Kinetic Theory of Gases," pp. 162–181, McGraw-Hill Book Company, Inc., New York, 1938.
[2] Eucken, *Physik. Z.*, **14**:324 (1913).

From the expression for the mean molecular speed, therefore,

$$V_s = \bar{v}\sqrt{\frac{\gamma\pi}{8}}$$

Thus the Mach number becomes

$$\text{Ma} = \frac{V}{V_s} = \frac{V}{\bar{v}}\sqrt{\frac{8}{\gamma\pi}} \tag{10-29}$$

Thus the Mach number is related directly to the ratio of the macroscopic velocity $V$ to the mean molecular velocity $\bar{v}$.

At ordinary pressures the intermolecular distances, even in gases where the intermolecular distances are very large when compared with solids and liquids, are much smaller than the body dimensions with which we are usually concerned in heat-transfer calculations, and therefore the notion of a continuous gas is well fulfilled. At low pressures, with corresponding low densities, the molecular mean free path becomes comparable with body dimensions, and then the effect of the molecular structure becomes a factor in flow and heat-transfer mechanisms.

The relative importance of effects due to the rarefaction of a fluid may be indicated by a comparison of the magnitude of the mean free molecular path in the fluid with some significant body dimension. Hence if $l$ is some body dimension which is a characteristic dimension in the flow field, the effects of rarefaction phenomena on flow and heat transfer will become important as soon as the ratio $\lambda/l$ can no longer be neglected. This ratio is dimensionless and is referred to as the *Knudsen number* Kn. The Knudsen number, which is thus of direct interest in the study of rarefied gas flow and heat transfer, is expressible in terms of the Mach and Reynolds numbers.

$$\text{Kn} = \frac{\lambda}{l} = \sqrt{\frac{\gamma\pi}{2}}\frac{\text{Ma}}{\text{Re}} \tag{10-30}$$

which can be evaluated approximately for a diatomic gas $\gamma = 1.4$ to yield

$$\text{Kn} \approx 1.48\frac{\text{Ma}}{\text{Re}}$$

Tsien[1] early proposed the division of fluid mechanics into various flow regimes based on characteristic ranges of the appropriate Knudsen number. For flows of low viscosity (i.e., large Reynolds numbers), the characteristic dimension of greatest importance is the boundary-layer thickness $\delta$. Since the boundary-layer thickness for laminar flow is related

---

[1] H. S. Tsien, *J. Aeronaut. Sci.*, **13**:653–654 (1946).

to the significant body dimension by

$$\frac{\delta}{l} \sim \frac{1}{\sqrt{\text{Re}}}$$

the corresponding Knudsen number becomes

$$\text{Kn} = \frac{\lambda}{\delta} \sim \frac{\text{Ma}}{\sqrt{\text{Re}}} \qquad (10\text{-}31)$$

Conventional gas dynamics hence will exist when $\text{Re} \gg 1$ and when the ratio $\text{Ma}/\sqrt{\text{Re}}$ is negligibly small. For slow flow, or Stokes-type flow, the significant flow dimension is the body dimension $l$. Thus the Knudsen number is based on $l$ to yield

$$\text{Kn} \sim \frac{\text{Ma}}{\text{Re}}$$

Thus continuum-type flow can be realized when $\text{Ma}/\text{Re}$ is negligibly small. In flows for which the Reynolds number is moderately large and the Mach number is large, the Knudsen number becomes of such a size that it no longer can be neglected, indicating the existence of rarefaction effects in the flow. In continuum flow a usual boundary condition at the interface between a fluid and a solid surface is that the fluid adjacent to the surface assumes both the velocity and the temperature of the surface. One of the more interesting effects of gas rarefaction in flow is that the gas adjacent to a solid surface no longer takes the velocity and temperature of the surface. The gas at the surface has a finite tangential velocity; it "slips" along the surface. The temperature of the gas at the surface is finitely different from the surface temperature; there is a "jump" in temperature between the surface and the adjacent gas. These effects are related to the molecular mean free path and parameters called *accommodation* and *reflection coefficients* which describe the statistical surface-molecule interaction. The flow regime just described is called the *slip-flow regime* and occurs for small but not negligible values of the Knudsen numbers.

For extremely low densities the mean free path $\lambda$ is very much larger than any significant body dimension $l$. In such a case the molecules leaving the body surface do not collide with freestream molecules until very far away from the surface. As a first approximation in such a case the molecular distribution in the region away from the surface can be assumed to be undistorted, that is, Maxwellian,[1] so that the flow near the body can be considered on the basis of the interaction between free molecules and the surface. This flow regime is named *free molecule flow* or *Knudsen flow*.

---

[1] Kennard, *op. cit.*, p. 45.

The transition from the gas-dynamics regime to the free-molecule-flow regime is gradual, showing no discontinuity experimentally, but for purposes of analysis it is convenient to define the flow regimes in an arbitrary manner. Tsien gave some such limits as

Gas dynamics: $\dfrac{\text{Ma}}{\sqrt{\text{Re}}} < 0.01$

Free molecule flow: $\dfrac{\text{Ma}}{\text{Re}} > 10$

The region in between is occupied by the *slip-flow* and *transition-flow* regimes. The transition-flow regime lies between the slip flow and free-molecule flows and is characterized by the fact that the molecular mean free path is approximately the same length as the characteristic dimension of the body. Collisions between molecule and surface and molecule and

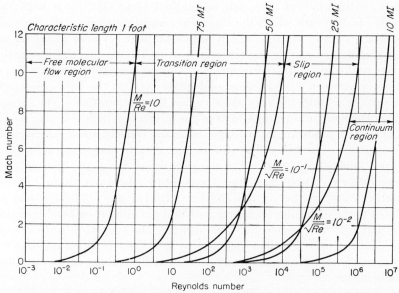

Fig. 10-10. Defined flow regions of gas dynamics.

molecule are frequent and of equal importance. Analysis in this region is difficult and rare. Present information is meager and mostly empirical. These flow regions are shown graphically in Fig. 10-10 with the corresponding altitude in miles above sea level, with the body dimension taken as 1 ft for the purpose of the altitude parameter. It is clear from Fig. 10-10 that for even moderately high Mach numbers a missile traveling at an altitude of 20 miles will experience slip-flow effects. Compressible boundary layers of $M > 4$ will show slip effects for Reynolds numbers smaller than about $10^5$. It is a certainty that hypersonic wind tunnels will show slip effects unless the reservoir stagnation pressure is very high.

*Slip Flow.* A completely satisfactory formulation of the flow and energy equations which might yield results on skin friction and heat transfer in a slightly rarefied gas, the slip-flow state, is lacking. Any of the attempts made are too formidable to be presented in a text of this scope, but some pertinent remarks may be in order. The most generally accepted formulation is due to Burnett,[1] Chapman and Cowling,[2] and others following the developments of Hilbert and Enskog in obtaining the first three terms in a perturbation series solution of the fundamental Maxwell-Boltzmann equation for the kinetic theory of gases. These solutions yield, in order of complication, statements of the *Euler* equation for incompressible, inviscid fluids; the *Navier-Stokes* equations for viscous, compressible fluids; and ultimately the *Burnett* equations. The first two equation sets with the corresponding energy equation for the second are well known in the practice of conventional fluid mechanics. The Burnett equations are clearly of higher order in the derivatives than are the Navier-Stokes equations, just as the Navier-Stokes equations have higher-order derivatives than the Euler equations for ideal fluids. The relative importance of these higher-order terms in the Burnett formulation as compared, say, with terms in the Navier-Stokes equations is determined by the magnitude of the rarefaction parameters as discussed earlier in defining the flow regimes. For example, the *Burnett stress term* is

$$\frac{\mu^2}{p} \frac{\partial u}{\partial x} \frac{1}{2} \left( \frac{\partial u}{\partial y} + \frac{\partial v}{\partial x} \right)$$

Dividing by the *Navier-Stokes stress term*

$$2\mu \frac{1}{2} \left( \frac{\partial u}{\partial y} + \frac{\partial v}{\partial x} \right)$$

we obtain

$$\frac{\mu}{p} \frac{\partial u}{\partial x} \sim \frac{\mu}{p} \frac{u}{l} \sim \frac{M^2}{\text{Re}}$$

Hence this term has an order of magnitude depending on Mach and Reynolds numbers and from earlier considerations is not negligible in the slip-flow region. Other Burnett terms have similar measure.

The foregoing notions are strictly applicable to a monatomic gas in which the molecules possess only the three translational degrees of freedom. For diatomic and polyatomic gases the distribution of internal energy over all degrees of freedom does not occur until the elapse of a

---

[1] D. Burnett, *Proc. London Math. Soc.*, **40**:382 (1935).

[2] S. Chapman and T. G. Cowling, "The Mathematical Theory of Non-uniform Gases," Cambridge University Press, London, 1939.

"relaxation time" following any abrupt change of state in the gas. Internal energy is stored first in the translational degrees of freedom, and only after a sufficient number of collisions will it be stored in the rotational and vibrational degrees of freedom. The number of collisions required varies from a few in the case of air to a thousand or more in the case of $CO_2$. Shock-wave thicknesses, for example, are almost completely determined by the higher-order equations and present extreme difficulties.

In spite of, and in many ways because of, the great complications arising from the attempts to formulate the slip-flow problems, a considerable body of literature exists having to do with slip-flow effects in fluid flow. This state of affairs is analogous to the treatment of the Navier-Stokes equations, for which no general solution and, indeed, only a few particular solutions exist. It is represented, however, by a vast literature of approximate solutions which enable the practical utilization of the ideas involved. The slip-flow studies are approximations, in many ways crude ones, in that the analysis has been based on the Navier-Stokes equations, usually for incompressible flow. The solutions have been approximate, with the slip-flow effects presumed to be concentrated in the contribution of additional terms in the tangential velocity and temperature boundary conditions; i.e., the velocity slip and the temperature jump at the gas-surface interface. In spite of the crudity of the assumptions regarding the slip effects, the results have been checked experimentally to a surprising accuracy. Another interesting and at the same time surprising result which has almost always been obtained in slip-flow analysis is that the results are found to correspond closely to free-molecule-flow results if the Knudsen numbers are taken as very large. Since slip flow is supposedly a phenomenon related to small Knudsen numbers, the exact significance of this is unknown; however, successful use of the fact has been made in semiempirical analysis to correlate experimental data over the entire flow regime from gas dynamics to free molecule flow.

*Approximate Slip-flow Analysis.* Since direct solutions for the flow and energy equations for this slip-flow region are unavailable at the moment, the attack on the problem has been made by using continuum formulations for the flow and energy equations, relegating the rarefaction effects to the boundary conditions. Two major effects have been observed in the slip-flow phenomena. First, it was noticed theoretically by Maxwell[1] and experimentally by Kundt and Warburg[1] that for flow conditions near a boundary the flow velocity is not zero but *slips* along the wall with a finite velocity. Second, a temperature discontinuity, or *jump*, was postulated by Poisson[1] during heat transfer from a surface to a

---

[1] Kennard, *op. cit.*, p. 311.

rarefied gas, and this effect was found to exist for static systems by the experiments of Smoluchowski.[1] These effects can be represented mathematically by

$$u_{y=0} - u_w = \frac{2 - f_s}{f_s} \lambda \left(\frac{\partial u}{\partial y}\right)_{y=0} \qquad (10\text{-}32)$$

and
$$t_{y=0} - t_w = \frac{2 - a}{a} \frac{2\gamma}{\gamma + 1} \frac{\lambda}{\Pr} \left(\frac{\partial T}{\partial y}\right)_{y=0} \qquad (10\text{-}33)$$

where $f_s$ is the specular reflection coefficient, or the fraction of the tangential momentum of the impinging molecules that is transmitted to the wall. $a$ is the thermal accommodation coefficient, or a measure of the extent to which the molecules upon being reflected (or absorbed and reemitted) from a surface have had their mean energy adjusted to the energy characteristic of the surface. Thus $a$ can be expressed as

$$a = \frac{E_r - E_i}{E_w - E_i} \qquad (10\text{-}34)$$

which for monatomic, perfect gases can be written exactly and for other gases in good approximation as

$$a = \frac{t_r - t_i}{t_w - t_i} \qquad (10\text{-}35)$$

where in the Eq. (10-34) $E_r$ is the energy of the molecules reflected (or reemitted) from the surface; $E_i$ is the energy of the molecules from the free-stream incident upon the surface; and $E_w$ is the energy of the molecules adjusted to the energy characteristic of the surface. In Eq. (10-35) $t_r$ is the temperature of the molecular stream reflected (or reemitted) from the surface; $t_i$ is the temperature of the incident molecular stream; and $t_w$ is the surface, or wall, temperature. Both the specular reflection coefficient $f_s$ and the accommodation coefficient $a$ have been treated as experimentally determined parameters rather than as variables. Table 10-1 gives some values of the slip coefficient as reported by Millikan[2] and Blankenstein.[3] It can be seen from the table that a good first-order estimate of $f_s = 1$ is perhaps acceptable. Table 10-2 gives some values of the thermal accommodation coefficient $a$ as reported by Knudsen,[4] Oliver,[5] Weidmann and Trumpler,[6] and Roberts.[7] In the case of the accommo-

---

[1] M. Smoluchowski, *Sitzber. Wien. Akad.*, **107**:304 (1898); *Anz. Akad. Wiss. Krakau A*, **5**:129 (1910).
[2] R. A. Millikan, *Phys. Rev.*, **21**:230 (1923).
[3] E. Blankenstein, *Phys. Rev.*, **22**:582 (1923).
[4] M. Knudsen, "Kinetic Theory of Gases," Methuen & Co., Ltd., London, 1934.
[5] R. N. Oliver, "Thermal Accommodation Coefficients," C.I.T. thesis, 1950.
[6] M. L. Weidmann and P. R. Trumpler, *Trans. ASME*, **68**:57 (1946).
[7] J. K. Roberts, *Proc. Roy. Soc.*, **129**:146 (1930); **135**:192 (1932).

TABLE 10-1. SPECULAR REFLECTION COEFFICIENT, $f_s$

| Gas | Surface | $f_s$ |
|---|---|---|
| Air | Machined brass | 1.00 |
| $CO_2$ | Machined brass | 1.00 |
| Air | Old shellac | 1.00 |
| $CO_2$ | Old shellac | 1.00 |
| Air | Hg | 1.00 |
| Air | Oil | 0.90 |
| $CO_2$ | Oil | 0.92 |
| $H_2$ | Oil | 0.93 |
| Air | Glass | 0.89 |
| He | Oil | 0.87 |
| Air | Fresh shellac | 0.79 |
| Air | $Ag_2O$ | 0.98 |
| He | $Ag_2O$ | 1.00 |
| $H_2$ | $Ag_2O$ | 1.00 |
| $O_2$ | $Ag_2O$ | 0.99 |

TABLE 10-2. THERMAL-ACCOMMODATION COEFFICIENT, $a$

| Gas | Surface | $a$ |
|---|---|---|
| $H_2$ | Bright platinum | 0.32 |
| $H_2$ | Black platinum | 0.74 |
| $O_2$ | Bright platinum | 0.81 |
| $O_2$ | Black platinum | 0.93 |
| $N_2$ | Platinum | 0.50 |
| $N_2$ | Tungsten | 0.35 |
| Air | Flat lacquer on bronze | 0.88–0.89 |
| Air | Polished bronze | 0.91–0.94 |
| Air | Machined bronze | 0.89–0.93 |
| Air | Etched bronze | 0.93–0.95 |
| Air | Polished cast iron | 0.87–0.93 |
| Air | Machined cast iron | 0.87–0.88 |
| Air | Etched cast iron | 0.89–0.96 |
| Air | Polished aluminum | 0.87–0.95 |
| Air | Machined aluminum | 0.95–0.97 |
| Air | Etched aluminum | 0.89–0.97 |
| He | Tungsten | 0.025–0.057 |

dation coefficient a fairly wide range of values is reported, such that in practice the values would have to be determined for the system under investigation.

*Flat Plate.* Laminar slip flow over a flat plate at zero angle of attack can be treated in terms of the Rayleigh problem of an impulsively started plate. The inertia and viscous terms in the Navier-Stokes equations can be simplified in this case. We shall calculate first the drag on a plate and then the heat transfer from the same plate.

The flow can be considered in two dimensions, but for the impulsively started plate we can neglect the convective inertia terms such as $u(\partial u/\partial x)$ and the viscous terms such as $\nu(\partial^2 u/\partial x^2)$, which leaves

$$\frac{\partial u}{\partial \tau} = \nu \frac{\partial^2 u}{\partial y^2} \qquad (10\text{-}36)$$

The left-hand term can be transformed into

$$\frac{\partial u}{\partial \tau} = \frac{\partial u}{\partial x}\frac{\partial x}{\partial \tau} + \frac{\partial u}{\partial y}\frac{\partial y}{\partial \tau}$$

and if a uniform velocity $U$ is given to the plate in the direction $x$ only, the above reduces, for $\partial y/\partial \tau = 0$, to

$$\frac{\partial u}{\partial \tau} = U \frac{\partial u}{\partial x}$$

and Eq. (10-36) becomes

$$U \frac{\partial u}{\partial x} = \nu \frac{\partial^2 u}{\partial y^2} \qquad (10\text{-}37)$$

The appropriate boundary conditions are

$$u = U \qquad x = 0 \qquad y > 0$$
$$u = \lambda \left(\frac{\partial u}{\partial y}\right) \qquad y = 0 \qquad x > 0$$

The second boundary condition is written for the velocity slip condition for a specular reflection coefficient of unity.

A drag coefficient is defined as

$$C_D = \frac{\text{total drag force on both sides of plate per unit width}}{\rho U^2 l/2}$$

The solution of Eq. (10-37) in terms of such a drag coefficient becomes

$$C_D \, \text{Ma} = \frac{2.67}{X_1^2}\left(e^{X_1^2}\operatorname{erfc} X_1 - 1 + \frac{2}{\sqrt{\pi}} X_1\right) \qquad (10\text{-}38)$$

where $X_1 = \tfrac{2}{3}\sqrt{\text{Re}/\text{Ma}^2}$. A plot of Eq. (10-38) is shown in Fig. 10-11 in comparison with experimental data on the total drag of flat plates from measurements made in a low-density high-speed wind tunnel.[1]

---

[1] S. A. Schaaf, *Univ. Calif. Inst. Eng. Research Rept.* HE-150-66, 1950. F. S. Sherman, *Univ. Calif. Inst. Eng. Research Rept.* HE-150-70, 1950, and HE-150-81, 1951.

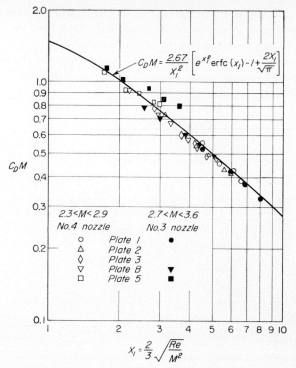

Fig. 10-11. Flat-plate drag coefficient. (*Courtesy of F. S. Sherman.*)

The heat transfer[1] from this plate can be calculated from the energy equation for the impulsively started plate. Again, in this case we can neglect heat-conduction terms such as $k(\partial^2 t/\partial x^2)$ as being small in comparison with $k(\partial^2 t/\partial y^2)$. We can neglect the convective term in $y$; thus for motion of the plate at constant velocity $U$ in the $x$ direction, the energy equation becomes

$$U \frac{\partial t}{\partial x} = \alpha \frac{\partial^2 t}{\partial y^2} \tag{10-39}$$

The appropriate boundary conditions are given in terms of the temperature-jump conditions at the wall and the state of the freestream.

$$t_{y=0} - t_W = \frac{2-a}{a} \frac{2\gamma}{\gamma+1} \frac{\lambda}{\Pr} \left(\frac{\partial t}{\partial y}\right)_{y=0} \qquad y=0 \quad x>0$$
$$t = t_f \qquad x=0 \qquad y>0$$

---

[1] R. M. Drake and E. D. Kane, "General Discussion on Heat Transfer," Institution of Mechanical Engineers and American Society of Mechanical Engineers, London, 1951; also J. R. Stalder, G. Goodwin, and M. O. Creager, "General Discussion on Heat Transfer," Institution of Mechanical Engineers and American Society of Mechanical Engineers, London, 1951.

Therefore, for $a = 0.8$, $\gamma = 1.4$, the solution of Eq. (10-39) becomes

$$\text{StMa} = \frac{0.38}{X_2^2}\left(e^{X_2^2}\operatorname{erfc} X_2 - 1 + \frac{2}{\sqrt{\pi}} X_2\right) \quad (10\text{-}40)$$

where $X_2 = \sqrt{\text{Re Pr}/6.9\text{Ma}^2}$ and St is the Stanton number

$$\text{St} = \frac{\text{Nu}}{\text{Re Pr}} = \frac{h}{U\rho c_p}$$

There are no experimental data with which to compare Eq. (10-40); however, Eq. (10-40) should yield results which are as good as Eq. (10-38).

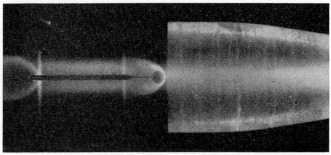

FIG. 10-12. Shock wave in front of sphere by nitrogen afterglow method. (*Courtesy of Low Pressures Research Project, University of California.*)

*The Sphere.* Heat transfer from spheres has received some attention,[1-3] since the spherical shape lends itself well to experimental procedures. The problem is complicated, however, by the fact that for the most part unless the sphere diameter is very small, the Mach number must be large in order that the Knudsen number will be large enough to establish the flow as slip flow. Ordinarily this means the Mach number is such that the flow is supersonic. In such a case a shock wave exists in front of the sphere (Fig. 10-12) and the conditions behind this shock wave must be considered before any sense can be made of the heat-transfer calculations. As one model we can consider a sphere in a supersonic, rarefied gas flow preceded by a normal shock wave behind which we can calculate the gas properties in order to determine the heat transfer.[4] As a second model we can consider a rarefied subsonic flow obtainable at very low densities such that the shock wave is not present and therefore offers

---

[1] R. M. Drake, Jr., F. M. Sauer, and S. A. Schaaf, *Univ. Calif. Inst. Eng. Research Rept.* HE-150-74, 1950.
[2] Drake and Kane, *op. cit.*
[3] R. M. Drake, Jr., and G. H. Backer, *Trans. ASME,* **74**:7 (1952).
[4] Drake, Sauer, and Schaaf, *op. cit.*

no real complication.[1] Kavanau has developed an expression for the rarefaction correction to the continuum solution for heat transfer from spheres that represents the experimental data that exist within 10 per cent. We shall consider this approach here.

If we consider the temperature jump in the slip-flow region as an effective thermal contact resistance at the gas-surface interface over and above the thermal resistance due to the viscous boundary layer, then the heat-transfer coefficient at low densities can be determined to the first order by a correction to the continuum heat-transfer coefficient at the same Reynolds number.

Consider a surface in continuum flow so that heat is transferred between the surface and the stream. The surface temperature is uniform, constant, and equal to $t_w{}^0$; the heat flux becomes

$$q = -k\left(\frac{\partial t}{\partial y}\right)_{y=0} = h^0(t_w{}^0 - t_r{}^0) \tag{10-41}$$

*where the zero superscript indicates the continuum state* and $t_r$ is the recovery temperature of the surface.

Now if the gas stream is rarefied while the Reynolds number is held constant, the additional resistance to heat transfer because of the temperature jump will appear, so that we can write a total convective heat-transfer coefficient incorporating both the thermal boundary-layer and temperature-jump resistances by writing

$$q = h(t_w - t_r) \tag{10-42}$$

The first-order temperature jump can be written as

$$t_{y=0} - t_w = \xi\left(\frac{\partial t}{\partial y}\right)_{y=0} \tag{10-43}$$

where $t_{y=0}$ and $(\partial t/\partial y)_{y=0}$ are the temperature and temperature gradient in the layer of gas immediately adjacent to the surface. The quantity $\xi$ is the temperature-jump distance defined from Eq. (10-33) as

$$\xi = 1.996\,\frac{2-a}{a}\,\frac{\gamma}{\gamma+1}\,\frac{\lambda}{\Pr}$$

or
$$\xi = \frac{\theta}{\Pr}\lambda \tag{10-44}$$

The first-order temperature-jump condition requires that $t_r = t_r{}^0$. It is now assumed that $q$, the heat transferred between the surface and the gas, is the same for the case with and without the temperature jump. This, in effect, assumes that Eq. (10-43) is valid when $t_{y=0} = t_w{}^0$. Com-

---

[1] L. L. Kavanau and R. M. Drake, Jr., *Univ. Calif. Inst. Eng. Research Rept.* HE-150-108, 1953. See also L. L. Kavanau, *Trans. ASME,* **77**:5 (1955).

bining Eqs. (10-43), (10-41), and (10-42) gives

$$\frac{h^0}{h} = 1 + \frac{\xi h^0}{k}$$

and substituting into this expression the value for $\xi$, we obtain with the aid of Eq. (10-30)

$$\frac{h^0}{h} = 1 + \frac{h^0 l}{k} \frac{\text{Ma}}{\text{Re Pr}} \sqrt{\frac{\gamma \pi \theta^2}{2}}$$

which can be written as

$$\frac{\text{Nu}^0}{\text{Nu}} = 1 + \sqrt{\frac{\gamma \pi \theta^2}{2}} \frac{\text{Ma}}{\text{Re Pr}} \text{Nu}^0 \quad (10\text{-}45)$$

If Eq. (10-45) is evaluated for a diatomic gas with $\gamma = 1.4$ and an accommodation coefficient $a = 0.8$, we obtain approximately

$$\frac{\text{Nu}^0}{\text{Nu}} = 1 + 2.59 \left(\frac{\text{Ma}}{\text{Re Pr}}\right) \text{Nu}^0$$

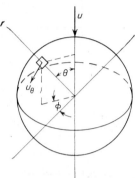

FIG. 10-13. Coordinate system for the sphere.

In order to interpret Eq. (10-45) it is necessary to have a solution for the continuum heat transfer from spheres. Such a solution has been obtained in virtue of rather gross approximations regarding the velocity distribution about the sphere.[1] Consider the energy equation for a spherical coordinate system, neglecting the compressibility and dissipation of mechanical-energy terms, in terms of the geometry shown in Fig. 10-13.

$$\rho c_p \left( u_r \frac{\partial t}{\partial r} + \frac{u_\theta}{r} \frac{\partial t}{\partial \theta} + \frac{u_\phi}{r \sin \theta} \frac{\partial t}{\partial \phi} \right) = k \left[ \frac{1}{r^2} \frac{\partial}{\partial r} \left( r^2 \frac{\partial t}{\partial r} \right) \right.$$
$$\left. + \frac{1}{r^2 \sin \theta} \frac{\partial}{\partial \theta} \left( \sin \theta \frac{\partial t}{\partial \theta} \right) + \frac{1}{r^2 \sin^2 \theta} \frac{\partial^2 t}{\partial \phi^2} \right] \quad (10\text{-}46)$$

As an approximation to the velocity slip condition let

$$u_r = u_\phi = 0 \qquad u_\theta = U$$

and from symmetry

$$\frac{\partial t}{\partial \phi} = \frac{\partial^2 t}{\partial \phi^2} = 0$$

and for small heat conduction in the $\theta$ direction compared with the con-

---

[1] Drake and Kane, *op. cit.*; also *Trans. ASME*, **74**:7 (1952). (This solution is also applicable for first-order slip-flow effects.)

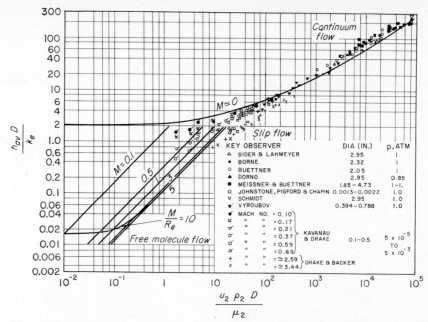

FIG. 10-14. Heat transfer from spheres (Reynolds numbers evaluated at condition 2 behind hypothetical normal shock for Ma > 1, and thermal conductivity evaluated at the equilibrium temperature). [*R. M. Drake, Jr., and G. H. Backer, Trans. ASME,* **74**:7 (1952).]

vective flow in that direction,

$$k \frac{\partial t}{\partial \theta} = k \frac{\partial^2 t}{\partial \theta^2} = 0$$

Equation (10-46) thus reduces to the abbreviated form

$$\rho c_p \frac{U}{r} \frac{\partial t}{\partial \theta} = k \left[ \frac{1}{r^2} \frac{\partial}{\partial r} \left( r^2 \frac{\partial t}{\partial r} \right) \right] \quad (10\text{-}47)$$

Equation (10-47) can be solved for the boundary conditions

$$t = t_w \quad r = r_0$$
$$t = t_f \quad r \to \infty$$

by means of the Laplace transform[1] to yield for the average heat-transfer coefficient in terms of the Nusselt number

---

[1] R. V. Churchill, "Modern Operational Mathematics in Engineering," McGraw-Hill Book Company, Inc., New York, 1944.

$$\overline{\mathrm{Nu}} = \frac{h_{\mathrm{av}} D}{k} = 2 + \frac{2}{\pi^2} \int_0^\infty \frac{(1 + e^{-\beta^2 \pi})(1 + \beta^4)^{-1} \, d\beta}{[J_1^2(\alpha_1 \beta) + Y_1^2(\alpha_1 \beta)]\beta} \quad (10\text{-}48)$$

where
$D$ = diameter of sphere
$\alpha_1 = \sqrt{2 \, \mathrm{Re} \, \mathrm{Pr}}$

$J_1(\alpha_1\beta)$ and $Y_1(\alpha_1\beta)$ = Bessel functions

Equation (10-48) is plotted in Figs. 10-14 and 10-15 as the line Ma $\sim 0$.

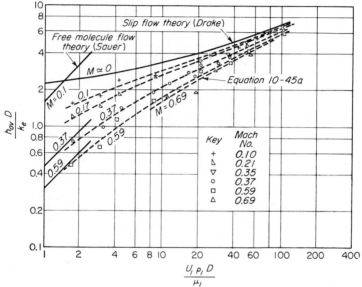

FIG. 10-15. Heat transfer from spheres at subsonic Mach numbers in a rarefied gas (Reynolds numbers evaluated at the freestream condition 1, and thermal conductivity evaluated at the equilibrium temperature). [L. L. Kavanau and R. M. Drake, Jr., *Report No. HE-150-108, University of California*, 1953. Also L. L. Kavanau, *Trans. ASME*, **77**:5 (1955).]

Using the combined results of Eqs. (10-45) and (10-48) we obtain the results as plotted on Fig. 10-15 in comparison with experimental heat-transfer data for small spheres. The constant 2.59 in Eq. (10-45) was adjusted empirically to a value 3.42 in order to represent the experimental data more accurately:

$$\overline{\mathrm{Nu}} = \frac{\overline{\mathrm{Nu}^0}}{1 + 3.42(\mathrm{Ma}/\mathrm{Re}\,\mathrm{Pr})\overline{\mathrm{Nu}^0}} \quad (10\text{-}45a)$$

The traces in Figs. 10-14 and 10-15 representing heat transfer in free molecule flow are the results of a kinetic-theory calculation made by F. M. Sauer.[1] The graphical results show the difference between taking

[1] F. M. Sauer, *J. Aeronaut. Sci.*, **18**:5 (1951).

the Reynolds number to zero in virtue of a zero velocity and taking it to zero in virtue of a zero density. The zero-velocity case for heat transfer from spheres becomes in the limit a case of radial conduction in a stagnant gas, whereas the approach to the zero-density case leads to the free-molecule-flow regime and much lower heat-transfer coefficients.

An interesting feature of these rarefied gas heat-transfer phenomena is shown in Fig. 10-16, a representation of the thermal-recovery factor plotted in relation to the rarefaction parameter. The recovery factors are seen to increase at large values of the Knudsen number in slip flow and

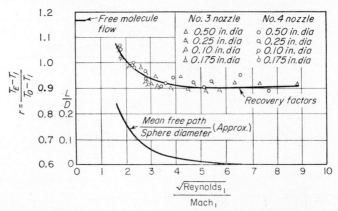

FIG. 10-16. Variation of thermal-recovery factor with rarefaction parameter. ($T_E$ is the equilibrium temperature attained by the high conductivity sphere, $T_O$ is the reservoir temperature, and $T_1$ is the free-stream temperature.) [R. M. Drake, Jr., and G. H. Backer, Trans. ASME, **74**:7 (1952).]

to approach a value greater than unity as predicted in the free-molecule-flow range. Another indication of this phenomenon, shown in Fig. 10-17, is the distribution of the thermal-recovery factor through the boundary layer in low-density flow, supersonic nozzles. This energy distribution is typical of those found in high-speed gas flows where the Prandtl number is less than unity. The fact that low-density flows have rather thick boundary layers makes this phenomenon more readily observable.

Heat transfer from cylinders in the slip-flow region produces effects similar to those described above for spheres. Sauer[1] has developed an analysis for cylinders, and Stalder[2] et al. have obtained experimental data for similar range of variables.

*Free Molecule Flow.* Heat transfer from a body to a rarefied gas stream in Maxwellian equilibrium can be calculated entirely from fundamental

[1] F. M. Sauer and R. M. Drake, Jr., *J. Aeronaut. Sci.*, **20**:3 (1953).
[2] J. R. Stalder, G. Goodwin, and M. O. Creager, *Natl. Advisory Comm. Aeronaut. Tech. Notes*, no. 2438, 1951.

notions of the kinetic theory of gases.[1,2] The results can be expressed as a function of the number of degrees of freedom $j$ of the molecule or the ratio of specific heats $\gamma$, which is directly related to $j$, so that the results become general in so far as the molecular structure of the gas is concerned. Furthermore, as has been shown by Oppenheim,[1] the determination of correlated results for a few fundamental shapes such as flat plates, horizontal circular cylinders, and spheres will enable the formulation of

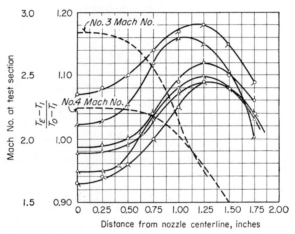

Fig. 10-17. Variation of thermal-recovery factor through laminar boundary layers in supersonic nozzles at very low pressures. ($T_E$ is the equilibrium temperature attained by the high conductivity spherical probe, $T_O$ is the reservoir temperature, and $T_1$ is the free-stream temperature.) No. 3 nozzle 5.25 in. diameter; No. 4 nozzle 5.5 in. diameter. [R. M. Drake, Jr. and G. H. Backer, Trans. ASME, **74**:7 (1952).]

results for more complicated cases by synthesis of the results of these simple shapes.

The convective heat transfer from a body in free molecule flow is established on the basis of an energy balance:

$$q = e_r - e_i \tag{10-49}$$

where $e_r = dE_R/dA$ = molecular-energy-transport rate per unit surface area of reemitted molecules

$e_i = dE_i/dA$ = energy-transport rate per unit surface area of molecules incident on the surface

$q = dQ/dA$ = heat-transfer rate per unit surface area to gas stream

The energy of the reemitted molecules depends upon the accommoda-

[1] A. K. Oppenheim, J. Aeronaut. Sci., **20**:1 (1953).
[2] J. R. Stalder and D. Jukoff, J. Aeronaut. Sci., **15**:7 (1948).

tion coefficient $a$ [Eq. (10-34)]. The combination of Eqs. (10-34) and (10-49) will yield

$$\frac{q}{a} = e_w - e_i \qquad (10\text{-}50)$$

Each energy term appearing in Eq. (10-50) is composed of translational energy plus part of the internal energy which is associated with degrees of freedom other than translation. From the principle of equipartition of energy, the internal energy of a gas molecule of $j$ degrees of freedom in equilibrium at temperature $T$ is

$$u = \frac{j}{2}\mathbf{k}T$$

where $\mathbf{k}$ is the Boltzmann constant ($\mathbf{k} = 5.66 \times 10^{-24}$ ft lb$_f$/molecule R). The translational energy of a molecule coming from the wall at temperature $T_w$ is

$$e_{t_w} = 2n\mathbf{k}T_w$$

The translational part of the internal energy is distributed among three degrees of freedom; therefore,

$$e_w = e_{t_w} + \frac{j-3}{2}n\mathbf{k}T_w = \frac{j+1}{2}n\mathbf{k}T_w \qquad (10\text{-}51)$$

where $n$ as before is the number of molecules striking a unit wall area per unit time.

The energy transport due to the incident molecules may then be

$$e_i = e_t + \frac{j-3}{2}n\mathbf{k}T_0 \qquad (10\text{-}52)$$

$e_t$ is the translational energy rate of the free stream striking a unit surface area of the wall per unit time and $T_0$ is the temperature of the freestream Substituting Eqs. (10-51) and (10-52) into Eq. (10-50) we obtain

$$\frac{q}{a} = \frac{j+1}{2}n\mathbf{k}T_w - \left(e_t + \frac{j-3}{2}n\mathbf{k}T_0\right) \qquad (10\text{-}53)$$

Since

$$\gamma = \frac{c_p}{c_v} = \frac{d(u+\mathbf{k}T)/dt}{du/dt} = 1 + \frac{2}{j}$$

it is possible to write Eq. (10-53) in terms of $\gamma$.

$$\frac{q}{a} = \frac{\gamma+1}{2(\gamma-1)}n\mathbf{k}T_w - \left[e_t + \frac{5-3\gamma}{2(\gamma-1)}n\mathbf{k}T_0\right] \qquad (10\text{-}54)$$

where $\gamma = \frac{5}{3}$ corresponds to a monatomic gas and $\gamma = \frac{7}{5}$ to a diatomic gas.

SPECIAL HEAT-TRANSFER PROCESSES

In order to calculate the heat-transfer rate from a body, expressions must be obtained for the number of molecules striking the surface per unit area per unit time $n$ and the translational energy $e_t$. These values as given by Oppenheim are

$$n = \frac{N\bar{v}}{2\sqrt{\pi}}\left[e^{-\eta^2} + \sqrt{\pi}\,\eta(1 + \mathrm{erf}\,\eta)\right] \qquad (10\text{-}55)$$

$$e_t = [(s^2 + \tfrac{5}{2})n - \Phi]kT_0 \qquad (10\text{-}56)$$

where $N$ = number of molecules per unit volume
$\bar{v}$ = mean molecular velocity
$\eta = U_\eta/\bar{v}$, with $U_\eta$ component of mass flow velocity $U$ normal to wall
$s = U/\bar{v} = \sqrt{\gamma/2}\,\mathrm{Ma}$ molecular speed ratio as related to Mach number
$\Phi = (N\bar{v}/4\sqrt{\pi})e^{-\eta^2}$.

Substitution of the values above into Eq. (10-54) results in the following dimensionless expression for the heat flux:

$$\frac{q}{ankT_0U} = \frac{\gamma+1}{2(\gamma-1)}\frac{n}{N\bar{v}s}\frac{T_w}{T_0} - \left(s^2 + \frac{\gamma}{\gamma-1}\right)\frac{n}{N\bar{v}s} + \frac{\Phi}{N\bar{v}s} \qquad (10\text{-}57)$$

In order to calculate the over-all heat transfer to a body in free molecule flow, the following surface integrals are defined:

$$\bar{G} = \frac{1}{A}\int_A \frac{e^{-\eta^2}}{2\sqrt{\pi}\,s}\,dA$$

$$\bar{F} = \frac{1}{A}\int_A \frac{\eta(1 + \mathrm{erf}\,\eta)}{2s}\,dA$$

where $A$ is the surface area of the body under investigation. Equation (10-57), when integrated over the surface area $A$, can be written in terms of these integrals.

$$\frac{\bar{Q}}{aNkT_0U} = \frac{\gamma+1}{2(\gamma-1)}\,(\bar{G}+\bar{F})\frac{T_w}{T_0} - \left(s^2 + \frac{\gamma}{\gamma-1}\right)(\bar{G}+\bar{F}) + \frac{1}{2}\bar{G} \qquad (10\text{-}58)$$

The heat transfer in free molecule flow is aptly represented in terms of the Stanton number and the thermal-recovery factor. For $\bar{Q} = 0$, Eq. (10-58) yields the expression for the wall recovery temperature in the instance where $T_w = \mathrm{const}$. This requires that the thermal resistance due to conduction within the body be small in comparison with the resistance due to fluid convection on the surface. In such a case,

$$\frac{T_r}{T_0} = 2\left(\frac{\gamma-1}{\gamma+1}\right)\left(\frac{\gamma}{\gamma-1} + s^2 - \frac{1}{2}\frac{\bar{G}}{\bar{G}+\bar{F}}\right) \qquad (10\text{-}59)$$

We can now write the heat flux in terms of the wall recovery temperature

$$\bar{Q} = a \frac{\gamma + 1}{2(\gamma - 1)} (\bar{G} + \bar{F})(T_w - T_r) N \mathbf{k} U$$

from which the Stanton number follows, noting that $N\mathbf{k}/\rho = R$, the gas constant.

$$\text{St} = \frac{h}{\rho c_p U} = \frac{\bar{Q}}{(T_w - T_r)\rho c_p U} = \frac{\gamma + 1}{2\gamma} a(\bar{G} + \bar{F}) \quad (10\text{-}60)$$

Making use of Eq. (10-27), the Eucken expression for the Prandtl number,

$$\text{St} = \frac{\text{Nu}}{\text{Re Pr}} = \frac{9\gamma - 5}{4\gamma} \frac{\text{Nu}}{\text{Re}}$$

The thermal-recovery factor can be expressed in terms of Eq. (10-59):

$$r = \frac{T_r - T_0}{T_s - T_0} = \frac{\gamma}{\gamma + 1} \left( 2 + \frac{1}{s^2} \frac{\bar{F}}{\bar{G} + \bar{F}} \right) \quad (10\text{-}61)$$

where $T_s$ is the stream stagnation temperature. The ratio of the stagnation temperature to the freestream temperature is given in terms of the molecular speed ratio $s$:

$$\frac{T_s}{T_0} = 1 + \frac{\gamma - 1}{2} \text{Ma}^2 = 1 + \frac{\gamma - 1}{\gamma} s^2 \quad (10\text{-}62)$$

The results of Eqs. (10-59) to (10-61) are shown in Figs. 10-18 to 10-20.

In addition to the shapes shown in the figures, other shapes can be easily deduced. For example, a flat plate inclined at an angle $\theta$ to the flow direction can be treated in terms of the flat plate normal to the flow.

$$\bar{G}_\theta(s) = \frac{e^{-\eta^2}}{2\sqrt{\pi} s} = \frac{\eta}{s} \frac{e^{-\eta^2}}{2\sqrt{\pi} \eta} = \bar{G}_n(\eta) \sin \theta$$

and
$$\bar{F}_\theta(s) = \bar{F}_n(\eta) \sin \theta$$

where $\eta = s \sin \theta$. The subscript $\theta$ indicates a flat plate at an angle of incidence $\theta$ with the onflow, and subscript $n$ indicates a flat plate normal to the onflow. Thus

$$\text{St}_\theta(s) = \text{St}_n(\eta) \sin \theta$$

$$r_\theta(s) = \frac{2\gamma}{\gamma + 1} \cos^2 \theta + r_n(\eta) \sin^2 \theta$$

It should be noted also that the flat-plate results apply to any surface making a constant angle with the flow. Therefore these results apply to wedges and cones in comparison with flat plates at angles of incidence with the flow, where the angle of incidence of the plate corresponds to half the opening angle of the wedge or cone. The zero-angle flat plate

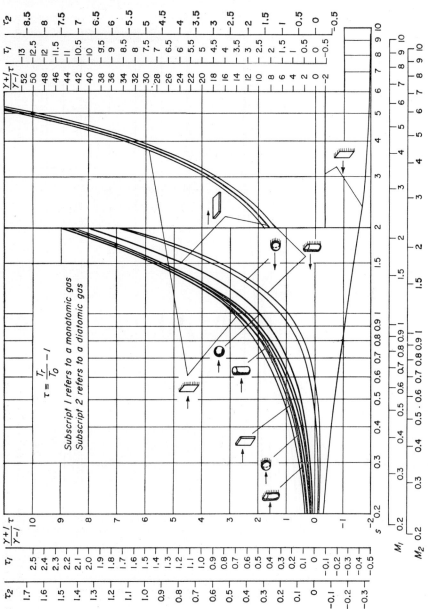

FIG. 10-18. Ratio of the wall-recovery temperature and the freestream temperature for a flat plate, a sphere, and a transverse cylinder in free molecule flow. [A. K. Oppenheim, J. Aeronaut. Sci., **20**:1 (1953).]

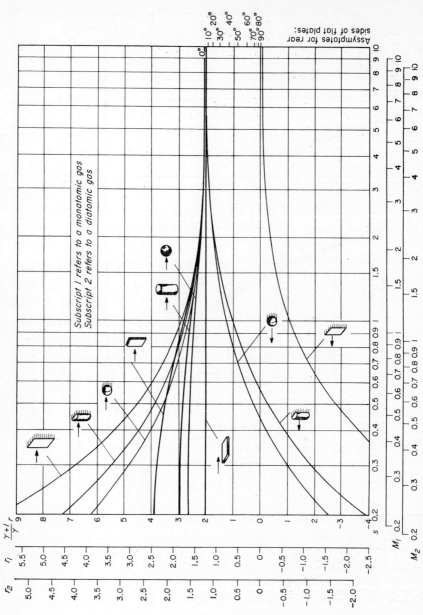

Fig. 10-19. Thermal-recovery factors for a flat plate, a sphere, and a transverse circular cylinder in a free molecule flow. [A. K. Oppenheim, *J. Aeronaut. Sci.*, **20**:1 (1953).]

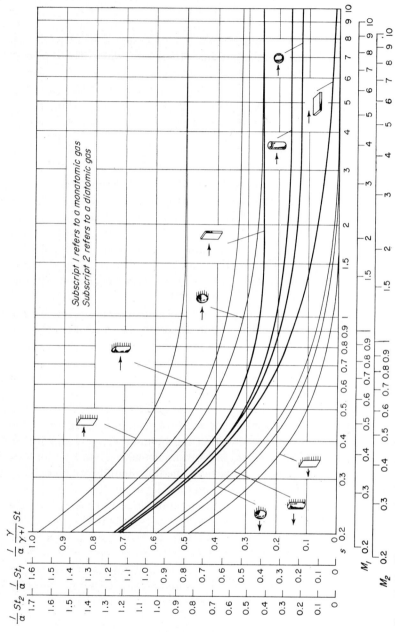

Fig. 10-20. Heat transfer from a flat plate, a sphere, and a transverse circular cylinder in a free molecule flow. [A. K. Oppenheim, *J. Aeronaut. Sci.*, **20**:1 (1953).]

corresponds in conditions to those occurring on a cylinder of any cross-sectional shape in axial flow.

**10-4. Heat Transfer in Liquid Metals.** The present section deals with heat transfer in liquid metals, because there are some characteristic features connected with this type of heat transfer which are not present in other fluids. The heat-conductivity values for liquid metals are very much larger than for any other fluids, and correspondingly the Prandtl numbers are very small. They cover a range of approximately 0.005 to 0.03. As a consequence of the high heat conductivity the dominating factor in the combined conduction and convection process is heat conduction.

For laminar flows through ducts the solutions which have been obtained in Sec. 7-7 apply for liquid metal as well as for other fluids in so far as the derivation of these equations makes no restriction as to the Prandtl number. The Nusselt numbers for developed flow conditions according to these relations are constant. The values of this constant are different for constant wall temperature and for constant heat flow. However, the entrance length for liquid-metal flows is quite short because of the low value of the Prandtl number. This means that the rearrangement of the temperature field takes place in a short section, and consequently, the axial temperature gradients in the flow direction may in some cases be of the same order of magnitude as the radial temperature gradients. This will apply especially for low Reynolds numbers, and for such values the solutions in Sec. 7-7 will have to be corrected because longitudinal conduction was neglected in deriving those equations.

Bounday-layer type of flow has been investigated in Sec. 7-3. The calculations in that section were made with the assumption that the flow boundary layer is thicker than the thermal boundary layer, and it was found that this condition is fulfilled for fluids with Prandtl numbers larger than 1. In this range, the ratio of the flow boundary-layer thickness to the thickness of the thermal boundary layer was found to be proportional to the cube root of the Prandtl number. Extrapolating this result in a qualitative way to low Prandtl numbers leads to the conclusion that for liquid metals the thermal boundary layer will be much thicker than the flow boundary layer. The calculation presented in Sec. 7-3 can also be carried out for this situation. It is, however, quite simple to obtain an approximate solution to the boundary-layer energy equation itself on the basis of the following consideration, which has been suggested by R. J. Grosh: The major portion of the temperature field in a liquid-metal boundary layer will be in a range where the velocity is equal to the stream velocity. This results because the thermal boundary layer is expected to be considerably thicker than the flow boundary layer. Correspondingly, one has to expect that the temperature field in the boundary layer can be calculated with good approximation on the basis of the assumption that in the whole field up to the plate surface the velocity is equal to the stream velocity. In this way the boundary-layer energy equation given in Sec. 7-6 simplifies to the following form:

$$u_s \frac{\partial t}{\partial x} = \alpha \frac{\partial^2 t}{\partial y^2}$$

The stream function is given by the relation $\psi = u_s y$ and the dimensionless stream function $f = 2\eta$. With this expression the temperature field simplifies to

$$\frac{t - t_w}{t_s - t_w} = \theta' = 2\sqrt{\frac{\Pr}{\pi}} \int_0^\eta e^{-\Pr \eta^2} d\eta$$

The temperature gradient at the plate surface is

$$\left(\frac{\partial t}{\partial y}\right)_w = \left(\frac{d\theta'}{d\eta}\right)_w \frac{1}{2} \sqrt{\frac{u_s}{\nu x}} (t_s - t_w) = \sqrt{\frac{\Pr}{\pi}} \sqrt{\frac{u_s}{\nu x}} (t_s - t_w)$$

The heat-transfer coefficient is

$$h = \frac{-k}{t_w - t_s} \left(\frac{\partial t}{\partial y}\right)_w = k \sqrt{\frac{\Pr}{\pi}} \sqrt{\frac{u_s}{\nu x}}$$

and the local Nusselt number is

$$\mathrm{Nu}_x = \frac{hx}{k} = \frac{1}{\sqrt{\pi}} \sqrt{\mathrm{Re}_x \Pr} = 0.565 \sqrt{\mathrm{Re}_x \Pr} \qquad (10\text{-}63)$$

R. J. Grosh and R. D. Cess[1] solved numerically the equation describing the temperature field in Sec. 7-6 and obtained values which were from 7 to 12 per cent smaller than the ones calculated from the equation above in a Prandtl-number range between 0.005 and 0.025. They also showed that for values $\mathrm{Re}_x \Pr > 50$ neglect of the longitudinal heat conduction is justified except in the immediate neighborhood of the plate leading edge.

The heat transfer around the circumference of a cylinder with its axis normal to the flow direction can be calculated if, again, the decrease of the velocity within the flow boundary layer is neglected. Boussinesq[2] in an early paper showed that the energy equation applying to heat conduction in a two-dimensional inviscid flow field can be simplified when it is transformed from the geometric coordinates $x$, $y$ to the flow potential $\varphi$ and the stream function $\psi$ as independent coordinates. It becomes

$$u_s \frac{\partial t}{\partial \varphi} = \alpha \left(\frac{\partial^2 t}{\partial \varphi^2} + \frac{\partial^2 t}{\partial \psi^2}\right)$$

---

[1] R. J. Grosh and R. D. Cess, *Trans. ASME*, **80**:667–676 (1958); see also E. M. Sparrow and J. L. Gregg, *J. Aeronaut. Sci.*, **24**, 852 (1957).

[2] K. J. Boussinesq, Calcul du pouvoir refroidissant des courrants fluides, *J. math.*, **1**:285 (1905).

If, as before, longitudinal conduction is neglected, then the term $\partial^2 t/\partial \varphi^2$ becomes zero and the equation assumes the same form as that for a flat plate. The Nusselt number is then obtained by transforming the solution back to the geometric coordinates $x$ and $y$. In this way Grosh and Cess calculated that the average Nusselt number for a single circular cylinder is

$$\overline{\mathrm{Nu}}_d = 1.015(\mathrm{Re}_d \, \mathrm{Pr})^{\frac{1}{2}} \tag{10-64}$$

The calculation assumes that the heat transport connected with the vortex motion in the back of the cylinder is negligibly small as compared with heat transported by conduction. Comparison with measurements indicates that this assumption is justified for values of the parameter Re Pr smaller than 500.

Turbulent duct flow of liquid metals can be calculated in a simple way when the heat transport by turbulent mixing can be considered small as compared with the conductive transport. The turbulent-velocity profile can be approximated by one of constant velocity with sufficient accuracy. Such a flow is usually referred to as *slug flow*. The calculations which have been discussed in Sec. 7-7 can easily be modified for this simplified velocity distribution. One finds in this way that the Nusselt number for constant heat flow is

$$\mathrm{Nu}_d = 8$$

and for constant wall temperature

$$\mathrm{Nu}_d = 5.8 \tag{10-65}$$

Actually the equation describing heat transfer for slug flow is identical with the equation for unsteady heat conduction in a solid body. Heat transfer in the entrance section of a tube can, therefore, easily be obtained from solutions to the unsteady-heat-conduction process into a rod with circular cross section.[1]

Heat exchange between a liquid metal and a solid surface through a turbulent boundary layer can be treated in exactly the same way as for a laminar boundary layer when the turbulent contribution to the heat exchange is negligible. This means that the relations developed for laminar flow should hold for turbulent flow as well.

Measurements on liquid metals have indicated that for turbulent flow through circular ducts, the heat exchange caused by turbulent mixing becomes appreciable and cannot be neglected when the parameter $\mathrm{Re}_d \, \mathrm{Pr}$ exceeds a value of approximately 50. Much used is the equation

$$\mathrm{Nu}_d = 7 + 0.025(\mathrm{Re}_d \, \mathrm{Pr})^{0.8} \tag{10-66}$$

[1] Heat transfer for turbulent flow through ducts with various cross sections is calculated this way by J. P. Hartnett and T. F. Irvine, *AIChE Journal*, **3**:313–317 (1957).

as developed by Lyons on the basis of calculations by Martinelli. The equation describes heat transfer for thermally developed conditions, turbulent flow through a tube, and constant heat flux at the wall. The results of experiments for the same condition scatter considerably and are on an average about 50 per cent lower than the predictions of the above equation. Comparisons with measurements on tube bundles indicate that calculations which consider only heat conduction agree with experimental values within 16 per cent for $Re_d$ Pr smaller than 500.[1]

For laminar flow through circular tubes several experimental investigations have obtained heat-transfer coefficients which are smaller than the values calculated from the equations in Sec. 7-7 for developed flow conditions. No satisfactory explanation of this fact is available. It has been suggested that nonwetting of the tube walls by the liquid metal may be responsible for the small heat-transfer coefficients and that nonwetting may be connected with gas entrainment in the metal flow.[2]

**10-5. Transpiration and Film Cooling.** Special cooling methods have been developed in an attempt to protect certain structural elements in turbojet and rocket engines from the influence of hot gases, like combustion chamber walls, exhaust nozzles, or gas turbine blades. These cooling methods will be discussed in their fundamental aspect in this section.

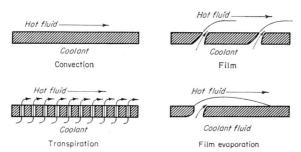

Fig. 10-21. Sketches of convection, transpiration, and film cooling.

Figure 10-21 shows sketches of these cooling methods. The upper left-hand sketch indicates the standard convection cooling. The upper right-hand arrangement shows the film cooling process in which a stream of coolant is blown through a series of slots in a direction that is tangential to the surface. In this way a layer is created which insulates the wall from the hot gases. The coolant film is gradually destroyed by mixing with the hot gases, so that its effectiveness decreases in the downstream direction. This disadvantage can be avoided by the process called

---

[1] Grosh and Cess, op. cit., pp. 677–682.
[2] Much useful information on heat transfer of liquid metals is found in "Liquid Metals Handbook," 2d ed., Atomic Energy Commission, Department of the Navy, Washington, D.C., 1952.

302    HEAT TRANSFER BY CONVECTION

transpiration cooling which is indicated in the lower left-hand sketch. In this method the wall is manufactured from a porous material and the coolant is blown through the pores. The coolant film on the hot gas side is, therefore, continuously renewed, and the cooling effectiveness can be made to stay constant along the surface. In the discussion up to now it was assumed that the coolant as well as the hot medium are either both gases or both liquids. When the hot medium is a gas, the cooling effectiveness can be very much increased by using a liquid as a coolant. This method is shown in the lower right-hand sketch of the figure. In this case a liquid film is created on the hot gas side of the wall, the liquid is evaporated on its surface, and the heat is absorbed by the evaporation process, thus substantially increasing the effectiveness of this cooling

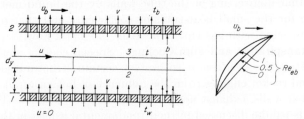

Fig. 10-22. Couette flow with ejection and suction through the porous walls. $t_b$ assumed larger than $t_w$.

method. It will be called evaporative film cooling or evaporative transpiration cooling, depending on whether the coolant is discharged through slots or through a porous wall.[1]

*Transpiration cooling* will now be considered in terms of its physical processes. In this cooling method a counterflow is created between the coolant ejected from the wall and the heat flux which moves from the hot gas toward the cooler surface. This reduces the heat-transfer coefficient as compared with conditions on a solid surface. We shall attempt to obtain an expression which approximately describes this reduction in heat transfer. For such a purpose the actual situation can be satisfactorily replaced by a considerably simplified model. As such, we shall use a Couette flow as it exists between two parallel walls, one of which is stationary whereas the other one moves with constant velocity within its plane. Exact solutions of the laminar boundary-layer equations with fluid ejection from the wall surface will be presented later in Sec. 16-2.

*Couette Flow with Transpiration Cooling.* Figure 10-22 presents a

[1] A very serious problem in the development of missiles, satellites, and spaceships is the excessive heating of the skin of these vehicles by friction in the high-velocity air stream. A method which appears very effective is called **ablation cooling**. The skin is manufactured of such a material that it sublimes, decomposes, or melts when the temperature increases by aerodynamic heating. In this way a mass flow away from the skin is created, which reduces heat transfer as in transpiration cooling.

sketch of the Couette flow between a stationary wall 1 and a moving wall
2. A coordinate system with the axes $x$ and $y$ is used as indicated in the
figure. Fluid will be carried along in the $x$ direction between the two
walls by the shear originating from the moving plane 2. It is well known
that the flow velocity $u$ in the $x$ direction increases linearly from the value
zero at wall 1 to the value $u_b$ with which wall 2 moves when the position
considered is sufficiently far from the entrance and when the pressure is
constant in the $x$ direction. If fluid is additionally injected into the flow
field through the porous wall 1 with velocity $v$ and correspondingly
removed through the wall 2, then the velocity field will be different.
With the pressure being constant in the $x$ and $y$ directions and the
velocity $v$ being constant along the walls, the velocities $u$ and $v$ change in
the $y$ direction only. Continuity additionally demands that the velocity
$v$ is also constant in the $y$ direction when the properties are assumed constant. A momentum balance in the $x$ direction on the small volume element 1-2-3-4 has to consider the following forces: a shear stress in plane
1-2 and one in plane 3-4. In addition, $x$ momentum is transported
through both of these planes. Only equal pressure forces occur in planes
1-4 and 2-3, and the momentum entering through 1-4 is equal to the
momentum leaving the element through plane 2-3. This results in the
following equation:

$$\mu \frac{d^2 u}{dy^2} - \rho v \frac{du}{dy} = 0$$

The boundary conditions are

For $y = 0$: $\qquad u = 0$
For $y = b$: $\qquad u = u_b$

A simple integration gives the following expression for the velocity component $u$:

$$\frac{u}{u_b} = \frac{e^{vy/\nu} - 1}{e^{vb/\nu} - 1}$$

This equation can be written in the following way by introducing a
Reynolds number $\text{Re}_{eb}$ based on the ejection velocity and the length
$b$ ($\text{Re}_{eb} = bv/\nu$)

$$\frac{u}{u_b} = \frac{e^{(y/b)\text{Re}_{eb}} - 1}{e^{\text{Re}_{eb}} - 1}$$

The shapes of the velocity profiles described by this equation for several
values of the ejection Reynolds number are indicated in Fig. 10-22.

It will now be assumed that the two walls are at two different temperatures $t_w$ and $t_b$ as indicated in the figure. The heat flow from surface 2
through the flowing fluid to surface 1 will be calculated. The tempera-

ture in the flow field will be a function of $y$ alone if the two surface temperatures are constant and if the flow is at a sufficient distance from the entrance. A heat balance on the volume element 1-2-3-4 has to consider the heat fluxes by conduction and convection through planes 1-2 and 3-4. No heat flux by conduction will occur through planes 1-4 and 2-3, and the entering and leaving convection fluxes will be equal. Heat may also be generated internally in the volume element. The heat generated per unit volume and time will be denoted by $\Phi$. This heat can, for instance, be generated by internal friction

$$\Phi = \mu \left(\frac{du}{dy}\right)^2$$

The heat balance results in the following equation:

$$k\frac{d^2t}{dy^2} - \rho v c \frac{dt}{dy} + \Phi = 0$$

The internal heat generation can be eliminated from our consideration by defining the heat-transfer coefficient with a recovery temperature in the same way as demonstrated in Sec. 10-1. The differential equation which describes the temperature field in a flow without heat generation (for low velocities)

$$k\frac{d^2t}{dy^2} - \rho v c \frac{dt}{dy} = 0$$

with the boundary conditions

For $y = 0$: $\qquad t = t_w$
For $y = b$: $\qquad t = t_b$

is completely analogous to the above equation describing the flow field. Its solution is

$$\frac{t - t_w}{t_b - t_w} = \frac{e^{\rho v c y/k} - 1}{e^{\rho v c b/k} - 1} = \frac{e^{(y/b)\operatorname{Re}_{eb}\operatorname{Pr}} - 1}{e^{\operatorname{Re}_{eb}\operatorname{Pr}} - 1}$$

The heat flow conducted into the stationary wall is obtained by a differentiation of this equation:

$$q_w = -k\left(\frac{dt}{dy}\right)_{y=0} = \frac{-k(t_b - t_w)\operatorname{Re}_{eb}\operatorname{Pr}}{b(e^{\operatorname{Re}_{eb}\operatorname{Pr}} - 1)}$$

If we define a heat-transfer coefficient $h$ by the equation

$$q_w = h(t_w - t_b)$$

and compare it with the heat-transfer coefficient $h_0$ in Couette flow without ejection ($v = 0$)

$$q_w = \frac{k}{b}(t_w - t_b) = h_0(t_w - t_b)$$

we obtain

$$\frac{h}{h_0} = \frac{\mathrm{Re}_{eb}\,\mathrm{Pr}}{e^{\mathrm{Re}_{eb}\mathrm{Pr}} - 1} \qquad (10\text{-}67)$$

This last equation gives us the reduction in the heat-transfer coefficient connected with the ejection process.

Heat generation by internal friction, important at high flow velocities, can be accounted for by replacing the wall temperature $t_b$ by the recovery temperature connected with this specific flow process. The above equation can be used as a first estimate of heat transfer in a laminar boundary layer on a flat plate with ejection provided the product $\mathrm{Re}_{eb}\,\mathrm{Pr}$ is written in the following way:

$$\mathrm{Re}_{eb}\,\mathrm{Pr} = \frac{\rho v c b}{k} = \frac{\rho v c}{h_0} = \frac{v}{u_s}\frac{1}{\mathrm{St}_0} \qquad (10\text{-}68)$$

where $v$ = ejection velocity at plate surface
$u_s$ = stream velocity
$\mathrm{St}_0$ = Stanton number on solid flat plate

Mickley[1] showed that the actual reduction in heat transfer decreases somewhat with increasing Prandtl number but is always larger than the one given by the Couette flow equation, so that this equation then gives a conservative estimate for laminar-boundary-layer flow. Heat transfer coefficients obtained from laminar-boundary-layer theory for $\mathrm{Pr} = 0.7$ and 1 are presented in Fig. 16-4.

W. D. Rannie[2] used the Couette flow model to describe conditions in the laminar sublayer of a turbulent boundary layer. This means that in the calculation presented above the distance of the two surfaces is replaced by the thickness of the laminar sublayer, the velocity $u_b$ of the moving plane by the velocity in the border between the sublayer and the turbulent boundary layer, and the temperature $t_b$ by the border temperature. Rannie also assumed that the turbulent mixing process in the boundary layer and the boundary-layer and sublayer thicknesses are not changed by the ejection. The resulting relation becomes somewhat complex and will not be presented here. Mickley showed that the simpler relation given by the Couette flow equation agrees quite well with his own experiments for turbulent boundary-layer flow of air over a flat plate with air ejection.

The relation for the temperature-recovery factor in Couette flow cannot be used to describe conditions in a boundary layer. Figure 10-23 pre-

---

[1] H. S. Mickley, R. C. Ross, A. L. Squyers, and W. E. Stewart, Heat Mass and Momentum Transfer for Flow over a Flat Plate with Blowing or Suction, *Natl. Advisory Comm. Aeronaut. Tech. Notes*, no. 3208, 1954.

[2] W. D. Rannie, A Simplified Theory of Porous Wall Cooling, *Calif. Inst. Technol., Jet Propulsion Lab., Progr. Rept.* 4-50, 1957.

sents the results of calculations for a laminar boundary layer.[1] It was calculated for air in the main flow as well as for the coolant. The change of recovery factor with ejection rate is smaller in a turbulent boundary layer.

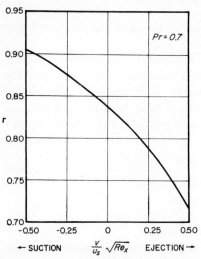

FIG. 10-23. Temperature-recovery factor $r$ for laminar boundary layer flow over a flat plate with ejection and suction. (*According to E. R. G. Eckert and J. N. B. Livingood.*) $Re_x$ based on stream velocity and distance $x$ from leading edge.

The temperature which the stationary wall assumes on the gas-flow side can be obtained by a heat balance on a control volume as indicated in Fig. 10-24. Steady state, no internal heat generation, and no heat transfer to the wall surfaces by radiation will be postulated. The surface 1-2 of the volume is located just outside the wall surface. The surface 3-4 is displaced from the wall surface on the coolant entry side by such an amount that it is located outside the boundary layer which exists on this side of the wall. It will be assumed that no heat flows through the surfaces 1-4 and 2-3. Heat is transported by conduction and by convection through the plane 1-2 and by convection only through the plane 3-4. The heat fluxes per unit area are indicated in the figure. It is postulated that the cooling air leaves the porous wall with the wall surface temperature $t_w$. This will agree with reality when the pores are sufficiently small.

A heat balance on the control volume results in the equation

$$h(t_s - t_w) = \rho v c(t_w - t_c) \tag{10-69}$$

The heat-conduction flux $k(dt/dy)_w$ has been replaced by $h(t_s - t_w)$. From this equation, the wall surface temperature $t_w$ can be calculated as soon as the coolant entry temperature $t_c$, the temperature $t_s$ in the stream outside the boundary layer, and the film heat-transfer coefficient $h$ are known. The temperature of the coolant on its way through the porous wall and the temperature in the porous material itself can be obtained by a calculation similar to the one presented above for Couette flow. Such a calculation has been published by L. Green,[2] who also included heat

---

[1] E. R. G. Eckert and T. N. B. Livingood, *Natl. Advisory Comm. Aeronaut. Tech. Notes*, no. 2733, 1952.

[2] L. Green, Jr., *J. Appl. Mech.*, **19**:173–178 (1952).

sources to cover conditions which may arise in nuclear reactors. Calculations have also been published describing the flow and heat-transfer process in ducts with porous walls. The results have been summarized and correlated by Eckert and Donoghue.[1]

It will now be attempted to obtain analytically expressions which describe heat transfer in *film cooling*. For such a purpose the significant physical processes influencing this cooling method have to be singled out. The development of the thermal boundary layer is dictated by the fact that cold fluid is added at the location of the slots. In addition, heat may be added to or removed from the boundary layer along the entire surface of the wall. These effects are also obtained in a model which has concentrated line heat sinks arranged at the location of the slots and in addition distributed heat sources or sinks along the wall surface.

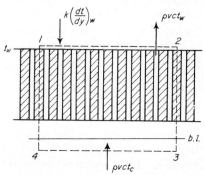

Fig. 10-24. Energy balance for transpiration cooling.

The thermal effects on the boundary layer should be well described by this model if one does not consider the immediate neighborhood of the slots. It neglects, on the other hand, the fact that in the actual arrangement the hydrodynamic boundary layer will also be influenced by the ejection of the fluid through the slots. It has been shown, however, in several of the preceding sections that local changes in the flow field have only a secondary effect on the heat-transfer process. Mathematically, the adoption of our model means that the equation for the velocity boundary layer is, for constant fluid properties, the same as on a solid wall and that the arrangement of heat sinks and sources is prescribed to the energy boundary-layer equation. The latter equation is linear when the fluid properties are constant. This means that a solution of the energy equation can be obtained by superposition of two solutions, one of which has the concentrated heat sinks only as boundary condition whereas the other one is obtained for the distributed sources or sinks. This latter solution will be identical with the ones which have been obtained before on solid surfaces for the appropriate heat-flux distribution. The heat transfer will, therefore, be described by the heat-transfer coefficients $h_0$ on a solid surface, and the heat flux from the wall will be given by the following equation:

$$q_w = h_0(t_w - t_r) \qquad (10\text{-}70)$$

[1] E. R. G. Eckert, P. L. Donoghue, and B. J. Moore, *Natl. Advisory Comm. Aeronaut. Tech. Notes*, no. 4102, 1957.

in which the recovery wall temperature $t_r$ is that temperature which the wall surface assumes when it is cooled by the concentrated heat sinks only. This temperature $t_r$ is obtained from the solution of the energy equation which considers the concentrated heat sinks as boundary conditions.

The method described above was proposed by the author[1] at a symposium held at the University of Michigan in 1952, and the analysis has been performed by M. Tribus and G. Klein.[2] They obtained the following relation for a flat plate with a turbulent boundary layer and with a line heat sink at its leading edge, assuming the properties to be the same for the main stream and for the coolant.

$$\frac{t_r - t_s}{t_c - t_s} = 5.77 (\text{Pr})^{2/3} (\text{Re}_s)^{0.2} \left(\frac{u_s x}{u_c s}\right)^{-0.8} \qquad (10\text{-}71)$$

The nomenclature in this equation is evident from Fig. 10-25. $\text{Re}_s$ is a Reynolds number based on the velocity $u_c$ at the slot exit and on the slot width $s$. K. Wieghardt[3] performed experiments on a geometry comparable to the model used by Tribus with air in the main flow and as the coolant. The adiabatic wall temperature and the temperature profiles in the boundary layer for $x/s > 100$ are shown in Fig. 10-25. Wieghardt presented his results in the following equation:

$$\frac{t_r - t_s}{t_c - t_s} = 21.8 \left(\frac{u_s x}{u_c s}\right)^{-0.8} \qquad (10\text{-}72)$$

valid for $u_c/u_s \leq 1$, $\text{Re}_x = 10^6$ to $10^7$, $x/s > 100$, $\text{Re}_s = 3{,}700$ to $12{,}000$. It can be observed that this relation is very similar to the one derived by Tribus. Introduction of the values for the slot Reynolds number into Eq. (10-71) leads to a numerical constant between 24 and 30. The difference between this value and the one obtained by Wieghardt can readily be explained by the fact that Wieghardt had not located the cooling air slot at the plate leading edge, so that the boundary layer has already grown to a certain thickness at the slot location. It may also reflect the fact that the velocity boundary layer is influenced by the coolant ejection. A number of experimental investigations have been conducted at various laboratories. The various results show considerable disagreement, so

---

[1] E. R. G. Eckert, Transpiration and Film Cooling, in "Heat Transfer" (a symposium), Engineering Research Institute, University of Michigan, Ann Arbor, Mich., 1953.

[2] M. Tribus and G. Klein, Forced Convection from Non-isothermal Surfaces, in "Heat Transfer" (a symposium), Engineering Research Institute, University of Michigan, Ann Arbor, Mich., 1953.

[3] K. Wieghardt: "Hot Air Discharge for De-icing," AAF-Translation No. F-TS919 RE, December, 1956.

SPECIAL HEAT-TRANSFER PROCESSES 309

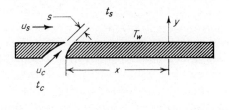

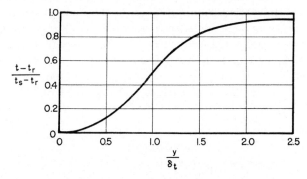

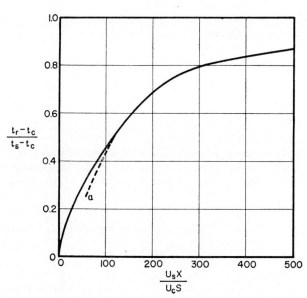

Fig. 10-25. Temperature profile in boundary layer and wall temperature for film cooling. $\delta_t$ is the wall distance where the temperature ratio assumes the value 0.5. (*According to K. Wieghardt.*) Curve *a* calculated with Eq. (10-72).

that additional work is necessary to clarify the heat-transfer process in film cooling.

Experiments on the *evaporative film cooling* indicate that the wall surface temperature stays below the evaporation temperature on all locations which are covered by a liquid film. The extent of this film can be calculated by a heat balance when the heat transfer from the hot gas to the evaporating surface is known. From the fact that the evaporation creates a mass flow from the interface into the main stream, one would conclude that these heat-transfer coefficients should be lower than for comparable conditions on a solid surface. Experiments performed at the National Advisory Committee for Aeronautics,[1] however, yielded heat-transfer coefficients which were considerably larger than those for a solid wall. Wrinkling of the liquid surface and increased turbulence may be responsible for this situation.

### PROBLEMS

**10-1.** Calculate the distribution of the static and total temperature in a gas flowing through a duct built with two parallel walls. Assume that the flow is steady, laminar, and fully developed and the duct walls are insulated.

**10-2.** Make the same calculation as in the preceding problem, however, for flow through a pipe with circular cross section.

**10-3.** Calculate and plot the static and total temperatures for laminar boundary-layer flow over an adiabatic flat plate for a fluid with Prandtl number 0.7 and 10.

**10-4.** Calculate the temperature which the skin of a missile assumes when it is in thermal equilibrium receiving heat by convection from the atmosphere and losing heat by radiation into the universe (0 R). The missile flies with a Mach number 6 at 20 miles (10 miles) altitude. Assume that the skin of the missile can be considered a flat plate, and make calculation for a location 1 ft from the leading edge. Assume that surface is black for the emitted radiation.

**10-5.** Derive an equation for the heat-transfer coefficient in transpiration cooling using Rannie's approach as described on page 305. Use the relations in Sec. 8-1 to describe conditions in the turbulent part of the flow field and the Couette flow model to describe conditions in the laminar sublayer. Develop the equation for turbulent tube flow and for a turbulent boundary layer.

[1] G. R. Kinney, Jr., and J. L. Sloop, Internal Film Cooling Experiments in a 4-inch Duct with Gas Temperatures to 2000 F, *Natl. Advisory Comm. Aeronaut. Tech. Mem.* E-50-F-19, 1950.

CHAPTER 11

# FREE CONVECTION

Free- or natural-convection flow arises in various ways, for instance, when a heated object is placed in a fluid, otherwise at rest, the density of which varies with temperature. Heat is transferred from the surface of the object to the fluid layers in its neighborhood. The density decrease which in a normal fluid is connected with a temperature increase causes these layers to rise and create the free-convection flow which now transports heat away from the object. Physically such a flow is described by stating that it is caused by body forces. In the specific example described above the body forces are gravitational forces. Free-convection flows under the influence of gravitational forces have been investigated most extensively because they are encountered frequently in nature as well as in engineering applications. Flows can be caused by other body forces as well. In a rotating system, for instance, centrifugal and Coriolis forces exist as body forces. Flow of cooling air through passages in the rotating blades of gas turbines is an example of flow under the influence of such body forces. In the boundary layers which surround missiles flying with high supersonic speed, temperatures may be so high that the air is ionized, which means that the atoms and molecules carry electrical charges. In this case electric or magnetic forces may arise which as body forces influence the flow. Little has been done to study free convection under the influence of forces other than gravitational. For a first estimate the relations available for gravitational free convection can be used to obtain information on flow under the influence of other body forces. The gravitational acceleration $g$ has then to be replaced in the Grashof number—a dimensionless term determining free-convection flow —by the acceleration corresponding to the body force in consideration, for instance, by the centrifugal acceleration $V^2/r$. The answer obtained in this way can usually be considered only as an approximate one because the field for the accelerations belonging to the various body forces is often different from the gravitational-acceleration field.

In this chapter only gravitational free convection will be considered on a number of typical geometries: a vertical flat plate, a horizontal circular cylinder with constant surface temperature, and a fluid enclosed between two plane walls or two concentric cylinders with different temperatures. Finally the situation where forced flow convection is influenced by body forces (mixed free and forced convection) will be briefly discussed.

**11-1. Laminar Heat Transfer on a Vertical Plate and Horizontal Tube.** Free-convection flow which is caused by temperature differences in a gas or liquid also builds boundary layers on the surfaces of solid bodies. Because of the low velocities encountered, these boundary layers are thicker than those in the examples treated up to now. The integrated boundary-layer equations for momentum and heat flow can be used again to calculate the heat transfer in free convection. In this section at first the simplest problem will be solved, namely, that of the vertical plate.[1] If such a plate is heated, the gas or liquid in its immediate neighborhood experiences an increase in its temperature as a result of heat transfer from the plate and begins to flow in an upward direction. In this way a boundary layer arises with zero thickness at the lower edge and with an increasing thickness in the upward direction. The distance along the surface from the lower edge may be called $x$, the distance from the plate surface $y$. Within the boundary layer, the temperature decreases from the value on the plate surface to the uniform value which the gas or liquid has outside the heated region. The temperature of the plate may be constant over the whole surface and steady. The difference between this value and the temperature outside the boundary layer is designated as $\vartheta_w$. The difference between the temperature at any point within the boundary layer and the temperature outside it may be denoted by $\vartheta$. The temperature field in the neighborhood of the plate must have the shape sketched in Fig. 11-1. In the figure, $\delta$ is the boundary-layer thickness. The velocity $u$ must be zero on the plate surface and outside the boundary layer. Therefore, the velocity field must have the shape sketched in Fig. 11-1. To solve the boundary-layer equation, the temperature profile is approximated by a parabola:

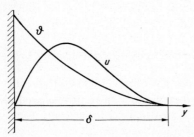

FIG. 11-1. Temperature and velocity profile in free-convection flow on a vertical wall.

$$\vartheta = \vartheta_w \left(1 - \frac{y}{\delta}\right)^2 \tag{11-1}$$

This equation already fulfills the boundary conditions $\vartheta = \vartheta_w$ for $y = 0$

---

[1] An analogous calculation for gases has been made by Squire in S. Goldstein, "Modern Developments in Fluid Dynamics," Oxford University Press, New York, 1938.

and $\vartheta = 0$ for $y = \delta$. A velocity profile similar to the one shown in Fig. 11-1 is represented by the equation

$$u = u_1 \frac{y}{\delta}\left(1 - \frac{y}{\delta}\right)^2 \tag{11-2}$$

with $u_1$ indicating an arbitrary function with the dimension of velocity. The maximum value of the velocity $u$ arises, according to this equation, at the distance $y = \delta/3$ from the wall. Its value is $u_{\max} = 4/27 u_1$. The boundary-layer thickness is assumed to be the same for the temperature and the velocity. This assumption has its justification in that it keeps the computational work small and that the results of the calculation performed with it agree quite well with experimental results.[1] Recently an analysis has been performed in which different boundary-layer thicknesses were assumed for the velocity and the temperature field.[2] The results deviated only moderately from those presented here.

The momentum equation (6-8) must be expanded by a summand which expresses the buoyancy force. This force is $g\rho\beta\vartheta$ per unit volume (see page 235). There acts on a volume element of height $dx$ and length $l$ (greater than the boundary-layer thickness, Fig. 7-2) the force $dx\, g\rho\beta \int_0^l \vartheta\, dy$. If this summand is added to Eq. (6-8) and it is considered that at the small velocities connected with free convection the pressure can be assumed constant and that the velocity $u$ outside the boundary layer is zero, there is obtained

$$\frac{d}{dx}\int_0^l u^2\, dy = g\beta \int_0^l \vartheta\, dy - \nu \left(\frac{du}{dy}\right)_w \tag{11-3}$$

The heat-flow equation (7-2) remains unchanged.

$$\frac{d}{dx}\int_0^l u\vartheta\, dy = -\alpha \left(\frac{d\vartheta}{dy}\right)_w \tag{11-4}$$

The temperatures are replaced here by the temperature difference $\vartheta$. The integrals in Eqs. (11-3) and (11-4) can be solved by introducing the temperature and velocity profiles (11-1) and (11-2).

$$\int_0^l u^2\, dy = \frac{u_1^2 \delta}{105} \qquad \int_0^l \vartheta\, dy = \frac{\vartheta_w \delta}{3} \qquad \int_0^l \vartheta u\, dy = \frac{1}{30} u_1 \vartheta_w \delta$$

Now the boundary-layer equations assume the form

$$\frac{1}{105}\frac{d}{dx}(u_1^2 \delta) = \frac{1}{3} g\beta \vartheta_w \delta - \nu \frac{u_1}{\delta}$$

$$\frac{1}{30} \vartheta_w \frac{d}{dx}(u_1 \delta) = 2\alpha \frac{\vartheta_w}{\delta}$$

[1] Y. S. Touloukian, G. A. Hawkins, and M. Jacob, *Trans. ASME*, **70**:13 (1948).
[2] E. M. Sparrow, *Natl. Advisory Comm. Aeronaut. Tech. Note* 3508, 1955. See also E. M. Sparrow and J. E. Gregg, *Trans. ASME*, **78**:435–440 (1956).

To solve these equations, let us try $u_1$ and $\delta$ as exponential functions of $x$:
$$u_1 = C_1 x^m \qquad \delta = C_2 x^n$$
Introducing these expressions we get
$$\frac{2m+n}{105} C_1{}^2 C_2 x^{2m+n-1} = g\beta\vartheta_w \frac{C_2}{3} x^n - \frac{C_1}{C_2} \nu x^{m-n}$$
$$\frac{m+n}{30} C_1 C_2 x^{m+n-1} = \frac{2\alpha}{C_2} x^{-n}$$

Since these equations must be valid for any value $x$, the exponents must have the same value for every summand:
$$2m + n - 1 = n = m - n$$
$$m + n - 1 = -n$$

This gives
$$m = \tfrac{1}{2} \quad \text{and} \quad n = \tfrac{1}{4}$$

Introducing these values into the above equation gives
$$\frac{C_1{}^2 C_2}{84} = g\beta\vartheta_w \frac{C_2}{3} - \frac{C_1}{C_2} \nu$$
$$\frac{C_1 C_2}{40} = \frac{2\alpha}{C_2}$$

or
$$C_1 = 5.17 \nu \left(\frac{20}{21} + \frac{\nu}{\alpha}\right)^{-1/2} \left(\frac{g\beta\vartheta_w}{\nu^2}\right)^{1/2}$$
$$C_2 = 3.93 \left(\frac{20}{21} + \frac{\nu}{\alpha}\right)^{1/4} \left(\frac{g\beta\vartheta_w}{\nu^2}\right)^{-1/4} \left(\frac{\nu}{\alpha}\right)^{-1/2}$$

Now the maximum velocity within the boundary layer is
$$u_{\max} = \frac{4}{27} u_1 = 0.766 \nu \left(0.952 + \frac{\nu}{\alpha}\right)^{-1/2} \left(\frac{g\beta\vartheta_w}{\nu^2}\right)^{1/2} x^{1/2} \qquad (11\text{-}5)$$

and the boundary-layer thickness
$$\delta = 3.93 \left(\frac{\nu}{\alpha}\right)^{-1/2} \left(0.952 + \frac{\nu}{\alpha}\right)^{1/4} \left(\frac{g\beta\vartheta_w}{\nu^2}\right)^{-1/4} x^{1/4} \qquad (11\text{-}6)$$

Again it is useful to make the boundary-layer thickness dimensionless by dividing it by the distance $x$ from the lower edge of the plate. In this equation there appears the Grashof number $\text{Gr}_x = g\beta\vartheta_w x^3/\nu^2$, based on the temperature difference between the plate and gas or liquid and the distance $x$ [Eq. (9-19)], and the Prandtl number $\text{Pr} = \nu/\alpha$:

$$\frac{\delta}{x} = 3.93 \, \text{Pr}^{-1/2}(0.952 + \text{Pr})^{1/4}(\text{Gr}_x)^{-1/4} \qquad (11\text{-}7)$$

The specific heat flow from the plate surface is represented by the equation
$$q = -k \left(\frac{d\vartheta}{dy}\right)_w$$

From Eq. (11-1) there is obtained

$$q = \frac{2k\vartheta_w}{\delta}$$

On the other hand, the equation by which the film-heat-transfer coefficient is defined reads

$$q = h\vartheta_w$$

Therefore
$$h = \frac{2k}{\delta} \qquad (11\text{-}8)$$

and in dimensionless form

$$\frac{hx}{k} = \mathrm{Nu}_x = 2\frac{x}{\delta}$$

By introducing the boundary-layer thickness one obtains

$$\mathrm{Nu}_x = 0.508\,\mathrm{Pr}^{1/2}(0.952 + \mathrm{Pr})^{-1/4}(\mathrm{Gr}_x)^{1/4} \qquad (11\text{-}9)$$

The local film heat-transfer coefficient decreases according to Eqs. (11-6) and (11-8) with increasing distance $x$. It is inversely proportional to the fourth root of $x$. By integration over the distance the average heat-transfer coefficient is found to be

$$\bar{h} = \tfrac{4}{3}h$$

This means that the average heat-transfer coefficient of a vertical plate with a height $x$ is $\tfrac{4}{3}$ the local value at the point $x$. For ideal gases the relationship $\beta = 1/T$ holds true. As long as the temperature differences are small, the expansion coefficient can be written $\beta = 1/T_0$, where $T_0$ is the absolute temperature in the gas outside the boundary layer. For air with a Prandtl number $\mathrm{Pr} = 0.714$,

$$\mathrm{Nu}_x = 0.378(\mathrm{Gr}_x)^{1/4} \qquad (11\text{-}10)$$

For this medium the heat transfer was calculated exactly by E. Pohlhausen in collaboration with E. Schmidt and W. Beckmann.[1] This resulted in a numerical value of 0.360 instead of 0.378 as in Eq. (11-10). Therefore, the approximate treatment agrees quite well with the much longer exact calculation. A comparison between the values calculated here and those measured and calculated by E. Schmidt and W. Beckmann is presented in Figs. 11-2 and 11-3. In the calculations, the property values were introduced at the plate temperature. S. Ostrach[2] has recently solved the laminar free-convection boundary-layer equations for a vertical plate on an electronic computer for several Prandtl numbers.

---

[1] E. Schmidt and W. Beckmann, *Tech. Mech. Thermodynam.*, **1**:1–24 (1930).
[2] S. Ostrach, *Natl. Advisory Comm. Aeronaut. Tech. Note* 2635, 1952.

316  HEAT TRANSFER BY CONVECTION

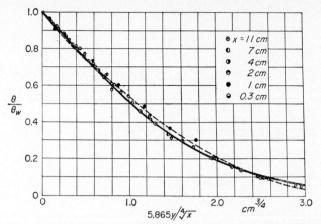

Fig. 11-2. Comparison between the measured and calculated temperature profile in free-convection flow on a vertical plate. (*Measured values according to E. Schmidt and W. Beckmann, solid-line exact solution by E. Schmidt and E. Pohlhausen, dashed-line approximate solution according to Squire.*)

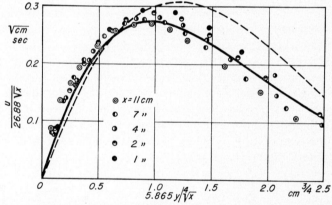

Fig. 11-3. Comparison between the measured and calculated velocity profile in free-convection flow on a vertical plate. (*Measured values according to E. Schmidt and W. Beckmann, solid-line exact solution by E. Schmidt and E. Pohlhausen, dashed-line approximate solution according to Squire.*)

The resulting velocity and temperature profiles are shown in Figs. 11-4 and 11-5. The Nusselt numbers obtained from this analysis agree with those calculated from Eq. (11-9) within 10 per cent in the Prandtl-number range from 0.01 to 1,000.

The agreement is much better than with the formula $\mathrm{Nu}_x = C(\mathrm{Gr}_x\,\mathrm{Pr})^{1/4}$ which is often used and which was derived by a generalization of the experiments with air on the assumption that inertia forces can be neglected in this type of flow [Eq. (9-19)]. Experiments by E. Schmidt

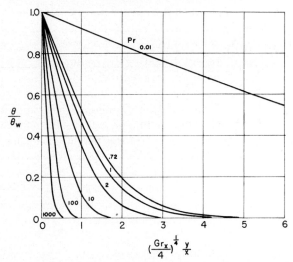

Fig. 11-4. Dimensionless temperature profiles for laminar free convection on a vertical plate. (*From S. Ostrach, Natl. Advisory Comm. Aeronaut. Tech. Note* 2635, 1952.)

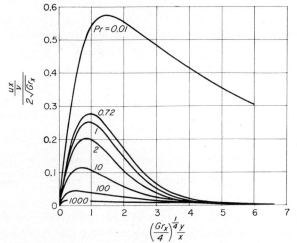

Fig. 11-5. Dimensionless velocity profiles for laminar free convection on a vertical plate. (*From S. Ostrach, Natl. Advisory Comm. Aeronaut. Tech. Note* 2635, 1952.)

and W. Beckmann, by E. R. G. Eckert and E. Soehngen[1] with air, and by H. H. Lorenz[2] and O. A. Saunders[3] with oil and mercury are also in fair agreement with the theoretical values.

[1] E. R. G. Eckert and E. Soehngen, *USAF Tech. Rept.* 5747, Wright-Patterson Air Force Base, Dayton, Ohio, 1948.
[2] H. H. Lorenz, *Z. tech. Physik*, **9**:362–366 (1934).
[3] O. A. Saunders, *Proc. Roy. Soc. (London)*, **A72**(948): 55–71, (1939).

Besides the vertical wall, the horizontal tube is important for industrial applications.[1] The boundary-layer equations were solved for this problem by R. Hermann.[2] The solutions agree well with the experimental results found by K. Jodlbauer.[3] The result can be summarized in the following way: The average film heat-transfer coefficient on a circular tube with the diameter $d$ has the same value as the average film heat-transfer coefficient on a vertical wall with the height $2.5d$ (see Fig. 11-6).

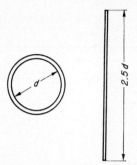

FIG. 11-6. Horizontal tube and vertical plate with equal average film heat-transfer coefficient in free convection.

For thin wires, the assumption that the boundary-layer thickness is small compared with the diameter does not hold true. The boundary-layer calculations of R. Hermann are based on this assumption, however. To take into account this fact we shall follow an idea first developed by Langmuir[4] and taken up by Rice.[5] We idealize the boundary layer as a stationary cylindrical air layer around the wire through which the heat given off by the wire must flow by conduction. Equations (3-6) and (3-7) therefore determine this heat flow, with $r_i$ equal to the radius of the wire (in this paragraph the diameter $d = 2r_i$ is used), $t_i$ the wire temperature, and $t_o$ the outside air temperature. On the other hand the definition equation

$$Q = \bar{h} 2\pi r_i l (t_i - t_o)$$

holds. Equating this expression to Eq. (3-7) gives

$$\bar{h} = k \frac{1}{r_i \ln(r_o/r_i)}$$

and

$$\overline{\mathrm{Nu}}_d = \frac{\bar{h} d}{k} = \frac{2}{\ln(r_a/r_i)} = \frac{2}{\ln(1 + 2b/d)}$$

where $b = r_a - r_i$ is the thickness of the idealized boundary layer. This thickness is so determined that for a boundary layer whose thickness is small as compared with the wire diameter, the equation transforms into an expression derived from Eq. (11-10). By changing to the average film heat-transfer coefficient and introducing the diameter $d = 2.5x$, we get

$$\overline{\mathrm{Nu}}_d = 0.400 (\mathrm{Gr}_d)^{1/4}$$

[1] Laminar free convection on a vertical cylinder was calculated by K. Millsaps and K. Pohlhausen, *J. Aeronaut. Sci.*, **25**:357–360 (1958).
[2] R. Hermann, *Z. angew. Math. Mech.*, **13**:433 (1933).
[3] K. Jodlbauer, *Forsch. Gebiete Ingenieurw.*, **4**:157 (1933).
[4] I. Langmuir, *Phys. Rev.*, **34**:401 (1912).
[5] C. W. Rice, *Trans. AIEE*, **42**:653 (1923).

For small values of $b/d$ the denominator $\ln(1 + 2b/d)$ transforms to $2b/d$ and therefore $\overline{\mathrm{Nu}}_d$ into $d/b$. Therefore $d/b = 0.400(\mathrm{Gr}_d)^{1/4}$, and introducing this into the above equation gives

$$\overline{\mathrm{Nu}}_d = \frac{2}{\ln\{1 + [2/0.400(\mathrm{Gr}_d)^{1/4}]\}} \tag{11-11}$$

This equation gives good agreement with experimental results.[1]

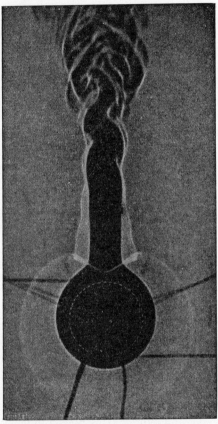

FIG. 11-7. Visualization of the boundary layer and film heat-transfer coefficient around a horizontal tube in free-convection flow by a schlieren photograph. (*From E. Schmidt, Forsch. Gebiete Ingenieurw.*, **3**:81–89, 1932.)

The dimensions of the boundary layer and the temperature field can be readily observed in schlieren or interference photographs. A schlieren method, which can be realized by very simple means, was described by

[1] Free convection to cylinders and spheres has been studied experimentally by J. K. Kyte, A. J. Madden, and E. L. Piret, *Chem. Eng. Progr.*, **49**:653–662 (1953). Experiments in rarefied air extended the Gr range to $10^{-7}$.

E. Schmidt.[1] An example of this method is shown in the photograph (Fig. 11-7), demonstrating free convection on a horizontal cylinder. A light source arranged at a great distance from the heated cylinder so that its light rays are parallel to the cylinder axis produces on a screen, without other means, the picture shown in Fig. 11-7. In passing through a field with a density gradient normal to its direction, a light ray is bent through a certain angle. Therefore, the light rays which traverse the heated boundary layer are deflected away from the cylinder, and the dark zone surrounding the contour in Fig. 11-7 indicates the boundary layer. It was shown by E. Schmidt that the distance of the heart-shaped bright line from the cylinder contour is proportional to the heat-transfer coefficient. The principle of an interferometer according to Zehnder and Mach[2] is shown in Fig. 11-8. Light from a monochromatic source $A$ is made parallel by a condensing lens $B$. It then falls on a plane glass plate $C$ whose surface is coated in such a way that half the light is reflected and the other half transmitted. The reflected part strikes the mirror $D$, is reflected by it, and falls on the glass plate $F$, which is coated in the same way as plate $C$. That part of the light transmitted through this plate is united with the second light beam, which is transmitted through plate $C$ and reflected by mirror $E$ and plate $F$. If the four plates $C$, $D$, $E$, $F$ are exactly parallel to one another, all light rays in each of the two beams have the same path length. On a screen $G$ there can then be observed a dark or a bright field, depending on whether the path lengths of the two beams between mirrors $CDF$ or $CEF$, respectively, vary by half a wavelength or an odd multiple of it, thus extinguishing each other, or by a full wavelength or a multiple, thus augmenting each other.

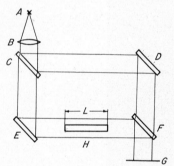

Fig. 11-8. Principle sketch of an interferometer according to Zehnder and Mach.

If a heated body, e.g., a cylinder $H$, is placed in one of the light beams, the temperature and therefore the density of the air are changed in the neighborhood of the body. Since the wavelength of light varies with density, there are now rays in the two beams which differ by one or more full wavelengths and others which differ by an odd number of half wavelengths. As a consequence of this, a field of dark and bright bands can

---

[1] E. Schmidt, *Forsch. Gebiete Ingenieurw.*, **3**:181–189 (1932).
[2] H. Schardin, *Z. Instrumentenk.*, **53**:396–403, 424–436 (1933).

now be observed on the screen $G$ around the shadow of cylinder $H$. The sharpness of this shadow can be improved if a lens is placed between plate $F$ and the screen, so that the lens creates an image of the central plane of the cylinder on the plane in which the screen $G$ is located. Figures 11-9 and 11-10 show photographs made in this way. Figure 11-9 represents the free convection around a horizontal tube, and Fig. 11-10 free convection along a vertical plane. The dark interference bands are lines of constant air density, and as the pressure can be assumed constant, these lines are isotherms. The temperature difference between any two successive bands can be easily calculated from the length of the cylinder or plate in the light-ray direction and the characteristics of the interferometer. From the photograph the temperature gradient in the air normal to the surface and therefore the heat-transfer coefficient can be determined. As the boundary-layer thickness is also seen, this method is very useful for investigating heat-transfer problems. The interferometer was first described by Zehnder and Mach. It was used for aerodynamic studies by T. Zobel.[1] The first studies in heat transfer were made by R. B. Kennard[2] and E. Eckert and E. Soehngen.[3]

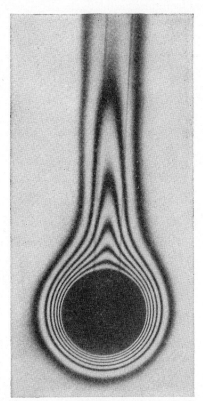

Fig. 11-9. Isotherms around a horizontal tube in free-convection flow as revealed by an interference photograph. (*Photograph by E. Eckert and E. Soehngen.*)

In the examples considered so far, the boundary layers were always laminar. The boundary layers which arise by free convection also change to turbulent flow when they have reached a certain thickness. In air the change occurs at a critical Grashof number around $Gr_x = 10^9$. This corresponds to a Reynolds number $Re_{c_x} = u_{max}\delta/\nu = 550$. The transition therefore takes place at a lower Reynolds number in free con-

[1] T. Zobel, *Z. Ver. deut. Ingr.*, **81**:503 (1937).
[2] R. B. Kennard, *J. Research Natl. Bur. Standards*, **8**:787 (1932).
[3] E. Eckert and E. Soehngen, *USAF Tech. Rept.* 5747, Wright-Patterson Air Force Base, Dayton, Ohio, 1948.

vection than in forced flow. Figure 11-11 is an interference photograph of a heated vertical plate of sufficient length to produce a turbulent boundary layer (3 ft). The interference lines again represent isotherms. The figures indicate the distance from the end of the lower plate in inches. It can be seen that the flow is smooth and laminar near the lower end.

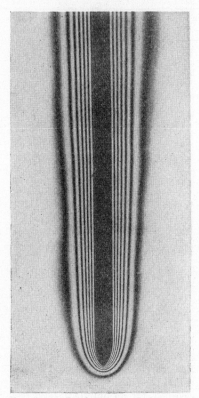

Fig. 11-10. Isotherms around a vertical plate in free-convection flow as revealed by an interference photograph. (*Photograph by E. Eckert and E. Soehngen.*)

At some distance from the lower end, however, waves of long wavelength start, since at a certain boundary-layer thickness the boundary-layer flow becomes unstable for transverse oscillations. These waves increase their amplitude in the downstream direction, become irregular, roll up into vortices, and break up into a coarse turbulence. The region shown in the photograph comprises only the first part of the transition zone to turbulent flow. It gives a very clear impression of the beginning of turbulence. It can be shown experimentally that even strong disturbances do not influence the laminar flow near the lower end of the plate but start waves at some distance from it. The fluctuations are felt through the whole boundary layer up to the plate surface and cause the heat transfer to fluctuate periodically with time.[1]

**11-2. Turbulent Heat Transfer on a Vertical Plate.** The integrated momentum and energy equations (11-3) and (11-4) can also be used to calculate free-convection heat transfer in that part of the surface of a vertical plate which is covered by a turbulent boundary layer.[2] Measured velocity and temperature profiles in such a boundary layer can be well approximated by the

---

[1] E. Eckert, "Heat Transfer and Fluid Mechanics Institute," American Society of Mechanical Engineers, Berkeley, Calif., 1949. E. Eckert, E. Soehngen, and P. J. Schneider in "50 Jahre Grenzschicht-theorie," ed. by H. Görtler and W. Tollmien, Vieweg-Verlag, Brunswick, Germany, 1955.

[2] E. R. G. Eckert and T. W. Jackson, *Natl. Advisory Comm. Aeronaut. Rept.* 1015, 1951.

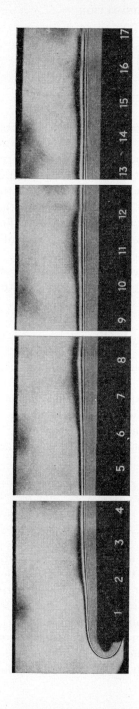 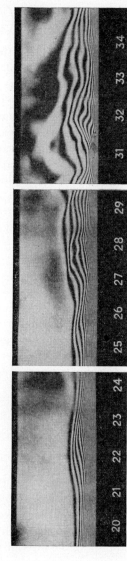

Fig. 11-11. Laminar and turbulent free-convection flow on a vertical plate revealed by interference photographs. Numbers give distance from the end of the lower plate in inches. *(Photograph by E. Eckert and E. Soehngen.)*

following equations:

$$\vartheta = \vartheta_w \left[ 1 - \left(\frac{y}{\delta}\right)^{1/7} \right] \quad (11\text{-}12)$$

$$u = u_1 \left(\frac{y}{\delta}\right)^{1/7} \left(1 - \frac{y}{\delta}\right)^4 \quad (11\text{-}13)$$

As for the case of forced flow, these equations cannot be expected to give the shear stress and the heat flow by their gradient at the surface. Accordingly, the last terms in Eqs. (11-3) and (11-4) have to be replaced by $\tau_w/\rho$ and by $q_w/\rho c_p$, and then experimental values for those parameters have to be found. It is reasonable to expect that close to the wall the relations connecting wall shear stress and heat flow with the temperatures and velocities in this range are the same for forced flow and for free-convection flow. Equation (11-13) has been intentionally chosen such that for the region close to the surface the equation has the same form $u = u_1(y/\delta)^{1/7}$ as Eq. (6-32) for turbulent forced flow. Accordingly, Eq. (6-33) will be assumed to describe the shear stress at the wall. A corresponding equation for heat flow at the wall is

$$q_w = 0.0228 \rho c_p u_1 \vartheta_w \left(\frac{\nu}{u_1 \delta}\right)^{1/4} (\text{Pr})^{-2/3} \quad (11\text{-}14)$$

This relation is obtained from Reynolds analogy with a correction term $(\text{Pr})^{-2/3}$ to account for the deviations from Reynolds analogy at Prandtl numbers different from 1. With these expressions, the solution of the integrated momentum and energy equations proceeds exactly in the same way as for the laminar boundary layer. For the details of this calculation the reader is referred to the paper cited.[1] The result is the following equation:

$$\text{Nu}_x = 0.0295(\text{Gr}_x)^{2/5}(\text{Pr})^{7/15}[1 + 0.494(\text{Pr})^{2/3}]^{-2/5} \quad (11\text{-}15)$$

An empirical relation is also often used

$$\text{Nu}_x = C(\text{Pr } \text{Gr}_x)^{1/3} \quad (11\text{-}16)$$

The constant $C$ has the value 0.10 for air and 0.17 for water. The range of $\text{Gr}_x$ and $\text{Pr}$ in the experiments is not sufficient to determine which of the two preceding equations gives better agreement with experiments.

Experiments on free-convection heat transfer from the surface of a sphere to a fluid filling the interior were made by E. Schmidt.[2] The fluid absorbs the heat by an increase of its mean temperature in time, so that the over-all process is unsteady. It can, however, still be considered as quasi-steady as far as the boundary layer on the surface and the film heat-

[1] *Ibid.*
[2] E. Schmidt, *Chem. Ingr. Tech.*, 1956, p. 175.

transfer coefficients are concerned. Alcohol, glycol, and water were used as fluids, and a range from $3 \times 10^5$ to $5 \times 10^{11}$ of the product $Gr_d\,Pr$ was covered. The results were summarized by the following equation:

$$Nu_d = 0.098(Gr_d\,Pr)^{0.345}$$

It is seen that the heat-transfer coefficients are quite close to the ones for turbulent free convection on a vertical plate [Eqs. (11-15) and (11-16)].

Calculation procedures similar to the ones developed in Secs. 11-1 and 11-2 have also been used to obtain information on local heat transfer existing on the surfaces of rotating disks.[1]

The effects of property variations on free-convection heat transfer in gases can be accounted for by introducing the properties into the previous relations at a reference temperature

$$t^* = t_w - 0.38(t_w - t_0)$$

according to E. M. Sparrow[2] ($t_0$ is the temperature outside the boundary layer).

**Example 11-1.** A vertical wall is heated from one side by steam to a temperature of 200 F. On the other side is air at 14.5 lb/in.² pressure and 68 F temperature. The local heat-transfer coefficient for free convection at a distance of 8 in. from the lower edge is to be calculated.

From the Appendix,

$$\nu = 24.21 \times 10^{-5}\,ft^2/sec \qquad k = 0.0181\,Btu/hr\,ft\,F \qquad Pr = 0.694$$

The expansion coefficient is $\beta = 1/T_0 = 1/528\,R$. Therefore

$$Gr = \frac{g\beta\vartheta_w x^3}{\nu^2} = \frac{32.2 \times 132 \times (8/12)^3}{528 \times 24.21^2 \times 10^{-10}} = 40.7 \times 10^6$$

The boundary layer therefore is laminar. Equation (11-7) gives

$$\frac{\delta}{x} = 3.93 \frac{1}{0.694^{1/2}}(0.952 + 0.694)^{1/4} \times 40.7^{-1/4} \times 10^{-6/4} = 0.0670$$

The boundary-layer thickness is $\delta = 0.536$ in. The film heat-transfer coefficient is, according to Eq. (11-8),

$$h = 2\frac{k}{\delta} = 2\frac{0.0181 \times 12}{0.536} = 0.81\,Btu/hr\,ft^2\,F$$

The average heat-transfer coefficient over the length of 8 in. is

$$\bar{h} = 4h/3 = 1.08\,Btu/hr\,ft^2\,F$$

[1] E. C. Cobb and O. A. Saunders, *Proc. Roy. Soc. (London),* **A**:343 (1956). R. L. Young, *Trans. ASME,* **78**:1163 (1956).

[2] E. M. Sparrow, *Trans. ASME,* **80**:879 (1958). The expansion coefficient $\beta$ has to be introduced at $t_0$.

The average heat-transfer coefficient on a horizontal tube with a diameter $d = 8/2.5 = 3.2$ in. has the same value. The maximum velocity within the boundary layer at a distance 8 in. from the bottom edge can be obtained from Eq. (11-5). By introducing the Grashof number, this equation becomes

$$u_{max} = 0.766 \frac{\nu}{x} (0.952 + \text{Pr})^{-\frac{1}{2}} (\text{Gr})^{\frac{1}{2}}$$

$$= 0.766 \frac{24.21 \times 10^{-5}}{\frac{8}{12}} (0.952 + 0.694)^{-\frac{1}{2}} \times 40.7^{\frac{1}{2}} \times 10^{\frac{5}{2}} = 1.38 \text{ fps}$$

As a consequence of the low velocity, small additional air velocities increase the heat transfer considerably. This fact must be kept in mind when considering industrial applications.

**11-3. Derivation of the Boundary-layer Equations.** In Secs. 6-3 and 7-2 the boundary-layer equations were derived for forced flow. This means that the effect of body forces within the fluid was neglected. In natural convection, however, the movement of the fluid is caused exclusively by the action of body forces. We shall consider the body force to be caused by gravitational acceleration. If the component of the gravitational acceleration in $x$ direction is denoted by $g_x$, then the boundary-layer differential equation (6-16), with the inclusion of the gravitational acceleration term and for steady flow, is

$$\rho \left( u \frac{\partial u}{\partial x} + v \frac{\partial u}{\partial y} \right) = -\frac{\partial p}{\partial x} + \frac{\partial}{\partial y} \left( \mu \frac{\partial u}{\partial y} \right) + g_x \rho \qquad (11\text{-}17)$$

The Navier-Stokes equation in the direction normal to the surface reduces by the order of magnitude analysis to

$$\frac{\partial p}{\partial y} = \rho g_y$$

indicating that in this case the pressure on normals to the surface is not constant.

Equation (11-17) will be transformed by adding and subtracting to the right-hand side the term $\rho_s g_x$:

$$\rho \left( u \frac{\partial u}{\partial x} + v \frac{\partial u}{\partial y} \right) = -\frac{\partial p}{\partial x} + \frac{\partial}{\partial y} \left( \mu \frac{\partial u}{\partial y} \right) + \rho_s g_x + (\rho - \rho_s) g_x$$

In free-convection boundary-layer flow the velocity $u$ outside the boundary layer is zero. Specifying the above equation for the outer edge of the boundary layer, therefore, results in

$$-\frac{\partial p}{\partial x} + \rho_s g_x = 0$$

This equation expresses the well-known fact that within a stagnant fluid in a gravitational field the pressure decreases in the vertical direction. In the preceding equation, therefore, the two terms $-\partial p/\partial x$ and $\rho_s g_x$ cancel out and the momentum differential equation for the free-convection boundary layer has the form

$$\rho\left(u\frac{\partial u}{\partial x} + v\frac{\partial u}{\partial y}\right) = \frac{\partial}{\partial y}\left(\mu\frac{\partial u}{\partial y}\right) + (\rho - \rho_s)g_x \qquad (11\text{-}18)$$

To this equation has again to be added the continuity equation and the energy equation, which equations have the same form as in forced flow. Solutions of this system of differential equations have been obtained for a number of geometries. In these solutions, the properties, including the density, have been considered as constant in all terms except the last one in the momentum equation, where the difference between the density locally within the boundary layer and the density outside the boundary layer at the same position $x$ occurs. When the density depends only on temperature, or as long as the pressure increase throughout the boundary layer is small, the last term in Eq. (11-18) can be transformed to

$$(\rho - \rho_s)g_x = -\beta\rho_s(t - t_s)g_x$$

E. Pohlhausen[1] obtained the solution for laminar boundary-layer flow of air along a heated vertical plate by finding a transformation which reduces the partial differential equations to total ones. The new parameters which make the transformation possible are

$$\eta = C\frac{y}{\sqrt[4]{x}} \qquad \vartheta' = \frac{t - t_s}{t_w - t_s} \qquad f = \frac{\psi}{4\nu C x^{\frac{3}{4}}}$$

$\psi$ is again the stream function ($u = \partial\psi/\partial y$, $v = -\partial\psi/\partial x$).

$$C = \sqrt[4]{\frac{\beta g_x(T_w - T_s)}{4\nu^2}}$$

the expansion coefficient $\beta$ was set equal $1/T_s$, and the dependent variables $\vartheta'$ and $f$ are assumed to be functions of $\eta$ only. The results of this calculation check quite well the results of measurements of the velocity and temperature profile. Figures 11-2 and 11-3 show the profiles calculated in this way and the agreement with measurements. Laminar free-convection flow along a vertical plate has also been calculated for a large range of Prandtl numbers by H. Schuh[2] and by S. Ostrach.[3] Schuh showed that the transformation to total differential equations is possible not only for the flat plate but for a group of surface geometries. One of these is the flow near the stagnation point of a horizontal cylinder.

[1] E. Schmidt and W. Beckmann, *Tech. Mach. Thermodynam.*, **1**:1–24 (1930).
[2] H. Schuh, Rept. and Transl. No. 1007, AVA monographs, 1948.
[3] S. Ostrach, *Natl. Advisory Comm. Aeronaut. Tech. Note* 2635, 1952.

## 11-4. Free Convection in a Fluid Enclosed between Two Plane Walls.

*Horizontal Walls.* No free-convection currents occur in a fluid which is enclosed between two parallel horizontal plates as long as the temperature of the upper plate is higher than the temperature of the lower one. Heat flows in this case from the upper to the lower plate, and the temperature in the fluid is constant in horizontal layers and increases in an upward direction. In a normal fluid for which the density decreases with temperature, such a temperature field is connected with a situation in which less dense layers are located above denser ones, a completely stable situation which does not cause any convection currents. Heat will be transported by conduction only (not considering radiation), and the temperature profile will be linear. This situation can be disturbed only near the edge of the plates.

The situation is different when a fluid is enclosed between two horizontal surfaces of which the upper surface is at a lower temperature than the lower one. Heat flow occurs now through the fluid in the direction from the lower toward the upper surface, and as a consequence, the fluid between the two plates assumes such temperatures that colder fluid particles are situated above warmer ones. For fluids whose density decreases with increasing temperature, this leads to an unstable situation. It does not give rise to convection currents as long as the product of Grashof (based on distance and temperature difference of the plates) and Prandtl numbers is low. However, when this parameter reaches a value around 1,700, then a peculiar free-convection flow pattern arises which can be observed in Fig. 11-12. (The figure was obtained by H. Siedentopf, and the flow is made visible by tiny aluminum particles in the fluid.) The flow field has a cellular structure with more or less regular hexagonal cells. In the interior of these cells the flow moves in an upward direction, and along the rim of the cell it returns downward. This flow situation is maintained up to a value of the product Grashof times Prandtl of around 47,000. Above this value, it changes to irregular turbulence. The lower critical Reynolds number at which this flow pattern is established was calculated theoretically by Rayleigh (1916) after H. Benard observed it for the first time (1900). Correspondingly, the parameter Grashof times Prandtl, as used in connection with this type of flow, is often referred to as Rayleigh number.

*Vertical Walls.* When the fluid layer is enclosed between two vertical walls, then it is expected that the height of the walls is an important geometric parameter as well as their distance. In a dimensionless presentation, accordingly, the height-to-distance ratio appears in addition to the Grashof number.

As long as $Gr_b < 124(Pr)^{-2}(20/21 + Pr)(L/b)$ with $L$ the height of layer and $b$ its thickness, the temperature decreases in the fluid layer linearly in a direction normal to the two walls over the major portion of the layer.[1] Only near the lower and upper ends of the air space do deviations from this linear temperature drop occur. They extend over a height which is approximately equal to the distance $b$ of the two walls. Except for the region near the end, the heat transfer is equal to that which can be calculated for pure conduction.

Distinct boundary layers can be observed for large values of the product $Gr_b\,Pr$. Observation of the temperature field with a Zehnder-Mach interferometer revealed a configuration as sketched in Fig. 11-13. A boundary layer grows in the upward direction on the hotter wall, while

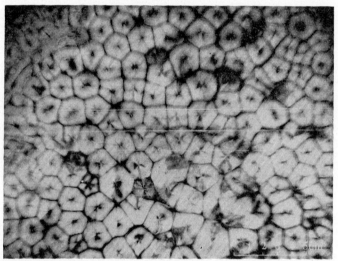

Fig. 11-12. Cellular flow pattern by free convection in horizontal air layers. (*According to H. Siedentopf, from L. Prandtl, "Strömungslehre," Vieweg-Verlag, Brunswick, Germany, 1949.*)

the one along the wall with the lower temperature grows in a downward direction. In the central core $b$ the temperature is found to be constant in a direction normal to the two walls. In the region of the two boundary layers the temperature varies in a manner quite similar to the variation found in a free-convection boundary layer on a single vertical surface. The two boundary-layer thicknesses and heat-transfer coefficients were also found to be not more than 20 per cent different from those calculated for a single plate with the same surface temperature and with a temperature at the outer edge of the boundary layer equal to the temperature in the fluid core $b$ at the height under consideration. This situation is maintained with decreasing values of $Gr_b\,Pr$ until the two boundary

[1] W. O. Carlson, Ph.D. thesis, University of Minnesota, Minneapolis, 1956.

layers almost touch each other. In contrast to the condition on a single heated or cooled plate in an infinite fluid, however, it is found that the temperature is not constant in the core $b$ in the space between the two vertical walls but increases in an upward direction.

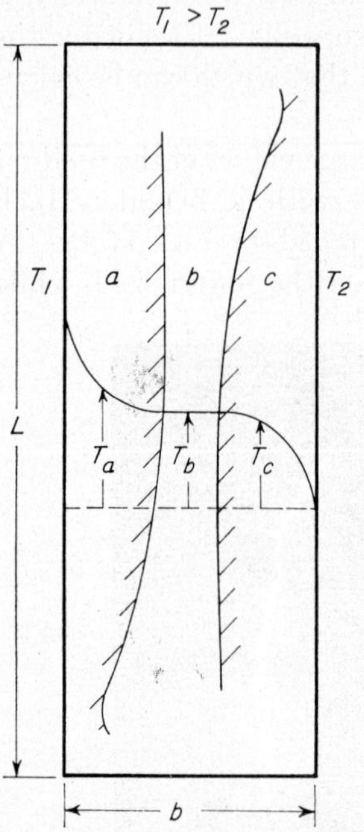

An interesting observation has been reported by E. Schmidt.[1] It was found that the heat transfer from the hot to the cold surface through a vertical fluid layer sometimes increases when the space between the two vertical walls is subdivided into smaller regions by thin horizontal walls. The above-mentioned fact that boundary layers develop at sufficiently large values of $Gr_b$ Pr explains this observation. The warm and cold boundary layers start essentially at each one of the horizontal partitions, and the boundary layers on both walls remain thinner than under conditions where they move along the full height of the walls.

FIG. 11-13. Sketch of boundary-layer development and temperature profile for free convection through a vertical fluid layer between a hot and a cold wall.

For engineering design calculations, interest is mainly concentrated on a determination of the total heat flow through fluid layers of various geometries from a hot to a cold surface. Relations for this heat flux will be discussed for fluid layers between parallel plane walls and for fluid layers enclosed between two concentric cylinders. Kraussold[2] showed that the average heat flow per unit area can as a first approximation be represented by a single relation for cylindrical layers as well as for plane horizontal or vertical ones. The average heat flow $q_m$ per unit time and area is for such layers conveniently characterized by an "equivalent heat conductivity $k_e$" defined by

$$q_m = \frac{k_e}{b}(t_h - t_c)$$

where $b$ = thickness of layer
$t_h$ = temperature of hot surface
$t_c$ = temperature of cold surface

For the cylindrical layer the heat flow $q_m$ is based on the mean cross section of the layer [at radius $r_m$]. Figure 11-14 shows the ratio $k_e/k$ of the

[1] E. Schmidt, *Chem. Ingr. Tech.*, 1956, pp. 175–180.
[2] H. Kraussold, *Forsch. Gebiete Ingenieurw.*, **2**:186–191 (1936).

equivalent to the actual heat conductivity plotted versus the product $Gr_b$ Pr in which $Gr_b$ is based on the thickness $b$ of the layer and on the temperature difference $t_h - t_c$. The experiments performed by Beckmann and Sellschopp on cylindrical layers and by the others on plane layers correlate quite well with the full line. For the range in which $k_e/k = 1$, heat transport through the layer takes place by conduction only. The deviation from $k_e/k = 1$ indicates the contribution of convection to the total heat flow. The correlation represented by the line can also be used for other geometries as a first approximation.

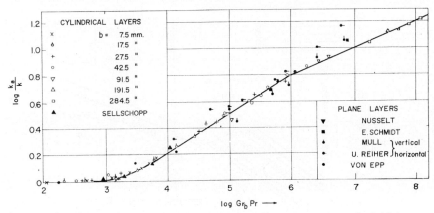

Fig. 11-14. Equivalent heat conductivity of free convection through fluid layers. (*H. Kraussold, Forsch. Gebiete Ingenieurw.*, **5**:186–191, 1934.)

**11-5. Mixed Free and Forced Convection.** In Chaps. 7 through 10 on forced convection the influence of body forces on flow and heat transfer has been neglected. Actually buoyancy forces are present in a normal fluid in which the density varies with temperature in any forced-convection flow, and it is of interest to know when they can be neglected and when they have to be accounted for. Such an investigation is made difficult by the large number of influencing parameters. In addition to Reynolds and Prandtl numbers, Grashof numbers as well as parameters describing the geometry of the boundaries and the orientation of the flow relative to the gravitational field are important. The situation which apparently has been investigated most intensively is forced flow through a circular pipe with its axis parallel to the direction of the gravitational acceleration and thus with the mean flow direction either parallel or opposite to it.[1]

It has to be expected that for large Reynolds values (and correspondingly large flow velocities) and small Grashof numbers, the influence of

[1] A thorough study of this case was made by T. M. Hallman, "Combined Forced and Free Convection in a Vertical Tube," Ph. D. thesis, Purdue University, 1958.

free convection on the heat transfer can be neglected. On the other hand, for large Grashof values and small Reynolds numbers the free convection should be the dominating factor. This expectation is borne out in Fig. 11-15, which summarizes theoretical and experimental investigations on vertical tubes with various length-to-diameter ratios ($L/d$) and for fluids with various Prandtl numbers.[1] The forced flow through the tube has the same direction as the buoyancy forces in the fluid near the tube wall. It can be observed that the mixed flow region, defined as the region in which the heat transfer differs by more than 10 per cent from

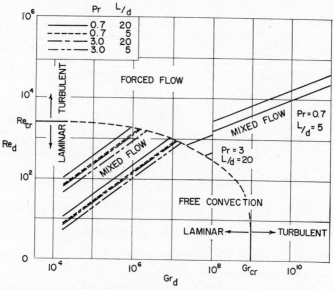

FIG. 11-15. Regimes for forced, mixed, and free convection. (*From E. R. G. Eckert and A. J. Diagula, Trans. ASME,* **75**:497–504, 1954.)

the one obtained with forced flow or free-convection relations, respectively, is actually quite small. McAdams proposed a rule by which heat transfer in the mixed flow region can be determined: One calculates the heat-transfer coefficient from forced-convection and free-convection relations and uses the larger value. Measured values obtained for flow through vertical tubes did not deviate by more than 25 per cent from the values calculated with McAdams' rule.

## PROBLEMS

**11-1.** Calculate laminar free-convection heat transfer along a flat plate inclined at an angle $\alpha$ to the vertical direction using the integrated boundary-layer equations and the same velocity profiles as in Sec. 11-1.

[1] E. R. G. Eckert and A. J. Diagula, *Trans. ASME,* **76**:497–504 (1954).

**11-2.** Calculate laminar free-convection heat-transfer coefficients around the periphery of a horizontal cylinder with circular cross section using the integrated boundary-layer equations and the same velocity profiles as in Sec. 11-1. Assume that the boundary-layer thickness is small as compared with the cylinder diameter.

**11-3.** Calculate the local heat-transfer coefficients which occur on the surface of a cooling air passage in a hollow gas turbine blade by the free-convection effect caused by centrifugal forces. Assume that the forced flow of air at atmospheric pressure and 500 F temperature through the passage is small so that its influence on the convective heat transfer can be neglected. The surface temperature of the passage may be assumed as 1000 F. The cooling passage is arranged with its axis normal to the axis of rotation, is 2 in. long, and is a distance of 15 in. from the axis. The turbine wheel to which the blade is attached rotates with 10,000 rpm. Assume for this calculation that the passage surface can be considered a flat plate with the centrifugal force acting as body force. Neglect the influence of Coriolis forces.

**11-4.** Derive Eq. (11-16) for turbulent free-convection heat transfer on a vertical flat plate using the procedure as briefly outlined in Sec. 11-2.

**11-5.** The heat conductivity of gases is often measured by Sutherland's method in which a platinum wire is arranged in the axis of capillary tube filled with the gas to be investigated. The wire is heated electrically, the tube is cooled on its outside, and the heat flow as well as the wire and tube temperatures are measured. The heat conductivity is calculated assuming that heat flows radially from the wire to the tube by conduction. Determine what diameter the capillary tube must have in order to avoid errors caused by free convection when it is desired to measure the heat conductivity of water vapor up to 150 atm pressure (see figures in Appendix for properties). Use Fig. 11-14 for this calculation. A temperature difference of 10 F is maintained between wire and tube.

**11-6.** The plane wall of a container kept at a temperature of 200 F is insulated with three layers of aluminum foil, one on top of the wall and the others at a distance of $\frac{1}{2}$ in. each, producing two air spaces. The outside wall of the insulation is cooled to 100 F by the surrounding air. What temperature does the medium sheet of foil assume? What is the heat flow per hour and square foot through the insulation? Neglect heat exchange by radiation.

**11-7.** In Prob. 11-3, at what velocity of the cooling air does the free-convection effect on heat transfer become negligible? (Assume that the hydraulic diameter of the air passage is 0.4 in.)

## CHAPTER 12

## CONDENSATION AND EVAPORATION

**12-1. Condensation.** If vapor is contained in a vessel whose walls have a temperature lower than the saturation temperature, the vapor condenses on the wall. The heat liberated by this process must be led away through the walls. The heat transfer connected with this process was calculated in 1916 by W. Nusselt[1] using his liquid-film theory. Nusselt starts with the conceptions that the condensed liquid forms a continuous film on the wall and that the heat flow is determined by the thermal resistance of this film. The simplest case will be calculated, namely, the condensation on a vertical wall at a constant temperature $t_w$. The condensed liquid builds up a film which flows by gravity downward and is continuously renewed by newly condensing vapor. The film has zero thickness at the upper edge of the wall, and the thickness increases in a downward direction as is indicated in Fig. 12-1. At a distance $x$ from the upper edge the film thickness may be $\delta$. For laminar flow, the velocity profile within the film can be assumed to be a parabola:

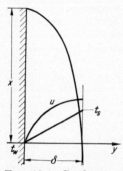

FIG. 12-1. Condensate-film thickness, velocity and temperature profile for film condensation on a vertical wall.

$$u = U\left(2\frac{y}{\delta} - \frac{y^2}{\delta^2}\right) \tag{12-1}$$

In this equation, $y$ is the distance from the wall and $U$ is the downward velocity of the liquid at the border adjacent to the vapor. As the heat of condensation flows normal to the velocities in the film, the temperature profile must be essentially linear, with a temperature $t_w$ at the wall and the saturation temperature $t_s$ at the border of the film adjacent to the vapor. It is surprising, perhaps, that the temperature field has here another shape than in the boundary layer of a liquid or gas (Sec. 7-3). This is explained by the fact that in the previously investigated problems the entire amount of heat flow to the wall originates in the boundary

[1] W. Nusselt, *Z. Ver. deut. Ingr.*, **60**:541 (1916).

layer whereas here all the heat is generated at the outer border of the liquid film by condensation of the vapor.

The heat flow through an element of the film with the height $dx$ and unit dimension in the horizontal direction along the wall is therefore

$$dQ = \frac{k}{\delta} dx\, (t_s - t_w) \tag{12-2}$$

Again a film heat-transfer coefficient $h$ is defined by the equation

$$dQ = h\, dx\, (t_s - t_w)$$

and therefore
$$h = \frac{k}{\delta} \tag{12-3}$$

The velocity with which the liquid film flows down the plate can be calculated from the equilibrium condition for the element of film under consideration. The weight of this film element must be counteracted by the shearing stress $\tau_w$ at the wall:

$$g\rho\delta\, dx = \tau_w\, dx$$

Acceleration and pressure forces are neglected as small.[1] By introducing Newton's law (6-2) for the shear stress and solving the above equation for the velocity gradient at the wall, there is obtained

$$\left(\frac{du}{dy}\right)_w = \frac{2U}{\delta} = \frac{g\rho\delta}{\mu}$$

and therefore
$$U = \frac{g\rho}{2\mu}\delta^2$$

Equation (12-1) therefore changes to

$$u = \frac{g\rho}{2\mu}\delta^2 \left(2\frac{y}{\delta} - \frac{y^2}{\delta^2}\right)$$

The average velocity $u_m$ at the cross section $x$ is two-thirds the maximum value, since the velocity profile is a parabola:

$$u_m = \frac{g\rho}{3\mu}\delta^2 \tag{12-4}$$

The mass of liquid flowing through the cross section at $x$ is

$$G = \rho u_m \delta = \frac{g\rho^2}{3\mu}\delta^3$$

Through the cross section at a distance $dx$ in the downward direction the mass flow is greater by the amount

$$dG = \frac{g\rho^2}{\mu}\delta^2\, d\delta \tag{12-5}$$

[1] A calculation that considers these effects is that of E. M. Sparrow and J. L. Gregg, *Trans. ASME*, series C, Journal of Heat Transfer, **81**:13 (1959).

This amount must be generated by condensation of vapor. With the equation $dQ = i_{fg} \, dG$, in which $i_{fg}$ is the heat of condensation per pound of liquid, there is obtained from Eq. (12-2)

$$dG = \frac{k}{i_{fg}\delta} \, dx \, (t_s - t_w) \tag{12-6}$$

and by equating (12-5) and (12-6) the formula

$$\delta^3 \frac{d\delta}{dx} = \frac{\mu k}{i_{fg} g \rho^2} (t_s - t_w)$$

is obtained. Integration gives

$$\frac{\delta^4}{4} = \frac{\mu k}{i_{fg} g \rho^2} (t_s - t_w) x$$

If the dynamic viscosity $\mu$ is replaced by the kinematic viscosity $\nu$, there is obtained for the thickness of the liquid film

$$\delta = \sqrt[4]{\frac{4\nu k (t_s - t_w) x}{g \rho i_{fg}}} \tag{12-7}$$

and for the film heat-transfer coefficient

$$h = \frac{k}{\delta} = \sqrt[4]{\frac{k^3 g \rho i_{fg}}{4 \nu (t_s - t_w) x}} \tag{12-8}$$

In dimensionless form it can be written

$$\mathrm{Nu}_x = \frac{hx}{k} = \frac{x}{\delta} = \sqrt[4]{\frac{g \rho i_{fg} x^3}{4 \nu k (t_s - t_w)}} \tag{12-9}$$

Equation (12-8) gives the local heat-transfer coefficient at a distance $x$ from the upper edge of the plate. The average value along this distance is found by integration:

$$\bar{h} = \tfrac{4}{3} h \tag{12-10}$$

According to Eq. (12-9), the Nusselt number depends on the dimensionless value on the right-hand side of the equation. It is sometimes referred to as the *Sherwood number*. The property values of the condensate must be introduced into the equation.

The liquid film changes to turbulent flow when its thickness exceeds a certain value. Heat transfer through the turbulent film can be calculated in the same way as for turbulent flow of liquids or gases along a plate. Such a calculation was done by U. Grigull.[1] The result shows that in the turbulent film the Nusselt number is a function of the dimensionless value in Eq. (12-9) and of the Prandtl number of the condensate. The equation for heat transfer in a turbulent film is rather complex. It is therefore better to take the heat-transfer coefficient from Fig. 12-2. In

[1] U. Grigull, *Forsch. Gebiete Ingenieurw.*, **13**:49–57 (1942).

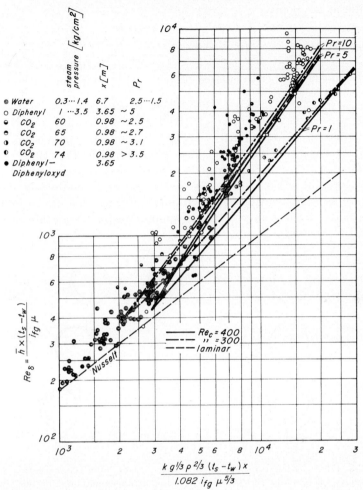

Fig. 12-2. Film condensation with laminar and turbulent film. (*From U. Grigull, Forsch. Gebiete Ingenieurw.*, **13**:49–57, 1942.)

this figure the Reynolds number of the film is plotted over the dimensionless value $kg^{1/3}\rho^{2/3}(t_s - t_w)x/1.082 i_{fg}\mu^{5/3}$ with the Prandtl number as a parameter. The Reynolds number is based on the film thickness $\delta$ and the average velocity $u_m$ in the film:

$$\mathrm{Re}_\delta = \frac{u_m \delta}{\nu} \tag{12-11}$$

From this, the average heat-transfer coefficient can be calculated by the fact that the heat flowing to the wall over the height $x$ originates from the heat of condensation of the condensate passing through a cross section of the film at the point $x$:

$$\bar{h}x(t_s - t_w) = \rho u_m i_{fg} \delta$$

Therefore
$$\mathrm{Re}_\delta = \frac{u_m \delta}{\nu} = \frac{\bar{h}x(t_s - t_w)}{i_{fg}\mu} \qquad (12\text{-}12)$$

The Reynolds number calculated from Eq. (12-9) for laminar flow is

$$\mathrm{Re}_\delta = \left[\frac{kg^{1/3}\rho^{2/3}x(t_s - t_w)}{1.082 i_{fg}\mu^{5/3}}\right]^{3/4} \qquad (12\text{-}13)$$

This equation is represented in Fig. 12-2 by the dashed line. For turbulent flow, two bundles of lines are inserted in Fig. 12-2. One is calculated with a critical Reynolds number for transition to turbulence $\mathrm{Re}_c = 400$; the other with $\mathrm{Re}_c = 300$. The great number of inserted points representing experimental results indicates that $\mathrm{Re}_c = 300$ conforms better with reality. In the laminar region, Nusselt's theory is verified by the experiments.

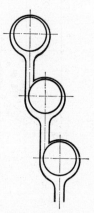

FIG. 12-3. Film condensation on three tubes above one another.

FIG. 12-4. Film condensation on a tube arrangement used in the Ginabat condenser.

Nusselt also calculated the heat transfer for condensation on a horizontal circular tube. The result can be expressed simply so that the average film heat-transfer coefficient on a tube with diameter $d$ has the same magnitude as the average film heat-transfer coefficient on a vertical wall with the height $x = 2.5d$. If several tubes are arranged one above the other, the condensate from the highest tube flows to the next underlying one and so on (Fig. 12-3). By this action the heat transfer of each successive tube is diminished. Nusselt's calculations showed that the heat flow of the second tube is only 60 per cent of the first. This decrease in the heat flow can be greatly diminished by an arrangement proposed by Ginabat (Fig. 12-4).

Most experiments have confirmed Nusselt's liquid-film theory. Sometimes, however, much higher values have been measured. E. Schmidt, W. Schurig, and W. Sellschopp[1] showed that this is because the vapor sometimes condenses not as a continuous film but in the form of small droplets which increase in size with time until they are so heavy that they run down the wall. In so doing they sweep their paths free of droplets. On these places the formation of new tiny droplets begins immediately. Figure 12-5 shows a photograph of this type of condensation. It can be easily understood that heat transfer for this type of condensation is much greater than for filmwise condensation. A great number of experiments have been conducted, especially in the United States, which analyze the conditions under which the condensation will be filmwise or dropwise. Pure vapor on clean surfaces always condenses as a film. Impurities in the vapor or on the wall, especially of fatty acids, lead to condensation in the form of droplets. The formation of droplets is easier on rough surfaces than on smooth ones. On steel or aluminum tubes under ordinary conditions one must always expect filmwise condensation. Since it is difficult to predict with certainty the formation of dropwise condensation, it is recommended that calculations be made with the equation for film condensation.

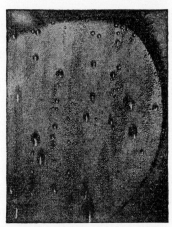

FIG. 12-5. Dropwise condensation. (*Photograph by E. Schmidt, W. Schurig, and W. Sellschopp, Tech. Mech. Thermodynam.*, **1**:53, 1930.)

Superheated vapor has practically the same film heat-transfer coefficients as saturated vapor. If a noncondensing gas is present together with the vapor, an additional resistance to the heat flow arises by concentration of the gas near the liquid-film surface. The heat transfer is thus considerably decreased. This problem is discussed in Part D, which deals with mass transfer.

**Example 12-1.** Saturated water vapor condenses as a film on a vertical wall. The film heat-transfer coefficient at a distance of 3 in. from the upper edge is to be calculated. The steam pressure is 1.43 $lb_f/in.^2$; the corresponding saturation temperature $t_s = 114$ F. The wall temperature is $t_w = 105$ F.

From steam tables take the heat of evaporation $r = 1029$ Btu/lb and the density $\rho = 62$ lb/ft$^3$. The property values $k$ and $\nu$ can be found in the Appendix:

$$\nu = 0.654 \times 10^{-5} \text{ ft}^2/\text{sec}$$

[1] E. Schmidt, W. Schurig, and W. Sellschopp, *Tech. Mech. Thermodynam.*, **1**:53 (1930).

$k = 0.367$ Btu/hr ft F. The dimensionless value in Eq. (12-9) is

$$\frac{g\rho x^3 i_{fg}}{4\nu k(t_s - t_w)} = \frac{32.2 \times 62 \times 1029 \times 3^3 \times 3600}{12^3 \times 4 \times 0.654 \times 10^{-5} \times 0.367 \times 9} = 1.34 \times 10^{12}$$

and the Nusselt number is Nu = $\sqrt[4]{1.34 \times 10^{12}} = 1.08 \times 10^3$. The local film heat-transfer coefficient is

$$h = \frac{0.367 \times 12}{3} \times 1.08 \times 10^3 = 1580 \text{ Btu/hr ft}^2 \text{ F}$$

and the average value along the height of 3 in. is

$$h = \tfrac{4}{3} \times 1580 = 2110 \text{ Btu/hr ft}^2 \text{ F}$$

The average film heat-transfer coefficient on a horizontal tube with a diameter $d = 3/2.5 = 1.2$ in. has the same magnitude. It can be seen that the heat-transfer coefficients connected with condensation are very great.

**12-2. Evaporation.** Evaporation of a liquid can occur in such a way that the vapor is produced at the liquid-vapor interface. This happens, for instance, when the heat of evaporation is transported into the fluid right at its surface, as, for example, by radiation directed toward and absorbed at the surface of the liquid. This evaporation process is quite similar to the melting process discussed in the chapter on conduction and can be calculated by essentially the same methods.

In most engineering situations, heat is added to the liquid from a submerged solid surface and the vapor is produced in the form of bubbles which start and grow at the heating surface, separating when they have reached a certain size and rising through the liquid. This mode of evaporation is called boiling. It will be discussed in this section. The complicated nature of the boiling process has for a long time restricted our knowledge to a collection of empirical data. Recently, however, considerable progress has been made in this area by intense experimental study and by development of models which sufficiently simplified the actual process so that dimensional analysis and analytical methods could be applied. There is good reason to expect that in the near future boiling heat transfer can be predicted in many cases by analysis just like the other modes of heat exchange which have been discussed in the previous sections.

For a discussion of the essential features of a boiling process, let us consider evaporation of a liquid from a submerged heating surface. Let us assume that the liquid is originally well below the evaporation temperature. Then it can be observed that some time after heat has been applied through the heating surface, vapor bubbles appear on that surface. These bubbles separate from the surface after having grown to a certain size and move into the bulk of the liquid. After a short path, however, they vanish again. Obviously they condense in the liquid, which in its bulk has not yet reached evaporation temperature. Only after the liquid has been heated sufficiently will the vapor bubbles travel all the way to the liquid surface and in this way start a flow of vapor into the space above the surface. Both of these processes have found engineering applications. The process in which the vapor is recondensed near the heating surface is called *local boiling* or *subcooled boiling*. The process in which the vapor bubbles penetrate the liquid surface is called *boiling with net evaporation*.

It has been established and can be readily understood that the boiling process is also influenced by the movement of the liquid near the heating surface. If this surface is submerged in a big container, then movement of the liquid is generated only by free convection or by the stirring action of the bubbles. This type of boiling is classified as *pool boiling*. If on the other hand the boiling occurs on the walls of a tube through which the liquid flows with considerable velocity, then this velocity makes itself felt and influences the bubble growth and separation. Boiling under such circumstances is called *forced-convection boiling*.

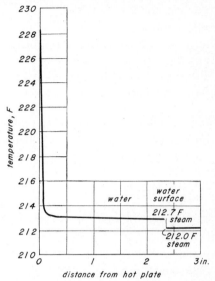

Fig. 12-6. Temperature profile in boiling water above a horizontal heating plate. (*According to M. Jakob and W. Fritz, from M. ten Bosch, " Die Wärmeübertragung,"* 3d ed., Springer-Verlag, Berlin, 1936.)

The most intensively investigated boiling process is *pool boiling with net evaporation*. This process will, therefore, be discussed in detail in the following paragraphs. The discussion will then be somewhat extended to local forced-convection boiling, and a few remarks will be devoted to forced-convection boiling with net evaporation.

Figure 12-6 presents the temperature field in boiling water measured for pool boiling above a horizontal heating surface with net evaporation. The space above the water surface was occupied with vapor and kept at atmospheric pressure (corresponding to a vapor temperature of 212 F). It is evident from the figure that the bulk of the water is at a fairly uniform temperature, but slightly superheated (by 0.7 F). Near the heating surface, however, the water temperature rises in a thin layer with a very steep gradient to a surface temperature of 228.3 F. In its details the temperature field depends on the magnitude of the heat flux, pressure, geometry, and boiling liquid. Qualitatively, however, the features discussed here appear in the same way in all boiling processes.

For an understanding of this temperature condition and of the heat transfer connected with boiling, it is necessary to consider in some detail the process of bubble formation in a liquid.[1] For this purpose some consideration will at first be given to a vapor bubble which is in thermal equilibrium within a liquid of uniform temperature. Under this condition, a bubble with a certain radius can be maintained, rather than causing it either to collapse by condensation of its vapor or to grow by evaporation from the liquid surface into its vapor space. In order to study the dynamic equilibrium of the bubble shown in Fig. 12-7, we imagine the bubble, with a radius $r$, to be cut into two halves and consider the equilibrium of one half. Forces which act on the surface within the plane 1-1 are the vapor pressure $p_v$ acting from the inside of the bubble, the liquid pressure $p_l$ from the outside of the bubble, and the surface tension within the surface of the bubble itself. If we denote this force per unit length of the surface by $\sigma$, then the force caused by surface tension is $2\pi r \sigma$. The vapor pressure and the liquid pressure generate a force in a direction normal to the plane 1-1 equal to the pressure difference times the cross-sectional area of the bubble. This gives the following equation:

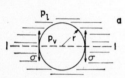

FIG. 12-7. Sketch of a vapor bubble in liquid.

$$\pi r^2 (p_v - p_l) = 2\pi r \sigma$$

The pressure difference itself is

$$p_v - p_l = \frac{2\sigma}{r}. \qquad (12\text{-}14)$$

The thermal equilibrium requires that there is no temperature difference between the vapor in the bubble and the liquid surrounding it. The vapor temperature within the bubble must also be equal to the saturation temperature belonging to the pressure $p_v$.[2] The pressure in the liquid is smaller, and therefore, the liquid must be superheated in order to allow a vapor bubble to be in equilibrium within it. The amount of superheat is larger the smaller the radius of the bubble. If the equation were valid

---

[1] F. Bosnjakovic: *Tech. Mech. Thermodynam.*, **1**:358-62 (1930).

[2] Thermodynamics shows that the saturation pressure $p_v$ of a vapor in equilibrium with a liquid at a certain temperature and separated by a plane interface is somewhat larger than the saturation pressure on a curved concave surface. This difference, however, is small compared with the pressure difference in Eq. (12-14) as long as the evaporation pressure is not close to the critical pressure of the boiling fluid.

down to a radius equal to zero, then this would mean that an infinitely large superheat is required to start a bubble within the liquid. Actually for very small bubble sizes considerations have to be made on the basis of the molecular structure rather than on a continuum basis from which the above equation is derived. However, the above equation makes it understandable that a large superheat is required to start the formation of new bubbles. The formation of new bubbles is made easier when gases are contained in the form of small gas bubbles within the liquid. Such gases serve as nuclei for the bubble formation. The process of the formation of new bubbles is referred to as the *nucleation process*. A vapor bubble grows when the liquid temperature is larger than the saturation temperature belonging to the pressure $p_v$. In this case, heat is conducted in the liquid toward the bubble surface, which is at saturation temperature. On the basis of a consideration of this conduction process, H. K. Forster and N. Zuber[1] were able to calculate this growth history of a bubble in good agreement with the results of measurements. Nucleation and growth of bubbles in a liquid occur in the homogeneous nuclear reactors where the heat of evaporation is supplied by the fission process of the fissionable atoms distributed in the liquid. The birth of new bubbles within the liquid is that part of the boiling process which up to now cannot be predicted analytically.

If boiling occurs on a solid surface through which heat is transported into the fluid, then the large superheat required for the growth of small bubbles is provided within the thin layer attached to the heating surface (Fig. 12-6). Places of especially large superheat will occur in grooves within the surface, and it is very probable that the formation of new bubbles occurs at such places. Actually, the conditions for the dynamic equilibrium of a bubble attached to a solid surface are more complicated than for the bubble in the interior of the fluid. Figure 12-8 shows a sketch of a bubble attached to a surface. Under this condition three surface tensions have to be considered. One, $\sigma_{lv}$, is determined by the properties of the vapor and the liquid and is acting in the liquid vapor interface. One, $\sigma_{ls}$, is determined by the properties of the solid surface and the liquid and is acting along the solid surface on the liquid side, and one, $\sigma_{vs}$, is determined by the properties of the vapor and the solid surface and is acting along the solid surface on the vapor side. The equilibrium in the horizontal direction requires the following relation among these surface tensions:

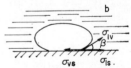

Fig. 12-8. Sketch of a vapor bubble on a surface.

$$\sigma_{vs} - \sigma_{ls} = \sigma_{lv} \cos \beta \qquad (12\text{-}15)$$

[1] H. K. Forster and N. Zuber, *J. Appl. Phys.*, **25**:474 (1954).

The angle $\beta$ depends on the relative magnitude of the different surface tensions. It may be smaller or larger than 90°. In the first case the liquid is said to *wet* the surface. When the angle $\beta$ is larger than 90°, the surface is considered to be *nonwetted*. It is frequently assumed that a thin vapor film exists between a liquid and a solid nonwetted surface even if no boiling occurs. It is understandable that the condition of surface wetting is of great influence in bubble formation.

It has been found that a surface initially requires a smaller temperature difference to transmit a certain amount of heat into the boiling fluid and that the required temperature difference gradually increases with time. It is assumed that the gases which originally were adsorbed at the surface facilitate the bubble formation until they are gradually removed with the bubbles. The considerations discussed make it understandable that the microstructure of the surface as well as its condition as far as impurities or adsorbed gas films are concerned have a considerable influence on boiling. On the other hand, the shape of the surface should be of minor importance, since the thickness of the thermal boundary layer on the surface is in most cases determined by the stirring action of the bubbles, a process concentrated in a comparatively small space. The growth and separation of a bubble from the surface are also more complicated processes than those inside a liquid, since the bubble grows into a boundary layer with a steep temperature gradient. It is also probable that the separation of the bubble is influenced by inertia forces, especially at high heat fluxes which cause a rapid growth rate and at higher pressures. Observations by Ellion[1] demonstrate this, since they show that bubbles are propelled into the fluid with considerable velocity from the underside of a horizontal, electrically heated ribbon. The complexity of this problem has up to now prevented a satisfactory analytical formulation.

We shall now consider the heat-transfer process connected with boiling. Figure 12-6 indicates that the thermal resistance in the evaporation process is concentrated essentially in a thin thermal boundary layer at the heated surface. Attention must, therefore, be fixed upon this region in an analysis of boiling heat exchange.

Two definitions are found in the literature for a film heat-transfer coefficient in boiling. In one of them, the heat flux $q$ at the heating surface is divided by the difference $\Delta t$ between the surface temperature and the saturation temperature belonging to the pressure in the liquid. The other definition uses the difference between the surface temperature and the fluid bulk temperature. This last definition agrees better with the one customarily used in convection without change of phase. The first one has the advantage in that the saturation temperature can be obtained

[1] M. E. Ellion, Ph.D. thesis, California Institute of Technology, 1953.

more easily and accurately from a pressure measurement. It is, therefore, more frequently used. The difference between both heat-transfer coefficients is, by the way, small. This becomes evident from Fig. 12-6. Figure 12-9 presents heat-transfer coefficients (according to the second definition) measured in pool boiling with net evaporation of water at atmospheric pressure. It can be observed that the heat-transfer coefficient increases at first at a moderate rate with the temperature difference $\Delta t$. At approximately $\Delta t = 10$ F, the rate of increase becomes considerably larger until at approximately $\Delta t = 40$ F, a maximum is reached.

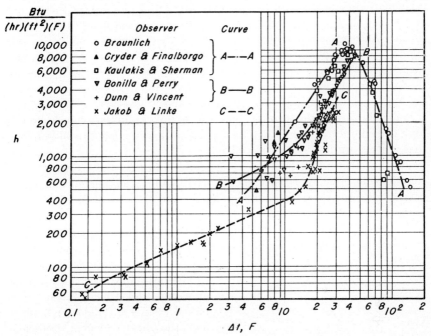

FIG. 12-9. Heat-transfer coefficients for water boiling on horizontal tubes ($A$) and horizontal plates ($B$ and $C$) at 1 atm. (*From W. H. McAdams, "Heat Transmission," 2d ed., McGraw-Hill Book Company, Inc., New York, 1942.*)

Beyond this value of $\Delta t$, the heat-transfer coefficient drops off. Measurements which extended the temperature difference beyond the range presented in Fig. 12-9 showed that the heat-transfer coefficient reaches a minimum at a $\Delta t$ of approximately 200 F and then starts rising again. Visual observation provided the explanation for this peculiar behavior. It could be observed that in the region $0 < \Delta t < 10$ F bubbles occur only on a few selected spots of the heating surface and rise in widely separated columns. It was reasoned by M. Jakob that the small number of bubbles can have only a minor influence on heat transfer and that the heat-transfer coefficient should be determined mainly by the free convection set up in the liquid by the temperature differences. Evaporation in this region is, therefore, called *free-convection boiling*. The number of spots on which bubbles start becomes larger with increasing $\Delta t$, and at a temperature difference beyond $\Delta t = 10$ F the heating surface is so densely populated with bubbles that their separation should cause a considerable stirring action in the fluid which increases the heat transfer. This is considered the reason for the steeper increase of the heat-transfer coefficient in the region $10 \text{ F} < \Delta t < 40 \text{ F}$ which is referred to as *nucleate boiling*. For the temperature region beyond 200 F in which the heat-transfer coefficient rises again with increasing temperature difference, the evaporation rate has become so high that a continuous vapor blanket covers the surface and the vapor bubbles grow out of this blanket. Boiling in this region is called *film boiling*. The low conductivity of the vapor film through which the heat must be transported explains the lower values of the heat-transfer coefficient for this region. At the highest temperature differences the energy transport is increased by radiation. The fact that a vapor film covers the heating surface at high heat fluxes has been known for a considerable time. Each reader has probably made the observation that when a drop of water is sprayed onto a glowing surface, it does not touch the surface but floats on a vapor film and that the droplet takes a considerable time to evaporate. This is called the Leidenfrost effect because it was reported and explained by Leidenfrost in 1756. The range $40 \text{ F} < \Delta t < 200 \text{ F}$ in which the heat-transfer coefficient decreases with increasing temperature difference is a transition region in which part of the surface is covered by bubbles and part by a film. Evaporation in this region is referred to as *transition boiling* or *partial film boiling*.

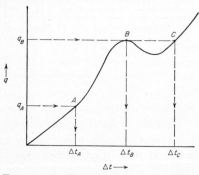

Fig. 12-10. Heat flux $q$ for boiling as function of temperature difference $\Delta t$.

The heat flow $q$ per unit surface area and unit time can be determined from Fig. 12-9 by multiplying the heat-transfer coefficient with the temperature difference $\Delta t$. When this is done, it is found that in the temperature range $40 \text{ F} < \Delta t < 200 \text{ F}$ the heat flow $q$ also drops with increasing $\Delta t$. As illustrated in Fig. 12-10, this has an important consequence in engineering applications. In many of those the heat flow $q$ is the quantity which is controlled. Let us assume that the heat flux is adjusted to the value $q_A$ in Fig. 12-10. The difference between the temperature of the heating surface and that of the liquid is $\Delta t_A$. When, now, the heat flux is gradually increased, then the temperature difference $\Delta t$ rises steadily until the value $q_B$

and with it $\Delta t_B$ are reached. A slight increase of the heat flux beyond this value, however, makes the temperature rise until a new equilibrium condition slightly above point $C$ is established. The corresponding temperature difference $\Delta t_C$ is usually so high that the temperature of the heating surface is beyond the melting point of most metals, in which case the surface is destroyed. Point $B$ is therefore referred to as *burnout point*. The trend to use high heat fluxes causes the engineer to design equipment so that it operates close to the burnout point, and the exact knowledge of this point has, therefore, great engineering importance.

If the assumption is correct that heat transfer in the *free-convection boiling* region is determined by the free-convection currents in the liquid, then it must be possible to calculate the heat-transfer coefficient in this region by relations of the form

$$\mathrm{Nu} = f(\mathrm{Gr},\mathrm{Pr}) \qquad (12\text{-}16)$$

describing free-convection heat transfer. The shape of the function $f$ will depend on the geometry of the heating surface. M. Jakob[1] has demonstrated that the known relations for free convection correlate measured heat-transfer coefficients for free-convection boiling satisfactorily.

In the *nucleate boiling* region it has been assumed that the stirring action of the bubbles is the determining factor for the heat-transfer process. This assumption was supported by experiments by F. C. Gunther and F. Kreith[2] in which it was found that the bubbles left the surface with velocities up to 15 fps. In addition, it has been established that the major portion of the heat flows in this region from the heating surface into the liquid and from there into the vapor bubbles. This means that the nature of the heat transfer at the heating surface can be considered as one by forced convection in a liquid where the convection is supplied by the movement of the bubbles and that the heat-transfer coefficient should be described by a relation of the form $\mathrm{Nu} = f(\mathrm{Re},\mathrm{Pr})$. Recent attempts at a theoretical treatment of heat transfer in nucleate boiling have, therefore, concentrated on the effort to express the stirring action by the growth of the bubbles.[3,4] The convection will, however, also depend on the number of bubbles generated per unit time and unit surface area, and the connection of this value with the microstructure of the surface makes its prediction difficult. Semiempirical correlations have been presented by Rohsenow[3] and by Forster and Zuber.[4] Both

---

[1] M. Jakob, "Heat Transfer," John Wiley & Sons, Inc., New York, 1949.

[2] F. C. Gunther and F. Kreith, Heat Transfer and Fluid Mechanical Institute, Berkeley, Calif., 1949.

[3] W. M. Rohsenow, *Trans. ASME*, **74**:969 (1952).

[4] H. K. Forster and N. Zuber, *Am. Inst. Chem. Engrs. J.*, **1**:531–535 (1955).

start from the assumption that heat transfer from the heating surface to the stirred liquid should be presented by a correlation of the form $Nu = f(Re, Pr)$. The length and velocity in these parameters are then related to bubble diameter, bubble velocity, and amount of vapor generated in the bubbles as a measure of bubble number. In this way Rohsenow obtained the following correlation:

$$\frac{c_l \Delta t}{i_{fg}} = a \left[ \frac{q}{\mu_l i_{fg}} \sqrt{\frac{\sigma}{g(\rho_l - \rho_v)}} \right]^{1/3} (Pr_l)^{1.7} \qquad (12\text{-}17)$$

where  $\Delta t$ = difference between surface temperature and saturation temperature
$i_{fg}$ = heat of evaporation
$q$ = heat flux at the heating surface
$c_l, \mu_l, \rho_l, Pr_l$ = specific heat, viscosity, density, and Prandtl number of liquid at saturation temperature

$\rho_v$ is the vapor density and $\sigma$ the liquid-vapor surface tension at saturation temperature. The constant $a$ depends on all the factors which influence nucleation. From an evaluation of available experimental results Rohsenow found values of $a$ between 0.0027 and 0.013, and from their own experiments on heat transfer in forced convection cooling with net evaporation Piret and Isbin[1] determined values of the constant $a$ between 0.0022 and 0.015. The wide range of this parameter indicates that considerable research is still required before a completely satisfactory correlation for nucleate boiling can be developed. The agreement between the constants $a$ which have been measured in pool boiling and forced-convection boiling also indicates that in the nucleate boiling range the stirring action of the bubbles is so great that the forced convection usually contributes little.

Experiments on boiling of liquid metals demonstrated the strong effect of surface wetting. It appears that with a nonwetting liquid *nucleate boiling* does not occur at all.

For heat transfer connected with *film boiling*, L. A. Bromley[2] developed a relation on the basis of a model which is essentially identical with Nusselt's film theory for condensation. The vapor film adjacent to the heating surface is assumed to rise under the influence of buoyancy forces, and heat is transferred by conduction through this film. The resulting relation for a vertical wall is identical with Eq. (12-9) with the only difference that the density $\rho$ in this equation has to be replaced by the difference $\rho_l - \rho_v$ of the liquid and vapor density. Otherwise, the vapor

---

[1] E. L. Piret and H. S. Isbin, *Chem. Eng. Progr.*, **50**:305 (1954).
[2] L. A. Bromley, *Chem. Eng. Progr.*, **46**:221–227 (1950).

properties have to be introduced into the equation. A relation for a horizontal tube analogous to Eq. (12-9) was found to agree with experimental results when a correction was applied to account for the facts that heat transfer by radiation through the vapor film increases the thickness of the vapor film and that the liquid exerts some frictional force on the moving vapor film.

It is of special importance to have a relation which predicts the peak nucleate heat flux occurring at the burnout point. Based on previous work by Kutateladze[1] and Rohsenow and Griffith,[2] N. Zuber[3] was able to develop such an equation analytically by a stability consideration on the interface between the vapor film and the liquid under the condition that they move relative to each other. Considering the two fluids as inviscid and moving under the influence of gravity and surface tension, he obtained the following equation for the maximum heat flux $q_B$ at burnout:

$$q_B = \frac{\pi}{24} i_{lv} \rho_v \left[ \frac{\sigma g(\rho_l - \rho_v)}{\rho_v^2} \right]^{1/4} \left( \frac{\rho_l + \rho_v}{\rho_l} \right)^{1/2} \qquad (12\text{-}18)$$

where $i_{lv}$ = heat of evaporation
$\rho_v$ = density of vapor
$\rho_l$ = density of liquid
$\sigma$ = vapor-liquid surface tension

Figure 12-11 compares this equation in a dimensionless presentation with experimental results obtained by various investigations and with various fluids. The agreement, which can be considered as very satisfactory, appears to indicate that surface condition does not influence heat transfer at the burnout point. Zuber's relation is valid for pool boiling with net evaporation. Experiments established the fact that in local boiling, the maximum heat flux is considerably higher (up to four times as large). Forced convection also increases the maximum heat flux.

The fact that heat-transfer coefficients from a surface into a boiling liquid are very large makes this process an effective one for keeping a wall cool. It is, for instance, used in the regenerative cooling of the walls of the combustion chambers in rockets. In this application, one of the fuels is ducted around the wall of the combustion chamber. Usually the heat flow is so large that locally the saturation temperature of the fluid is exceeded and bubble formation occurs in the neighborhood of the wall. However, in the core of the coolant the temperature is lower than the

[1] S. S. Kutateladze, *Izvest. Akad. Nauk. U.S.S.R., Otdel. Tekh. Nauk.* 4, 1951, p. 529.
[2] W. M. Rohsenow and P. Griffith, *Preprint* 9, American Society of Mechanical Engineers and American Institute of Chemical Engineers, Heat Transfer Symposium, Louisville, Ky., March, 1955.
[3] N. Zuber, *Trans. ASME*, **80**:711–720 (1958).

350    HEAT TRANSFER BY CONVECTION

evaporation temperature, and the bubbles collapse as soon as they separate from the surface and penetrate into the interior of the fluid. This heat-transfer process has to be classified as *forced-convection local boiling* and has been studied intensively in recent years. If, for instance, water is ducted through a pipe, and if the pipe is heated from the outside, then the heat flow per unit area of the pipe wall follows the law indicated in Fig. 12-12 which presents the results of experiments by Rohsenow and

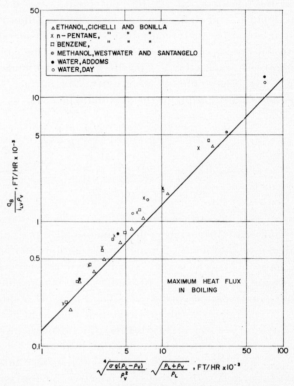

Fig. 12-11. Correlation for heat flux at burnout point and comparison with experimental results. (*From N. Zuber, ASME paper 57-HT-4, 1957.*)

Clark.[1] It increases at first with increasing difference between wall temperature and bulk temperature at the same rate as for forced flow without evaporation (Colburn line). As soon as bubble formation at the surface occurs, however, the increase in heat flow becomes very rapid and the temperature difference required is almost independent of the amount of heat flow. At the same time, it is found that various curves in the figure are obtained, depending on the magnitude of the saturation

[1] W. M. Rohsenow and J. A. Clark, Heat Transfer and Fluid Mechanics Institute, Stanford University Press, Stanford, Calif., pp. 193–207, 1951.

temperature. Rohsenow and Clark pointed out that these curves can be brought together to form a single line by plotting the heat flow per unit area over the difference between wall temperature and saturation temperature instead of wall temperature minus bulk temperature. This

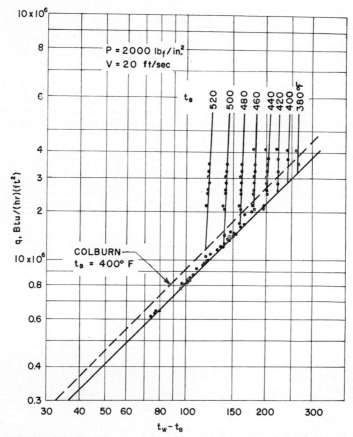

FIG. 12-12. Local boiling heat transfer for turbulent forced flow through a tube ($d = 0.18$ in., Re $\approx 3 \times 10^5$). (*From W. M. Rohsenow, Heat Transfer with Evaporation, in "Heat Transfer" (a symposium), University of Michigan Press, Ann Arbor, Mich., 1952.*)

shows that the main resistance to the heat flow is concentrated near the surface and is influenced by the saturation temperature rather than by the bulk temperature. The fact that in the nucleate boiling region forced flow has little effect on heat transfer has been pointed out before.

Special conditions arise when the boiling occurs inside long tubes as found in steam boilers or evaporators. The mixture of liquid and vapor is forced through the tubes either in forced- or in natural-convection flow.

Pressure drop and heat transfer follow an involved and complicated pattern. For a more extensive discussion and for empirical relations describing this two-phase flow and the other areas of evaporative heat transfer reference is made to various texts.[1]

## PROBLEMS

**12-1.** A condenser is to condense 20,000 lb/hr of water vapor at a pressure of 1.4 $lb_f/in.^2$ The heating surface consists of horizontal brass tubes with a 1.2-in. ID and a 0.04-in. wall thickness. The tube length is 9 ft. The amount of cooling water is 800,000 lb/hr, its entrance temperature is 80 F, and its velocity in the tubes is 6 fps. How many tubes have to be provided in the condenser?

**12-2.** Calculate the thickness of the condensate film on a horizontal tube, assuming the film thickness to be small as compared with the tube diameter, by applying the method used in Sec. 12-1.

[1] W. H. McAdams, "Heat Transmission," 3d ed., McGraw-Hill Book Company, Inc., New York, 1954, or Max Jakob, "Heat Transfer," vol. 1, John Wiley & Sons, Inc., New York, 1949.

PART C

# THERMAL RADIATION

**Basic Concepts.** All substances, solid bodies as well as liquids and gases, at normal and especially at elevated temperatures emit energy in the form of radiation and are also capable of absorbing such energy. All heat-transfer processes are, therefore, more or less accompanied by a heat exchange by radiation. In some cases, radiative heat exchange is so small that it can be neglected or it is of such a nature that it is simply included in the conductive heat transfer. This is generally the case in solids and liquids. In gases, however, the radiative transport is so different in nature from the conductive and convective heat transport that it has to be considered separately. Heat exchange by radiation will be the subject of discussion in Part C.

Calculations of heat transfer by conduction or convection in Parts A and B started always with an energy balance on a small-volume element, and it turned out that it was sufficient for such a balance to know the conditions in the immediate neighborhood of the element. These conditions could be expressed by gradients of the state parameters involved, especially of the temperature, and as a consequence the energy balance could be written down as a differential equation. In radiative heat exchange the situation is usually more complicated. Let us consider, for example, that we want to study radiative heat exchange within the combustion chamber of a jet-engine combustor or a rocket. For this purpose, an energy balance on a small arbitrarily located volume element within the combustor has to be made. When the energy gain of the gas contained in this volume element is considered, the gas is found to absorb radiative energy which has been emitted from other gas elements or from elements of the solid walls even when these are located at a considerable distance. Therefore, the energy balance depends not only on the conditions in the immediate neighborhood of the volume element under consideration, and the equation describing this balance is an integrodifferential equation. Such an equation, however, is quite difficult to solve, and practically all engineering calculations are based on simplifying assumptions. A second factor, which makes radiative heat exchange more complicated than conductive or even convective heat transfer, is connected with the fact that the radiative properties of the various substances encountered in engineering are more difficult to describe than, for instance, the heat conductivity, knowledge of which is required for calculation of a heat-conduction process. Therefore, Chap. 13 will be devoted to a discussion of the radiative properties of solids, liquids, and gases. Radiative heat exchange will be discussed in Chap. 14.

The radiative energy emitted from a substance increases very strongly with temperature. As a consequence, radiative heat exchange becomes very large and dominating at high temperatures. Many processes in new engineering areas occur at high temperatures, and a knowledge of radiative heat transfer becomes very important for the design of the pertinent equipment. Nuclear power plants, gas turbines, and the various propulsion devices for aircraft, missiles, satellites, and space vehicles are examples of such engineering areas. A fusion power plant with temperatures of millions of degrees will pose very severe cooling problems caused by radiative heat transfer.

CHAPTER 13

# THERMAL RADIATION PROPERTIES

Heat rays, as they travel through space, exhibit all the features characteristic of waves. They show "interference" phenomena when bundles which originated from the same source but have traveled different paths are united again. They also can be polarized, that is, made to oscillate only in a certain direction when they are transmitted through suitable filters. This indicates that the rays are transverse waves oscillating in a direction normal to the path along which they travel. The rays are an oscillating electric and magnetic field. Their nature is the same as that of other electromagnetic waves, such as X rays, visible light, and the waves used in wireless transmission. All these rays travel with approximately the same velocity. In a vacuum the velocity for all electromagnetic waves is

$$c_v = 3 \times 10^{10} \text{ cm/sec} = 186{,}000 \text{ miles/sec}$$

In other media the velocity is somewhat slower and can be calculated from the refraction index

$$n = \frac{c_v}{c}$$

with $c$ denoting the velocity in the medium under consideration. The various electromagnetic waves differ only by the wavelength $\lambda$. The scale shown in Fig. 13-1 indicates how the waves are classified according to wavelength. The wavelength is measured in this figure in various scales. $1\mu = 10^{-4}$ cm is a convenient scale for heat radiation. Molecular dimensions (diameter of a molecule, distance of molecules) have numbers of order 1 when expressed in angstroms. The human eye is sensitive to rays in the range from 0.40 up to $0.70\mu$. Radiation emitted from heated bodies comprises rays with wavelengths 0.3 to $10\mu$ and more.

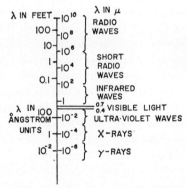

FIG. 13-1. Wavelengths of various electromagnetic waves.

It can be observed that a large part of thermal radiation has waves outside the visible range. This range with wavelengths larger than those of light is called *infrared*.

In a vacuum and in a homogeneous medium radiation travels in straight paths. However, when a radiation ray impinges on an interface between two different media, it is split into two portions. One part is reflected from the interface, and the other part penetrates into the interior of the second medium. The reflection occurs in such a way that the angle $\beta_r$ between the reflected beam and the normal to the surface equals the angle $\beta_i$ made by the impinging beam with the same normal, and that incident beam, reflected beam, and interface normal are in one plane

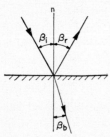

FIG. 13-2. Reflection and refraction of a ray on a surface.

(Fig. 13-2). The beam traveling into the interior of the second medium is bent (refracted) in such a way that the ratio of the sine of the angle $\beta_i$ to the sine of the angle $\beta_b$ between the penetrating beam and the surface normal is equal to the ratio of the refraction index of the second medium to the refraction index of the medium in which the beam traveled before impinging on the interface. The reflection and refraction laws hold also for bundles consisting of a large number of rays when they impinge on a perfectly plane surface. This reflection is called *specular* or *regular*. On rough surfaces, however, the incident bundle of rays is distributed into all directions after reflection. This kind of reflection is called *diffuse* reflection.

In heat-exchange calculations, we are especially interested in determining the amount of energy which is transported by radiation. The amount of energy carried along with the radiation is not changed as the radiant rays travel through a vacuum. When radiation travels through other media, the energy generally decreases in flow direction because it is partly absorbed or scattered. The energy flow may also increase in the direction of the rays when radiation is emitted by the molecules of which the medium consists. The absorption and emission of infrared radiation are, however, very weak for air and many other gases, so that such gases in most cases can be considered with good approximation as nonabsorbing and nonemitting media.

We shall now consider what happens to radiant energy impinging on an interface between two media. Radiation may travel through a medium 1 toward the interface. The energy flow impinging per unit time on a unit area of the interface is called *incident radiation* or *irradiation* and may be denoted by $H$. Part of the radiation will be reflected back into medium 1. The energy leaving the interface with the reflected radiation per unit time and area will be denoted by $\rho H$. $\rho$ indicates the ratio of the reflected to the incident radiation and is called *reflectivity*. The rest of the radiation penetrates through the interface into medium 2. If medium 2 does absorb radiation, then part of this radiation will be absorbed while in medium 2 and its amount may be denoted by $\alpha H$. $\alpha$, the ratio of the absorbed to the incident radiation, is called *absorptivity*. The rest of the radiant energy will leave medium 2 somewhere through its surfaces. This amount may be denoted by $\tau H$. $\tau$, the ratio of the transmitted to the incident radiation, is called *transmissivity*. From the law of conservation of energy it follows that

$$\rho + \alpha + \tau = 1 \tag{13-1}$$

The parameters $\rho$, $\alpha$, and $\tau$ depend on the nature of the two media 1 and 2 and on the structure of the interface. In the following discussion of these properties medium 1 will always be air. Replacement of air by another gas changes the values of these parameters very little.

Solid and liquid bodies absorb practically the entire infrared radiation penetrating through their surfaces within a very thin layer. Electric conductors need a layer of the order of magnitude $1\mu$ for the absorption. Nonconductors, with very few exceptions, absorb practically the entire radiation in layers of order 0.1-in. thickness.[1] The thickness of materials used in engineering is almost always greater than these values, and Eq. (13-1) can, in this case, be simplified to

$$\rho + \alpha = 1 \tag{13-2}$$

This equation then describes the behavior of almost all solid and liquid materials for infrared radiation. In a simplified way one talks often about an absorbing surface, although in reality the absorption occurs in a layer of small but finite thickness.

Gases, on the other hand, are found to reflect very little of the radiation impinging on their interface with air or another gas. However, they usually transmit an appreciable portion of the radiation even in layers of fairly large thickness. For gases, therefore, $\rho$ can usually be neglected as small and the equation

$$\alpha + \tau = 1 \tag{13-2a}$$

expresses their behavior against foreign radiation.

When they are at a finite absolute temperature, most media will emit thermal radiation. When we consider the medium 2 a radiating one, then a radiative energy flux will travel through the interface into medium 1 in a direction opposite the incident radiation. The energy flux through a unit area of the interface per unit time generated by the radiation emitted from medium 2 is called *emissive power* and is denoted by $e$. It

[1] E. Eckert, *Forsch. Gebiete Ingenieurw.*, **9**:251–253 (1938).

will be shown that the laws governing the radiation emitted by a body are especially simple when no radiation is reflected from its surface ($\rho = 0$, $\alpha = 1$). Such a body is called a *black body*. The name comes from the fact that a surface which absorbs all light rays appears black to the eye. A surface can absorb practically all thermal radiation without absorbing all light rays and therefore without appearing black to the eye. For instance, a whitewashed wall is nearly black for infrared radiation.

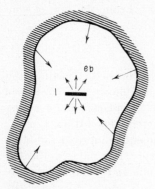

Fig. 13-3. Derivation of Kirchhoff's law.

Totally black surfaces do not exist in nature, since a certain percentage of the incident radiation is always reflected, but there are surfaces which reflect only a very small portion of the incident radiation. A surface which is very nearly black to infrared radiation is snow, with an absorptivity of 0.985 for thermal radiation from a body whose temperature is not too high. Although no totally black surfaces exist in nature, the concept of a black body is a useful one because the laws governing its radiation are comparatively simple and because it can be proved that no surface radiates more heat than a black one. This is the content of Kirchhoff's law.

In order to prove Kirchhoff's law, we shall consider a hollow space (Fig. 13-3) whose walls have a uniform temperature. The space enclosed by the walls may be a vacuum. The walls of this enclosure will emit radiation, and this radiation will be partially reflected and in this way will travel back and forth through the enclosure. Now we place a small body 1 with the shape of a disk in the interior of the enclosure. If the disk is small enough, then we can assume that the radiation in the enclosure is changed only by an infinitesimally small amount through the presence of the disk. We shall assume that the disk has black surfaces and is at a temperature equal to the temperature of the enclosure walls. The amount of heat which is radiated per unit time from a unit surface of the black body 1 will be denoted by $e_b$. We can immediately state that the amount of heat impinging on the surface of body 1 per unit time and area is also $e_b$ for the following reason: In a vacuum or a system in thermal equilibrium the body 1 can exchange heat with its surroundings by radiation only. If the amount of heat impinging on the surface of body 1 and absorbed by it ($\alpha = 1$) were different from the amount of heat which the body emits, then the body would either cool off or heat up. Both temperature changes, however, are forbidden by the second law of thermodynamics, which states that no temperature differences can arise spontaneously in a closed system when it is in thermal equilibrium. We can now change the position and the orientation of the black disk 1 and in this way prove that in a black enclosure of uniform temperature the radiation impinging on a unit surface is independent of the location and of the direction of that surface. We can call such a radiation *isotropic*.

Now we remove body 1 and replace it by body 2, which is also shaped as a small disk but consists of a material with an absorptivity $\alpha$. The radiative energy impinging on the surface of this body is again $e_b$. The body absorbs the amount $\alpha e_b$. The emissive power $e$ of body 2 must have the same magnitude according to the second law of thermodynamics:

$$e = \alpha e_b \tag{13-3}$$

This is the mathematical expression for Kirchhoff's law. In words we can state that the ratio of the emissive power $e$ of a body to the emissive power $e_b$ of a black body at the same temperature equals the absorptivity of the first body. The ratio of the emissive power of any body to the emissive power of a black body of equal temperature is called emissivity $\epsilon$. Kirchhoff's law can also be written in the form

$$\epsilon = \alpha \tag{13-4}$$

$\alpha$ is always smaller than 1. Therefore, the emissive power $e$ is always smaller than the emissive power of a black body at equal temperature. Restrictions on the law as expressed by Eq. (13-4) will be discussed in Sec. 13-2.

Radiators occurring in nature simultaneously emit radiation of various wavelengths. Accordingly, in our calculations we have usually to deal with radiation which contains rays of a wide range of wavelengths. It is, however, often necessary to start the considerations with a radiation the rays of which have all the same wavelength. Such a radiation is called *monochromatic*. Experimentally, very nearly monochromatic radiation can be produced by appropriate filters. We can imagine that the walls of the hollow space in Fig. 13-3 emit black radiation of a certain wavelength only. In this way, Eq. (13-4) is derived for monochromatic radiation. Equation (13-4) holds also for radiation which is composed of several wavelengths, in this case, however, only with certain limitations which will be pointed out later on.

Kirchhoff's law can be demonstrated by the following simple experiment. A cross is painted with ink or with black paint on a piece of chromium- or nickel-plated metal sheet. The cross appears dark on the metal background because the absorptivity of the paint for incident visible radiation is higher than the absorptivity of the metal. If, however, the sheet is heated in a bunsen burner until it starts glowing, then the cross appears brighter than the background, indicating that the emissivity of the paint is also higher than the emissivity of the metal. This, however, is just what is stated by Kirchhoff's law.

Considerable information on the radiation properties of materials (reflectivity, absorptivity, emissivity) can be obtained from macroscopic considerations employing Maxwell's electromagnetic wave theory. For a complete knowledge, however, it is necessary to study the interaction between the radiation and the material on a molecular scale. It has been found that radiation behaves then as if it consisted of finite particles (photons).

From his electromagnetic theory, Maxwell predicted in 1865 that radiation impinging on a surface exerts a pressure on that surface, and he calculated the magnitude of that pressure $p_r$ exerted on a perfect mirror (a surface with $\rho = 1$) as

$$p_r = \frac{u}{3} \tag{13-5}$$

where $u$ denotes the radiation density (the amount of radiative energy contained per unit volume in the space above the surface). An interesting analogy exists between the radiation pressure and the pressure which an ideal gas exerts on the walls of a container. Gas kinetic theory shows that the gas pressure $p$ is $p = \frac{1}{3}\rho v^2$, with $\rho$ expressing the density of the gas and $v$ the mean velocity of the molecules. If we assume that the photons contained in a unit volume have a mass $\rho_r$ and move with light velocity $c$ and write in analogy to the gas $p_r = \frac{1}{3}\rho_r c^2$, then a comparison of the two equations for the radiation pressure results in

$$u = \rho_r c^2$$

which is Einstein's equation expressing the equivalence of energy and mass. A radiation field is sometimes referred to as a "photon gas" because of this equivalence. The radiation pressure is usually very small. Solar radiation, for instance, exerts at the surface of the earth a pressure of 0.4 mg/m² in metric units. Nevertheless, in 1901 Lebedow was able to demonstrate this pressure experimentally.

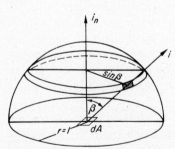

FIG. 13-4. Determining the radiation in different directions.

**13-1. The Black Body.** A surface element of a black body radiates in all directions. The heat flow per unit surface area leaving the body per unit time by this radiation is its emissive power $e_b$. The heat radiation from a unit surface in a certain direction can be measured by determining the heat flow $dq$ radiated per unit area of the radiating body and passing through an element of the surface of a sphere constructed with the radius $r = 1$ around the radiating surface element (Fig. 13-4). The area of such a surface element on this sphere has the same numerical value as the solid angle $d\omega$. The radiation in the contemplated direction is defined as

$$i = \frac{dq}{d\omega} \tag{13-6}$$

and is called *radiation intensity*. In the preceding section it has been pointed out that black-body radiation as created in an isothermal enclosure is isotropic, that is, independent of direction. From this fact, it can be easily demonstrated that the radiation intensity $i_b$ of a black body is given by Lambert's cosine law

$$i_b = i_{bn} \cos \beta \tag{13-7}$$

where $i_{bn}$ is the radiation in the direction of the normal to the surface element and $\beta$ is the angle between the radiation $i_b$ and the normal. The total heat flux radiated per unit area of the black surface element, its emissive power, can be found by integration:

$$e_b = \int i_b \, d\omega = 2\pi i_{bn} \int_0^{\pi/2} \cos \beta \sin \beta \, d\beta = \pi i_{bn} \tag{13-8}$$

A solid angle, corresponding to a ring-shaped surface on the sphere, $2\pi \sin \beta \, d\beta$, could be used in carrying out the integration, since the radiation intensity $i_b$ is constant in all directions comprised in a conical surface with the opening angle $\beta$ (Fig. 13-4). The total heat radiated from the unit surface area of a black body is therefore $\pi$ times the heat radiated in normal direction.

The heat radiated from a body is always comprised of rays of different wavelengths. The amount of heat $di$ radiated within an infinitesimally small wavelength range is also an infinitesimally small quantity. The fraction

$$i_\lambda = \frac{di}{d\lambda} \tag{13-9}$$

however, is a finite quantity and is called monochromatic radiation intensity. In order to calculate the heat radiated by a black body as a function of the wavelength, Max Planck formulated the concept of his famous quantum theory, which became the origin of modern physics. Planck's law for the monochromatic thermal radiation intensity of a black body in normal direction is

$$i_{\lambda bn} = \frac{2C_1}{\lambda^5 (e^{C_2/\lambda T} - 1)} \tag{13-10}$$

where $T$ is the absolute temperature. Without the factor 2 the equation gives the radiation intensity of polarized black radiation. The constants $C_1$ and $C_2$ are contained together with other radiation constants in Table 13-1. It may be observed that these constants are known with very high accuracy. Equation (13-10) is represented in Fig. 13-5 as a function of the wavelength $\lambda$ with the absolute temperature $T$ as parameter. From Fig. 13-5 it can be seen that at any temperature, a black

TABLE 13-1. RADIATION CONSTANTS IN PLANCK'S, WIEN'S, AND STEFAN-BOLTZMANN'S EQUATIONS*

| | | |
|---|---|---|
| $C_1$ | $0.18892 \times 10^8$ Btu $\mu^4$/hr ft$^2$ | $0.59525 \times 10^{-12}$ watt cm$^2$ |
| $C_2$ | 25,896 $\mu$ R | 1.4387 cm K |
| $C_3$ | 5215.6 $\mu$ R | .28976 cm K |
| $\sigma$ | $0.1714 \times 10^{-8}$ Btu/hr ft$^2$ R$^4$ | $4.876 \times 10^{-8}$ kcal/m$^2$ hr K$^4$ † |

* According to N. W. Snyder, *Trans. ASME*, **76**:537–539 (1954).
† 1 kcal = 760 kw-hr.

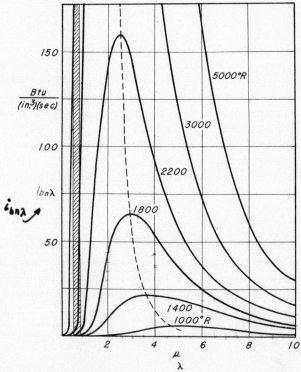

FIG. 13-5. Black-body radiation as a function of the wavelength for different temperatures.

body radiates heat in a wide range of wavelengths. The maximum value for the radiation intensity moves with increasing temperature to shorter wavelengths. This is described by Wien's law:

$$\lambda_{\max} T = C_3 \tag{13-11}$$

where $\lambda_{\max}$ is the wavelength for the maximum radiation intensity. The visible range of wavelengths is indicated in Fig. 13-5 as a shaded band. It can be seen that with increasing temperature an increasingly greater part of the whole radiation falls into the visible range. As the human eye can sense very small energies, the small radiation in the visible range at 1800 R, or 1340 F, is seen as dark glowing red. With increasing temperature, the visible part of the radiation becomes greater and therefore the brightness increases. But even at a temperature of 5000 R this visible part is comparatively small and the radiant energy is almost completely contained in infrared rays. The filament of a tungsten lamp has this temperature; therefore the efficiency of such a lamp is poor. Only at the temperature of the sun (approximately 10,000 R) does the maximum value of the monochromatic radiation intensity lie within the visible range.

Equation (13-10) can be brought into the following form:

$$\frac{i_{\lambda b n}}{T^5} = \frac{2C_1}{(\lambda T)^5 (e^{c_2/\lambda T} - 1)} \tag{13-12}$$

which indicates that $i_{\lambda b n}/T^5$ is a function of the product $\lambda T$ and of the various constants only. This function is found tabulated in various handbooks.[1] Table 13-2 tabulates Eq. (13-12) with the constants taken from Table 13-1.

The radiation intensity in the whole range of wavelengths, which is of most interest for engineering calculations, can be found from Planck's law by integration:

$$i_{bn} = \int_0^\infty i_{b\lambda n} \, d\lambda$$

The integration is possible by introducing $x = 1/\lambda$:

$$i_{bn} = 2C_1 \int_0^\infty x^3 [e^{(c_2/T)x} - 1]^{-1} \, dx$$
$$= 2C_1 \int_0^\infty x^3 [e^{-(c_2/T)x} + e^{-(c_2/T)2x} + e^{-(c_2/T)3x} + \cdots] \, dx$$

Each single integral $\int_0^\infty x^3 e^{-(c_2/T)ix} \, dx$ ($i = 1, 2, 3, \ldots$) can be solved by

---

[1] See, for instance, E. Jahnke and F. Emde, "Tables of Functions," p. 46, Teubner Verlagsgesellschaft, Leipzig, 1933.

TABLE 13-2. Monochromatic Radiation Intensity $i_{bn\lambda}$ of a Black Body in Normal Direction* as a Function of Absolute Temperature $T$ and Wavelength $\lambda$†

| $\lambda T$, $\mu$ R | $\dfrac{i_{bn\lambda}}{T^5} \times 10^{15}$ Btu hr ft² $\mu$ R⁵ | $\lambda T$, $\mu$ R | $\dfrac{i_{bn\lambda}}{T^5} \times 10^{15}$ Btu hr ft² $\mu$ R⁵ |
|---|---|---|---|
| 1000 | 0.0001977 | 10400 | 25.82 |
| 1200 | 0.005941 | 10600 | 24.71 |
| 1400 | 0.05991 | 10800 | 23.65 |
| 1600 | 0.3109 | 11000 | 22.64 |
| 1800 | 1.039 | 11200 | 21.68 |
| 2000 | 2.585 | 11400 | 20.76 |
| 2200 | 5.210 | 11600 | 19.88 |
| 2400 | 8.987 | 11800 | 19.04 |
| 2600 | 13.81 | 12000 | 18.25 |
| 2800 | 19.43 | 12200 | 17.49 |
| 3000 | 25.49 | 12400 | 16.76 |
| 3200 | 31.67 | 12600 | 16.07 |
| 3400 | 37.66 | 12800 | 15.41 |
| 3600 | 43.22 | 13000 | 14.79 |
| 3800 | 48.17 | 13200 | 14.18 |
| 4000 | 52.43 | 13400 | 13.62 |
| 4200 | 55.95 | 13600 | 13.07 |
| 4400 | 58.73 | 13800 | 12.55 |
| 4600 | 60.78 | 14000 | 12.12 |
| 4800 | 62.18 | 14200 | 11.59 |
| 5000 | 63.00 | 14400 | 11.13 |
| 5200 | 63.33 | 14600 | 10.71 |
| 5400 | 63.08 | 14800 | 10.30 |
| 5600 | 62.51 | 15000 | 9.89 |
| 5800 | 61.63 | 16000 | 8.194 |
| 6000 | 60.48 | 17000 | 6.824 |
| 6200 | 59.12 | 18000 | 5.720 |
| 6400 | 57.60 | 19000 | 4.827 |
| 6600 | 55.96 | 20000 | 4.099 |
| 6800 | 54.23 | 21000 | 3.522 |
| 7000 | 52.44 | 22000 | 3.006 |
| 7200 | 50.62 | 23000 | 2.589 |
| 7400 | 48.79 | 24000 | 2.248 |
| 7600 | 46.95 | 25000 | 1.957 |
| 7800 | 45.14 | 26000 | 1.711 |
| 8000 | 43.36 | 27000 | 1.505 |
| 8200 | 41.61 | 28000 | 1.330 |
| 8400 | 39.91 | 29000 | 1.174 |
| 8600 | 38.25 | 30000 | 1.0437 |
| 8800 | 36.65 | 40000 | 0.3718 |
| 9000 | 35.10 | 50000 | 0.1636 |
| 9200 | 33.60 | 60000 | 0.08279 |
| 9400 | 32.17 | 70000 | 0.04611 |
| 9600 | 30.79 | 80000 | 0.02759 |
| 9800 | 29.46 | 90000 | 0.01756 |
| 10000 | 28.19 | 100000 | 0.01154 |
| 10200 | 26.98 | ∞ | 0 |

* Calculated using a table published by R. V. Dunkle, *Trans. ASME*, **76**:549 (1954).
† 1 $\mu$ = $10^{-4}$ cm.

partial integration. It gives $6T^4/C_2^4 i^4$. Therefore

$$i_{bn} = \frac{12C_1}{C_2^4} T^4 \left(1 + \frac{1}{2^4} + \frac{1}{3^4} + \cdots \right) = \frac{12C_1}{C_2^4} \frac{\pi^4}{90} T^4 = K_n T^4 \quad (13\text{-}13)$$

This is Stefan-Boltzmann's law, found experimentally by Stefan and theoretically by Boltzmann. According to this law, black-body radiation is proportional to the fourth power of the absolute temperature. The emissive power of a black body is $\pi$ times the radiation intensity as presented by Eq. (13-13). Therefore, it can be written

$$e_b = \sigma T^4 \quad (13\text{-}14)$$

The numerical value of the Boltzmann constant $\sigma$ has been determined by numerous measurements. The resulting value $\sigma = 0.173 \times 10^{-8}$ Btu/hr ft$^2$ R$^4$ is slightly greater than the value resulting from Eq. (13-13) which is contained in Table 13-1.

In nature an exactly black surface does not exist. It can be approached, however, to any desired degree in the following way. Kirchhoff's law states that a body which absorbs all impinging radiation radiates as a black body. We can realize a black body therefore by using a hollow space in which there is a small opening (a *hohlraum*) (Fig. 13-6). The walls of the space must be kept at uniform temperature. Radiation which streams through the opening into the hollow space is reflected repeatedly before it leaves the space again through the opening. If the surface of the space has a great absorptivity, then at every reflection the major part of the radiation is absorbed, so that after several reflections almost all radiation is absorbed. The opening of the hollow space therefore acts like a black surface. If the size of the opening is decreased, the black body can be approximated as closely as is desired. If the hollow space is heated, then the opening emits black-body radiation according to Kirchhoff's law.

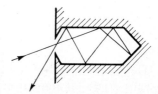

Fig. 13-6. Radiation of a hollow space.

In the rest of this chapter, the derivation of the laws for black-body radiation will be discussed. Historically, Boltzmann derived the fourth-power law in 1884 from classical thermodynamics. Wien then developed the displacement law (13-11) in 1896 on the same basis. For a derivation of the law describing the monochromatic radiation intensity it was necessary to leave the ground of classical physics. M. Planck succeeded in deriving this law in 1900 through his quantum concept.

We shall first derive Stefan-Boltzmann's law. As a preparation it is necessary to establish the connection between radiation density $u_b$ and the radiation intensity $i_b$ of a *hohlraum* whose walls are at constant temperature. The hollow space may, for our consideration, have the shape of a sphere, and we shall calculate the radiation density in the center of this sphere (Fig. 13-7). The radiation density in such a *hohlraum* must be independent of location, for otherwise the radiation pressure would vary with location and it would be possible to obtain mechanical work by a device which is exposed to radiation pressure at various locations. This, however, would violate the second law of thermodynamics. We consider the radiation emitted by an element $dA_s$ of the spherical surface and passing through an area element $dA$ located at the center of the sphere. This radiation passing through $dA$ in a time $d\tau$ can be expressed as

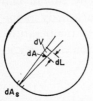

Fig. 13-7. Radiation in an enclosure of uniform temperature.

$$i_{bn} \frac{dA_s\, dA}{r^2} d\tau$$

where the radiation intensity of the enclosure in normal direction is $i_{bn}$. Since radiation moves with the velocity of light $c$, the radiation passing through $dA$ in the time element $d\tau$ must be contained in a column behind $dA$ of length $dL$ where $dL$ is fixed by the equation

$$dL = c\, d\tau$$

Introducing this expression into the above relationship results in the following expression for the radiation coming from $dA_s$ and contained in $dV = dA\, dL$:

$$i_{bn} \frac{dA_s}{r^2 c} dV$$

The total amount of radiation contained in the volume $dV$ will be found when the total amount of radiation originating from the whole enclosure surface is calculated. This is obtained by integrating the above expression over the spherical surface.

$$i_{bn} \frac{dV}{c} \int \frac{dA_s}{r^2}$$

$dA_s/r^2$ is nothing else than the solid angle under which the area $dA_s$ is seen from the center of the sphere. Therefore, the integral $\int dA_s/r^2$ must be equal to $4\pi$. The total radiation contained per unit volume, the radiation density $u_b$ is

$$u_b = \frac{4\pi}{c} i_{bn} = \frac{4}{c} e_b \qquad (13\text{-}15)$$

It is a function of temperature only.

After this preparation, Stefan-Boltzmann's law can be derived. For this purpose imagine that the hollow space of uniform temperature which has been considered quite frequently has now the shape indicated in Fig. 13-8. It is a cylinder closed by a piston, the surface of which is assumed to be a perfect mirror ($\rho = 1$). The space within the cylinder may be assumed a perfect vacuum, and the cylinder surfaces are kept at a temperature $T$. Accordingly, the surfaces will radiate heat and the interior of the cylinder is filled with heat radiation whose density is $u_b$. This density is a function of the temperature of the cylinder walls. It will also exert a pressure of the amount $p_r = u_b/3$ [Eq. (13-5)] on the wall of the piston. Now imagine the following cycle performed with the cylinder. The piston is moved to the right so that the volume of the cylinder changes from $V_1$ to $V_2$. The temperature is kept constant during this time. This means that the radiation density is constant. Since the volume increases, the total amount of energy contained within the cylinder will increase during this period and has to be furnished by addition of heat $Q$ from the surrounding to the cylinder walls.

FIG. 13-8. Heat engine operated on radiation pressure.

The radiation pressure also stays constant during this period and will perform work on the cylinder of the magnitude

$$W = p_r(V_2 - V_1) = \frac{u_b}{3}(V_2 - V_1)$$

The internal energy contained within the space of the cylinder increases by an amount

$$\Delta U_b = u_b(V_2 - V_1)$$

The first law of thermodynamics states that the amount of heat added during this process must be

$$Q = \Delta U_b + W = \tfrac{4}{3} u_b(V_2 - V_1)$$

After the volume $V_2$ is reached, the cylinder is still moved on by an infinitely small amount $dV$, now without addition of heat. Correspondingly the density of radiation within the cylinder will decrease and the temperature and the radiation pressure will drop. This is indicated in the lower half of the diagram. After a temperature $T - dT$ is reached, the movement of the cylinder is reversed. The cylinder is moved to the left, at first in such a way that the temperature is kept constant and finally adiabatically, until the original state is reached again. The total work done on the cylinder during the cycle will be

$$dW = dp_r(V_2 - V_1) = \frac{du_b}{3}(V_2 - V_1)$$

The process which we consider is a reversible one if it is carried out slowly enough so that the heat addition and extraction occur without temperature drop. The cycle has been performed between two temperature limits $T$ and $T - dT$, and according to the second law of thermodynamics, all reversible cycles performed between two fixed temperature levels have the same thermal efficiency $(dT/T)$, namely, the efficiency of a Carnot process. Accordingly, the thermal efficiency of the process considered here is

$$\eta = \frac{dT}{T} = \frac{dW}{Q} = \frac{du_b}{4u_b}$$

It follows that

$$\frac{du_b}{u_b} = 4\frac{dT}{T}$$

This differential equation can be integrated:

$$\ln u_b = 4 \ln T + \ln C$$

or
$$u_b = CT^4 \qquad (13\text{-}16)$$

According to Eq. (13-15) the last equation can also be written in terms of the emissive power of a black body and thus states Stefan-Boltzmann's law of radiation. Wien's law was derived in a similar manner.

Toward the end of the nineteenth century, there were a considerable number of attempts also to predict theoretically the monochromatic emissive power. These, however, were unsuccessful in the sense that the calculations resulted in laws which were not in agreement with experimental results. Nevertheless, one of the predictions will be described, since it leads to the step which Planck made in order to derive his law. This attempt was made by Rayleigh and Jeans (1900). It is based on the concept that in an enclosure of uniform temperature in which black radiation exists, waves of a large number of wavelengths travel through space, being in equilibrium with the surrounding walls by the continuous emission and absorption of energy by the molecules of which the wall material is composed. The simplest arrangement by which electromagnetic waves of a certain wavelength can be created is an arrangement of two atoms having a plus and minus electric charge and interconnected by elastic forces so that they are able to perform oscillations. Such a harmonic oscillator will then be able to emit and absorb radiation. There must be a large number of such oscillators, with different frequencies, arranged along the wall of the enclosure to produce radiation of a black body. According to the law of equipartition of energy, which had been derived from statistical mechanics, it has to be expected that the oscillators for each wavelength have, on an average, stored at a certain temperature the same amount of energy, namely, $\mathbf{k}T/2$ per degree of freedom. An oscillator stores kinetic energy as well as potential energy, and therefore its energy is

$$E = \mathbf{k}T$$

where **k** is the gas constant per molecule (Boltzmann's constant). If we knew, in addition, the number of oscillators for each wavelength, then we would be able to calculate the energy contained in the arrangement of oscillators, and this again, according to the equipartition law, has to be equal to the energy of the black-body radiation which is in equilibrium with these oscillators. Rayleigh and Jeans in 1900 calculated the number $dZ$ of waves in a frequency range between $\nu$ and $\nu + d\nu$ which are contained in a certain volume $V$ from the concept that the waves must be standing waves with a whole number of waves within the confines of the volume. They derived the following equation:[1]

$$dZ = \frac{8\pi\nu^2}{c^3} V \, d\nu$$

The radiation density $u_\nu \, d\nu$ of the radiation field in the volume $V$ can then be obtained by multiplying the average energy of a specific oscillator which corresponds to the average energy of a wave with the number of such waves. This results in

$$V u_\nu \, d\nu = E \, dZ = \frac{8\pi\nu^2}{c^3} V \mathbf{k} T \, d\nu$$

In experimental physics, the energy is usually expressed as a function of wavelength rather than of frequency. The equation connecting the wavelength with the frequency is

$$\nu \lambda = c$$

with $c$ expressing the velocity of light. The amount of energy, therefore, can also be written

$$u_\lambda \, d\lambda = u_\nu \, d\nu$$

if the differential of the frequency is described by

$$d\nu = \frac{c}{\lambda^2} d\lambda$$

Introducing those expressions into the above equation and dividing by the volume $V$ give for the monochromatic radiation density

$$u_\lambda = \frac{8\pi}{\lambda^4} \mathbf{k} T$$

This equation is Rayleigh-Jean's law for monochromatic black radiation. According to this law the radiation will increase indefinitely when the wavelength goes toward zero, which is in contradiction to experiments.

[1] For a derivation see, for instance, W. P. Allis and M. A. Herlin, "Thermodynamics and Statistical Mechanics," pp. 199–214, McGraw-Hill Book Company, Inc., New York, 1952.

However, it was found that for large wavelengths, it describes the actual monochromatic black radiation quite well.

At this point, Max Planck stepped in with his bold idea that energy can be exchanged on a molecular basis not in any arbitrary quantity but only in multiples of a certain energy element $\epsilon$. Therefore, the quantity $n\epsilon$, with $n$ being a whole number, describes energy exchange on a molecular basis.

Planck showed that the average energy which an oscillator in equilibrium at the temperature $T$ assumes on this basis is no longer described by the equipartition law as stated in the previous equation but by the following relationship:

$$E = \frac{\epsilon}{e^{(\epsilon/kT)} - 1}$$

For small values of $\epsilon$, the exponent in the denominator becomes small. In this case the $e$ function can be developed into a series of which only the first two terms are considered. This procedure results in the following equation:

$$\lim_{\epsilon \to 0} E = \frac{\epsilon}{1 + \frac{\epsilon}{kT} - 1} = kT$$

For small values of $\epsilon$, therefore, Planck's law of the energy of an oscillator transforms into the equipartition law. For increasing values of $\epsilon$, however, the energy $E$ drops off to smaller values and converges rapidly toward zero.

Planck specified the energy quantum additionally by assuming that it is proportional to the frequency of the oscillator according to the following equation:

$$\epsilon = h\nu \qquad (13\text{-}17)$$

In this way he obtained the result that the energy drops off at large frequencies or at small wavelengths, which was necessary in order to bring the calculation into agreement with experimental results. The term $h$ has been found to be a universal constant, the so-called *Planck's quantum*. If the new equation for the average energy stored in a linear oscillator is now introduced into the previous equation instead of $kT$, then the following equation is obtained for the energy density at a certain wavelength:

$$u_\lambda = \frac{8\pi hc}{\lambda^5 (e^{hc/\lambda kT} - 1)}$$

If in addition we change from the energy density to the radiation intensity we obtain finally Planck's famous law for black-body radiation intensity:

$$i_{bn\lambda} = \frac{2hc^2}{\lambda^5 (e^{hc/\lambda kT} - 1)}$$

This equation contains only one unknown constant, namely, the quantum $h$, which could be determined from the measurement of black-body radiation intensity. The equation is in excellent agreement with the results of experimental determinations of the radiation intensity of a black body. Its derivation led to the concept that energy consists of finite particles or quanta in much the same way as matter which can be broken up only into certain particles of finite size: molecules, atoms, electrons, and so on. This concept became the basis for the development of modern physics.

**13-2. Solids and Liquid Bodies.** The radiative behavior of solids and liquids in the infrared wavelength range is characterized by the fact that almost all of them absorb practically all incident energy in very thin layers immediately below their surfaces and that emitted radiation leaving their surface also comes from such a thin layer. As a consequence, the equation

$$\rho + \alpha = 1 \qquad (13\text{-}2)$$

gives the connection between reflectivity $\rho$ and absorptivity $\alpha$.

All solid and liquid bodies reflect a fraction of impinging radiation and therefore, according to Kirchhoff's law, radiate less energy than the black body. With respect to their thermal radiation properties in the infrared wavelength range, there is a principal difference between electric conductors and nonconductors. The conductors reflect the major part of impinging radiation and therefore radiate comparatively little heat. Nonconductors absorb the major part of impinging radiation and radiate comparatively much heat. The rule does not apply to radiation in the visible or shorter wavelength range. For both groups, the reflectivity, absorptivity, and emissivity change with wavelength. Figure 13-9a shows the reflectivity $\rho_\lambda$ of monochromatic radiation for aluminum as an example of metals, and Fig. 13-9b for some nonconducting surfaces of engineering importance, according to W. Sieber.[1] Figure 13-9c presents the monochromatic absorptivity $\alpha_\lambda$ of water layers measured by Aschkinass.[2] Generally, reflectivity and absorptivity depend on surface temperature. All curves in the figure hold for a surface at ambient temperature.

The polished aluminum surface has a smaller reflectivity for shorter than for longer wavelengths. The trend for all metals is the same as that for aluminum. Equation (13-2) shows, on the other hand, that the absorptivity of metals is greater at shorter wavelengths. The variation of the reflectivity of electric nonconductors with wavelength is more irregular. The oxide film of the anodized aluminum surface (Fig. 13-9a) corresponds more to the nonconductors. A characteristic feature of nonconductors is the fact that the reflectivity is generally lower at longer

---

[1] W. Sieber, *Z. tech. Physik,* **22**:130–135 (1941).
[2] E. Aschkinass, *Wien. Ann.,* **55**:404 (1895).

wavelengths. Surfaces which have a bright color have high reflectivity at short wavelengths. Figure 13-9c presents the absorptivity of thin layers of water. The difference between the two curves $a$ and $b$ indicates that a layer with a thickness of 0.01 mm still transmits a measurable part

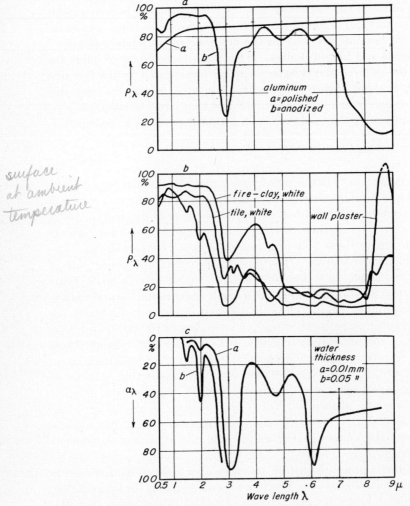

FIG. 13-9. Monochromatic reflectivity and absorptivity as a function of wavelength for different materials. [*From W. Sieber, Z. tech. Physik*, **22**:130–135 (1941).]

of incident radiation. Therefore, in this case, Eq. (13-2) cannot be used to calculate the reflectivity.

A surface whose absorptivity is the same for all wavelengths is called a gray surface. The surfaces represented in Fig. 13-9 approximate a gray surface but imperfectly. Good approximations of a gray surface are slate, tarboard, and dark linoleum. These materials absorb from 85 to 92 per cent of the incident radiation in the range of wavelengths between 0.5 and 9μ.

Kirchhoff's law

$$\epsilon_\lambda = \alpha_\lambda \qquad (13\text{-}18)$$

can be used to calculate the monochromatic emissivity from absorptivity measurements.

For engineering calculations, the data for the total radiation as obtained by integration over the entire wavelength range are most important. For this reason, total radiation properties of surfaces are usually listed in engineering handbooks. The use of these values may, however, lead to considerable inaccuracies if the definition of the total properties is not carefully kept in mind.

The total emissivity of a surface at a temperature $T_s$ is defined as the ratio of the emissive power $e$ of the surface to the emissive power $e_b$ of a black surface at the same temperature.

$$\epsilon = \frac{e(T_s)}{e_b(T_s)}$$

The same definition holds for the monochromatic values

$$\epsilon_\lambda = \frac{e_\lambda(T_s)}{e_{b\lambda}(T_s)}$$

With the equations

$$e = \int_0^\infty e_\lambda \, d\lambda \qquad e_b = \int_0^\infty e_{b\lambda} \, d\lambda$$

one obtains

$$\epsilon = \frac{\int_0^\infty \epsilon_\lambda(T_s) e_{b\lambda}(T_s) \, d\lambda}{\int_0^\infty e_{b\lambda}(T_s) \, d\lambda} = f(T_s) \qquad (13\text{-}19)$$

In this equation it is indicated which values are functions of the surface temperature. The emissivity $\epsilon_\lambda$ depends in addition on the nature of the surface.

The total absorptivity of a surface is the ratio of absorbed radiation $\alpha H$ to the impinging irradiation $H$. The latter can be expressed as

$$H = \int_0^\infty H_\lambda \, d\lambda$$

With the monochromatic absorptivity $\alpha_\lambda$ defined as the ratio of monochromatic absorbed radiation to incident monochromatic irradiation, the total absorptivity is

$$\alpha = \frac{\int_0^\infty \alpha_\lambda(T_s) H_\lambda \, d\lambda}{\int_0^\infty H_\lambda \, d\lambda} \qquad (13\text{-}20)$$

The total absorptivity depends, therefore, not only on the nature and temperature of the absorbing surface but also on the wavelength distribution of the incident radiation. The same situation exists for the total reflectivity. Such values cannot be tabulated unless the nature of the incident radiation is clearly specified.

When the incident radiation comes from a black body at temperature $T_i$ whose emissive power is $e_b(T_i)$, then the total absorptivity is

$$\alpha = \frac{\int_0^\infty \alpha_\lambda(T_s) e_{b\lambda}(T_i) \, d\lambda}{e_b(T_i)} = f(T_s, T_i) \qquad (13\text{-}21)$$

This absorptivity depends on the nature and temperature of the absorbing surface as well as on the temperature of the incident black radiation. This value can be calculated as soon as the monochromatic absorptivity and the temperature $T_i$ are known. For impinging nonblack radiation the values may be considerably different. A comparison of Eqs. (13-19) and (13-20) shows that Kirchhoff's law [Eq. (13-4)] does not hold generally for the total absorptivity and emissivity of a surface. Only when the impinging radiation is black and when its temperature is the same as that of the absorbing surface does Eq. (13-19) become identical with Eqs. (13-20) and (13-21).

The integrals in the above equations are usually determined numerically or graphically. To obtain the absorptivity for impinging black radiation, for instance, each ordinate of the pertinent curve in Fig. 13-5 is multiplied with the corresponding $\alpha_\lambda$ (taken, for example, from Fig. 13-9). The resulting curve must be planimetered and divided by the area under the radiation curve in Fig. 13-5. In this way the absorptivity and reflectivity values in Fig. 13-10 were determined by W. Sieber. In the curves, the difference between electric conductors (represented by aluminum) and nonconductors can be observed. For nonconductors, the absorptivity decreases with temperature; for conductors, it increases. Industrial radiators have temperatures between 500 and 5000 R. For such radiation, the absorptivity of nonconductors is greater than the values for conductors. The sun has a temperature of 10,000 R. At this temperature nonconductors with white surfaces absorb less of the radiation than metal surfaces. Only a few metals, e.g., silver, also have very small absorptivity in the range of visible radiation. Silver has a value of $\rho = 0.96$ for the reflectivity in this range.

THERMAL RADIATION PROPERTIES 375

Only for certain types of surfaces does a relation similar to Kirchhoff's law exist between the total absorptivity and emissivity. For a gray surface $\alpha_\lambda$ is independent of wavelength and Eq. (13-20) simplifies to

$$\alpha = \alpha_\lambda(T_s) = \epsilon_\lambda(T_s) = \epsilon(T_s) \qquad (13\text{-}22)$$

The absorptivity is equal to the emissivity of the surface at the surface temperature. Unfortunately, not too many surfaces exist in nature

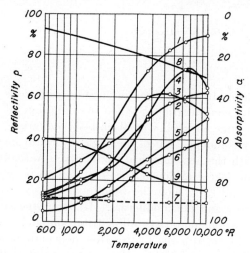

FIG. 13-10. Total reflectivity and absorptivity of different materials for incident black radiation of the indicated temperature. (1) White fire clay. (2) Asbestos. (3) Cork. (4) Wood. (5) Porcelain. (6) Concrete. (7) Roof shingles. (8) Aluminum. (9) Graphite. [*According to* W. Sieber, *Z. tech. Physik*, **22**:130–135 (1941).]

which can be considered gray. For a surface whose monochromatic absorptivity is independent of temperature and for incident black radiation, Eq. (13-21) becomes, considering that $\alpha_\lambda = \epsilon_\lambda$,

$$\alpha = \frac{\int_0^\infty \epsilon_\lambda e_{b\lambda}(T_i)\, d\lambda}{e_b(T_i)} = \epsilon(T_i) \qquad (13\text{-}23)$$

The absorptivity is equal to the emissivity of the surface, however, at the temperature of the incident black radiation. For a metal surface at $T_s$ and for incident black radiation at $T_i$ it can be demonstrated[1] from electromagnetic theory that the following relation holds:

$$\alpha = \epsilon(T_m) \qquad (13\text{-}24)$$

with $$T_m = \sqrt{T_s T_i} \qquad (13\text{-}25)$$

Equations (13-22) to (13-25) are useful for an estimate of total absorp-

[1] E. Eckert, *Forsch. Gebiete Ingenieurw.*, **7**:265–270 (1936).

tivities from tabulated emissivities. For an accurate calculation, however, the knowledge of monochromatic values is indispensable.

The emissivity was up to now defined as a ratio of emissive powers. One can also define it in terms of radiation intensities. In order to distinguish between the two values, the first one is called *hemispherical emissivity*

$$\epsilon = \frac{e}{e_b} \tag{13-26}$$

the second one *directional emissivity*

$$\epsilon_\beta = \frac{i}{i_b} \tag{13-27}$$

In the last equation both radiation intensities have to be introduced for the angle $\beta$ under consideration. When the radiation intensity of a surface follows the same cosine law as black-body radiation, then the directional emissivity is independent of the angle $\beta$ toward the surface normal and identical with the hemispherical emissivity. Actually, all materials exhibit a certain degree of dependence of the emissivity on the angle $\beta$.

Figures 13-11 and 13-12 show the directional emissivity for a number of electric conductors and nonconductors. If the cosine law were strictly fulfilled by the various materials, all distribution curves would be semicircles. It can be seen in the figures that nonconductors have smaller emissivities for emission angles in the neighborhood of 90° than would be expected from the cosine law. For conductors, the emissivity first increases and afterward decreases as the emission angle approaches 90°. This decrease occurs at very small angles and is, therefore, not seen in Fig. 13-11. In Table A-10 the hemispherical emissivity and the value $\epsilon_n$ in the direction of the surface normal are compiled. Some theoretical information on the ratio of the two emissivities will be presented later on.

The distribution curves in Figs. 13-11 and 13-12 can also be interpreted in a different way. The brightness under which a radiating surface appears to the eye depends on the radiative flux originating from a unit area of the radiating body as it appears to the eye. Correspondingly, the apparent brightness $b$ is defined as the amount of radiation emitted per unit time from a surface divided by the area of the projection of this surface on a plane normal to the direction from which the surface is viewed. If we consider a surface element $dA$ located at the center of the hemisphere in Fig. 13-4, then we can write for the radiative flux $i\,dA$. The projection of $dA$ on a plane normal to the beam indicated as $i$ in the figure is $dA \cos \beta$. Therefore, the apparent brightness under which the area $dA$ appears as viewed under an angle $\beta$ is

$$b = \frac{i}{\cos \beta} \tag{13-28}$$

If we multiply numerator and denominator of this equation by $i_{bn}$, we obtain

$$b = \frac{i}{i_b} i_{bn} = \epsilon_\beta i_{bn}$$

The apparent brightness is therefore proportional to the directional emissivity. A radiating black surface appears to have the same brightness

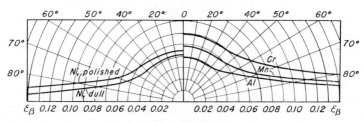

Fig. 13-11

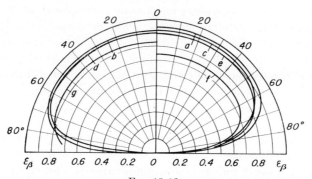

Fig. 13-12

Figs. 13-11 and 13-12. Emissivity of materials in different directions. (a) Wet ice. (b) Wood. (c) Glass. (d) Paper. (e) Clay. (f) Copper oxide. (g) Aluminum oxide. The temperature of the radiating metal surfaces was around 300 F; that of the nonmetallic surfaces between 32 and 200 F. [From E. Schmidt and E. Eckert, Forsch. Gebiete Ingenieurw., **6**:175–183 (1935).]

regardless of from which angle it is viewed. A radiating metal surface appears brighter when viewed under a large angle (except very close to 90°), and the opposite holds of a radiating nonconductor. A radiating sphere appears as a disk with uniform brightness when its surface is black, it appears with a brighter rim when it has a metal surface, and it has a brighter center field when its surface is a nonconductor.

Table A-10, in which emissivities $\epsilon_n$ and $\epsilon$ are listed for various surface conditions, indicates a difficulty which is inherent in tabulating such values. The emissivity depends, sometimes to a considerable degree, on the surface condition, and it is often difficult to describe this condition precisely enough.

Emissivity values depend also on temperature. Values for the emissivity of various materials as a function of temperature can be found in the reference books, periodicals, and research reports. From the hemispherical emissivity $\epsilon$ the heat radiated per unit surface area at the absolute temperature $T$ and the emissive power $e$ can be calculated by the formula

$$e = \epsilon \sigma T^4 \qquad (13\text{-}29)$$

The reflectivity $\rho$ also depends on the direction of the incident radiation, and a reflectivity for hemispherical and directional irradiation should be distinguished in the same way as for the absorptivity. According to Eq. (13-1) either reflectivity describes the total reflected energy regardless of its distribution in space.

With respect to the spatial distribution of reflected radiation, one distinguishes between specular and diffuse reflection. These two concepts are only approximated in reality. Figures 13-13 and 13-14 represent measured distribution curves for black-body radiation at 500 F impinging almost normally on different surfaces. The incident-ray bundle had an opening angle of 6°. From Fig. 13-13 it can be seen that surfaces which appear specular to the eye reflect much more thermal radiation in the proximity of the regularly reflected ray than in the other directions. On the other hand, it can be stated that specular and completely diffuse reflection are limiting ideal cases which are only approached by actual surfaces.[1]

The information on the radiation properties of solids and liquids that can be obtained from electromagnetic theory will now be briefly reviewed. This theory considers radiation as a combination of electric and magnetic waves and calculates reflectivity and absorptivity values of a material by applying the laws for reflection of such waves on the interface of substances which differ in their electrical properties. It is found that predictions obtained by this theory are in fairly good agreement with experimental results, so that they can be used as guides to estimate radiation properties of materials for which measured values are not available as long as the radiation considered is primarily located in the infrared range. The latter condition is fulfilled in practically all engineering applications dealing with radiation produced on earth. The highest temperatures obtained by combustion (in rockets) are of the order of 5000 R, and even for this temperature the radiation of a black body occurs mainly in the infrared region according to Fig. 13-5. Solar radiation, however, is to a

[1] New measurements by B. Münch on spatial distribution of reflected thermal radiation are contained in "Mitteilungen aus dem Institut für Thermodynamik und Verbrennungsmotoren der Eidgenössischen Technischen Hochschule, Zurich, Switzerland," Publisher, Leemann, Zurich.

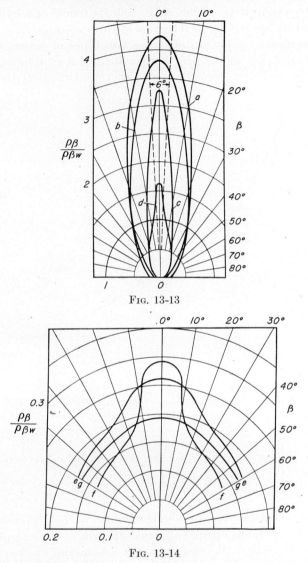

Fig. 13-13

Fig. 13-14

Figs. 13-13 and 13-14. Reflectivity of materials for black-body radiation. (*a*) Aluminum paint. (*b*) Iron scraped. (*c*) Black iron. (*d*) Copper oxide. (*e*) Cast iron. (*f*) Clay. (*g*) Wood. $\rho_\beta$ measured reflectivity, $\rho_{\beta w}$ reflectivity of a completely diffuse white surface with $\rho = 1$, both measured under an angle $\beta$ toward the surface normal. [*From* E. Eckert, *Forsch. Gebiete Ingenieurw.*, **1**:265–270 (1936).]

considerable extent in the visible range, and the following relations should not be used for absorption or reflection of this radiation.

Electromagnetic theory distinguishes two types of materials: electric conductors and nonconductors. For conductors, it establishes a relation between the radiation properties and the electric resistivity $\rho_e$. The

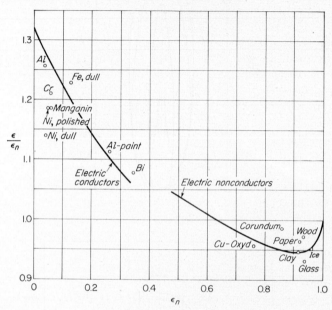

Fig. 13-15. Theoretical and experimental values for the ratio of hemispherical to normal emissivity. [*From E. Schmidt and E. Eckert, Forsch. Gebiete Ingenieurw.*, **6**:175–183 (1935).]

following equation was derived for the emissivity in a direction normal to the surface:

$$\epsilon_n = 0.576 \sqrt{\rho_e T} - 0.124 \rho_e T \tag{13-30}$$

The electric resistivity has to be introduced into this equation in ohm-centimeters, the temperature in degrees Kelvin. This relation agrees within 20 per cent with measurements. A useful relation was also obtained by E. Schmidt and E. Eckert[1] for the ratio $\epsilon/\epsilon_n$. The curves in Fig. 13-15 show as full lines the prediction of the theory for electric conductors and nonconductors. The points indicate measured results. It can be seen that fairly good agreement exists between experimental results and those calculated from electromagnetic theory. The calculation shows also that, for large angles $\beta$, the radiation emitted or absorbed by a metal surface is strongly polarized. Waves fluctuating in a direction

---

[1] E. Schmidt and E. Eckert, *Forsch. Gebiete Ingenieurw.*, **6**:175–183 (1935).

parallel to the incidence plane (plane containing the incident ray and the surface normal) are by far dominating. This can be easily understood in a qualitative way. An electric conductor is characterized by the fact that its electrons are quite mobile. If an electric wave impinges in a normal direction on the surface, then the electrons can easily follow the fluctuating electric field, and this will prevent a field from arising in the interior of the conductor. This means that the wave has to be reflected from the surface. The same situation will still exist when the wave impinges under an angle $\beta$ as long as the plane in which it fluctuates is normal to the incidence plane. When it is within the incidence plane, however, the electrons near the surface would have to penetrate the surface in order to follow the fluctuating electric field. Since they cannot do this, it follows that more of the impinging wave will penetrate into the interior of the conductor.

For electric nonconductors, the theory relates radiation properties to the refraction coefficient. This relation is not too useful because for engineering materials refraction coefficients are usually not known. However, the theory also permits a calculation of the spatial distribution of emitted radiation which is in good agreement with measured results (Fig. 13-12) and in this way predicts the ratio indicated in Fig. 13-15. The rule for obtaining the absorptivity of a metal surface for incident black radiation from its emissivity which is expressed by Eqs. (13-24) and (13-25) comes also from electromagnetic theory.

**13-3. Gases.** Gases generally absorb and radiate energy by increasing or releasing some of the energy internally stored within the molecules. Such energy changes may be caused by changes in the rotation of the molecules or in the vibration of the atoms within the molecule, by changes in the electron orbits, and finally by changes in the nuclear arrangements. They occur as quanta in finite steps, the amount of energy per step being smallest for changes in rotation and largest for changes in the nuclear arrangement. According to Planck's equation (13-17), the radiation connected with small energy quanta has a small frequency or large wavelength. The larger the energy steps, the shorter the wavelength of the equivalent radiation. Radiation is called *thermal radiation* when thermal equilibrium is closely approximated at least locally in the gas. Changes in the energy state of the molecules are caused by collisions which have sufficient kinetic energy. The higher the temperature of the gas, the larger the energy quanta which can be released in the molecular collisions and thus the shorter the wavelength of the released radiation. At temperatures in the range up to, say, 5000 F, as they occur in normal engineering applications, the bulk of radiation is connected with changes in the rotational or vibrational energy of the molecules.

The corresponding radiation has wavelengths in the range 1 to $30\mu$

**382** THERMAL RADIATION

(Fig. 13-5). Changes in the electron orbits move the radiation into the visible range. This radiation contributes, however, essentially to the total energy only at very high temperatures like the temperature of the sun. Changes in the nuclei finally shift the radiation to extremely short waves (Fig. 13-1). In this section only the infrared radiation will be considered. The atoms in their rotational or vibrational movement act as small transmitters of electromagnetic waves, since electric charges are usually connected with them.

Elementary gases, consisting of atoms all of which are the same kind, have no free electric charges connected with the atom. Such gases as

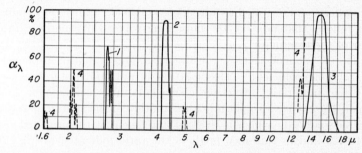

Fig. 13-16. Absorption bands of carbon dioxide. (1) 5-cm, (2) 3-cm, (3) 6.3-cm, (4) 100-cm layers. (*Measured by E. F. Barker, G. Hertz, C. Schaefer, and B. Philipps.*)

hydrogen, oxygen, and nitrogen radiate no heat and are perfectly transparent to foreign heat radiation. For engineering calculations, the radiation of carbon dioxide and of water vapor are most important, since these gases are good radiators and are present in combustion gases in high concentration. Carbon monoxide, sulfur dioxide, and methane also radiate well but are generally present only in small concentrations. Figures 13-16 and 13-17 show the absorption spectra of carbon dioxide and water vapor. It can be seen that gases are different in their behavior from solid and liquid bodies in so far as they radiate and absorb radiation only within certain limited wavelength ranges. For water vapor these ranges lie comparatively close together. The radiation takes place essentially at wavelengths longer than 1 $\mu$; therefore, it is invisible. From the figures it can also be seen that gases, in contrast to solid and liquid bodies, need great thickness to absorb the major part of incident radiation.

When a monochromatic beam has an intensity $B_\lambda$ after having traveled a distance $s$ in an absorbing gas, then the amount $dB_\lambda$ by which the intensity decreases through absorption on the next path length $ds$ is described by the equation

$$\frac{dB_\lambda}{B_\lambda} = -a_\lambda \, ds \tag{13-31}$$

The exponent $a_\lambda$ is called the *absorption coefficient*. It may be expected that $a_\lambda$ is to a first approximation proportional to the number of molecules per unit volume and that it increases with temperature. Figures 13-16 and 13-17 indicate that it also depends strongly on wavelength. The number of molecules at a constant temperature is proportional to the pressure. Actually it has been found that the absorption coefficient increases somewhat faster than linearly with pressure. In a mixture of

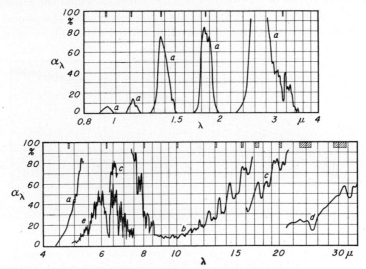

FIG. 13-17. Absorption bands of water vapor. (*a*) 127 C and 109 cm. (*b*) 127 C and 104 cm. (*c*) 127 C and 32.4 cm. (*d*) 81 C and 32.4 cm at 4 cm atm. (*e*) Room temperature and 220 cm at 7 cm atm. [*From E. Schmidt, Forsch. Gebiete Ingenieurw.,* **3**:57 (1932).]

an absorbing gas with a nonabsorbing one, the absorption coefficient would again be expected to be proportional to the partial pressure of the absorbing gas.

Equation (13-31) can be integrated when the temperature and, in a gas mixture, the concentration of the absorbing gas are constant. The result is $B_\lambda = B_{\lambda 0} e^{-a_\lambda s}$, where $B_{\lambda 0}$ indicates the intensity with which the radiant beam entered the gas (at $s = 0$).

According to our definition, the transmissivity of a gas layer of thickness $s$ is

$$\tau_\lambda = e^{-a_\lambda s} \qquad (13\text{-}32)$$

and the absorptivity is

$$\alpha_\lambda = 1 - e^{-a_\lambda s} \qquad (13\text{-}33)$$

The same equation also gives the emissivity $\epsilon_\lambda$ according to Kirchhoff's law.

It may have been noticed that the radiation properties $\tau_\lambda$, $\alpha_\lambda$, and $\epsilon_\lambda$ have been defined for the gas in a slightly different way from solids and liquids. For those they were referred to the radiant energy $H$ incident on the surface, whereas in Eqs. (13-32) and (13-33) they are referred to the radiant energy $B$ traveling from the interface into the gas. The difference, however, is very small because the reflectivity of a gas-vacuum or gas-gas interface is very small.

According to the discussion of the parameters on which the absorption coefficient depends, one expects the monochromatic transmissivity, absorptivity, and emissivity to depend, at a fixed temperature, primarily

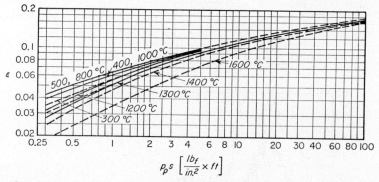

Fig. 13-18. Emissivity of carbon dioxide in mixture with nitrogen or air. $s$ is the thickness of the gas layer; $p_p$ is the partial pressure of $CO_2$. Total pressure is 1 atm. (*From E. Eckert, VDI-Forschungsheft 387, 1937.*) Dashed lines indicate extrapolated values.

on the product of thickness $s$ times pressure $p$ or partial pressure $p_p$ of the absorbing and radiating gas. This statement is sometimes referred to as *Beer's law*. The following discussion will show that this law can be considered as an approximation only.

For engineering heat-transfer calculations, the total radiated heat integrated over all wavelengths is most important. Therefore, a number of extensive experimental investigations were carried out in this field after A. Schack[1] had emphasized the importance of gas radiation for heat transfer in furnaces and steam boilers. Measurements on carbon dioxide were made by H. C. Hottel[2] and coworkers and by Eckert.[3] Figure 13-18 shows the emissivity of carbon dioxide for different temperatures (in degrees centigrade) and thicknesses and for a total pressure of 1 atm. The emissivity is determined in the same way as for solid bodies by comparing the radiation of the gas with black-body radiation at the same temperature (Eq. 13-29). E. Eckert made measurements with carbon

[1] A. Schack, *Z. tech. Physik*, **5**:267–278 (1924).
[2] H. C. Hottel and H. G. Mangelsdorf, *Trans. Am. Inst. Chem. Engrs.*, **31**:517–549 (1935). H. C. Hottel and V. C. Smith, *Trans. ASME*, **57**:463–470 (1935).
[3] E. Eckert, *VDI-Forschungsheft* 387, 1937.

dioxide–nitrogen mixtures of various partial pressures and thicknesses and found that Beer's law holds for carbon dioxide. Hottel and Mangelsdorf used carbon dioxide–air mixtures. Both experimental results agree fairly well. Only the extrapolated values for higher temperatures show greater differences, Hottel's values being smaller. Since a theoretical investigation by Schwiedessen showed also that the emissivity decreases at higher temperatures, the curve for 1600 C in Fig. 13-18 is taken from Hottel's values.

The first measurements of water vapor were made by E. Schmidt.[1] He investigated various thicknesses of pure water vapor. Experiments by Hottel and coworkers[2] with water-vapor–air mixtures of 22-in. thickness agree with Schmidt's values at high partial pressures of the water vapor but are much lower at low partial pressures. Measurements by E. Schmidt and E. Eckert[3] explained these differences. They showed that Beer's law is not valid for water vapor, that the emissivity of this gas at a constant product of partial pressure times thickness is smaller for small partial pressures. This can be accounted for if the emissivity of pure water vapor is multiplied by a reduction factor less than 1 as soon as the partial pressure is less than 1 atm. Figure 13-19 shows in the large diagram the emissivity of pure water vapor according to the measurements of Schmidt for various temperatures (in Celsius degrees) and at atmospheric pressure. The small diagram gives the reduction factor $f$ by which the values from the large diagram have to be multiplied when the partial pressure of water vapor is smaller than 1. This factor is a mean value derived from a paper by Schwiedessen,[4] who compared all measurements and extrapolated them to higher temperatures. They also take into account the results of new measurements.[5] The measurements by Schmidt and Eckert indicated a small dependence of the reduction factor on gas temperature. This has been neglected in the small diagram of Fig. 13-19.

All measurements were carried out at a total pressure of the mixtures equal to 1 atm. At higher total pressures the radiation increases because the number of radiating molecules in a given thickness of the layer increases. It appears, however, that the radiation of any molecule also increases with pressure. Hottel's chapter on radiation in McAdams' book[6] contains diagrams from which the influence of total pressure can be estimated.

[1] E. Schmidt, *Forsch. Gebiete Ingenieurw.*, **3**:57–70 (1932).
[2] Hottel and Mangelsdorf, *loc. cit.* Hottel and Smith, *loc. cit.*
[3] E. Schmidt and E. Eckert, *Forsch. Gebiete Ingenieurw.*, **8**:87 (1937).
[4] H. Schwiedessen, *Arch. Eisenhüttenw.*, **14**:9–14, 145–153, 207–210 (1940).
[5] H. C. Hottel and R. B. Egbert, *Trans. Am. Inst. Chem. Engrs.*, **38**:531–565 (1942). J. E. Ebarhardt and H. C. Hottel, *Trans. ASME*, **58**:185–193 (1936).
[6] W. H. McAdams, "Heat Transmission," 3d ed., McGraw-Hill Book Company, Inc., New York, 1954.

In engineering calculations one has often to consider mixtures of water vapor and carbon dioxide with nonradiating gases. Generally, the monochromatic absorptivity of a mixture of two gases 1 and 2 with the

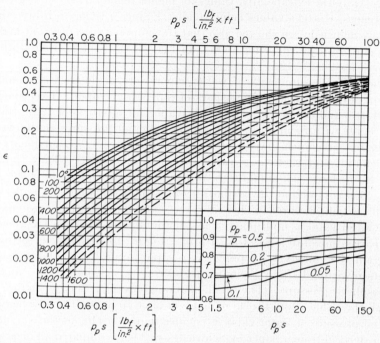

FIG. 13-19. Emissivity of water vapor in mixture with nitrogen or air. $s$ is the thickness of the gas layer; $p_p$ is the partial pressure of water vapor, total pressure $p = 1$ atm. When the partial pressure $p_p$ is different from 1, the emissivities read in the big figure have to be multiplied by $f$ read from the small diagram. Dashed lines indicate extrapolated values. (*According to measurements by E. Schmidt, H. Hottel, H. Mangelsdorf, and E. Eckert.*)

absorption coefficients $a_{\lambda 1}$ and $a_{\lambda 2}$ is

$$\alpha_\lambda = 1 - e^{-(a_{\lambda 1} + a_{\lambda 2})s} \tag{13-34}$$

assuming that the two gases do not influence each other in their absorption. The resulting absorptivity $\alpha_\lambda$ is smaller than the sum of the absorptivities of the two gases individually. The same statement holds for the emissivity. Table 13-3 gives corresponding corrections $\Delta\epsilon$ which have to be subtracted from the sum of the individual emissivities of water vapor and carbon dioxide to obtain the radiation of a mixture of both gases.

In determining absorptivities from the total emissivities for gases, the same difficulties are encountered as for solid materials. Kirchhoff's law (13-4) gives strictly the total absorptivity of a gas only for incident radiation from a black body with a temperature which is equal to the gas temperature. Absorptivities calculated with Eq. (13-4) can be used with good approximation to calculate radiative exchange of a gas with a black surface as long as the gas temperature is higher than the surface temperature of the radiating body. Some discussion on the reverse situation is found in Hottel's chapter in McAdams.[1] A more accurate

TABLE 13-3. CORRECTION $\Delta\epsilon$ FOR SIMULTANEOUS RADIATION OF $H_2O$ AND $CO_2$*

$$\epsilon_{H_2O+CO_2} = \epsilon_{H_2O} + \epsilon_{CO_2} - \Delta\epsilon$$

| $p_p s$ of $CO_2$, $lb_f/in.^2$ ft | $p_p s$ of $H_2O$, $lb_f/in.^2$ ft | | | Temp. of mixture, C |
|---|---|---|---|---|
| | 2.5 | 15 | 50 | |
| 2.5 | 0.000 | 0.014 | 0.034 | 400 |
| | 0.003 | 0.012 | 0.018 | 800 |
| | 0.007 | 0.020 | 0.035 | 1200 |
| 15 | 0.000 | 0.017 | 0.043 | 400 |
| | 0.008 | 0.024 | 0.032 | 800 |
| | 0.018 | 0.047 | 0.052 | 1200 |
| 50 | 0.000 | 0.017 | 0.043 | 400 |
| | 0.013 | 0.025 | 0.033 | 800 |
| | 0.030 | 0.049 | 0.054 | 1200 |

* According to E. Eckert, *VDI-Forschungsheft* 387, 1937.

determination of absorptivities and transmissivities would require a knowledge of monochromatic emissivities.

Equation (13-33) is concerned with radiation traveling through the gas in a certain direction. When the absorptivity is set equal to the emissivity, then Eq. (13-33) has to be interpreted as the ratio of intensities

$$\epsilon_{\lambda,\beta} = \frac{i_\lambda}{i_{b\lambda}} = 1 - e^{-a_\lambda s} \tag{13-35}$$

In calculations on heat exchange by radiation, one is primarily interested in the energy flux which arrives at a certain location $dA$ on the boundary of a gas body and which comes from all directions in space. The calculation will be made for monochromatic radiation. The radiative flux arriving per unit area and time, the emissive power, is given by the equation

$$e_\lambda = \int i_\lambda \, d\omega$$

where $d\omega$ indicates a small solid angle.

[1] "Heat Transmission," 3d ed., McGraw-Hill Book Company, Inc., New York, 1954.

For a spherical coordinate system with its origin at $dA$, the solid angle $d\omega$ can be expressed by the two angles $\beta$ and $\varphi$. When additionally Eq. (13-35) is introduced, then the emissive power can be written

$$e_\lambda = \int_{\varphi=0}^{\varphi=2\pi} \int_{\beta=0}^{\beta=\pi/2} (1 - e^{-a_\lambda s}) i_{b\lambda} \sin \beta \, d\beta \, d\varphi$$

One can define a monochromatic hemispheric emissivity as the ratio of the emissive power $e_\lambda$ to the emissive power of a black surface at the same temperature:

$$\bar{\epsilon}_\lambda = \frac{e_\lambda}{e_{b\lambda}} = \frac{1}{\pi} \int_{\varphi=0}^{\varphi=2\pi} \int_{\beta=0}^{\beta=\pi/2} (1 - e^{-a_\lambda s}) \sin \beta \cos \beta \, d\beta \, d\varphi \quad (13\text{-}36)$$

This equation can be integrated for a specific shape of the radiating gas

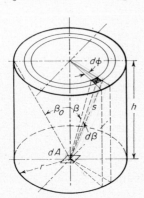

FIG. 13-20. Calculation of radiation of cylindrical gas body.

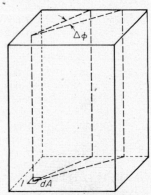

FIG. 13-21. Calculation of radiation of gas body.

body. Integrations of Eq. (13-36) have been performed by Nusselt, Jakob, E. Schmidt, Hottel, and Eckert. As an example, the integration will be carried out for the cylindrical gas shape illustrated in Fig. 13-20, the solution for which was given by E. Schmidt.[1]

**Example 13-1.** The problem is to calculate the amount of radiation from a gas body with the shape of a cylinder and with uniform temperature and concentration into the area element $dA$ located at the center of its base (Fig. 13-20). The integration over the angle $\varphi$ can be carried out immediately because of the rotational symmetry of the configuration. This integration of Eq. (13-36) gives

$$\bar{\epsilon}_\lambda = 2 \int_0^{\pi/2} (1 - e^{-a_\lambda s}) \sin \beta \cos \beta \, d\beta$$

The integration over the angle $\beta$ has to be performed in two parts: from $\beta = 0$ to $\beta = \beta_0$ and from $\beta_0$ to $\beta = \pi/2$. For the first part, the emissivity is

$$\bar{\epsilon}_\lambda = -2 \int_0^{\beta_0} [1 - e^{-a_\lambda (h/\cos \beta)}] \cos \beta \, d(\cos \beta)$$

[1] E. Schmidt, *Z. Ver. deut. Ingr.*, **77**:1162–1164 (1933).

We introduce a new variable

$$\zeta = \frac{a_\lambda h}{\cos \beta}$$

and obtain

$$\bar{\epsilon}_\lambda = 2(a_\lambda h)^2 \int_{ah}^{ah/\cos\beta_0} (1 - e^{-\zeta}) \frac{d\zeta}{\zeta^3}$$

The integral can now be solved by partial integration

$$\bar{\epsilon}_\lambda = (a_\lambda h)^2 \left| -\frac{1}{\zeta^2} + \frac{R^{-\zeta}}{\zeta^2} - \frac{e^{-\zeta}}{\zeta} - Ei(-\zeta) \right|_{a_\lambda h}^{a_\lambda h/\cos\beta_0}$$

The function

$$Ei(-\zeta) = \int_0^z \frac{e^{-x}}{x} dx$$

has been tabulated and is called integrallogarithm.[1] The integration of the second portion can be carried out in an analogous manner. The sum of the two emissivities obtained in this way gives the emissivity for the cylindrical gas shape.

This value can be utilized to obtain the emissivity of a gaseous body shaped like a parallelepiped into any location of its surface. This is illustrated in Fig. 13-21. Assume that the emissivity of the gas body into the location 1 is to be determined. Then the gas body can be divided up into wedge-shaped parts in the manner indicated in the figure. The emissivity for any wedge is obtained from the one for the cylinder presented in Fig. 13-20 when the emissivity for the cylinder is multiplied by the ratio $\Delta\varphi/2\pi$. A summation or integration over all wedges of which the gas body in Fig. 13-21 is composed results in the local emissivity at the location $dA$ for the parallelepiped.

Another integration has to be performed to obtain the mean emissivity which describes the radiative exchange between a gas body and a finite area of its surface. Such a mean emissivity indicates, for instance, the radiative exchange between the gas body shown in Fig. 13-21 and any one of the rectangles of which the surface is composed.

A very simple expression can be obtained for the mean emissivity of a gas body of arbitrary shape and its surrounding surface under the condition that the self-absorption of the gas is negligibly small. This requires either the absorption coefficient $a_\lambda$ or the dimensions of the gas body to be small. In this case, the $e$ function in the expression

$$\epsilon_\lambda = 1 - e^{-a_\lambda s}$$

can be developed in a series, and the series can be terminated after the second term. This results in the following equation describing the emissivity of a gas with negligible self-absorption

$$\epsilon_\lambda = a_\lambda s \tag{13-37}$$

Figure 13-22 indicates such a gas body with the volume $V$ and surface area $A$. We will at first derive an expression which describes that part of

[1] Jahnke-Emde, "Tables of Functions," p. 78, Springer-Verlag, Berlin, 1933.

the radiation with wavelength $\lambda$ originating in the volume element $dV$ which arrives at the surface area $dA$. If the volume element is seen from the area $dA$ under the solid angle $d\omega$, then the heat radiated into the element $dA$ per unit time can be written as

$$d^2q_\lambda = di_\lambda \, d\omega \, dA$$

The term $di_\lambda$ indicates the energy emitted by the volume element $dV$ per

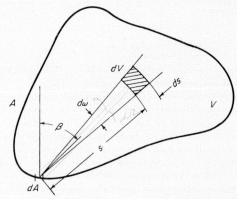

Fig. 13-22. Calculation of mean hemispherical emissivity of a gas body with negligible self-absorption.

unit solid angle in the direction of $d\omega$. With the emissivity

$$d\epsilon_\lambda = a_\lambda \, ds$$

of the volume element with the thickness $ds$, this equation becomes

$$d^2q_\lambda = i_{b\lambda} \, d\epsilon_\lambda \, d\omega \, dA = \frac{e_{b\lambda}}{\pi} \cos\beta \, a_\lambda \, ds \, d\omega \, dA$$

The volume element itself can be expressed as

$$dV = s^2 \, d\omega \, ds$$

Therefore, the radiative heat flux into the surface element becomes

$$d^2q_\lambda = \frac{e_{b\lambda}}{\pi} a_\lambda \, dV \, \frac{\cos\beta \, dA}{s^2}$$

The term $\cos\beta \, dA/s^2$ appearing in this equation can be interpreted as the solid angle $d\Omega$ under which the area $dA$ is seen from the location of $dV$. Therefore,

$$d^2q_\lambda = \frac{e_{b\lambda}}{\pi} a_\lambda \, dV \, d\Omega$$

The total energy exchange between the volume element $dV$ and the total surface area $A$ of the gas body results from an integration of the above equation over the whole space surrounding $dV$.

$$dq_\lambda = \frac{e_{b\lambda}}{\pi} a_\lambda \, dV \, 4\pi = 4a_\lambda e_{b\lambda} \, dV$$

By another integration, the monochromatic radiation emitted by the total volume $V$ and arriving at its surface area $A$ is obtained.

$$q_\lambda = 4a_\lambda e_{b\lambda} V \qquad (13\text{-}38)$$

With the mean hemispherical emissivity $\bar{\epsilon}_\lambda$ of the gas body into its surface area $A$, the radiative heat flux can also be expressed as

$$q_\lambda = \bar{\epsilon}_\lambda e_{b\lambda} A$$

This results in the relation

$$\bar{\epsilon}_\lambda = \frac{4V}{A} a_\lambda \qquad (13\text{-}39)$$

for the mean hemispherical emissivity of a gas body of negligible self-absorption into its surface area.

The local, hemispherical emissivity for a gas body with the shape of a hemisphere into the center of its base area can immediately be found by integration of Eq. (13-36). The path length $s$ is in this case equal to the radius $r$ of the hemisphere and is, therefore, independent of angle. The integration then results in the following expression for the monochromatic emissivity of the hemisphere:

$$\epsilon_{\lambda H} = 2 \int_0^{\pi/2} (1 - e^{-a_\lambda r}) \sin \beta \, d(\sin \beta) = 1 - e^{-a_\lambda r} \qquad (13\text{-}40)$$

For other shapes, the integration of Eq. (13-36) is usually quite tedious. For this reason it is very fortunate for engineering calculations that H. Hottel developed a simple procedure to obtain approximate values for the emissivity of gas bodies. He demonstrated that for each specific shape of a gas body, there can be found a hemisphere of an equivalent radius $L_e$ in such a way that the ratio $\bar{\epsilon}_\lambda/\epsilon_{\lambda H}$ is equal to 1 within approximately $\pm 10$ per cent for all values of $a_\lambda s$ which are of engineering importance (between 0.05 and infinity). Table 13-4 contains values of the equivalent radius for gas volumes of different shapes, and Fig. 13-23 shows the ratio $\bar{\epsilon}_\lambda/\epsilon_{\lambda H}$ as a function of $a_\lambda s$ for some shapes and for the value of $L_e$ in Table 13-4.

The equivalent radius for a gas volume with negligible self-absorption can be obtained from Eq. (13-39). The local emissivity of a gas body with the shape of a hemisphere, with radius $L_e$, and with negligible self-absorption into the center of its base area is

$$\epsilon_{\lambda H} = a_\lambda L_e$$

Setting this emissivity equal to the one expressed by Eq. (13-39) results in

$$L_e = \frac{4V}{A} \tag{13-41}$$

H. Hausen[1] and F. M. Port[2] found that this last equation can be used to obtain a good approximation for the equivalent radius of a gas body with

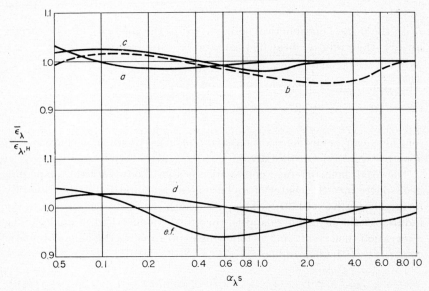

FIG. 13-23. Ratio of the mean hemispherical emissivity $\bar{\epsilon}_\lambda$ of various gas bodies to the emissivity $\epsilon_{\lambda H}$ of a hemisphere into the center of its base. The curves $a$ to $f$ belong to the first six geometries listed in Table 13-4 in alphabetical order. The length $s$ is the dimension $d$, $r$, or $l$, and $\epsilon_{\lambda H}$ is the emissivity of a hemisphere with radius $L_e$ as listed in the table. [From E. Eckert, *VDI-Forschungsheft* 387 (1937)].

the parameter $a_\lambda s$ in the range usually encountered in engineering calculations by multiplying the right-hand side of the equation with a factor 0.9. It can easily be checked that Eq. (13-41) multiplied with this factor checks the values for the equivalent radius contained in Table 13-4 very well, and it may also have been noticed that Eq. (13-41) is quite similar to the relation which describes the hydraulic diameter of a pipe.

An integration over all wavelengths has to be performed, using Eq. (13-36), when the total radiation of a gas body is needed. In using the

---

[1] According to communication in a letter.
[2] F. M. Port, "Heat Transmission by Radiation from Gases," Sc.D. thesis, Massachusetts Institute of Technology, 1939.

approximate procedure based on the equivalent radius, such an integration is not necessary since the equivalent radius holds for any value of the parameter $a_\lambda s$. The equivalent radius can, therefore, be used directly in connection with Figs. 13-18 and 13-19 to obtain the hemispherical

TABLE 13-4. EQUIVALENT RADII $L_e$ OF GAS VOLUMES OF DIFFERENT SHAPE*

| Shape of gas volume | $L_e$ |
|---|---|
| Circular cylinder: height = diameter = $d$ | |
|   Irradiation into the center of the base | $0.77d$ |
| Circular cylinder: height = $\infty$, diameter = $d$ | |
|   Irradiation into the convex surface | $0.95d$ |
| Cylinder: height = $\infty$, base semicircle with radius $r$ | |
|   Irradiation into the center of the plane rectangular surface | $1.26r$ |
| Sphere: diameter = $d$ | |
|   Irradiation into the surface | $0.65d$ |
| Volume between two infinite planes separated by distance $l$ | |
|   Irradiation into the planes | $1.8l$ |
| Circular cylinder: height = $\infty$, diameter = $d$ | |
|   Irradiation into the center of the base | $0.9d$ |
| Tube bundle:† | |
|   In triangular arrangement | |
|     $s = 2d$ | $3.0(s - d)$ |
|     $s = 3d$ | $3.8(s - d)$ |
|   In square arrangement | |
|     $s = 2d$ | $3.5(s - d)$ |
| Cube: length = $a$ | |
|   Irradiation into each surface plane | $0.66a$ |

\* According to H. C. Hottel and E. Eckert.
† $s$ = distance between tube centers
  $d$ = tube diameter

emissivity $\bar{\epsilon}$. The energy flux from a gas body with a temperature $T_g$ into a unit of the area $dA$ is then obtained from the equation

$$e = \bar{\epsilon}\sigma T_g^4 \tag{13-42}$$

## PROBLEMS

**13-1.** Calculate the total emissivity (integrated over all wavelengths) of a surface of anodized aluminum, assuming that the surface has a temperature of 300 F and that the absorptivity obtained from Fig. 13-9a represents the monochromatic emissivity of such a surface. (Assume that the emissivity for wavelengths longer than $9\mu$ is equal to the value at $9\mu$.)

**13-2.** Calculate the total absorptivity of the surface considered in Prob. 13-1 for impinging black radiation of 1800 R assuming that the surface has a temperature of 300 F. Assume that the monochromatic absorptivity obtained from Fig. 13-9a applies here.

**13-3.** Calculate the total absorptivity of the surface considered in Prob. 13-1 for impinging radiation coming from a fire-clay surface at 1800 F. Assume that the absorptivity obtained from the corresponding curve in Fig. 13-9b represents the emissivity of fire clay at that temperature.

**13-4.** Calculate the total hemispherical emissivity for glass from the distribution curve in Fig. 13-12.

**13-5.** Calculate the monochromatic emissivity of a gas layer between two parallel planes using Eq. (13-36), and verify the value $L_e = 1.8l$ given in Table 13-4.

## CHAPTER 14

## HEAT EXCHANGE BY RADIATION

**14-1. Black Bodies.** *Heat Exchange between Two Area Elements.* When different heated bodies with black surfaces are arranged so that they can see one another, every body radiates heat to the others and absorbs heat radiated from the other bodies. The hotter bodies lose more heat by radiation than they absorb. For the cooler bodies the opposite is true. In this way a heat flow from the hotter to the cooler bodies arises which will be calculated in the following paragraphs. In this section it is assumed that the surfaces of the radiating bodies are black. In Fig. 14-1, $dA_1$ and $dA_2$ represent surface elements of two radiating bodies. The distance between them is $s$, and the angles between any of the two normals to the surfaces and the connecting line $s$ are $\beta_1$ and $\beta_2$, respectively. Then the heat $d^2Q_{b1}$ radiated per unit time from surface $dA_1$ within the solid angle under which $dA_2$ is seen from $dA_1$ is, according to Eqs. (13-6) and (13-7),

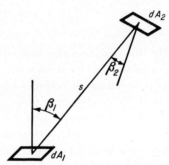

Fig. 14-1. Heat exchange by radiation of two small surfaces.

$$d^2Q_{b1} = i_{bn1} \cos \beta_1 \, d\omega_1 \, dA_1 \qquad (14\text{-}1)$$

where $i_{bn1}$ is the radiation intensity of $dA_1$ in the normal direction and $d\omega_1$ is the solid angle under which $dA_2$ is seen from $dA_1$. The expression

$$d\omega_1 = \frac{dA_2 \cos \beta_2}{s^2} \qquad (14\text{-}2)$$

gives this angle. Equation (14-1) becomes

$$d^2Q_{b1} = i_{bn1} \frac{\cos \beta_1 \cos \beta_2}{s^2} dA_1 \, dA_2$$

This heat is absorbed by the black surface $dA_2$. In the same way the heat $d^2Q_{b2}$ radiated from $dA_2$ to $dA_1$ is

$$d^2Q_{b2} = i_{bn2} \cos \beta_2 \, d\omega_2 \, dA_2 = i_{bn2} \frac{\cos \beta_1 \cos \beta_2}{s^2} dA_1 \, dA_2 \quad (14\text{-}3)$$

This amount of heat is absorbed by the black surface $dA_1$. The net flow from $dA_1$ to $dA_2$ originated by the exchange of thermal radiation is therefore

$$d^2Q_b = d^2Q_{b1} - d^2Q_{b2} = (i_{bn1} - i_{bn2}) \frac{\cos \beta_1 \cos \beta_2}{s^2} dA_1 \, dA_2$$

The radiation intensity $i_{bn1}$ of the black surface in a direction normal to the surface is given by

$$i_{bn1} = \frac{\sigma}{\pi} T_1^4 \quad (14\text{-}4)$$

An analogous equation holds for the radiation intensity $i_{bn2}$. Therefore

$$d^2Q_b = \frac{\cos \beta_1 \cos \beta_2}{s^2} dA_1 \, dA_2 \frac{\sigma}{\pi} (T_1^4 - T_2^4) \quad (14\text{-}5)$$

By this equation the heat exchange of two small surfaces with arbitrary positions can be calculated. By integration it is possible to calculate the heat exchange of surfaces whose sizes are not small when compared with the distances between them. In order to simplify these calculations it is useful to introduce a new concept, that of the *shape factor F*. The shape factor $dF_{1-2}$ of $dA_1$ with respect to $dA_2$ is the heat $d^2Q_{b1}/dA_1$ radiated per unit area from $dA_1$ to $dA_2$, divided by the emissive power $e_{b1}$ of $dA_1$. The latter is $e_{b1} = \pi i_{bn1}$. From Eqs. (14-1) and (14-2) there is obtained

$$dF_{1-2} = \frac{1}{\pi} \cos \beta_1 \, d\omega_1 = \frac{\cos \beta_1 \cos \beta_2}{\pi s^2} dA_2 \quad (14\text{-}6)$$

Therefore the shape factor is fixed by a purely geometrical relation. The heat radiated from $dA_1$ and intercepted by $dA_2$ can now be written as

$$d^2Q_{b1} = e_{b1} \, dF_{1-2} \, dA_1$$

In the same way, the heat radiated from $dA_2$ and intercepted by $dA_1$ is, according to Eqs. (14-2) and (14-3),

$$d^2Q_{b2} = i_{bn2} \cos \beta_1 \, d\omega_1 \, dA_1 = e_{b2} \, dF_{1-2} \, dA_1$$

and the net heat exchange is

$$d^2Q_b = (e_{b1} - e_{b2}) \, dF_{1-2} \, dA_1 = \sigma \, dF_{1-2} \, dA_1 (T_1^4 - T_2^4) \quad (14\text{-}7)$$

The calculations can be referred in the same way to the surface element $dA_2$, and thus is obtained

$$d^2Q_b = (e_{b1} - e_{b2}) \, dF_{2-1} \, dA_2 = \sigma \, dF_{2-1} \, dA_2 (T_1^4 - T_2^4) \quad (14\text{-}8)$$

The shape factor $dF_{2-1}$ of $dA_2$ with respect to $dA_1$ is

$$dF_{2-1} = \frac{1}{\pi} \cos \beta_2 \, d\omega_2 = \frac{\cos \beta_1 \cos \beta_2}{\pi s^2} dA_1$$

*Heat Exchange between an Area Element and a Finite Area.* The heat exchange by radiation between an area element and a surface whose size is not small compared with their distance apart can be determined by integration of Eqs. (14-5) and (14-7). The result is the equation

$$dQ_b = \sigma F_{1-2} \, dA_1 (T_1^4 - T_2^4)$$

in which the shape factor is

$$F_{1-2} = \int_{A_2} \frac{\cos \beta_1 \cos \beta_2}{\pi s^2} dA_2$$

Integrations of this equation, which are often tedious, have been performed for a number of configurations and are found in the various books dealing with heat transfer.[1] Table 14-1 contains a number of such shape factors.

As an example, the shape factor will be calculated which characterizes radiative exchange between an area element $dA_1$ and a circular area $A_2$ in a relative position as indicated in Fig. 14-2. Since the shape factor depends only on angular relations, one can always change the scale of the configuration so that one of the pertinent dimensions has the length 1. This has been done in the figure. From Fig. 14-2 one can show that

$$s^2 = h^2 + 1 + \rho^2 + 2\rho \cos \alpha \qquad dA_2 = \rho \, d\rho \, d\alpha$$

$$\cos \beta_1 = \frac{1 + \rho \cos \alpha}{s} \qquad \cos \beta_2 = \frac{h}{s}$$

With these expressions Eq. (14-6) becomes

$$dF_{1-2} = \frac{h}{\pi} \frac{1 + \rho \cos \alpha}{(h^2 + 1 + \rho^2 + 2\rho \cos \alpha)^2} \rho \, d\rho \, d\alpha$$

and

$$F_{1-2} = \frac{2h}{\pi} \int_0^r \rho \, d\rho \int_0^\pi \frac{1 + \rho \cos \alpha}{(h^2 + 1 + \rho^2 + 2\rho \cos \alpha)^2} d\alpha$$

The double integration results in

$$F_{1-2} = \frac{h}{2} \left[ \frac{h^2 + r^2 + 1}{\sqrt{(h^2 + r^2 + 1)^2 - 4r^2}} - 1 \right]$$

---

[1] For example, W. H. McAdams, "Heat Transmission," 3d ed., McGraw-Hill Book Company, Inc., New York, 1954; E. Eckert, "Technische Strahlungsaustauschrechnungen," VDI-Verlag, Berlin, 1937; P. H. Moon, "Scientific Basis of Illuminating Engineering," McGraw-Hill Book Company, Inc., New York, 1936.

## Table 14-1.

**RADIATING AREA ELEMENT**

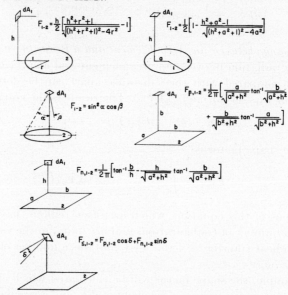

$$F_{1-2} = \frac{h}{2}\left[\frac{h^2+r^2+1}{\sqrt{(h^2+r^2+1)^2-4r^2}} - 1\right]$$

$$F_{1-2} = \frac{1}{2}\left[1 - \frac{h^2+a^2-1}{\sqrt{(h^2+a^2+1)^2-4a^2}}\right]$$

$$F_{1-2} = \sin^2\alpha \cos\beta$$

$$F_{p,1-2} = \frac{1}{2\pi}\left[\frac{a}{\sqrt{a^2+h^2}}\tan^{-1}\frac{b}{\sqrt{a^2+h^2}} + \frac{b}{\sqrt{b^2+h^2}}\tan^{-1}\frac{a}{\sqrt{b^2+h^2}}\right]$$

$$F_{n,1-2} = \frac{1}{2\pi}\left[\tan^{-1}\frac{b}{h} - \frac{h}{\sqrt{a^2+h^2}}\tan^{-1}\frac{b}{\sqrt{a^2+h^2}}\right]$$

$$F_{\delta,1-2} = F_{p,1-2}\cos\delta + F_{n,1-2}\sin\delta$$

**FINITE RADIATING AREA:**

*graph of this → in notes or Perry's p.487 fig. 19 & 20*

$$\bar{F}_{1-2} = \frac{1}{2}\left[\frac{h^2+1}{r^2} + 1 - \sqrt{\left(\frac{h^2+1}{r^2}+1\right)^2 - \frac{4}{r^2}}\right]$$

$$\bar{F}_{1-3} = \frac{2}{ab\pi}\left[b\sqrt{h^2+a^2}\tan^{-1}\frac{b}{\sqrt{a^2+h^2}} + a\sqrt{b^2+h^2}\tan^{-1}\frac{a}{\sqrt{b^2+h^2}}\right.$$
$$\left. - bh\tan^{-1}\frac{b}{h} - ah\tan^{-1}\frac{a}{h} - \frac{h^2}{2}\ln\frac{(h^2+a^2+b^2)h^2}{(a^2+h^2)(b^2+h^2)}\right]$$

$$\bar{F}_{2-3} = \frac{1}{\pi}\left[\tan^{-1}\frac{b}{h} + \frac{a}{h}\tan^{-1}\frac{b}{a} - \frac{\sqrt{a^2+h^2}}{h}\tan^{-1}\frac{b}{\sqrt{a^2+h^2}}\right.$$
$$\left. + \frac{h}{4b}\ln\frac{(a^2+b^2+h^2)h^2}{(a^2+b^2)(b^2+h^2)} + \frac{a^2}{4bh}\ln\frac{(a^2+b^2+h^2)a^2}{(a^2+b^2)(a^2+h^2)}\right.$$
$$\left. - \frac{b}{4h}\ln\frac{(a^2+b^2+h^2)b^2}{(a^2+b^2)(h^2+b^2)}\right]$$

$$F_{1-(2+3+4+5)} = F_{1-2} + F_{1-3} + F_{1-4} + F_{1-5}$$

In the same way the other shape factors in Table 14-1 can be calculated. The numerical integration can be replaced by a graphical procedure first mentioned by R. A. Hermann (1900). According to Eq. (14-6), the shape factor of a surface element $dA_1$ with respect to a finite surface $A_2$ is

$$F_{1-2} = \frac{1}{\pi} \int \cos \beta_1 \, d\omega_1 \qquad (14\text{-}9)$$

The expression under the integral sign $\cos \beta_1 \, d\omega_1$ is the projection of the

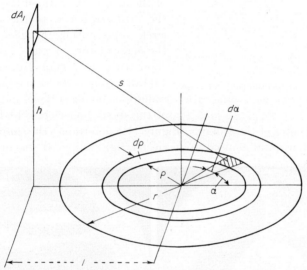

FIG. 14-2. Shape factor for radiative exchange between an area element and a circular disk.

solid-angle element $d\omega_1$ on the plane of the radiating surface $dA_1$. The integral is the sum of the projections of all solid-angle elements $d\omega_1$, or the projection of the solid angle corresponding to the whole surface $A_2$ on the plane of $dA_1$. From this the following construction arises: We project by central projection the surface $A_2$ (denoted by 1 in Fig. 14-3) onto the surface of a hemisphere with the radius $R$. This surface $1'$ must be projected once more by normal projection to the plane of the radiating surface $dA_1$. The area of this second projection $1''$ divided by the area of the circle $\pi R^2$ gives the shape factor $F_{1-2}$ as can be seen from Eq. (14-9). This graphical construction can be replaced by a simple mechanical integrator[1] or by optical projection.[2] For this purpose a point light

[1] H. Hottel, *Mech. Eng.*, **52** (1930); and V. H. Cherry, D. D. Davis, and L. M. K. Boelter, *Trans. Illum. Eng. Soc. N.Y.*, 1939.
[2] E. Eckert, *Z. Ver. deut. Ingr.*, **79**:1495–1496 (1935).

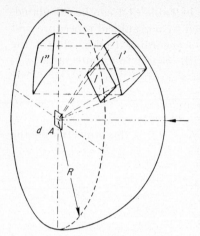

FIG. 14-3. Determination of the geometrical factor for a surface 1.

source is arranged in the center of the hemisphere in Fig. 14-3. The area 1, the geometrical factor of which is to be determined, is made of cardboard and arranged in the proper position. It is projected by the point light source as shadow on a milk-glass hemisphere. When this hemisphere is photographed from a great distance in the direction of the arrow, the ratio of the shadow of the cardboard area to the area of the circle representing the glass sphere is the shape factor. Figures 14-4 and 14-5 show two such photographs. Figure 14-4 determines the shape factor of the radiation cooling surfaces in the furnace of the steam boiler (Fig. 14-15) as seen from the center of the stoker. Figure 14-5 determines the shape factor of the tungsten spiral as used in an electric light bulb. It was produced by

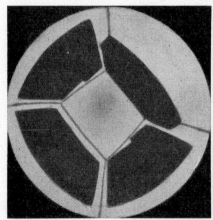

FIG. 14-4. Photographic determination of the geometrical factor of the radiation cooling surface in the furnace of a steam boiler as seen from the center of the stoker. [From E. Eckert, Z. Ver. deut. Ingr., **79**: 1495–1496 (1935).]

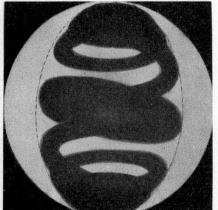

FIG. 14-5. Photographic determination of the geometrical factor of the tungsten coil in an electric bulb. [From E. Eckert, Z. Ver. deut. Ingr., **79**:1495–1496 (1935).]

positioning the model of the spiral a short distance from the point light source.[1]

From a study of the graphical construction shown in Fig. 14-3, analyt-

[1] *Ibid.*

ical expressions can be derived for the shape factor of many geometries without any integration. An example will amplify this statement. The shape factor with which an area element $dA$ irradiates a rectangle $A$ may be determined when $dA$ is arranged parallel to and below a corner of the rectangle $A$. Figure 14-6 shows this arrangement. Area $A$ is divided

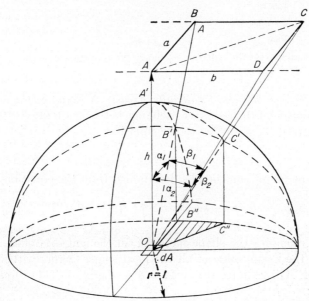

FIG. 14-6. Shape factor for radiative exchange between a rectangle and an area element.

into two triangles by the line $AC$, and the geometric construction is indicated for the triangle $ABC$. The primed figure $OB''C'''$, which indicates the shape factor, is the projection of the sector $OB'C'$ of a great circle. The area of the sector $OB'C'$ is $\beta_1/2\pi$ times the area of the circle or $\beta_1/2$ (the radius of the sphere is chosen to be 1). The area of the sector $OB''C'''$ is therefore $(\beta_1/2) \sin \alpha_1$. The shape factor is this area divided by the area of the base circle, or $(\beta_1/2\pi) \sin \alpha_1$. In the same way the shape factor of the triangle $ACD$ is found as $(\beta_2/2\pi) \sin \alpha_2$, and the shape factor of the rectangle is the sum of the shape factor of the two triangles, or

$$F_{1-2} = \frac{\beta_1 \sin \alpha_1 + \beta_2 \sin \alpha_2}{2\pi}$$

Replacement of the angles by the lengths $a$, $b$, and $h$ results in the relation contained in Table 14-1 for this configuration.

The equations in Table 14-1 can be used to determine shape factors for many more configurations. Figure 14-7 illustrates this in two examples. The validity of the relations contained in the figure is immediately evident from the geometric interpretation of the shape factor.

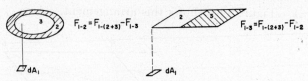

FIG. 14-7. Shape factor for radiative exchange between an area element and a finite area.

*Heat Exchange between Two Finite Areas.* A shape factor is also used for a calculation of the radiative exchange between a finite surface $A_1$ and a finite surface $A_2$ according to the equation

$$Q_b = \bar{F}_{1-2} A_1 \sigma (T_1^4 - T_2^4) \qquad (14\text{-}10)$$

The shape factor can then be determined from the equation

$$\bar{F}_{1-2} = \frac{1}{A_1} \int_{A_1} F_{1-2}\, dA_1 = \frac{1}{A_1} \int_{A_1} \int_{A_2} \frac{\cos \beta_1 \cos \beta_2}{\pi s^2}\, dA_2\, dA_1$$

where $F_{1-2}$ denotes the shape factor of an element $dA_1$ of $A_1$. $\bar{F}_{1-2}$ is, therefore, the mean value of all the local shape factors with which any location on $A_1$ irradiates the surface $A_2$. Equation (14-10) can also be written as

$$Q_b = \bar{F}_{2-1} A_2 \sigma (T_1^4 - T_2^4) \qquad (14\text{-}11)$$

with

$$\bar{F}_{2-1} = \frac{1}{A_2} \int_{A_2} F_{2-1}\, dA_2 = \frac{1}{A_2} \int_{A_2} \int_{A_1} \frac{\cos \beta_1 \cos \beta_2}{\pi s^2}\, dA_1\, dA_2$$

indicating the mean shape factor of area $A_2$ against $A_1$. The following relation obviously connects the two shape factors

$$\bar{F}_{1-2} A_1 = \bar{F}_{2-1} A_2$$

This is a convenient relation because sometimes one shape factor is simpler to calculate than the other. When, for instance, the surface $A_1$ is completely surrounded by the surface $A_2$, it can be stated at once that $F_{1-2}$ is equal to 1, since all rays emitted from $A_1$ will impinge on $A_2$. Table 14-1 contains expressions for a number of shape factors. For a graphical determination of the shape factor for a finite surface $A_1$ and a finite surface $A_2$, the surface $A_1$ must be divided into small areas of equal size and the construction shown in Fig. 14-3 be performed for the centers of every one of these areas. The average value of all the geometrical factors determined in this way is the geometrical factor $\bar{F}_{1-2}$.

## 14-2. Solid, Liquid, and Gaseous Bodies.

If the surfaces which exchange heat by radiation are not black, conditions become more involved, since part of the thermal radiation is reflected by the surfaces. Some of the radiation travels in this way back and forth between the surfaces until it is finally absorbed. The influence of this action on the heat flow can best be studied with two parallel surfaces whose distance apart is small compared with their size, so that practically all radiation emitted by one surface falls upon the second. The shape factor of either surface is therefore 1. We shall first consider monochromatic radiation. In Fig. 14-8 the travel of a heat ray emitted by the surface 1 is traced.

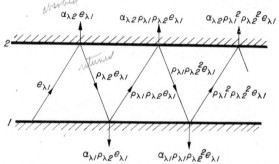

Fig. 14-8. Radiant heat exchange between two parallel surfaces.

The monochromatic emissive power of the radiation emitted from this surface per unit time and area is $e_{\lambda 1}$. From it a part $\alpha_{\lambda 2} e_{\lambda 1}$ is absorbed on surface 2 and a part $\rho_{\lambda 2} e_{\lambda 1}$ is reflected back to surface 1. Here a part $\alpha_{\lambda 1} \rho_{\lambda 2} e_{\lambda 1}$ is absorbed; $\rho_{\lambda 1} \rho_{\lambda 2} e_{\lambda 1}$ is reflected to surface 2; and so on. The amount of heat which left surface 1 per unit time and area is therefore

$$q_{\lambda 1} = (1 - \alpha_{\lambda 1}\rho_{\lambda 2} - \alpha_{\lambda 1}\rho_{\lambda 1}\rho_{\lambda 2}^2 - \alpha_{\lambda 1}\rho_{\lambda 1}^2\rho_{\lambda 2}^3 - \cdots)e_{\lambda 1}$$
$$= [1 - \alpha_{\lambda 1}\rho_{\lambda 2}(1 + \rho_{\lambda 1}\rho_{\lambda 2} + \rho_{\lambda 1}^2\rho_{\lambda 2}^2 + \cdots)]e_{\lambda 1}$$

The series within the parentheses gives $1/(1 - \rho_{\lambda 1}\rho_{\lambda 2})$, since $\rho_{\lambda 1}$ and $\rho_{\lambda 2}$ are smaller than 1. Therefore

$$q_{\lambda 1} = \left(1 - \frac{\alpha_{\lambda 1}\rho_{\lambda 2}}{1 - \rho_{\lambda 1}\rho_{\lambda 2}}\right) e_{\lambda 1}$$

For bodies opaque to thermal radiation, Eq. (13-2) holds. Expressing $\rho_\lambda$ by the absorptivity $\alpha_\lambda$ results in

$$q_{\lambda 1} = \left[1 - \frac{\alpha_{\lambda 1}(1 - \alpha_{\lambda 2})}{\alpha_{\lambda 1} + \alpha_{\lambda 2} - \alpha_{\lambda 1}\alpha_{\lambda 2}}\right] e_{\lambda 1} = \frac{\alpha_{\lambda 2}}{\alpha_{\lambda 1} + \alpha_{\lambda 2} - \alpha_{\lambda 1}\alpha_{\lambda 2}} e_{\lambda 1}$$

In the same manner, surface 2 emits radiation of emissive power $e_{\lambda 2}$. The amount $\alpha_{\lambda 1} e_{\lambda 2}$ is absorbed by surface 1. The rest is reflected, impinges on surface 2, and is there partly reflected back to surface 1. Here the part

$\alpha_{\lambda 1}\rho_{\lambda 1}\rho_{\lambda 2}e_{\lambda 2}$ is absorbed, the rest reflected, and so on. By summing up all radiation absorbed by surface 1 there is obtained

$$q_{\lambda 2} = \alpha_{\lambda 1}(1 + \rho_{\lambda 1}\rho_{\lambda 2} + \rho_{\lambda 1}{}^2\rho_{\lambda 2}{}^2 + \cdots)e_{\lambda 2}$$

which gives as before

$$q_{\lambda 2} = \frac{\alpha_{\lambda 1}}{1 - \rho_{\lambda 1}\rho_{\lambda 2}} e_{\lambda 2} = \frac{\alpha_{\lambda 1}}{\alpha_{\lambda 1} + \alpha_{\lambda 2} - \alpha_{\lambda 1}\alpha_{\lambda 2}} e_{\lambda 2}$$

The net heat flux from surface 1 to surface 2 per unit time and area is

$$q_\lambda = q_{\lambda 1} - q_{\lambda 2} = \frac{\alpha_{\lambda 2}e_{\lambda 1} - \alpha_{\lambda 1}e_{\lambda 2}}{\alpha_{\lambda 1} + \alpha_{\lambda 2} - \alpha_{\lambda 1}\alpha_{\lambda 2}} = \frac{\alpha_{\lambda 2}\epsilon_{\lambda 1}e_{b\lambda 1} - \alpha_{\lambda 1}\epsilon_{\lambda 2}e_{b\lambda 2}}{\alpha_{\lambda 1} + \alpha_{\lambda 2} - \alpha_{\lambda 1}\alpha_{\lambda 2}}$$

This equation actually is another derivation of Kirchhoff's law. If the two surfaces have equal temperatures, then $e_{b\lambda 1} = e_{b\lambda 2}$ and the heat flow $q$ must be zero according to the second law of thermodynamics. From the above equation it follows that

$$\alpha_{\lambda 2}\epsilon_{\lambda 1} = \alpha_{\lambda 1}\epsilon_{\lambda 2} \quad \text{or} \quad \frac{\alpha_{\lambda 1}}{\epsilon_{\lambda 1}} = \frac{\alpha_{\lambda 2}}{\epsilon_{\lambda 2}} = 1$$

and this ratio must be equal to 1, since the expression must also be valid for a black surface for which $\alpha_\lambda$ and $\epsilon_\lambda$ are 1.

With Kirchhoff's relation, the net heat exchange can be written

$$q_\lambda = \frac{1}{(1/\epsilon_{\lambda 1}) + (1/\epsilon_{\lambda 2}) - 1}(e_{b\lambda 1} - e_{b\lambda 2}) \qquad (14\text{-}12)$$

The total radiative heat exchange is obtained by integration over all wavelengths:

$$q = \int_0^\infty \frac{1}{(1/\epsilon_{\lambda 1}) + (1/\epsilon_{\lambda 2}) - 1}(e_{b\lambda 1} - e_{b\lambda 2})\,d\lambda \qquad (14\text{-}13)$$

For *gray* surfaces, for which the emissivities are independent of wavelength, Eq. (14-13) becomes

$$q = \frac{1}{(1/\epsilon_1) + (1/\epsilon_2) - 1}(e_{b1} - e_{b2}) = \frac{\sigma}{(1/\epsilon_1) + (1/\epsilon_2) - 1}(T_1{}^4 - T_2{}^4) \qquad (14\text{-}14)$$

One sometimes defines an interchange factor $\epsilon_{1-2}$ by the equation

$$\frac{1}{\epsilon_{1-2}} = \frac{1}{\epsilon_1} + \frac{1}{\epsilon_2} - 1 \qquad (14\text{-}15)$$

and can write

$$q = \epsilon_{1-2}\sigma(T_1{}^4 - T_2{}^4) \qquad (14\text{-}16)$$

For other geometries, the calculation of the radiative interchange becomes quite involved unless certain restrictions are placed on the radiative properties. For the following calculations, it will be assumed that the emitted radiation follows Lambert's cosine law and that the radiation properties are independent of wavelength (gray surfaces).

For two concentric spheres or cylinders whose surfaces reflect diffusely, a calculation analogous to the one presented at the beginning of this section leads to the equation

$$q = \frac{\sigma}{(1/\epsilon_1) + (A_1/A_2)[(1/\epsilon_2) - 1]} (T_1^4 - T_2^4) \qquad (14\text{-}17)$$

$q$ is the heat flow per unit area of the smaller surface $A_1$.

Equation (14-17) can be derived more rapidly in the following way.[1] Let us denote the total radiative flux (consisting of emitted and reflected radiation) which leaves surface 1 per unit time and area by $B_1$ and the total radiative flux leaving a unit area of a surface 2 by $B_2$. The letter $B$ is chosen in analogy to the practice in illumination engineering where the total light flux leaving a surface determines its brightness. The total radiative flux leaving surface 1 is then $A_1 B_1$, and the total radiative flux leaving surface 2 is $A_2 B_2$. Only part of this latter flux impinges on surface 1, namely, the amount $\bar{F}_{2-1} A_2 B_2$. The shape factor $\bar{F}_{2-1}$ with which surface 2 views surface 1 can easily be calculated, since generally $A_1 \bar{F}_{1-2} = A_2 \bar{F}_{2-1}$ and since $\bar{F}_{1-2} = 1$ because surface 1 is completely surrounded by surface 2. The net heat flow from surface 1 caused by the radiative interchange must be the difference between the total radiation leaving this surface and arriving at it, or

$$q A_1 = A_1 (B_1 - B_2)$$

The total radiative flux leaving surface 1 is composed of emitted and reflected radiation:

$$B_1 = e_1 + (1 - \alpha_1) B_2$$

The total radiation leaving surface 2 consists of emitted radiation, of reflected radiation coming from $A_1$, and of reflected radiation coming from $A_2$:

$$B_2 = e_2 + (1 - \alpha_2) F_{2-1} B_1 + (1 - \alpha_2)(1 - F_{2-1}) B_2$$

From the last three equations and considering $F_{2-1} = A_1/A_2$, the following expression is obtained:

$$q = \frac{\alpha_2 e_1 - \alpha_1 e_2}{\alpha_2 + (A_1/A_2)(\alpha_1 - \alpha_1 \alpha_2)} \qquad (14\text{-}18)$$

[1] This method and its extension discussed in the following section has apparently been used for the first time by G. Poljak, *Tech. Physics U.S.S.R.*, **1**:555 (1935).

Introducing $e_1 = \epsilon_1 e_{b1}$, $e_2 = \epsilon_2 e_{b2}$, $\alpha_1 = \epsilon_1$, and $\alpha_2 = \epsilon_2$ results in Eq. (14-17).

The calculation for nongray surfaces has to start with a consideration of monochromatic radiation. The total radiative heat exchange is finally obtained by integration over all wavelengths:

$$q = \int_0^\infty \frac{e_{b\lambda 1} - e_{b\lambda 2}}{(1/\epsilon_{\lambda 1}) + (A_1/A_2)[(1/\epsilon_{\lambda 2}) - 1]} \, d\lambda \qquad (14\text{-}19)$$

It is easily seen that Eqs. (14-12) to (14-16) describe the heat exchange between two concentric cylinders or spheres when the surfaces reflect specularly, because any ray starting from the inner surface $A_1$ is reflected back to $A_1$ from $A_2$. Rays originating from $A_2$ can be divided into two groups: one alternating between $A_2$ and $A_1$ and one which never reaches $A_1$.

For two surfaces whose size is small compared with their distance apart (Fig. 14-1), the fraction of the reflected radiation which falls back to the radiating surface becomes so small that it can be neglected. The interchange factor then becomes for gray surface elements

$$\epsilon_{1-2} = \epsilon_1 \epsilon_2 \qquad (14\text{-}20)$$

This equation generally gives a lower limit for the interchange factor. The upper limit is 1.

In a way similar to the radiation exchange between two parallel walls the interchange between a gas and its enclosing walls can be calculated if it is assumed that the walls have *gray* surfaces.[1] The interchange factor $\epsilon_{g-w}$ between the gas and the wall is

$$\frac{1}{\epsilon_{g-w}} = \frac{1}{\bar{\epsilon}_g} + \frac{1}{\bar{\epsilon}_{g\infty} \bar{\epsilon}_w} - \frac{1}{\bar{\epsilon}_{g\infty}} \qquad (14\text{-}21)$$

where $\bar{\epsilon}_g$ is the emissivity of the gas (see Figs. 13-18 and 13-19) and $\bar{\epsilon}_{g\infty}$ is the emissivity of the same gas with an infinitely thick layer. $\bar{\epsilon}_{g\infty}$ can be determined approximately in Figs. 13-18 and 13-19 by extrapolation. $\epsilon_w$ is the emissivity of the walls. The heat flow per unit wall area is

$$q = \epsilon_{g-w} \sigma T_g^4 - \epsilon_{g-w} \sigma T_w^4 \qquad (14\text{-}22)$$

Into the first summand, $\bar{\epsilon}_g$ and $\epsilon_w$ have to be introduced at gas temperature; into the second, at wall temperature. When the emissivity of the walls is great (in the neighborhood of 1), the interchange factor can be calculated with the simpler equation

$$\epsilon_{g-w} = \bar{\epsilon}_g \epsilon_w \qquad (14\text{-}23)$$

Equations (14-21) and (14-23) exactly describe the interchange factor

[1] E. Eckert, *VDI-Forschungsheft* 387, p. 19, 1937.

between the walls of a spherical or long cylindrical container of uniform temperature $T_w$ and a gas of temperature $T_g$ inside the container or for a gas layer of temperature $T_g$ between two plane walls with temperature $T_w$.

**Example 14-1.** The heat exchange by radiation between the two walls of a thermos bottle is to be calculated. The walls may be assumed silvered on the sides turned toward each other. The contents of the bottle have a temperature of $t_1 = 212$ F. The ambient temperature is $t_2 = 68$ F. The inner and outer walls, respectively, have nearly the same temperatures. The emissivity of silver is $\epsilon = 0.02$ (from Table $A$-10, assuming $\epsilon_n = \epsilon$). As the surfaces reflect specularly, Eq. (14-14) must be used to determine the interchange factor.

$$\frac{1}{\epsilon_{1-2}} = \frac{1}{0.02} + \frac{1}{0.02} - 1 = \frac{1}{0.01}$$

The heat flow per unit area of the inner wall is

$$q = 0.01 \times 0.173 \times 10^{-8}[(212 + 460)^4 - (68 + 460)^4] = 2.18 \text{ Btu/hr ft}^2$$

To get the same insulation effect by a cork layer, its thickness must be

$$b = \frac{k}{q}(t_1 - t_2) \qquad \text{with } k = 0.025 \text{ Btu hr ft F}$$

This gives

$$b = \frac{0.025(212 - 68)}{2.18} = 1.65 \text{ ft}$$

It can be seen that the insulating effect of the thermos bottle is excellent even though it must be kept in mind that some additional losses occur by conduction in the glass walls. The loss by radiation can be decreased by arranging walls silvered on both sides between the inner and outer wall. With $n$ additional walls the heat loss is decreased by the factor $1/(n+1)$.

The small radiation exchange between metal surfaces is utilized in Alfoil insulation. This insulation is built up from aluminum foils which surround the body to be insulated at approximately ½-in. distance from each other. The total heat flow through the insulation is by radiation, conduction through the air space between the foils, and convection. The convection parts are comparatively small as long as the distance between the foils is kept low enough, and the conduction is small, since air is a good insulator.

**14-3. Radiative Heat Exchange inside an Enclosure.** Frequently the following task is encountered in engineering calculations: An enclosure is filled with a nonradiating or radiating gas, and the walls, of which the enclosure is composed, are kept at various temperatures. The radiative heat exchange between the various surfaces and the gas in the enclosure is to be calculated. Usually some of the surfaces are kept at known temperatures by heating or cooling, whereas others can be assumed adiabatic to heat flow. The combustion chamber of the steam boiler shown in Fig. 14-15 is an example of such an enclosure. The stoker is covered with a glowing layer of coal, the temperature of which is usually considered known in such calculations. The temperature of the tube surfaces is practically equal to the evaporation temperature of the water inside the tubes. The temperature of the refractory walls, however, is determined by a balance of the heat received and given off by radiation and convection. The radiative heat exchange is usually dominating, so that a calculation neglecting the convective heat exchange is a good approximation. The interior of the chamber is filled with a gas which participates in the radiative exchange. To calculate the heat exchange and the temperatures of the walls and of the gas inside the enclosure is a formidable task, and considerable simplifications are required to solve the problem in a reasonable time. In reality the temperature varies locally on the refractory walls. For calculation purposes these walls are usually divided into a limited number of areas and the temperature is considered constant in each area. Also the temperature and composition within the flame or combustion gases filling the chamber vary locally. In such calculations either the space inside the chamber is assumed to be filled by a homogenous gas of uniform temperature or it is subdivided into a limited number of volumes and the temperature and composition are assumed constant in each.

*Nonradiating Gas, Black Surfaces.* In this section an enclosure will be considered which is filled with a gas which does not radiate or absorb heat. The enclosure consists of $n$ walls with diffusely reflecting surfaces. The emission of radiation is assumed to follow the cosine law. The temperature is assumed constant over each surface, and convective heat transfer is neglected. The temperature is described for some surfaces, and the remaining ones are considered adiabatic. The task is to calculate the radiative heat exchange between the various surfaces and the unknown adiabatic surface temperatures. The shape factors between any of the surfaces will be assumed as known.

As a first step we shall consider all surfaces to be black. Then the calculation is straightforward. The heat exchange between two arbitrary surfaces $i$ and $k$ is

$$Q_{i-k} = A_i F_{i-k}(e_{bi} - e_{bk})$$

The heat exchange between surface $i$ and the rest of the enclosure is obtained by summation

$$Q_i = \sum_k A_i F_{i-k}(e_{bi} - e_{bk}) \qquad (14\text{-}24)$$

The $n$ surfaces can be divided into two groups. In the first group the temperature $T_i$ and, therefore, $e_{bi}$ are known and the heat flux $Q_i$ is sought. In the second group the heat flux $Q_i = 0$ and the temperature is to be calculated. The problem contains $n$ unknown values which are obtained by a solution of the $n$ linear equations. V. Paschkis, and more recently

A. K. Oppenheim,[1] pointed out that electric analogues can be devised for this radiation exchange process. In Fig. 14-9 the network for the enclosure shown in part $a$ is sketched in part $b$. The enclosure consists of five surfaces of which surfaces 2, 3, and 5 may be kept at prescribed temperatures. A knot in the analogue corresponds to a surface in the enclosure. All knots are interconnected by wires with resistances proportional to $1/A_i F_{i-k}$. The knots corresponding to surfaces with known temperature are connected with electric potentials proportional to $e_{bi}$. The knots which correspond to the adiabatic surfaces 1 and 4 are without external connection. The currents from the outside connections to the knots correspond to $Q_i$, and the electric potential of any knot without outside connection corresponds to the emissive power of the respective adiabatic surface. Actually, an enclosure has to be considered as a configuration in space, and all its surfaces must be included in the calculation. In Fig. 14-9 the surfaces outside the plane of drawing have not been included so as not to obscure the sketch.

*Nonradiating Gas, Gray Walls.* When the surfaces of the enclosure are nonblack surfaces, then innumerable reflections occur within the enclosure and a calculation may at first glance appear hopeless. However, a generalization of the procedure which has been used on page 405 surmounts these difficulties. We shall first consider gray surfaces ($\alpha = \epsilon$). $B$ will indicate the total radiative flux leaving any surface per unit area, and $H$ the total flux impinging per unit area. We shall now fix our attention on surface $i$ and develop an expression for the heat exchange $Q_i$ between this area and the rest of the enclosure. The heat flux leaving the surface is $A_i B_i$. This term can also be written as $\sum_k A_i F_{i-k} B_i$. The heat flux directed from all other surfaces against surface $i$ is $A_i H_i = \sum_k A_i F_{i-k} B_k$. Therefore the heat exchange of area $i$ is

$$Q_i = \sum_k A_i F_{i-k} (B_i - B_k) \tag{14-25}$$

On the other hand

$$Q_i = A_i (B_i - H_i) \tag{14-26}$$

The heat flow leaving surface $i$ is composed of emitted and reflected radiation:

$$B_i = \epsilon_i e_{bi} + (1 - \epsilon_i) H_i \tag{14-27}$$

---

[1] V. Paschkis, *Elektrotech. u. Maschinenbau*, **54**:617 (1936); and A. K. Oppenheim, Radiation Analysis by the Network Method, *ASME Preprint Paper* 54-A-75.

410  THERMAL RADIATION

Elimination of $H_i$ from the last two equations results in

$$Q_i = A_i \frac{\epsilon_i}{1 - \epsilon_i}(e_{bi} - B_i)$$

Two equations of the form (14-25) and (14-28) can be written for the surfaces, as compared with one equation for the black surfaces. On the other hand the $B_i$ appear as additional unknown quantities, so that $2n$ equations are available for $2n$ unknowns and the remaining problem is to solve $2n$ simultaneous linear equations.

The electric analogue for the enclosure in Fig. 14-9a with gray surfaces

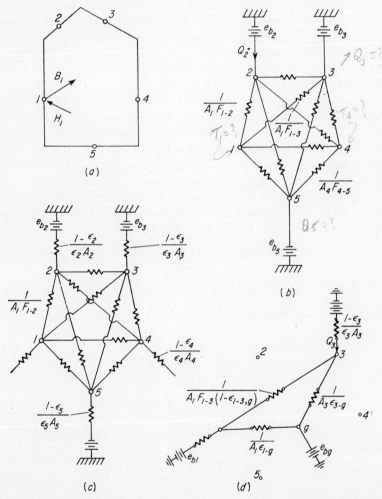

FIG. 14-9. Radiative exchange in an enclosure and an electric analogue. (*According to A. K. Oppenheim.*)

is illustrated in Fig. 14-9c according to Oppenheim. Equation (14-28) indicates that a resistance proportional to $(1 - \epsilon_i)/\epsilon_i A_i$ has to be inserted into each connection of a knot with an outside potential. An interesting conclusion can be drawn immediately from the analogue. Insertion of a resistance on the outside of a knot without outside connection (for instance, knot 4) obviously has no influence on the potential of this knot. This means that the temperature of an adiabatic wall is independent of the absorptivity of the surface of this wall.

It should be pointed out that the previous derivation rests on the assumption that all surfaces are irradiated uniformly over their areas. The subdivision of the enclosure into the $n$ surfaces must be made in such a way that this condition is reasonably approximated.

When the walls which exchange heat by radiation are nongray surfaces, then Eqs. (14-25) to (14-28) have to be written for monochromatic radiation. The total amount of heat which a surface $i$ loses by the radiative exchange is finally obtained by an integration over all wavelengths.

$$Q_i = \int_0^\infty Q_{\lambda i}\, d\lambda$$

The calculation becomes a fairly involved trial-and-error procedure, since for an adiabatic wall the condition $Q_i = 0$ applies only to the total and not to the monochromatic radiation.

*Radiating Gas.* The calculation presented in the last section will now be extended to the situation in which a radiating and absorbing gas is contained inside the enclosure. It will again be assumed that the walls of the enclosure radiate according to the cosine law, that they are diffuse reflectors and gray surfaces. The gas which fills the enclosure will at first also be postulated as a gray gas ($\alpha = \epsilon$). Later on this assumption will be relaxed. The gas will also be assumed to have a uniform temperature and concentration throughout the space inside the enclosure.

Equations (14-26) and (14-27) are valid for the present situation as well, and their combination results in the expression

$$Q_i = A_i \frac{\epsilon_i}{1 - \epsilon_i} (e_{bi} - B_i) \tag{14-29}$$

Equation (14-25), however, changes because part of the radiation coming from the surface $k$ is absorbed in the gas before it reaches surface $i$ and on the other hand the gas radiates energy toward the latter surface. We shall now derive the equation expressing these physical processes. The radiative flux leaving the surface $k$ and directed toward the surface $i$ is

$$A_i F_{i-k} B_k$$

The radiative flux coming from $k$ and arriving at $i$ can then be expressed as

$$A_i F_{i-k}(1 - \epsilon_{i-k,g}) B_k$$

In this equation, $\epsilon_{i-k,g}$ is the emissivity which for a gray surface, according to Kirchhoff's law, is equal to the absorptivity, and this value has to be determined for all radiant rays traveling between the surfaces $i$ and $k$. It has to be obtained from Eq. (13-36) in which the integration is not carried out over the whole hemisphere but only over the solid angle indicated by the shape factor $F_{i-k}$.

The radiative flux emitted by the gas and arriving at $i$ is

$$A_i \epsilon_{i-g} e_{bg}$$

where $\epsilon_{i-g}$ indicates the emissivity of the gas body into area $i$. Finally the radiative flux leaving $i$ is

$$A_i B_i$$

From these components the net radiative heat loss of surface $i$ can be obtained:

$$\begin{aligned} Q_i &= A_i[B_i - \Sigma F_{i-k}(1 - \epsilon_{i-k,g})B_k - \epsilon_{i-g}e_{bg}] \\ &= A_i \epsilon_{i-g}(B_i - e_{bg}) + \Sigma A_i F_{i-k}(1 - \epsilon_{i-k,g})(B_i - B_k) \end{aligned} \quad (14\text{-}30)$$

The net heat loss of the gas caused by the radiative exchange is

$$Q_g = \sum_i A_i \epsilon_{i-g}(e_{bg} - B_i) \tag{14-31}$$

Equations (14-29) to (14-31) are sufficient to calculate the unknown parameters in this problem—the heat loss for surfaces with known temperature, the temperature for adiabatic surfaces, and the heat loss or the temperature of the gas in the enclosure.

The electric analogue to the enclosure sketched in Fig. 14-9a is shown in Fig. 14-9d. Only the connections between knots 1, 3, and $g$ are inserted. Knot $g$ corresponds to the enclosed gas. Analogous connections have to be made between the other knots. Knot $g$ is either connected with a battery giving the potential $e_{bg}$ when the temperature of the gas is prescribed or is left floating when no heat sources or sinks are contained in the gas. For the first situation, the current from the battery into knot $g$ is proportional to the heat loss of the gas. In the second situation, the potential at knot $g$ is proportional to the emissive power of a black body at the gas temperature.

Referring to the section on gas radiation, it is evident that real gases differ considerably from a gray one. For a calculation of radiative heat exchange in an enclosure filled with a nongray gas, Eqs. (14-29) to (14-31) have to be applied to monochromatic radiation. The total heat loss is then found by an integration of the monochromatic heat flux over all wavelengths. This calculation becomes very cumbersome, and an approximation to the real conditions is therefore desirable. The concept of a *selectively gray gas*[1] is useful for this purpose. To approximate a real gas by a selectively gray one, the whole wavelength range is divided into two portions. One comprises the wavelength ranges in which the gas absorbs, and the other one consists of the rest. The calculation in the first range is then made in exactly the same manner as for a gray gas, assuming that in this range the absorptivity is independent of wavelength. For the second part, the calculation procedure developed in the preceding paragraph for a nonradiating gas is used.

The assumption that the temperature of the gas body in the enclosure is uniform is in many cases not a realistic one. An obvious improvement of the calculation procedure is obtained when the gas body is divided into a number of parts and the temperature and concentration are allowed to vary from one part to the next. As an example, the geometry in Fig. 14-10a will be investigated.[2]

An absorbing and radiating gray gas of locally uniform composition is enclosed between two infinite parallel walls, 1 and 2. The distance between the two walls is $b$. For our calculation, we shall divide the gas body into $n$ layers of equal thickness $\Delta b$.

For the calculation of the radiative heat exchange, the radiation parameters of a gas layer and of the walls have to be determined. The following values are needed and will be assumed independent of temperature.

$\epsilon_w$ = hemispherical emissivity of the wall

$\epsilon_g$ = hemispherical emissivity of a gas layer with thickness $\Delta b$

$\tau_{gw,k}$ = hemispherical transmissivity of a gas body (consisting of $k$ layers) for wall radiation

$\tau_{gg,k}$ = transmissivity of a gas body ($k$ layers) for gas radiation from one layer

$\alpha_{gw,k}$ = absorptivity of a gas layer with thickness $\Delta b$ for wall radiation with $k$ gas layers in between

$\alpha_{gg,k}$ = absorptivity of a gas layer for radiation from another gas layer (both with thickness $\Delta b$) with $k$ gas layers in between

These radiation parameters have to be determined by an integration process analogous to the one described in Sec. 13-3. Its calculation will not be discussed here.

The radiative flux arriving at wall 1 may be $H_1$ and the radiative flux leaving this wall is indicated by $B_1$. Equation (14-29) applies again to

[1] E. Eckert, *VDI-Forschungsheft* 387, 1937.
[2] This calculation was discussed by E. R. G. Eckert in a lecture on radiation at the Lewis Laboratory of the National Advisory Committee for Aeronautics, May, 1957.

414  THERMAL RADIATION

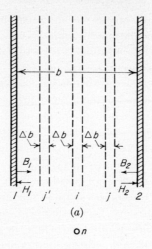

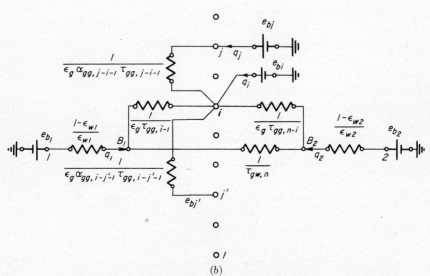

Fig. 14-10. Radiative exchange between a gas layer and two plane walls.

the present situation. It remains to express the net heat loss $q_1$ per unit area of wall 1 by the arriving and the leaving radiative fluxes and to describe the arriving flux from the various parts of which it consists. An arbitrary layer is denoted by $i$. The following equation holds:

$$q_1 = B_1 - \left(\tau_{gw,n}B_2 + \sum_{i=1}^{i=n} \epsilon_g \tau_{gg,i-1} e_{bi}\right)$$

This equation may now be written for the situation that the two walls and the enclosed gas are all at the same temperature. From Kirchhoff's law, it is known that in an enclosure of uniform temperature the radiative flux through any arbitrary plane is equal to the radiative flux coming from a black surface and that the net heat loss by radiation of wall 1 is zero. The above equation gives for this special situation

$$1 - \tau_{gw,n} - \sum_{i=1}^{i=n} \epsilon_g \tau_{gg,i-1} = 0$$

With this relation, the preceding equation can be transformed to

$$q_1 = \tau_{gw,n}(B_1 - B_2) + \sum_{i=1}^{i=n} \epsilon_g \tau_{gg,i-1}(B_1 - e_{bi}) \tag{14-32}$$

A corresponding equation holds for wall 2.

As a next step, the radiative loss of the gas layer $i$ will be calculated. This quantity is determined by the radiation emitted from the gas through both of its surfaces, by the radiative flux which comes from the two walls and is absorbed by the gas layer, and by the radiation originating in the two gas bodies on both sides of the layer and being absorbed in the layer. This energy balance can be expressed by the following equation:

$$q_i = 2\epsilon_g e_{bi} - \alpha_{gw,i-1}\tau_{gw,i-1}B_1 - \alpha_{gw,n-i}\tau_{gw,n-i}B_2$$
$$- \sum_{j=n+1}^{j=n} \epsilon_g \alpha_{gg,j-i-1}\tau_{gg,j-i-1}e_{bj} - \sum_{j'=1}^{j'=i-1} \epsilon_g \alpha_{gg,i-j'-1}\tau_{gg,i-j'-1}e_{bj'}$$

Specializing this equation for the condition of a uniform temperature results with the same consideration as above in the relation

$$2\epsilon_g = \alpha_{gw}\tau_{gw,i-1} + \alpha_{gw,n-i}\tau_{gw,n-i} + \sum_{j=n+1}^{j=n} \epsilon_g \alpha_{gg,j-i-1}\tau_{gg,j-i-1}$$
$$+ \sum_{j'=1}^{j'=i-1} \epsilon_g \alpha_{gg,i-j'-1}\tau_{i-j'-1}$$

In addition, it is evident that for each ray the directional emissivity must be equal to the directional absorptivity. Therefore,

$$\epsilon_g \tau_{gg,k} = \alpha_{gw}\tau_{gw,k}$$

The last two equations can be introduced into the expression for $q_i$. Then the following equation is obtained:

$$q_i = \epsilon_g \tau_{gg,i-1}(e_{bi} - B_1) + \epsilon_g \tau_{gg,n-i}(e_{bi} - B_2)$$
$$+ \sum_{j=i+1}^{j=n} \epsilon_g \alpha_{gg,j-i-1} \tau_{gg,j-i-1}(e_{bi} - e_{bj})$$
$$+ \sum_{j'=1}^{j'=i-1} \epsilon_g \alpha_{gg,i-j'-1} \tau_{gg,i-j'-1}(e_{bi} - e_{bj}) \quad (14\text{-}33)$$

Eqs. (14-29), (14-32), and (14-33), together with two corresponding equations for the wall 2, completely describe the problem. It can easily be checked that the number of equations is equal to the number of unknowns.

The electric analogue which corresponds to these equations is presented in Fig. 14-10b. The vertical row of knots, 1 to $n$, corresponds to the $n$ gas layers. The necessary connections are indicated between the knot $i$ and the knots 1, 2, $j$, and $j'$ respectively. The network would have to be completed by making all the other connections in the same way. The temperature of the various gas layers may be determined by internal heat sources or sinks. In this case the temperature of the layers has to be considered as prescribed. Accordingly, the potential of the knots 1 to $n$ is fixed by appropriate batteries. The temperature of some or all gas layers may also be determined solely by the radiative heat exchange. In this case, the corresponding knots are left floating and the potentials of the knots indicate their black-body emissive power and in this way their temperature.

For a nongray gas, the calculation has to be performed at first for monochromatic radiation with a succeeding integration over all wavelengths. This procedure can again be simplified by the concept of a selectively gray gas.

*Apparent Emissivity in a Cylindrical Hole.* In the preceding sections, the temperature and radiation properties have been considered as constant over finite areas and volumes, respectively. In reality, most parameters usually vary continuously over the surfaces and volumes considered. The mathematical formulation of the problem of radiative exchange under this condition leads to integral equations. As an example, the following situation will be discussed in this section:

A cylindrical hole as sketched in Fig. 14-11 is drilled in a material of constant temperature. The walls of the hole may be considered gray, and the emissivity of the wall material denoted by $\epsilon$. This property, together with the constant wall temperature, determines the radiative flux emitted from any location in the hole surface. The total flux $B$ leaving any location in the hole will be larger because the emitted radiation is supplemented by radiation reflected on the surface. We can define a *local apparent emissivity* of the wall by the following expression:

$$\epsilon_a(x) = \frac{B(x)}{e_b}$$

The radiative flux $B$ as well as the apparent emissivity have to be considered as functions of the distance $x$ from the end of the hole. The radiative flux arriving on a unit surface area at distance $x$ may be denoted

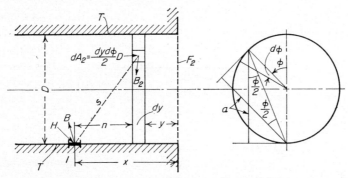

FIG. 14-11. Radiation from a cylindrical hole.

by $H$. The following equation relates the fluxes arriving and leaving such an area:

$$B = \epsilon e_b + (1 - \epsilon)H \tag{14-34}$$

In order to obtain an expression for the flux $H$, the radiative flux $dH$ starting from an area element $dA_2$ and arriving at 1 will be expressed first. The coordinates indicating the position of $dA_2$ are $y$ and $\varphi$. It follows that

$$dH = B_2 \, dF_{1-2} = \epsilon_a(y) e_b \frac{\cos \beta_1 \cos \beta_2}{\pi s^2} dA_2 = \epsilon_a(y) e_b \frac{a^2 \, dy \, D \, d\varphi}{2\pi s^4}$$

The following relations are evident from Fig. 14-11:

$$a = D \cos^2\left(\frac{\varphi}{2}\right) \qquad s^2 = (x-y)^2 + D^2 \cos^2\left(\frac{\varphi}{2}\right)$$

After introduction of these, the above equation changes to

$$dH = \epsilon_a(y) e_b \frac{D^3 \cos^4(\varphi/2)}{\pi[(x-y)^2 + D^2 \cos^2(\varphi/2)]^2} d\left(\frac{\varphi}{2}\right) dy$$

The total flux arriving at location 1 is obtained by integration over the periphery and over the length of the hole:

$$H = \frac{2D^3}{\pi} e_b \int_{y=0}^{\infty} \epsilon_a(y) \int_{\varphi/2=0}^{\pi/2} \frac{\cos^4(\varphi/2)}{[(x-y)^2 + D^2 \cos^2(\varphi/2)]^2} d\left(\frac{\varphi}{2}\right) dy$$

The integration over the angle $\varphi$ can be carried out and with the following substitutions

$$\frac{x}{D} = \xi \qquad \frac{y}{D} = \eta$$

results in

$$H = e_b \int_0^\infty \epsilon_a(\eta) \left\{ 1 - |\xi - \eta| \frac{2(\xi - \eta)^2 + 3}{2[(\xi - \eta)^2 + 1]^{3/2}} \right\} d\eta \qquad (14\text{-}35)$$

The integration has to be carried out in two parts: from $y = 0$ to $y = x$ and from $y = x$ to $y = \text{infinity}$. $|\xi - \eta|$ indicates that this expression

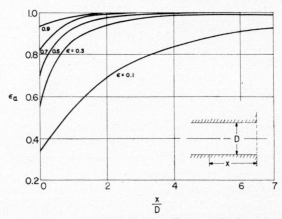

FIG. 14-12. Apparent emissivity of walls in a cylindrical hole.

must always be introduced as positive. Combination of Eqs. (14-34) and (14-35) gives the following equation:

$$\epsilon_a(\xi) = \epsilon + (1 - \epsilon) \int_0^\infty \epsilon_a(\eta) \left\{ 1 - |\xi - \eta| \frac{2(\xi - \eta)^2 + 3}{2[(\xi - \eta)^2 + 1]^{3/2}} \right\} d\eta \qquad (14\text{-}36)$$

The unknown parameter $\epsilon_a$ appears in this equation not only on the left-hand side, where it has to be taken at the location $\xi$, but also on the right-hand side in the integral where it has to be introduced at location $\eta$. Equation (14-36) is, therefore, an integral equation and in particular a homogeneous linear integral equation. An approximate solution of this equation was obtained by E. Eckert.[1] It is presented in Fig. 14-12. In the figure, it can be observed how the apparent emissivity in the hole increases with increasing distance $x$. The figure permits one to determine what distance from the entrance of the hole is required to have the apparent emissivity approach, to any desired degree, the value 1 which holds for a black body.

[1] E. Eckert, "Technische Strahlungsaustauschrechnungen," VDI-Verlag, Berlin, 1937.

**14-4. Radiation of Flames.** From the emissivity charts for carbon dioxide and water vapor, the radiation of combustion gases after completion of combustion can be calculated, e.g., the heat transfer by radiation within the tube bundles (convective surface) of water-tube boilers. The radiation of flames is usually considerably greater than the radiation as computed from the carbon dioxide and water molecules present.

Most fuels burn with a luminous flame. The yellow glow of a flame comes from the hydrocarbons which are contained in the fuel and are gradually broken down in the flame. Thereby are formed molecules which have an increasing percentage of carbon, and carbon particles occur which glow in the flame and give it the yellowish color. These particles also emit considerable thermal radiation. All bituminous fuels, e.g., wood, lignite, and the younger coals burn in this way with luminous flames. Only anthracite, coke, and some gases (generator gas, hydrogen, blast-furnace gas) have nonluminous flames. The faint bluish shine which these flames emit, and which is called *chemoluminosity*, arises from the chemical reactions within the gaseous components. It is not connected, however, with any thermal radiation worth mentioning as has been proved by several investigations. The emission of a luminous flame depends on the number of carbon particles, and this number varies greatly with the conditions under which the combustion occurs. It is influenced, for instance, by the mixing of air and the combustible gases and by the temperatures of both components. As long as these conditions cannot be predicted in a furnace, there is no possibility of calculating the radiation of a flame exactly.

Figure 14-13 shows[1] an example of the results of absorptivity measurements on a benzene-air flame of approximately 3200 F. The flame filled a tubular combustor of 2-in. diameter. The measurement was made along a beam in the direction of the tube diameter 5 in. from the fuel nozzle. The fuel-air ratio was approximately 0.01. Three pressure levels were maintained: 40, 150, and 450 in. Hg abs. It can be observed in the figure that with increasing pressure the flame becomes gradually luminous, and the luminous radiation fills the empty valleys more and more between the absorption bands caused by gaseous radiation. The wavelength range shown in the figure is the important one for the specific flame temperature. Black-body radiation at this temperature has its maximum at $2$-$\mu$ wavelength and drops to below 10 per cent of the maximum at 1 and 5 $\mu$.

For a first approximation, the flame is often considered as a black body, and the fact that this is not entirely true is corrected for by an empirical

---

[1] From E. C. Miller, A. E. Blake, R. M. Schirmer, G. D. Kittredge, and E. H. Fromm, "Radiation from Laboratory Scale Jet Combustor Flames," Phillips Petroleum Company Research Report 1526-56R on U.S. Navy Contract 52-132-C, 1956.

factor $p$ smaller than 1. This gives for the net heat flow $Q$ per unit time to the furnace wall the equation

$$Q = p\epsilon_w \sigma A (T_f^4 - T_w^4) \qquad (14\text{-}37)$$

assuming that the flame fills all the interior of the furnace, where $\epsilon_w$ is the emissivity (equal to the absorptivity) of the furnace walls, $T_f$ the absolute temperature of the flame, $T_w$ the absolute wall temperature, and $A$ the area of the wall.

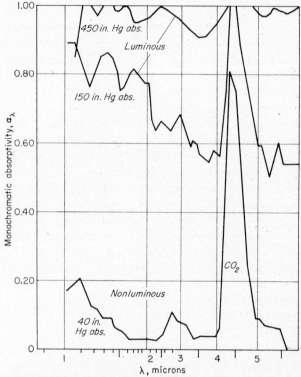

Fig. 14-13. Monochromatic absorptivity of luminous and nonluminous benzene-air flames. (*According to E. C. Miller, A. E. Blake, R. M. Schirmer, G. D. Kittredge, and E. H. Fromm.*)

The empirical factor $p$ depends mainly on the fuel and on the size of the furnace. For large furnaces $p$ is nearly unity. In water-cooled furnaces of steam boilers, the cooling surface receives, in addition to the radiation from the flame, radiation from the refractory-lined walls and from the fuel layer on the stokers and grates. The factor $p$ therefore is greater in furnaces in which only a small part is water-cooled than in furnaces whose walls are entirely lined with cooling surfaces. In furnaces of usual size,

$p$ lies between 0.6 and 1. If the net radiative heat flow to a cooling surface consisting of a row of tubes before a refractory wall is to be calculated with Eq. (14-37), the question arises as to which area is to be used for $A$. Investigations of this situation were made by H. C. Hottel[1] and E. Eckert.[2] According to them, the area $A$ is determined by multiplying the area of the furnace walls before which the tubes are arranged by a

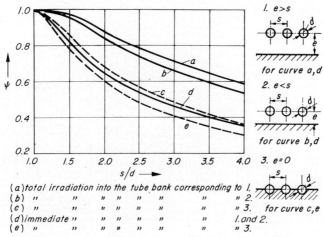

Fig. 14-14. Radiation factor $\psi$ for a row of tubes before a refractory wall. [From E. Eckert, Arch. Wärmewirtsch., **13** (1932).]

factor $\psi$. This factor is represented in Fig. 14-14. The $\psi$ values were calculated with the methods discussed in Secs. 14-1 and 14-3.

**Example 14-2.** The heat absorption by the cooling surface in the combustion chamber of a steam boiler is to be calculated. Figure 14-15 shows this cooling surface to consist of rows of steel tubes positioned in front of the refractory walls. The steam pressure of the boiler is assumed as 1,000 lb/in.[2] To this pressure there corresponds a saturation temperature 544.6 F. The external temperature of the tube walls is somewhat higher; it is assumed to be 600 F. The flame temperature is 2500 F. The spacings between centers of the tubes along the cooling surface are taken to be $2d$ ($d$ is the tube diameter).

First of all the surface area $A$ which is to be introduced into Eq. (14-37) must be calculated. From Fig. 14-14 on the line $a$ is found $\psi = 0.88$ (the distance of the tubes from the wall can be taken as $>s$). The area of the wall which is lined with cooling tubes is assumed to be $5.25 \times 14.75 + 5.25 \times 17.4 = 168.8$ ft². Then surface $A$ is $168.8 \times 0.88 = 148.5$ ft². The factor $p$ in Eq. (14-37) may be 0.87. The effective emissivity $\epsilon_w$ of the tube-wall arrangement is estimated from Table A-10 as $\epsilon_w = 0.8$, since the tube surface is covered with the scale in the rolling process and often with

---

[1] H. C. Hottel, *Mech. Eng.*, **52**:699 (1930); *Trans. ASME*, **53**:265 (1931).

[2] E. Eckert, *Hauptverein deut. Ingr. Mitt.*, 1931, pp 483–486; *Arch. Wärmewirtsch.*, **13**:241 (1932).

slag. Now Eq. (14-37) gives

$$Q = 0.87 \times 0.8 \times 0.173 \times 10^{-8} \times 148.5[(2500 + 460)^4 - (600 + 460)^4]$$
$$= 13{,}500{,}000 \text{ Btu/hr}$$

The heat of vaporization at 1,000 lb/in.$^2$ is 649.4 Btu/lb. Therefore the cooling surface generates $13{,}500{,}000/649.4 = 20{,}800$ lb/hr of steam. The surface area of the tubes is greater than the area of the wall which is lined with the tubes by the factor

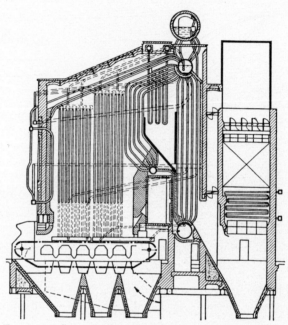

Fig. 14-15. Steam boiler with radiation heating surface.

$\pi d/s$. The tube surface therefore is $(\pi/2)168.8 = 265$ ft$^2$, and the steam generation per square foot of the tube surface is 78.5 lb/hr ft$^2$. The heat absorption per square foot of tube surface is $13{,}500{,}000/265 = 50{,}900$ Btu/hr ft$^2$.

**14-5. The Heat-transfer Coefficient for Radiation.** Often heat transfer occurs simultaneously by convection and radiation. As long as the radiation takes place between solid surfaces, neither mode of heat transfer interferes with the other. In heat transfer between a gas and a wall exchanging heat by radiation and convection, some interference between these two ways of heat exchange occurs, but this effect is very tedious to calculate. Therefore both parts of the total heat flow are usually calculated separately and summed. Since the heat flow by convection is expressed by a film heat-transfer coefficient, it is useful in some cases to define a heat-transfer coefficient $h_r$ for the radiated heat in the same way, i.e., by dividing the heat flow per unit area $q$ by the temperature difference:

$$h_r = \frac{q}{t_1 - t_2} = \epsilon_{1-2}\sigma F_{1-2} \frac{T_1^4 - T_2^4}{T_1 - T_2} = \epsilon_{1-2}\sigma F_{1-2} f(T_1, T_2) \quad (14\text{-}38)$$

Contrary to the character of the convection heat-transfer coefficient, $h_r$ is seen to depend very much on the temperatures $T_1$ and $T_2$.

The temperature function

$$f(T_1, T_2) = \frac{T_1^4 - T_2^4}{T_1 - T_2} = T_1^3 + T_1^2 T_2 + T_1 T_2^2 + T_2^3$$

is tabulated in Table 14-2. When $T_1$ is not too different from $T_2$, $f$ can well be approximated by $4\bar{T}^3$, where $\bar{T} = (T_1 + T_2)/2$. This function must be multiplied by the interchange factor $\epsilon_{1-2}$, by the Stefan-Boltzmann constant $\sigma$, and by the shape factor $F_{1-2}$ in order to get the radiation heat-transfer coefficient. This coefficient increases greatly with temperature.

**Example 14-3.** The combustion-chamber walls of modern steam boilers are lined with cooling tubes, which absorb heat radiated from the flame and decrease the flame temperature. Through these cooling tubes circulates the boiling water contained in the boiler. The interesting question arises as to how the total heating surface must be divided into this cooling surface and the convective surface in order to get the steam generation per unit heating surface as high as possible. To answer this question it is necessary to calculate the heat-transfer coefficients for the radiation cooling surfaces and the convective surface. As an example, the tubes of the boiler are of 2-in. diameter. For the convective surface these tubes are arranged on the corners of triangles with equal sides 4 in. long (Fig. 14-16). Let the velocity of the combustion gases at the entrance of the convective surface be 15 fps, and consider the gases to contain 14 volume per cent carbon dioxide and 7 volume per cent water. The surface temperature of the tube walls is taken as 600 F.

The heat absorption of one tube of the radiation surface is given by Eq. (14-37). The area $A$ for 1 ft of tube length is $\psi s$ ($s$ is the distance between tube centers). The heat-transfer coefficient by radiation is found by dividing the heat flow $Q$ by the tube surface area and by the difference between the flame temperature $T_f$ and the surface temperature $T_w$ of the tube:

$$h_r = \frac{p\epsilon_w \psi s \sigma}{\pi d} \frac{T_f^4 - T_w^4}{T_f - T_w}$$

The combustion chamber may be large; therefore $p = 1$. The effective emissivity of the tube-wall arrangement is assumed to be $\epsilon_w = 0.94$. Otherwise, the same numerical values are used as in the previous example:

$$h_r = \frac{0.94 \times 0.88 \times 2 \times 0.173 \times 10^{-8}}{\pi} \frac{(2{,}500 + 460)^4 - (600 + 460)^4}{2{,}500 - 600}$$
$$= 36.2 \text{ Btu/hr ft}^2 \text{ F}$$

Introducing other values for the ratio $s/d$ of tube distance to tube diameter and for the flame temperature $T_f$ determined the heat-transfer coefficients $h_r$ in Fig. 14-16.

Now the heat-transfer coefficient $h_b$ for the tubes within the convective surface must be calculated. This coefficient comprises heat transfer by gas radiation and by

Table 14-2. Temperature Factor $(T_1^4 - T_2^4)/(T_1 - T_2) \times 10^{-8}$ for the Radiation Heat-transfer Coefficient

| $t_2$ F \ $t_1$ F | −460 | −200 | 0 | 200 | 400 | 600 | 800 | 1000 | 1200 | 1400 | 1600 | 1800 | 2000 |
|---|---|---|---|---|---|---|---|---|---|---|---|---|---|
| −460 | 0 | | | | | | | | | | | | |
| −200 | 0.176 | 0.703 | | | | | | | | | | | |
| 0 | 0.973 | 2.011 | 3.894 | | | | | | | | | | |
| 200 | 2.875 | 4.628 | 7.246 | 11.50 | | | | | | | | | |
| 400 | 6.361 | 9.041 | 12.56 | 17.87 | 25.44 | | | | | | | | |
| 600 | 11.91 | 15.72 | 20.29 | 26.81 | 35.75 | 47.64 | | | | | | | |
| 800 | 20.00 | 25.16 | 30.95 | 38.85 | 49.35 | 62.95 | 80.00 | | | | | | |
| 1000 | 31.12 | 37.83 | 44.99 | 54.43 | 66.61 | 82.05 | 101.2 | 124.5 | | | | | |
| 1200 | 45.74 | 54.20 | 62.90 | 74.03 | 88.08 | 105.5 | 126.8 | 152.5 | 183.0 | | | | |
| 1400 | 64.35 | 74.78 | 85.18 | 98.17 | 114.2 | 133.9 | 157.5 | 185.7 | 218.9 | 257.4 | | | |
| 1600 | 87.42 | 100.0 | 112.3 | 127.3 | 145.5 | 167.5 | 193.6 | 224.4 | 260.4 | 302.0 | 349.7 | | |
| 1800 | 115.4 | 130.4 | 144.7 | 161.9 | 182.5 | 206.9 | 235.7 | 269.3 | 308.3 | 353.0 | 404.0 | 461.6 | |
| 2000 | 148.9 | 166.4 | 182.9 | 202.4 | 225.5 | 252.6 | 284.2 | 320.8 | 362.8 | 410.8 | 465.3 | 526.5 | 595.6 |

convection. In order to calculate the radiation of carbon dioxide and water in the combustion gases, the equivalent length $L_e$ of the gas layer must be known. From Table 13-4 take $L_e = 3(s - d) = (3 \times 2)/12 = 0.5$ ft for the arrangement of tubes used here. The partial pressure of the carbon dioxide is equal to $p = 0.14$ atm for a total pressure of 1 atm; therefore the product $pL_e$ is 0.07 ft atm. or 1.03 (lb$_f$/in.²) ft. The gas temperature may be assumed to be 1800 F. From Fig. 13-18 the emissivity of carbon dioxide is $\epsilon_{CO_2} = 0.06$ at 1800 F and $\epsilon_{CO_2} = 0.055$ at 600 F. The emissivity of the tube wall is $\epsilon_w = 0.8$; therefore, by Eq. (14-23) the interchange factor becomes

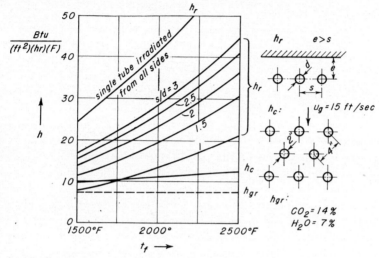

Fig. 14-16. Comparison of radiation heat-transfer coefficients $h_r$ and convection heat-transfer coefficients $h_c$ as functions of the gas temperature.

$0.8 \times 0.06 = 0.048$ at 1800 F and $0.8 \times 0.055 = 0.044$ at 600 F. Now Eq. (14-22) gives the specific heat flow by the carbon dioxide radiation:

$$q_{CO_2} = 0.173 \times 10^{-8} \times 0.048(1{,}800 + 460)^4 - 0.173 \times 10^{-8} \times 0.044(600 + 460)^4$$
$$= 2070 \text{ Btu/hr ft}^2$$

For water vapor, the partial pressure is $p_p = 0.07$ atm; therefore the product $pL_e = 0.035$ ft atm or 0.51 ft lb$_f$/in.². From Fig. 13-19 we obtain in the small diagram the reduction factor $f = 0.67$ for $p_p = 0.07$ atm and $p_pL_e = 0.035$ ft atm. This gives the emissivity $\epsilon_{H_2O} = 0.0176$ at 1800 F and $\epsilon_{H_2O} = 0.0481$ at 600 F. The interchange factors are $0.8 \times 0.0176 = 0.0140$ and $0.8 \times 0.0481 = 0.0385$. The specific heat flow by water radiation is

$$q_{H_2O} = 0.173 \times 10^{-8} \times 0.0140(1{,}800 + 460)^4 - 0.173 \times 10^{-8} \times 0.0385(600 + 460)^4$$
$$= 548 \text{ Btu/hr ft}^2$$

and therefore the total heat flow by gas radiation is

$$548 + 2{,}070 = 2618 \text{ Btu/hr ft}^2$$

and the heat-transfer coefficient by gas radiation is

$$h_{r,g} = \frac{2618}{1{,}800 - 600} = 2.18 \text{ Btu/hr ft}^2 \text{ F}$$

The heat-transfer coefficient by convection can be calculated with Eq. (9-28) and the values in Table 9-3. It is found to be

$$h_c = 8.14 \text{ Btu/hr ft}^2 \text{ F}$$

Therefore, the total heat-transfer coefficient for the convective surface is $h_b = 2.18 + 8.14 = 10.32$ Btu/hr ft² F for a gas temperature of 1800 F. For other gas temperatures and tube arrangements the values are found in Fig. 14-16.

A comparison between the values $h_r$ for the radiation cooling surface and the values $h_b$ for the convective surface in Fig. 14-16 shows that only for very small ratios of $s/d$ (defined on page 244) and flame temperatures under 1800 F does the convective surface have greater heat-transfer coefficients than the radiation cooling surface. It is therefore advantageous to line the whole combustion chamber with narrowly spaced cooling tubes as long as other considerations (difficulties in ignition, etc.) do not prevent this.

**14-6. Radiation Error in Temperature Measurements.** Temperature measurements in a solid or fluid by an immersed probe (thermometer, thermocouple, etc.) are subject to systematic errors, and an estimate of the magnitude of these errors should be made in each installation. The procedure for such a calculation which determines the conduction error was developed in Sec. 3-4. The heat capacity of the thermometer causes an error when the temperature changes with time (Sec. 4-1). When the temperatures are measured in a gas, an additional error may arise owing to the fact that the thermometer is in radiative heat exchange with surrounding solid surfaces which may have a temperature different from the gas. The radiation error will be calculated in this section.

Assume a thermometer positioned in a nonradiating gas of temperature $t_g$ in an enclosure, the walls of which have a temperature $t_w$. A balance between the heat transferred per unit surface area from the gas to the thermometer by convection and from the thermometer to the walls by radiation gives

$$h_c(t_g - t_t) = h_r(t_t - t_w)$$

when $t_t$ = thermometer temperature
$h_c$ = convective heat-transfer coefficient
$h_r$ = radiative heat-transfer coefficient

The second coefficient is obtained from Eq. (14-38) which simplifies to $h_r = \epsilon_t \sigma f(T_t, T_w)$ with $\epsilon_t$ indicating the emissivity of the thermometer surface, since the area of the thermometer is small compared with the area of the walls and the angle factor of the thermometer against the wall is 1. The relative error $E$ is defined as

$$E = \frac{t_g - t_t}{t_g - t_w} = \frac{1}{1 + h_c/h_r} \qquad (14\text{-}39)$$

If, for instance, the air temperature is measured by a mercury thermometer in a heated room with walls which are cooler than the air, then heat exchange between air and thermometer is by free convection. Convective heat-transfer coefficients are then of the order of 2 Btu/ft² hr F. Radiative heat-transfer coefficients at temperatures near atmospheric temperature are found to be of order 0.5 Btu/ft² hr F. The relative radiation error then becomes approximately 20 per cent! This means that no accurate temperature measurement is possible with an ordinary thermometer under these circumstances.

The radiation error can be substantially reduced by a radiation shield. Figure 14-17 is a sketch of a thermometer arranged to measure the gas temperature $t_g$ in an enclosure with walls of temperature $t_w$. The shield $s$ must be designed so that it protects the temperature-sensing element of the thermometer from the radiation coming from the walls but does not inhibit convective heat transfer at the thermometer surface. For instance, a short, hollow cylinder open on both ends as indicated in Fig. 14-17 is a desirable configuration. For a calculation of the radiation error, values referring to the thermometer have the index $t$ and values pertaining to the shield the index $s$. $A_s$ is the surface area of one side of the shield and $A_t$ the thermometer surface area. Heat balances can be written for the thermometer and the shield.

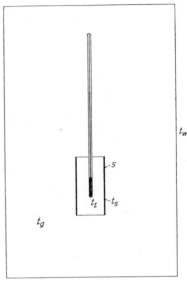

Fig. 14-17. Calculation of the radiation error of a thermometer with a radiation shield.

$$h_{r,t}(t_t - t_s) = h_{c,t}(t_g - t_t)$$
$$2A_s h_{c,s}(t_g - t_s) + A_t h_{r,t}(t_t - t_s) = A_s h_{r,s}(t_s - t_w)$$

Elimination of $t_s$ and calculation of the relative error results in

$$E = \frac{t_g - t_t}{t_g - t_w} = \frac{h_{r,t}}{h_{r,t} + h_{c,t} + 2\dfrac{h_{r,t} + h_{c,s}}{h_{r,s}} + 2\dfrac{h_{c,s}h_{c,t}}{h_{r,s}} + \dfrac{A_t}{A_s}\dfrac{h_{r,t} + h_{c,t}}{h_{r,s}}}$$

The last term in the denominator can usually be neglected, since $A_t/A_s \ll 1$. The relative error simplifies then to

$$E = \frac{t_g - t_t}{t_g - t_w} = \frac{1}{[1 + (h_{c,t}/h_{r,t})][1 + 2(h_{c,s}/h_{r,s})]} \qquad (14\text{-}40)$$

The symbols are the same as for the unshielded thermometer. Equation

(14-40) shows that all radiation terms must be kept small in order to reduce the error. This can be done by using a material of low emissivity for the outside of the shield. The inside of the shield should be blackened in order to avoid reflection of wall radiation onto the thermometer. If possible, the surface of the thermometer should be given a coating of low emissivity.

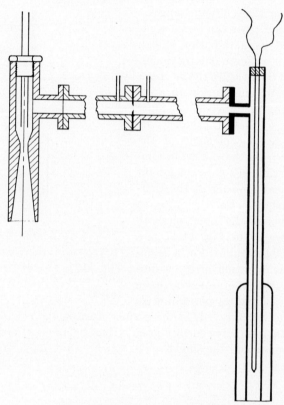

Fig. 14-18. Suction thermometer.

Under extreme conditions (for instance, when measuring gas temperatures in furnaces) shielding is not sufficient to reduce radiation errors, and one must then try to increase the convective heat transfer by decreasing the diameter of the thermometer (very thin wire thermocouples) or by artificially increasing the gas velocity past the thermometer. This can be done by sucking the gas through the radiation shield. Such a thermometer is called a suction thermometer. It is shown in Fig. 14-18.

A special situation arises when the total temperature is to be measured in a high-velocity, high-temperature gas stream. The diffusor thermometer (Fig. 10-6) is the most common instrument for this purpose. Three errors—conduction, incomplete temperature recovery, and radiation—have to be considered in steady state, and in addition, heat-capacity errors arise for unsteady temperatures. By multiple shielding and careful optimization of conditions with respect to all errors satisfactory instruments can be designed.[1]

**14-7. Pyrometry.** All thermometers have an upper limit to which they can be used. A platinum–platinum-rhodium thermocouple, for instance, can measure temperatures up to 2700 F. There are other metal combinations known which can be used up to 4500 F. However, they are difficult to handle. Beyond these limits temperature is measured by the intensity of emitted radiation. This method is called pyrometry. With this technique, the international temperature scale is extended beyond the temperature of melting gold (2405 R) by measurement of the radiation intensity of a black surface. Pyrometers have an additional advantage, namely, that the measurement can be made from a distance. They are, therefore, used even for lower temperatures and, when properly designed and applied, are very accurate instruments.

In the following paragraphs the most commonly used methods for measuring the temperatures of the surfaces of solids or liquids will be discussed. Two classes of instruments are in existence. One class measures the total (integrated) radiation emitted from a surface calorimetrically. The second class compares by observation with the eye the monochromatic radiation intensity of the object in the visible-wavelength range with a known monochromatic radiation intensity.

Each pyrometer must be calibrated by viewing with it objects with known radiation intensity. These are, for instance, created by immersing black bodies (ceramic pieces with cavities) in baths of molten metal with known temperature. In its use the instrument has one basic disadvantage, which is connected with the fact that the radiation intensity of a solid or liquid surface depends not only on its temperature but also on its emissivity. Either the emissivity of the surface has to be known or it has to be increased artificially to the value 1 when the true temperature is to be determined from a pyrometer reading. Otherwise, an apparent temperature is read which does not agree with the true temperature. A number of differently defined apparent temperatures are used in pyrometry. The *total radiation temperature* $T_t$ is the temperature of a black body when it emits the same total radiation intensity as the surface toward which the pyrometer is directed. From this definition follows the relation between $T_t$ and the true surface temperature $T$:

$$\sigma T_t^4 = \epsilon \sigma T^4 \qquad \text{or} \qquad T = T_t / \sqrt[4]{\epsilon} \qquad (14\text{-}41)$$

---

[1] F. D. Werner and R. E. Keppel, An Improved Multiple Shield Gas Temperature Probe, "Proceedings of the Third Midwestern Conference on Fluid Mechanics," pp. 463–478, University of Minnesota Book Store, Minneapolis, 1953.

with $\epsilon$ indicating the emissivity of the surface. For a metal with an emissivity $\epsilon = 0.05$, the apparent total radiation temperature is only approximately half the true temperature.

The *apparent monochromatic temperature* ($T_\lambda$) is the temperature of a black body which at a defined wavelength in the visible range (usually at $\lambda_r = 0.665\ \mu$) has the same radiation intensity as the surface under consideration. According to this definition

$$\frac{\lambda_r^{-5}}{e^{C_2/\lambda_r T_\lambda} - 1} = \epsilon_{\lambda_r} \frac{\lambda_r^{-5}}{e^{C_2/\lambda_r T} - 1}$$

At temperatures occurring in engineering applications, the first term in the denominator of the above equation is large compared with 1, and the 1 can be neglected without causing an error which would exceed the obtainable accuracy. This equation then can be transformed to the following relation:

$$\frac{1}{T} = \frac{1}{T_\lambda} + \frac{\lambda_r}{C_2} \ln \epsilon_{\lambda_r} \qquad (14\text{-}42)$$

$\epsilon_\lambda$ denotes the monochromatic emissivity of the surface. The equation shows that the apparent monochromatic temperature is always lower than the true temperature. The same is true for the total radiation temperature. For metals which have especially low emissivities $\epsilon_{\lambda_r}$ is usually higher than $\epsilon$. The monochromatic temperature then is closer to the true temperature than the total radiation temperature.

The *color temperature* $T_c$ is defined as the temperature at which a black body has the same ratio of radiation intensities at two defined wavelengths in the visible range as the surface in question. Usually the wavelengths of red light ($\lambda_r = 0.665\ \mu$) and of green light ($\lambda_g = 0.544\ \mu$) are used. Broadly speaking, the color temperature is that temperature at which a radiating black body appears to the eye to have the same color as the surface under consideration. From the definition it follows that

$$\frac{1}{T} = \frac{1}{T_c} + \frac{\lambda_r \lambda_g}{(\lambda_r - \lambda_g) C_2} \ln \left( \frac{\epsilon_{\lambda_g}}{\epsilon_{\lambda_r}} \right) \qquad (14\text{-}43)$$

The color temperature for a gray surface for which the emissivity does not vary with wavelength ($\epsilon_{\lambda_g} = \epsilon_{\lambda_r}$) is therefore equal to the true temperature. For most surfaces the color temperature comes closer to the true temperature than the other apparent temperatures.

Figures 14-19 and 14-20 show the two most commonly used pyrometers. The total radiation pyrometer (Fig. 14-19) concentrates by a front gold-plated mirror $a$ the radiation coming from a surface $s$ on a thermopile $b$ and measures the electric potential generated in the thermopile by a millivoltmeter $c$. Diaphragms $d$ prevent stray radiation from impinging on the thermopile. This pyrometer, therefore, measures the apparent total radiation temperature. The optical pyrometer with vanishing filament (Fig. 14-20) is essentially a telescope. Its objective $a$ when directed on a radiating object $b$ creates an image $c$ in the plane where

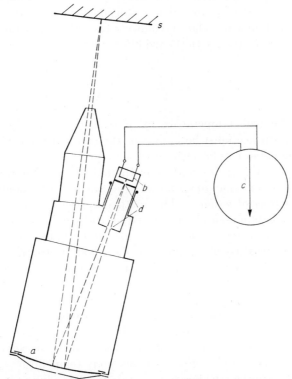

FIG. 14-19. Pyrometer for measuring total radiation. (*According to E. Schmidt.*)

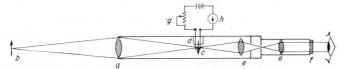

FIG. 14-20. Optical pyrometer.

the filament $d$ of a small lamp is located. Through the ocular $e$ and a color filter $f$, the eye observes the image $c$ of the object and the lamp filament simultaneously. The temperature of the filament is changed by adjustment of the resistance $g$ in the circuit which heats the lamp. It is adjusted until the radiation intensity of the filament is identical with the apparent intensity of the image. The heating current for the filament read on instrument $h$ is, when calibrated, an indication of the radiation intensity of the object. By a smoked glass the apparent intensity can be adjusted to that value for which the human eye is most sensitive. Usually the color filter is a red filter ($\lambda_r = 0.665\ \mu$), and the instrument reads the apparent monochromatic temperature. The color temperature can be measured when the pyrometer is additionally equipped with a green filter $\lambda_g$. The following relation connects the color temperature $T_c$ with the temperature $T_{\lambda_r}$ read with the red filter and the value $T_{\lambda_g}$ read with the green filter

$$\frac{1}{T_c} = \frac{(1/\lambda_r T_{\lambda_r}) - (1/\lambda_g T_{\lambda_g})}{(1/\lambda_r) - (1/\lambda_g)} \qquad (14\text{-}44)$$

To obtain the true temperature by pyrometry, either the emissivity values of the surface must be known or the emissivity must be made to approach the value 1. This can often be done by drilling a hole into the surface and by viewing the bottom of the hole with the pyrometer.

Temperatures in flames or radiating gases are most commonly measured with the *line reversal method*. This method is based on Kirchhoff's law. The gas is made luminous at the desired location by introducing traces of sodium. When the gas is viewed by a spectroscope, the sodium lines can be observed as bright-yellow lines. A black body is arranged behind the gas or flame. It produces a continuous spectrum in the spectroscope. The sodium lines appear as bright lines on this background when the gas temperature is higher than that of the black body and as dark lines when the black-body temperature is higher. The lines disappear when gas and black body have the same temperature, because in this case the gas absorbs just as much of the background radiation as it radiates (Kirchhoff's law). A measurement of the black-body temperature then reveals the gas temperature at the location containing the sodium traces which is viewed by the pyrometer.

**14-8. Solar Radiation.** A large energy source is available to us in the form of radiation coming from the sun, and considerable efforts are being made to find ways by which this energy can be utilized effectively. For instance, the possibility of heating homes by this source is being studied at present at various research centers. This section is devoted to a brief discussion of solar radiation. The sun radiates very nearly like a "black" circular disk with a temperature of 10,000 R. Rays drawn from a point on the earth toward two opposite points on the circumference of the sun include an angle of 32 min or 0.00931 radian. As a consequence of the high temperature, the maximum radiation intensity is found at 0.5-$\mu$ wavelength and approximately one-half of the radiation occurs in the visible-wavelength range, the rest in the infrared up to approximately 3 $\mu$. Part of the solar radiation which is directed toward the earth is absorbed, reflected, or refracted away by the atmosphere, and the rest reaches the surface of the earth. In a yearly average the earth absorbs approximately 43 per cent of the radiation coming from the sun (27 per cent directly and 16 per cent as diffuse sky radiation), 42 per cent is reflected or refracted back to space from clouds and the air and reflected from the earth's surface, and 15 per cent is absorbed in the atmosphere.

The amount of solar radiation which impinges on a unit area of a surface normal to the sun's radiation and located outside the atmosphere

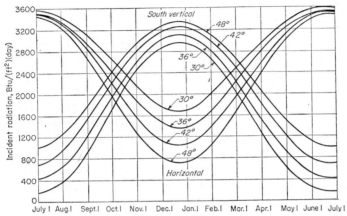

FIG. 14-21. Daily solar irradiation on a horizontal and a south-facing vertical surface at various north latitudes outside the atmosphere. [*According to R. C. Jordan and J. L. Threlkeld, Trans. Am. Soc. Heating Ventilating Engrs.*, **60**:177–238 (1954).]

does not depend on the location on the earth or on the time of the day and is therefore often called the *solar constant*. It changes somewhat during the year, however, because it depends on the distance from the sun. Its amount is 406 Btu/ft² hr in January and 433 Btu/ft² hr in July. Solar absorbers will often have a position which is fixed relative to the earth's surface. Solar radiation impinges on such a surface under an angle which changes during the day and the year. If absorption of radiation in the atmosphere is neglected, then the amount of radiation impinging on a unit area of such a surface can readily be calculated from the angle between the surface normal and the solar rays. Figure 14-21 presents the results of such a calculation for a horizontal surface and for a vertical one facing south as a function of geographic latitude and time of the year.[1] Two facts are interesting to note. During the winter months, when solar energy is needed for heating, a vertical surface in the range of latitudes covered by the United States receives a considerably larger energy flux

[1] R. C. Jordan and J. L. Threlkeld, *Trans. Am. Soc. Heating Ventilating Engrs.*, **60**:177–238 (1954).

than a horizontal one. This energy flux also does not vary appreciably with geographic latitude.

Actually this amount of energy is not available because part of the solar radiation is absorbed in the atmosphere or reflected and refracted back into space. Absorption is caused in clear air mainly by ozone and water vapor. Additional absorption is caused by dust particles. This latter, of course, depends considerably on location. The amount by which solar radiation is reduced by absorption depends also on the length of travel of the sun rays through the atmosphere, or on the angle between the sun rays and the earth's surface (solar altitude). Figure 14-22 gives, according to investigations by P. Moon,[1] the solar radiation incident on a surface normal to the sun rays and located on the earth's surface for representative conditions of the absorbing constituents in the atmosphere on a cloudless winter day.

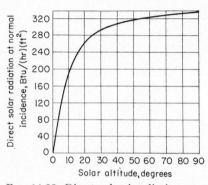

FIG. 14-22. Direct solar irradiation on a surface normal to the sun's rays at the earth's surface on a cloudless winter day. [According to R. C. Jordan and J. L. Threlkeld, Trans. Am. Soc. Heating Ventilating Engrs., **60**:177–238 (1954).]

Part of the sun's radiation reaches the earth as diffuse sky radiation. The amount of this radiation varies widely. On a cloudless day approximately 15 to 30 Btu/ft$^2$ hr of sky radiation impinges on a horizontal surface.

On cloudy days the solar radiation drops to very low values (1 per cent and less of the radiation on clear days).

The amount of solar radiation which a surface absorbs depends on its absorptivity. It has to be kept in mind that solar radiation has short wavelengths and that the absorptivity for such radiation may be considerably different from the absorptivity for long-wave radiation, which can, for instance, be obtained through use of Table A-10. Figure 13-10 shows, for instance, that the white surface 1 has considerably smaller absorptivities than the aluminum surface for solar radiation. Surfaces which are to be protected against the sun's radiation should be given a white, nonmetallic surface coat (e.g., roofs of cars). For solar absorbers one is interested in having a surface which has a high absorptivity for solar radiation but a low emissivity for long-wave radiation in order to minimize heat losses by radiation from the surface. Most surfaces with high absorptivity have the opposite trend (see Fig. 13-9). The National

[1] P. Moon, J. Franklin Inst., **230**:583–617 (1940).

Physical Laboratory in Israel was recently successful[1] in developing a surface with the desired properties by applying a thin black layer of a special material on a metal surface. Short-wave radiation is absorbed in the layer, whereas long-wave radiation is transmitted and reflected from the metal surface.

Solar energy has the advantage that it is readily available; on the other hand, its flux density is too small for many purposes. In such cases, collectors have to be used which concentrate the energy at the target. Parabolic mirrors are generally used for this purpose. In this paragraph, we shall determine what flux density can be obtained by the use of such a mirror, which is shown in Fig. 14-23.[2] Solar radiation approaches it with an angle $\alpha = 32$ min. The bundle of rays approaching the mirror along its axis is reflected in itself. At a distance $f$ from the apex, which is equal to the focal length of the mirror, it creates an image of the sun of diameter $d = f\alpha$ and with an area $\pi f^2 \alpha^2/4$. Onto this area is also collected radiation which is incident on the other parts of the mirror surface. There

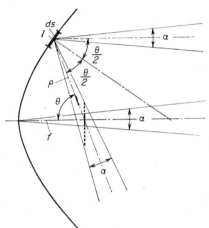

Fig. 14-23. Parabolic mirror for a solar furnace.

are, however, certain losses connected with this process, because not all the incoming rays are concentrated on the sun's image. A bundle incident on point 1 of the surface is, for instance, reflected as a cone with the opening angle $\alpha$, the axis of which goes through the focal point. The plane of the sun's image cuts out of this cone an elliptical area

$$\frac{\alpha^2 \rho^2}{\cos \vartheta} \frac{\pi}{4}$$

The ratio of the area of the sun's image to this area indicates the ratio of the radiation impinging on the sun's image after reflection at point 1 to the radiation incident at point 1.

$$\frac{f^2}{\rho^2} \cos \vartheta \qquad (14\text{-}45)$$

The energy reflected under ideal conditions from a ring-shaped area $dA$ of

[1] *Sci. American*, **194**:102 (1956).
[2] N. K. Hiester, T. E. Tietz, E. Loh, and P. Duwez, *Jet Propulsion*, **27**:507–513 (1957).

the mirror with the length $ds$ in the meridional plane is

$$dq = C\, dA \cos \frac{\vartheta}{2}$$

when $C$ is the solar constant. With the expressions

$$ds = \frac{\rho\, d\vartheta}{\cos(\vartheta/2)} \qquad dA = 2\pi\rho \sin \vartheta\, ds = 2\pi\rho^2 \sin \vartheta\, \frac{d\vartheta}{\cos(\vartheta/2)}$$

which are evident from the figure, the reflected radiation becomes

$$dq = C 2\pi\rho^2 \sin \vartheta\, d\vartheta$$

The amount of this radiation collected at the sun's image is found by multiplying this expression with Eq. (14-45). The total energy collected on the sun's image by the mirror is then found by integration:

$$Q = 2\pi C f^2 \int_0^\vartheta \sin \vartheta \cos \vartheta\, d\vartheta = \pi C f^2 \sin^2 \vartheta$$

Up to this point it has been assumed that none of the energy approaching from the sun is lost in the collection process. In reality, a reduction of this energy will be caused by absorption in the atmosphere and by the fact that the mirror surface will have a reflectivity smaller than 1 and will also suffer losses caused by imperfect geometry. These three factors will be accounted for by two efficiency terms $\eta_a$ and $\eta_r$, respectively. The heat flux collected at the image of the sun is

$$Q = \pi C \eta_a \eta_r f^2 \sin^2 \vartheta \tag{14-46}$$

The flux density—the energy arriving per unit area—is obtained by dividing Eq. (14-46) by the area of the sun's image:

$$q = \frac{4}{\alpha^2} C \eta_a \eta_r \sin^2 \vartheta$$

This equation can be written in a different way which is more readily understandable. The solar constant can be expressed in the following way:

$$C = \frac{\alpha^2}{4} \sigma T_s^4 \tag{14-47}$$

in which $T_s$ indicates the black-body temperature of the sun (10,000 R). The term $\alpha^2/4$ is the shape factor of the sun from the location of the mirror. With this equation the heat flux density can be expressed as

$$q = \sigma T_s^4 \eta_a \eta_r \sin^2 \vartheta$$

When the mirror is made with an opening angle of 60° and when the values $\eta_a = 0.7$ and $\eta_r = 0.8$ are introduced, values which can be met in real installations, then the heat-flux density is 42 per cent of the value $\sigma T_s^4$ or 42 per cent of the heat-flux density from a black surface at the temperature of the sun (680,000 Btu/hr ft²). It can be seen that high flux densities can be obtained by a solar collector.

It is also of interest to calculate to what temperature $T_m$ a target located at the position of the sun's image can be heated by a solar collector under optimum conditions. A target will obviously assume its highest temperature when heat losses from its back surfaces are prevented and when no heat conduction or convection occurs on the front surface which is exposed to the solar radiation. Under these circumstances the heat loss will be by radiation alone. Equating this heat loss with the heat flux arriving from the sun gives the following relation:

$$\epsilon \sigma T_m^4 = \alpha \sigma T_s^4 \eta_a \eta_r \sin^2 \vartheta$$

For a gray surface the absorptivity $\alpha$ is equal to the emissivity $\epsilon$, and the temperature of the object becomes

$$T_m = T_s \sqrt[4]{\eta_a \eta_r} \sin^{1/2} \vartheta \qquad (14\text{-}48)$$

With the same values as before, $\vartheta = 60°$, $\eta_a = 0.7$, $\eta_r = 0.8$, one finds that the maximum temperature of the object is equal to 80 per cent of the temperature of the sun, or 8000 R.

## PROBLEMS

**14-1.** Calculate the shape factor for an area element exchanging radiation with an area having the shape of a circular disk in a position as indicated in Fig. 14-2 except that both areas are parallel and $r > 1$.

**14-2.** Develop equations describing the shape factor between an area element and a circular disk for any mutual angular position. Use for this purpose the procedure as outlined in the text for the example shown in Fig. 14-6.

**14-3.** Show that the factor $\psi$ read from curve $d$ in Fig. 14-14 for a value $s/d = 3$ is correct. $\psi$ on this curve is identical with the shape factor between the furnace wall and the row of tubes.

**14-4.** Determine the local shape factor between an area element of the furnace wall and the tubes in the arrangement belonging to curve $e$ in Fig. 14-14, and prove by averaging that the $\psi$ value read from this figure for $s/d = 3$ is correct.

**14-5.** Determine the shape factor for heat exchange between an area element and a circular disk oriented at an angle of 45° to the element when the disk diameter is equal to its distance from the element. Assume that the circular disk radiates like a black body but that the area element is an aluminum surface with a spatial distribution curve as given in Fig. 13-11. Compare with the shape factor for two black surfaces of equal shape.

**14-6.** Prove that Eq. (14-21) describing the interchange factor between a gas layer and gray walls is correct.

**14-7.** Calculate the heat exchange between the various surfaces surrounding the combustion chamber of the steam boiler in Fig. 14-15. Measure the dimensions of the various cooled and uncooled surfaces in this figure, and assume that the width of the combustion chamber normal to the plane of the drawing is equal to half of the stoker length. Use the information contained in the examples in Secs. 14-4 and 14-5, and assume that the gases and flame in the combustion chamber do not radiate or absorb heat and that the bed of coal on the stoker has a temperature of 3000 F. Consider the surface of the coal bed a black surface, and assume that the emissivity of the refractory lining equals 0.7. Use the method in Sec. 14-3 for the calculation.

**14-8.** Calculate the radiation error made in measuring the temperature of air at 1000 F and atmospheric pressure with the suction thermometer shown in Fig. 14-18. The thermocouple wires (iron-constantan) are 0.02 in. thick. The tube surrounding the thermocouple has a diameter of 0.5 in., and the outer radiation shield is 0.8 in. in diameter. Both are made of stainless steel. The air is sucked with 150-fps velocity through the thermometer.

**14-9.** Show analytically how the line reversal method works by tracing the intensity of a beam starting from the black body and traveling through the flame toward the spectroscope.

PART D

# THE TRANSFER OF MASS

In many engineering activities, especially in chemical engineering, processes are utilized which are classified as *mass-transfer processes*. This term refers to the situation that, in a mixture with local concentration differences, a component is transported from one location to another by diffusion or convection. Engineering-wise, the mass transport through an interface between various media or phases of the same medium is of special importance.

The differential equations describing such a mass-transfer process can be derived in very general form by the procedures developed by the science of irreversible thermodynamics. These equations will not be derived or discussed here, but only simpler forms which describe diffusion and convection in mixtures of ideal gases or of dilute liquid solutions. Sometimes chemical reactions are coupled with the mass-transfer process. Such reactions are also excluded from the treatment in this book. The surface on which the mass transfer is studied is assumed to be an interface between the gas and a solid material or a liquid, or an interface between a liquid and a solid material. Specific processes to which the relations developed in Part D can be applied are, for instance, the evaporation and condensation of one component of a gas mixture on a solid surface.

It will be assumed that the surface on which the mass transfer occurs is known and well defined in its geometry. Quite often this is not the case, for instance, in evaporation of a liquid in droplet form or when a gas moves in the form of bubbles through a liquid. Considerable research is still required before the performance of such processes can be satisfactorily predicted, and they will not be treated here. Also excluded is mass movement through porous materials as it occurs in drying processes.

As an introduction to the discussion of the transfer processes themselves, the basic equations connecting the important physical parameters in a gas mixture in equilibrium will be developed. The necessary relations for describing a gas-vapor mixture will then be discussed, the latter for the example of humid air.

CHAPTER 15

# RELATIONS FOR TWO COMPONENT MIXTURES

**15-1. Basic Equations for Two-component Gas Mixtures.** The equations describing mass transfer are expressed in the literature in a wide variety of parameters, and the task often arises of converting one form into the other. For this reason the parameters and their definitions will be discussed for a mixture of two components which will be identified by subscripts 1 and 2. It is assumed that the mixture is in equilibrium, that is, in a state of constant temperature and pressure, and also that the composition of the mixture is locally uniform. The masses $m_1$ and $m_2$ are contained in the volume $V$. The total mass of the mixture is $m = m_1 + m_2$, and the total density $\rho = m/V$. One can also define partial densities $\rho_1 = m_1/V$, $\rho_2 = m_2/V$. These values are also often referred to as *concentration*. The mixture may also be characterized by the *mass fractions* $w_1 = m_1/m$, $w_2 = m_2/m$. Instead of the mass, the mole can be used as a unit. If $M_1$ and $M_2$ denote the molecular weights, then the number of moles for the two components are $N_1 = m_1/M_1$, $N_2 = m_2/M_2$. The actual number of molecules is found by multiplying $N$ by Loschmidt's number ($6.03 \times 10^{23}$ molecules per g mole or $2.733 \times 10^{26}$ molecules per lb mole). Again, *total mole densities* $N/V$, ($N = N_1 + N_2$) and *partial mole densities*, or mole concentrations, $N_1/V$, $N_2/V$ can be defined, as well as *mole fractions* $N_1/N$, $N_2/N$.

In a mixture of ideal gases the composition is often expressed by the partial pressures. For a mixture of ideal gases which do not react chemically, Dalton's law applies: *Where different gases are present in the same space, every gas fills the whole available space as if the other gases were not present.* According to the ideal gas law, the pressure which any of the gases creates is its *partial pressure*. The total pressure of the mixture equals the sum of all partial pressures. On a molecular basis this means that no attractive forces are exerted among the molecules. If in a two-component gas mixture the partial pressures are denoted by $p_1$ and $p_2$, the total pressure by $p$, the gas constants by $R_1$ and $R_2$, and the common temperature by $T$, then the following equations express Dalton's law:

$$p_1 V = m_1 R_1 T = \frac{m_1}{M_1} \Re T = N_1 \Re T$$

$$p_2 V = m_2 R_2 T = \frac{m_2}{M_2} \Re T = N_2 \Re T$$

$$p = p_1 + p_2$$

with $\Re$ expressing the universal gas constant and $R_1 = \Re/M_1$, $R_2 = \Re/M_2$. The universal gas constant has the value $\Re = 1{,}545.4$ ft lb$_f$/lb mole $R$. By addition of the above equations the gas law for the mixture is obtained:

$$pV = mRT = N\Re T$$

with the individual gas constant

$$R = (m_1 R_1 + m_2 R_2)/m = w_1 R_1 + w_2 R_2 = w_1 R_1 + (1 - w_1) R_2$$

The various properties defined in the preceding paragraph can be expressed in terms of partial pressures:

$$\rho_1 = \frac{m_1}{V} = \frac{p_1}{R_1 T} \qquad w_1 = \frac{m_1}{m} = \frac{p_1 R}{p R_1} = \frac{R_2 p_1}{R_1 p - p_1 (1 - R_2)}$$

$$\frac{N_1}{V} = \frac{p_1}{RT} \qquad \frac{N_1}{N} = \frac{p_1}{p}$$

The variety in the expressions for mass transfer comes from the fact that the driving potential can be expressed as a gradient of any of the parameters: partial mass or mole density, mass or mole fraction, or partial pressure. Calculations for humid air are often based on still another parameter which will be introduced in the next section.

**15-2. Basic Equations and $s$-$i$ Diagram for Humid Air.** For calculations with humid air it is useful to base all equations on one pound of dry air as the water-vapor content changes. The mass of water vapor per pound of dry air is called *specific humidity* and is denoted by $s$.

From the above equations, it follows that

$$s = \frac{m_v}{m_a} = \frac{M_v}{M_a} \frac{p_v}{p_a} \qquad (15\text{-}1)$$

A subscript $a$ indicates air; a subscript $v$ indicates vapor. The molecular weight of water vapor is 18; the molecular weight of air is 28.96. Therefore

$$s = 0.622 \frac{p_v}{p_a} = 0.622 \frac{p_v}{p - p_v} \qquad (15\text{-}2)$$

At any given temperature, air can hold only a certain maximum content of water vapor. Air with this maximum content is said to be *saturated*. The saturation point is fixed by the condition that the partial pressure of the water vapor cannot be greater than the saturation pressure of water vapor corresponding to the air temperature. The saturation pressure belonging to different temperatures can be found in the $s$-$i$ diagram of Fig. 15-5 (page 448). The specific humidity of saturated air is fixed by the equation

$$s = 0.622 \frac{p_s}{p - p_s} \qquad (15\text{-}3)$$

where $p_s$ denotes the saturation pressure.

The humidity of air is often expressed by the ratio of the actual vapor pressure $p_v$ to the saturation pressure $p_s$ at the air temperature. This ratio is called *relative humidity* and denoted by $r$:

$$r = \frac{p_v}{p_s} \qquad (15\text{-}4)$$

Sometimes the ratio of the specific humidities is used instead of the ratio of the partial pressures. This ratio

$$r_s = \frac{s}{s_s} \qquad (15\text{-}5)$$

can be expressed by the partial pressures of the water vapor. At low temperatures, the partial pressure of the vapor is very small as compared with the total barometric pressure. It can then be written instead of Eq. (15-2)

$$s \approx 0.622 \frac{p_v}{p}$$

and for saturated air

$$s_s \approx 0.622 \frac{p_s}{p}$$

Therefore $$r_s = \frac{s}{s_s} \approx \frac{p_v}{p_s} = r$$

At low temperatures there is therefore no difference between the ratios $r$ and $r_s$.

An amount of water exceeding that at saturation can be held in the air only in liquid form as water droplets (fog) or in solid form as snow. The amount of water in 1 lb of dry air is $s - s_s$.

For the following calculations, the enthalpy of moist air is needed. As long as the specific humidity does not exceed the saturation point, the enthalpy of moist air consists of the enthalpy of the dry air and the enthalpy of the water vapor. Now there arises the difficulty that the temperature 0 F is commonly used as a datum for air; however, the melting point (32 F) of ice is used as a datum point for water vapor. We shall use here the temperature $t_0 = 32$ F as a datum for both air and water vapor. With the specific heat $c_{pa}$ of air at constant pressure, the enthalpy of the dry air is $i_a = c_{pa}(t - t_0)$. The enthalpy $i_v$ of water vapor at low temperatures can be expressed in the form $i_v = c_{pv}(t - t_0) + i_{v,0}$ with $c_{pv}$ indicating the specific heat of water vapor at constant pressure and $i_{v,0}$ the heat of vaporization at 32 F. The enthalpy of moist air is

$$i = i_a + s i_v = c_{pa}(t - t_0) + s[c_{pv}(t - t_0) + i_{v,0}] \quad (15\text{-}6)$$

with the numerical values $c_{pa} = 0.240$ Btu/lb F, $c_{pv} = 0.44$ Btu/lb F, and $i_{v,0} = 1076$ Btu/lb. When the air contains water also in liquid form, the enthalpy of the liquid water $i_f = c_f(t - t_0)$ must be added. ($c_f = 1$ Btu/lb F.) This gives

$$i = c_{pa}(t - t_0) + s_s[c_{pv}(t - t_0) + i_{v,0}] + (s - s_s)c_f(t - t_0) \quad (15\text{-}7)$$

The enthalpy of ice is $i_i = -i_{s,0} + c_i(t - t_0)$ with the heat of fusion $i_{s,0}$ equal to 143 Btu/lb, and the specific heat of ice $c_i$ is equal to 0.5 Btu/lb F. The enthalpy of ice has a negative value, since the enthalpy of liquid water at 32 F is fixed at zero. The enthalpy of air with ice fog is

$$i = c_{pa}(t - t_0) + s_s[c_{pv}(t - t_0) + i_{v,0}] - (s - s_s)[i_{s,0} - c_i(t - t_0)] \quad (15\text{-}8)$$

Equation (15-6) must be applied as long as $s < s_s$. For $s > s_s$ and $t > 32$ F, Eq. (15-7) must be used, and for $s > s_s$ and $t < 32$ F, Eq. (15-8) holds true.

In nonsaturated air, the water vapor is present in a superheated condition, since the vapor pressure is smaller than the saturation pressure corresponding to the air temperature. Under special conditions air may hold an amount of water vapor greater than that belonging to the saturation point. This is possible, for instance, if the air is cooled very rapidly (by adiabatic expansion, e.g., in a nozzle) or if the air does not contain any solid particles (dust, soot) on which the formation of droplets can start. This condition is called *supersaturation*. For engineering calculations this condition is usually of minor importance. We shall assume, therefore, in the following calculations that no supersaturation occurs.

Drying and moistening processes are usually studied with the help of the psychrometric chart of humid air. In this chart the temperature (dry-bulb temperature) is used as abscissa and the specific humidity as ordinate. R. Mollier[1] introduced another diagram which uses the enthalpy instead of the temperature. This diagram has the advantage that processes of mixing two or more air streams or of admixing moisture as water or water vapor to air can be solved exactly by simple geometrical constructions on it, whereas such constructions give only approximate solutions in the psychrometric chart. Moreover, such Mollier diagrams can be generalized for mixtures of two substances other than air and water. In the original Mollier diagram, the specific humidity is the abscissa and

[1] R. Mollier, *Z. Ver. deut. Ingr.*, **67**:869–872 (1923); **73**:1009–1013 (1929).

the enthalpy the ordinate. We shall turn it around so that the enthalpy is the abscissa and the specific humidity the ordinate. In this presentation the diagram is very similar to the psychrometric chart, the only difference being that the lines of constant temperature which are vertical in the psychrometric chart are straight lines, slightly inclined by varying angles, in the Mollier diagram. In the following pages this diagram for humid air will be developed.

If the specific humidity $s$ is used as ordinate and the enthalpy $i$ as the abscissa, Fig. 15-1a arises. In the figure are shown some isotherms, which are composed of straight lines. The isotherm for $t = 32$ F has, for

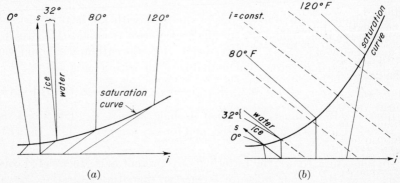

Fig. 15-1. Specific-humidity–enthalpy diagram for moist air. This diagram holds for a particular total pressure. (*According to R. Mollier.*)

example, in the range of unsaturated air the inclination 1076 Btu/lb toward the ordinate up to the saturation limit. For higher humidities $s$, the isotherm is a vertical line described by Eq. (15-7). If excess water is contained in the air as ice, the isotherm above the saturation limit is given by Eq. (15-8). It has therefore a negative inclination of 143 Btu/lb toward the ordinate. The isotherms for positive temperatures are inclined toward the right; the 32 F isotherm for foggy air ($s$ greater than $s_s$) is vertical. The isotherms for negative temperatures are inclined toward the left. By connecting all points which correspond to the saturation limit at different temperatures, the saturation curve in the diagram arises. As the saturation humidity depends on the total pressure according to Eq. (15-3), this curve and therefore the whole diagram hold only for a certain total pressure.

The saturation curve separates the regions of unsaturated and foggy states. At the point of intersection with the 32 F isotherm, the saturation curve has a slight break, since for temperatures above 32 F it is fixed by the equilibrium between water and its vapor and for temperatures below 32 F by the equilibrium between ice and water vapor. In the form shown in Fig. 15-1a, the diagram has the disadvantage that the region of unsaturated air is comparatively small. This can be avoided by using a system of oblique-angled coordinates. Mollier proposed to give the $s$ axis such a direction that the 32 F isotherm is vertical in the unsaturated range. Such a diagram is shown in Fig. 15-1b. The lines $i$ = const in this diagram are a group of parallel oblique lines. Such a diagram on an increased scale is added to the book as Fig. 15-5.

This diagram will now be used to work some examples. The condition of humid air at a certain pressure is given by two state functions, e.g., by the temperature $t_1$ and the specific humidity $s_1$. In Fig. 15-2 this fixes the point of state 1. We can immediately determine the relative humidity $\rho$ by following the isotherm $t_1$ up to the point of intersection with the saturation curve and by reading the saturation humidity $s_{s1}$. The value is $\rho_1 = s_1/s_{s1}$. For not too high temperatures, the relative humidity $r$ has the same value. For higher temperatures it must be determined from the partial pressures. The $s$-$i$ diagram (page 448) shows curves of $r$-const. When air at the state 1 is cooled, its specific humidity does not change as long as the air is not saturated.

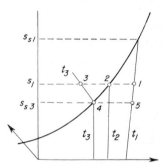

FIG. 15-2. Determination of the state of moist air in the Mollier $s$-$i$ diagram.

This means that in the diagram the state moves on a horizontal line. The relative humidity increases with decreasing temperature, and at point 2 saturation is reached. This point is called the *dew point* and the corresponding temperature, *dew-point temperature*. When the air is cooled to still lower temperatures, part of the water vapor condenses in form of small droplets (fog). At state point 3 the temperature $t_3$ is reached. Of the total specific humidity $s_1$, a part $s_{s3}$ is present as vapor, the rest as liquid. The vapor part $s_{s3}$ is fixed by the intersection of the isotherm through point 3 with the saturation line. The liquid part of the water can be removed from the air, for instance, by waiting until the droplets sink to the ground. After this has occurred, the humidity of the air will have decreased from $s_1$ to $s_{s3}$. The corresponding state point in the diagram is point 4. In this way, therefore, air can be dried. If, for instance, air with the state 1 must be dried to the specific humidity $s_{s3}$, the air must be cooled to the temperature $t_3$. By so doing, the dew point is first reached and afterward point 4 when the liquid water is removed. Now the air can be heated again to the initial temperature so that the final state 5 has the same temperature but a lower specific humidity than the original state 1. This method is applied very often in air conditioning.

As a second example, the state can be determined which arises by mixing two amounts of air with different temperatures and humidities. The mass of air $m_1$ may have a temperature $t_1$ and a specific humidity $s_1$; the mass of air $m_2$ may have a temperature $t_2$ and a specific humidity $s_2$.

By mixing, there results a mass of air $m$ with a temperature $t$ and a specific humidity $s$. The following equations hold true:

$$m_1 + m_2 = m \qquad \text{mass balance for air} \qquad (15\text{-}9)$$
$$m_1 s_1 + m_2 s_2 = ms \qquad \text{mass balance for water} \qquad (15\text{-}10)$$
$$m_1 i_1 + m_2 i_2 = mi \qquad \text{heat balance} \qquad (15\text{-}11)$$

where the values $i$ are the enthalpies of the moist air. They can be read from an $s$-$i$ diagram. Eliminating $m$ with the help of the first equation gives

$$m_1(s - s_1) = m_2(s_2 - s)$$
$$m_1(i - i_1) = m_2(i_2 - i)$$

and by division

$$\frac{s - s_1}{i - i_1} = \frac{s_2 - s}{i_2 - i} \qquad (15\text{-}12)$$

From this equation it follows that on an $s$-$i$ diagram the state point $M$ for the mixture must be situated on a straight line 1-2 connecting the two initial points (Fig. 15-3). The location of point $M$ on the line follows from the equation

$$\frac{m_1}{m_2} = \frac{s_2 - s}{s - s_1} \qquad (15\text{-}13)$$

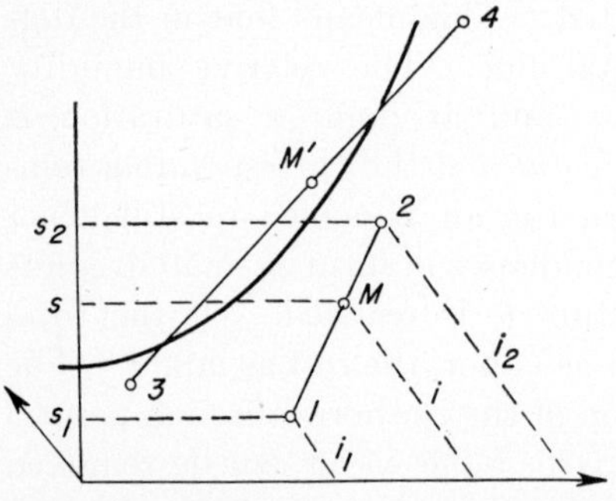

Fig. 15-3. Mixing of air streams in the Mollier $s$-$i$ diagram.

The ratio of the distance $M$-2 in Fig. 15-3 to the distance $M$-1 must therefore be equal to the ratio of the masses $m_1/m_2$. If air with the state 3 is mixed with air of the state 4, the resulting state of the mixture lies on line 3-4 (Fig. 15-3). It can be seen that the mixture may lie in the foggy range. A very common example of this fact is the visibility of the exhaled air in cold weather. The warm, wet air from the lungs mixes with the cold outside air, and mixtures arise whose water content is visible as fog.

Another example is the moistening of air with liquid water or with steam. The initial state of the air can be determined by its mass $m_1$, its temperature $t_1$, and its humidity $s_1$. The mass of water mixed with the air can be expressed by $m_w$ and its enthalpy by $i_w$. Then

$$m_1(s_2 - s_1) = m_w \quad \text{mass balance} \tag{15-14}$$
$$m_1(i_2 - i_1) = m_w i_w \quad \text{heat balance} \tag{15-15}$$

with the subscript 2 indicating the final state of the mixture. By dividing the equations there is obtained

$$\frac{i_2 - i_1}{s_2 - s_1} = i_w \tag{15-16}$$

By this equation, the direction in which the state of the air moves on the $s$-$i$ diagram during moistening is fixed. The ratio of the enthalpy change $(i - i_1)$ to the change $(s - s_1)$ in specific humidity must be equal to the enthalpy of $i_w$ of the water. This gives the angle $\alpha$ in Fig. 15-4 a certain value. In order to find this direction easily in such a Mollier diagram, a scale can be inserted along the edge (such as is done in Fig. 15-5). The rays of this scale originate at the origin ($i = 0$ and $s = 0$) and are denoted by the corresponding enthalpy $i_w$. The scale is indicated at the top of Fig. 15-4. A dashed line connecting the origin with the value $i_w$ on this scale determines the direction in which the state of the air changes. If a parallel line is drawn through point 1 in Fig. 15-4, which gives the state of the air before the humidification, then the state 2 of the air after humidification must lie on this line. The final state $s_2$ on this line is fixed by the amount of water admixed. Equation (15-14) gives

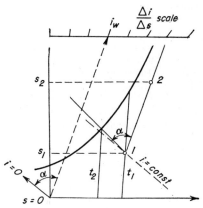

Fig. 15-4. Moistening of air with steam or water in the Mollier $s$-$i$ diagram.

$$s_2 - s_1 = \frac{m_w}{m_1} \tag{15-17}$$

and so fixes the final point of state. The enthalpy of steam has the order of magnitude 1000 Btu/lb. From the large $s$-$i$ diagram it can be seen that the direction of this change of state is approximately vertical. Therefore, if unsaturated air is humidified with steam, its temperature changes only by a small amount. Foggy air is heated when mixed with steam. The enthalpy of liquid water has values which are approximately equal to its temperature. The direction of a change of state produced by injecting water into air has a strong inclination to the left. In the unsaturated range air therefore cools by injection of water. It may seem strange at first that air can be cooled by injection of water when the water is warmer than the air. The reason is that the water evaporates and takes its heat of evaporation from the air. Other examples for the use of the diagram are given in the following paragraphs. For mixtures of two components other than air and water similar diagrams have proved very useful.[1]

## PROBLEMS

**15-1.** Saturated air with atmospheric pressure and 50 F temperature is mixed with air which has a specific humidity $s = 0.04$. What must be the minimum temperature of the second air in order to exclude any local formation of fog in the mixing process? (Use the enthalpy–specific-humidity diagram Fig. 15-5 to obtain the answer.)

**15-2.** The specific humidity $s$ of air with atmospheric pressure and a temperature of 140 F is to be reduced from 0.05 to 0.02. What sequence of cooling and heating processes can you suggest by which this result can be obtained. With what maximum temperature must water as coolant be available to make this process possible? What amount of heat has to be removed and added in this sequence per pound of dry air? (Use Fig. 15-5.)

**15-3.** Air at atmospheric pressure with a temperature of 150 F and a relative humidity $r = 0.1$ is to be saturated by a spray of water with 80 F temperature or saturated steam at atmospheric pressure. What is the temperature of the saturated air in both cases, and what amount of water or steam is necessary per pound of dry air? (Use Fig. 15-5.)

**15-4.** An enthalpy–specific-humidity diagram for fuel vapor-air mixtures is useful for a study of the physical processes occurring in the carburetor of a combustion engine. Design such a diagram for benzene-air mixtures using physical properties contained in various handbooks.

[1] A description of the design and use of such diagrams is found in F. Bosnjaković, "Technische Thermodynamik," vol. 2, Theodor Steinkopf, Verlagskuchhandlung, Dresden, 1937.

CHAPTER 16

MASS TRANSFER

**16-1. Diffusion.** Heat conduction in a gas is caused by the random movement of the molecules which by their mixing tend to equalize existing differences in their energy. By the same movement local differences in concentration of a gas mixture diminish in time even if no macroscopic mixing occurs. This process is called diffusion.

We shall derive the basic laws for the diffusion process from a gas-kinetic consideration similar to the one in Sec. 10-3. For this purpose we consider a mixture of two components enclosed in a vessel and being at uniform temperature and uniform pressure (Fig. 16-1). The total number of molecules per unit volume is $n$ (the number of molecules of the two components is $n_1$ and $n_2$; $n = n_1 + n_2$). Gas kinetics shows that under the condition of uniform temperature and pressure the number $n$ must be locally constant in the vessel. We assume that the numbers $n_1$ and $n_2$ vary in the $y$ direction, $n_1$ being greater in the lower part and $n_2$ in the upper part of the vessel. We fix our attention to the plane 0-0 where the numbers for the two components are $n_{1,0}$ and $n_{2,0}$. Molecules will continuously cross this plane in their random movement. All molecules 1 contained in a small volume element with a unit base area in the plane 0-0 and of height $v_{1\perp} \, d\tau$, with an average velocity $v_{1\perp}$ in the upward $y$ direction, will cross this unit area of plane 0-0 in a time $d\tau$. The number of molecules per unit volume in this volume element is somewhat greater than the number $n_{1,0}$ at the plane 0-0 if this plane coincides with the

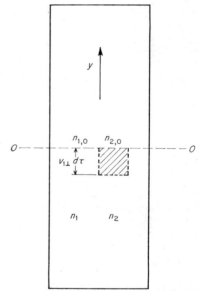

Fig. 16-1. Diffusion in a binary mixture.

upper base. It can be written as $n_{1,0} - j\lambda(dn_1/dy)$, because the molecules come on an average from a plane with a distance of $j\lambda$ from plane 0-0, where $\lambda$ denotes the mean-free-path length and $j$ a number of order of magnitude 1. In addition we can write $v_{1\perp} = i'v_1$, with $v_1$ indicating the mean molecular velocity of species 1 and $i'$ indicating a number of order 1. The number of molecules 1 crossing plane 0-0 per unit area and time in an upward direction is, therefore, $i'v_1[n_{1,0} - j\lambda(dn_1/dy)]$. The number of molecules 1 moving in a downward direction through plane 0-0 is $i'v_1[n_{1,0} + j\lambda(dn_1/dy)]$, and thus the net flow of molecules 1 through plane 0-0 is

$$\dot{n}_1 = -i_1 v_1 \lambda \frac{dn_1}{dy} \tag{16-1}$$

with $i_1 = 2i'j$ being a number of order 1. In the same way the net flow of molecules 2 through plane 0-0 is

$$\dot{n}_2 = -i_2 v_2 \lambda \frac{dn_2}{dy} \tag{16-2}$$

The mean free path will depend on the size of both molecules 1 and 2. Therefore, a diffusion coefficient $D_{12}$ for diffusion of a component 1 into a mixture of 1 and 2 is defined by the equation

$$\dot{n}_1 = -D_{12} \frac{dn_1}{dy} \tag{16-3}$$

and a diffusion coefficient $D_{21}$ of gas 2 into a mixture 1 and 2,

$$\dot{n}_2 = -D_{21} \frac{dn_2}{dy} \tag{16-4}$$

From the equations $n = n_1 + n_2$ and $n = $ const it follows that

$$\frac{dn_1}{dy} = -\frac{dn_2}{dy}$$

In order to preserve $n = $ const during the process, the total number of molecules crossing plane 0-0 must be zero $(\dot{n}_1 + \dot{n}_2 = 0)$. Therefore

$$D_{12} = D_{21}$$

The diffusion coefficient for diffusion of gas 1 is equal to the diffusion coefficient for diffusion of gas 2. In the following discussion we shall denote both simply by $D$. More extensive gas kinetic calculations, which include determination of the numerical factors $i$,[†] show that $D$ depends only slightly on the concentration of the components in the mix-

[†] S. Chapman and T. G. Cowling, "The Mathematical Theory of Non-uniform Gases," Cambridge University Press, London, 1952.

ture, that it is inversely proportional to the pressure, and that with temperature it increases proportional to $T^{1+b}$, where $b$ has a value between 0.5 and 1. Table A-9 in the appendix contains diffusion coefficients for various mixtures. The symbols $n$ and $\dot{n}$ in Eqs. (16-3) and (16-4) can also be interpreted as moles per unit volume or per unit area and time, respectively. For solids and liquids, diffusion coefficients are also defined by Eqs. (16-3) and (16-4). Information on such coefficients and their dependence on the various parameters is scarcer than on gases.

The diffusion equation can also be expressed in terms of concentration gradients. When both sides of Eq. (16-3) are multiplied with the molecular weight $M_1$, there is obtained

$$\dot{m}_1 = -D \frac{d\rho_1}{dy} \tag{16-5}$$

($\dot{m}_1$ is the mass flux per unit time and area; $\rho_1$ is the partial mass density of component 1). This equation is often referred to as *Fick's law*. In partial pressure gradients it reads

$$\dot{m}_1 = -\frac{D}{R_1 T} \frac{dp_1}{dy}$$

The diffusion process illustrated in Fig. 16-1 is called *equimolal counter-diffusion*. Actually there is a slight convective motion inside the container shown in Fig. 16-1. This can be seen in the following way: The net flow of moles $n_1$ through plane 0-0 upward is equal to the net flow of moles $n_2$ downward. The two mass flows, however, will not balance out when the molecular weights $M_1$ and $M_2$ are different. A mass flow occurs in Fig. 16-1 as long as the diffusion goes on. In order to set up equations for the combined action of diffusion and convection in a mass-transfer process, one has to know the diffusive flow of molecules through a plane through which no convective flow occurs (which moves along with the flow). Chapman and Cowling[1] derive the following expression for the number of moles $\dot{n}_1$ passing through a unit area of such a plane per unit time in a two-component mixture for the conditions that the pressure is constant, that no body forces are present, but that the temperature varies locally:

$$\dot{n}_1 = -\frac{M_2}{M^2} \rho D \left[ \frac{d(n_1/n)}{dy} + k_T \frac{1}{T} \frac{dT}{dy} \right] \tag{16-6}$$

In this equation $M$ is the molecular weight of the mixture and $\rho$ the mixture density. $k_T$ is called the *thermal-diffusion ratio*. The equation shows that diffusion occurs in a mixture under the presence of temperature gradients even when there are no concentration differences. This

[1] *Ibid.*

process is called *thermal diffusion*. The values of the thermal-diffusion ratio are such that an appreciable mass flow occurs only with extreme temperature differences. In normal engineering mass-transfer processes, thermal diffusion can be neglected, and this will be done here in the following considerations. The diffusive mass flow can be written in a simple way in terms of the mass ratio. Multiplying Eq. (16-6) by $M_1$ and replacing the mole ratio $n_1/n$ by the mass ratio $w_1$ result in

$$\dot{m}_1 = -\rho D \left( \frac{dw_1}{dy} \right) \tag{16-7}$$

when the thermal diffusion term is neglected. This expression can also be obtained by adding the convective mass flow to the diffusion flow through plane 0-0 in Fig. 16-1.

A combination of diffusive and convective mass transfer, which occurs frequently in engineering processes, will now be discussed in connection with the following simple experiment. A glass tube closed on one end is filled at the bottom with water (Fig. 16-2). Over the open end of the tube air is blown with a certain mass fraction $w_{1s}$ of water vapor. The air velocity will be assumed small. The total pressure within the tube is then constant and equal to the outside pressure. Also the temperature will be assumed as constant throughout the tube. The mass fraction $w_{1s}$ of the water vapor outside the tube is generally different from that over the water surface. The partial pressure of the water vapor over the water surface equals the saturation pressure at the surface temperature.[1] In this way the mass fraction $w_{1w}$ is prescribed. Gradients of the mass fraction $w_1$ and a diffusive flow of water vapor will exist in the tube according to Eq. (16-7).

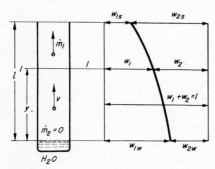

Fig. 16-2. Diffusion of water vapor through air.

Since the mass fraction of the water vapor $w_1$ and the mass fraction of the air $w_2$ add up to 1, a gradient of the vapor mass fraction corresponds to a gradient of the mass fraction of the air. As Eq. (16-7) must also be

---

[1] Theoretically the vapor pressure over the surface is slightly lower than the saturation pressure as soon as a mass flow from the surface occurs, but the difference is very small and can be neglected as long as the pressure is not very low. On the other hand the evaporation uses up heat, which must, in many cases, flow from inside the liquid to the surface. This results in a temperature gradient in the liquid near the surface and a surface temperature lower than the average liquid temperature.

valid for air, a mass flow of air must arise which is in a direction opposite to the vapor flow. No air can leave the tube at the bottom, however, as it is closed there. Therefore a convective flow in an upward direction must be present within the tube, which compensates for the diffusive air flow. The velocity of this convective flow is $v$. The amount of water vapor which is transported by this convective flow through the cross section 1-1 per unit area and time is $\rho_1 v$. Therefore the whole mass velocity of the water vapor at the cross section 1-1 is

$$\dot{m}_1 = -\rho D \frac{dw_1}{dy} + \rho_1 v \qquad (16\text{-}8)$$

The same relation must be valid for the resulting air flow. Since this flow is zero, the resulting equation is

$$\dot{m}_2 = -\rho D \frac{dw_2}{dy} + \rho_2 v = 0 \qquad (16\text{-}9)$$

This equation gives the velocity:

$$v = \frac{D}{w_2} \frac{dw_2}{dy} \qquad (16\text{-}10)$$

The mass fraction of the air can be expressed by

$$w_2 = 1 - w_1 \qquad \frac{dw_2}{dy} = -\frac{dw_1}{dy}$$

This gives

$$v = -\frac{D}{1-w_1} \frac{dw_1}{dy} \qquad (16\text{-}11)$$

The mass flow of water vapor through the tube is, according to Eq. (16-8),

$$\dot{m}_1 = -\rho D \frac{dw_1}{dy} - \frac{\rho_1}{1-w_1} D \frac{dw_1}{dy} = -\frac{\rho^2}{\rho_2} D \frac{dw_1}{dy}$$

For an integration of this equation, it is advantageous to change to partial pressures. With the relations

$$w_1 = \frac{p_1 R}{R_1 p} \qquad R = w_1(R_1 - R_2) + R_2$$

there is obtained

$$\frac{dw_1}{dy} = \frac{R^2}{R_1 R_2} \frac{1}{p} \frac{dp_1}{dy}$$

and

$$\dot{m}_1 = \frac{\rho R}{\rho_2 R_2} \frac{D}{R_1 T} \frac{dp_1}{dy} \qquad (16\text{-}12)$$

$$\dot{m}_1 = -\frac{D}{R_1 T} \frac{p}{p - p_1} \frac{dp_1}{dy} \qquad (16\text{-}13)$$

This equation is called *Stefan's law*.[1]

[1] Stefan, *Wien. Ber.*, **68**:385–425 (1874).

When the cross section of the tube is constant over $y$, Eq. (16-13) can be integrated, since $\dot{m}_1$ is then independent of $y$. By separating the variables, there results

$$\frac{dp_1}{p - p_1} = - \frac{\dot{m}_1}{D} \frac{R_1 T}{p} dy$$

and by integration between $y = 0$ and $y = l$,

$$\ln \frac{p - p_{1s}}{p - p_{1w}} = \dot{m}_1 \frac{l}{D} \frac{R_1 T}{p}$$

or
$$\dot{m}_1 = \frac{D}{l} \frac{p}{R_1 T} \ln \frac{p - p_{1s}}{p - p_{1w}} \tag{16-14}$$

With this equation, the mass velocity can be calculated from the partial pressures at both ends of the tube when the diffusion coefficient is known. Inversely, such an experiment can be used to determine the diffusion coefficient by measuring the partial pressures and the mass evaporated per hour. As long as the partial pressures of the diffusing gas at the wall and in the stream are small when compared with the total pressure, Eq. (16-14) can be simplified to

$$\dot{m}_1 = \frac{D}{l} \frac{1}{R_1 T} (p_{1w} - p_{1s}) \tag{16-15}$$

This equation has exactly the same form as Eq. (1-7) for heat conduction in a plane wall. In the same way Stefan's law [Eq. (16-13)] transforms into Fick's law [Eq. (16-5)] at small partial pressure differences and then has the same shape as the general equation for heat conduction [Eq. (2-2)].

For larger partial-pressure differences, Eq. (16-14) can be brought into a form which is very similar to Eq. (16-15) in the following way:

$$\dot{m}_1 = \frac{D}{l} \frac{p}{R_1 T} \frac{p_{1w} - p_{1s}}{p_{2s} - p_{2w}} \ln \frac{p_{2s}}{p_{2w}} = \frac{D}{l} \frac{1}{R_1 T} \frac{p}{p_{2m}} (p_{1w} - p_{1s}) \tag{16-16}$$

$p_{2m}$ is the log mean pressure and can be determined conveniently with the aid of Table 1-2. $p/p_{2m}$ is the factor by which the mass flow $\dot{m}_1$ is larger in the diffusion process indicated in Fig. 16-2 than in the one calculated by Eq. (16-15).

The preceding calculations were made for the evaporation of water into air in order to connect them with a concrete example. They can generally be applied, however, to the mutual diffusion of any two gases. Equation (16-8) and a corresponding equation for the second gas hold generally, whereas Eq. (16-13) is valid only for a surface which is impermeable to one of the two gases and for isothermal conditions, and Eq. (16-14) is further restricted to one-dimensional flow. All evaporation, condensation, and sublimation processes from a surface into a gas fulfill the condition of a semipermeable surface.

There are cases in which a mass flow of both components exists. For instance, in combustion of a solid carbon surface the oxygen diffuses toward the surface and the carbon monoxide or dioxide generated by the combustion process diffuses away from it. Equation (16-10) has in this case to be replaced by an equation which expresses the fact that the ratio of the two mass velocities at the surface is prescribed by the stoichiometric equation which describes the combustion process.

It is interesting to note that an enthalpy transport is connected with the diffusion process even in the isothermal field. For the example illustrated in Fig. 16-2 the water-vapor diffusion causes an enthalpy stream per unit area and time equal to $\dot{m}_1 c_{p1} T$. Even through a surface arranged in such a way in the gas that no net mass flow occurs through it ($\dot{m}_1 = \dot{m}_2$), an enthalpy flow is still present in the amount $\dot{m}_1(c_{p1} - c_{p2})T$. If in addition temperature differences are present, then the heat flow per unit time and area through a plane which moves with the flow is

$$q = -k\frac{dT}{dy} + \dot{m}_1(c_{p1} - c_{p2})T \qquad (16\text{-}17)$$

It is assumed that the direction $y$ is normal to the plane. The heat flow per unit time and area through plane 1-1 in Fig. 16-2 is

$$q = -k\frac{dT}{dy} + \dot{m}_1 c_{p1} T$$

when the temperature in the tube is not uniform. The following relation results from Eqs. (16-8) and (16-11):

$$\dot{m}_1 = -\rho D\frac{dw_1}{dy} + \rho_1 v = \rho(1 - w_1)v + \rho_1 v = \rho v$$

and
$$q = -k\frac{dT}{dy} + \rho v c_{p1} T$$

**Example 16-1.** A vertical tube with 0.775-in.$^2$ cross section is filled with water at the bottom. The distance from the water surface to the open end of the tube is 2.527 in. A flow of perfectly dried air is blown over the open end, and the tube is held at a constant temperature of 87.5 F. The water evaporated is measured by weighing and is found to be $5.165 \times 10^{-5}$ lb/hr. From this experiment the diffusion coefficient of water vapor in air is to be calculated.

The mass flux is

$$\dot{m}_1 = \frac{5.615 \times 10^{-5} \times 144}{0.775} = 1.043 \times 10^{-2} \text{ lb/hr ft}^2$$

The saturation pressure of water vapor at 87.5 F can be found in the steam tables. It is $p_{1w} = 0.6453$ lb$_f$/in.$^2$. The gas constant of water vapor is obtained from its molecular weight $M_1 = 18.016$ and the universal gas constant $\mathcal{R} = 1545.4$ ft lb$_f$/lb mole R:

$$R_1 = \frac{\mathcal{R}}{M_1} = \frac{1{,}545}{18.02} = 85.74 \text{ ft lb}_f/\text{lb R}$$

Solving Eq. (16-14) for $D$ gives

$$D = \frac{\dot{m}_1 l R_1 T}{p \ln [p/(p - p_{1w})]}$$
$$= \frac{1.043 \times 10^{-2} \times 2.527 \times 85.74 \times (460 + 87.5)}{12 \times 14.22 \times 144 \times \ln 14.22/(14.22 - 0.6453)} = 1.085 \text{ ft}^2/\text{hr}$$

Two things must be watched, since the occurrence of either may falsify the result of the experiment. The temperature of the water surface may be lower than that of the surrounding, and free convection may occur within the tube, for since humid air is not so heavy as dry air, the air in the bottom is lighter than that near the upper end. Both errors can be kept small by using a tube with a small diameter.

**16-2. Laminar Boundary Layer on a Flat Plate with Mass and Heat Transfer.** Mass transfer in a stagnant or nearly stagnant gas mixture has been studied in the previous section. Mass transfer in engineering applications is usually more complex because forced or free mass movements occur and forced or natural convection also contributes to the mass exchange. When mass is transferred from a solid surface into a fluid stream, the transfer process is essentially concentrated in the boundary layer. This process will be studied on a flat plate arranged in a stream of uniform velocity of such magnitude that a laminar boundary layer exists along the surface. In most situations a heat-transfer process is connected with the mass transfer. In evaporation of vapor from a wet surface or condensation on the surface, for instance, heat is absorbed or released at the surface by the change of phase, and this process usually creates temperature differences in the fluid and thus heat transfer. In many practical cases conditions will be such that the boundary-layer flow is turbulent. The laminar flow, however, is simpler to calculate and serves as a good model on which the basic phenomena can be studied.[1]

The complete set of boundary-layer equations which describe the combined momentum, mass, and energy transport in a two-dimensional, steady laminar boundary layer of a two-component mixture are[2]

Continuity:

$$\frac{\partial(\rho u)}{\partial x} + \frac{\partial(\rho v)}{\partial y} = 0$$

---

[1] Turbulent mass transfer is studied analytically by R. G. Deissler, *Natl. Advisory Comm. Aeronaut., Tech. Notes* 3145, 1954.

[2] For a derivation of these equations see Chapman and Cowling, *op. cit.*

Momentum:

$$\rho u \frac{\partial u}{\partial x} + \rho v \frac{\partial u}{\partial y} = \frac{\partial}{\partial y}\left(\mu \frac{\partial u}{\partial y}\right) - \frac{\partial p}{\partial x}$$

Diffusion:

$$\rho u \frac{\partial w_1}{\partial x} + \rho v \frac{\partial w_1}{\partial y} = \frac{\partial}{\partial y}\left(\rho D \frac{\partial w_1}{\partial y}\right)$$

Energy:

$$\rho u c_p \frac{\partial t}{\partial x} + \rho v c_p \frac{\partial t}{\partial y} = \frac{\partial}{\partial y}\left(k \frac{\partial t}{\partial y}\right) + \mu \left(\frac{\partial u}{\partial y}\right)^2 + u \frac{\partial p}{\partial x} + \rho D(c_{p2} - c_{p1})\frac{\partial t}{\partial y}\frac{\partial w_1}{\partial y}$$

The diffusion equation is written in terms of the mass fraction $w_1$ of the component which moves away from the surface. In the energy equation a new term appears as the last term on the right-hand side. It describes the effect that has been pointed out in the previous section, namely, that energy transport is connected with diffusion even when no temperature gradients occur. Actually diffusion also occurs in the presence of temperature differences without concentration gradients. This process is called thermal diffusion. Correspondingly, in this case additional terms should be included in the mass- and energy-transport equations. However, it appears that thermal diffusion contributes essentially to the overall process only when the temperature gradients are extremely large and that it can be neglected in normal engineering mass-transfer processes.

The above equations are very difficult and tedious to solve and will be simplified by the assumption that the fluid properties can be considered constant. In the present situation this means not only that the properties of each component are independent of temperature and pressure but also that the properties of the two components in the fluid differ very little from each other. In addition the flow velocities may be sufficiently small (smaller than, say, one-half the sound velocity) so that frictional heating can be neglected. We also consider a constant velocity outside the boundary layer. The above system of equations then simplifies to

Continuity: $$\frac{\partial u}{\partial x} + \frac{\partial v}{\partial y} = 0 \qquad (16\text{-}18)$$

Momentum: $$\rho u \frac{\partial u}{\partial x} + \rho v \frac{\partial u}{\partial y} = \mu \frac{\partial^2 u}{\partial y^2} \qquad (16\text{-}19)$$

Diffusion: $$u \frac{\partial w_1}{\partial x} + v \frac{\partial w_1}{\partial y} = D \frac{\partial^2 w_1}{\partial y^2} \qquad (16\text{-}20)$$

Energy: $$u \frac{\partial t}{\partial x} + v \frac{\partial t}{\partial y} = \alpha \frac{\partial^2 t}{\partial y^2} \qquad (16\text{-}21)$$

To these equations belong a set of boundary conditions:

At $y = 0$: $\quad u = 0 \quad v = v_w \quad w_1 = w_{1w} \quad t = t_w$ (16-22)
At $y \to \infty$: $\quad u = u_s \quad w_1 = w_{1s} \quad t = t_s$ (16-23)

The mass fraction $w_{1w}$ and temperature $t_w$ at the wall will be assumed constant (independent of $x$). The condition $v = v_w$ accounts for the fact that a convective flow is generally connected with mass transfer from a surface. It was mentioned in the preceding paragraph that such a convective flow is always present when the surface is impermeable for one of the components (evaporation, condensation). It also occurs when both components pass through the surface unless the two opposing mass flows are just equal.

This convective velocity $v_w$ then appears as an additional parameter in the above set of differential equations, and we shall discuss later on how it is connected with the over-all transfer process. Otherwise, for the constant-property fluid, the flow process is independent of the mass- and heat-transfer process. In addition there exist pronounced similarities of the three differential equations, which will help in obtaining solutions. The partial differential equations will now be transformed to total differential ones by introduction of the variable $\eta = (y/2)\sqrt{u_s/\nu x}$ and $f$ used in Sec. 6-6 and of the following dimensionless parameters describing the mass fraction and temperature:

$$\varphi' = \frac{w_1 - w_{1w}}{w_{1s} - w_{1w}} \qquad \theta' = \frac{t - t_w}{t_s - t_w}$$

The transformation results in the following equations:

$$\frac{d^3f}{d\eta^3} + f\frac{d^2f}{d\eta^2} = 0$$

$$\frac{d^2\varphi'}{d\eta^2} + (\text{Sc})f\frac{d\varphi'}{d\eta} = 0$$

$$\frac{d^2\theta'}{d\eta^2} + (\text{Pr})f\frac{d\theta'}{d\eta} = 0$$

In the third equation, the dimensionless ratio $\text{Pr} = \nu/\alpha$ is found. In the same way a dimensionless ratio $\nu/D$ appears in the transformation process in the second equation. It is referred to as the *Schmidt number* $\text{Sc} = \nu/D$.

The boundary conditions become

At $y = 0$:

$$\frac{df}{d\eta} = 0 \qquad \varphi' = 0 \qquad \theta' = 0 \qquad f = f_w = \frac{v_w}{u_s}\sqrt{\text{Re}_x} = \text{const}$$

At $y \to \infty$: $\qquad \dfrac{df}{d\eta} = 1 \qquad \varphi' = 1 \qquad \theta' = 1$

The boundary condition $f_w = $ const, which is a necessary condition for the transformation, implies that the ratio $v_w/u_s = $ const$/\sqrt{Re_x}$ or that $v_w/u_s$ is inversely proportional to $\sqrt{x}$. Fortunately this restriction is not too serious, since it is found to agree in most cases with the condition of a constant temperature and mass fraction at the surface. The boundary conditions on $df/d\eta$, $\varphi'$, and $\theta'$ are identical. Also, the three differential equations are identical relations among these three parameters when

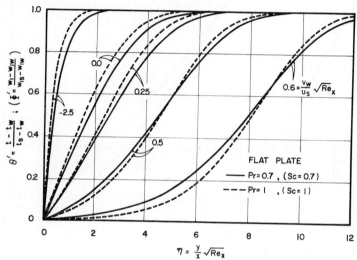

FIG. 16-3. Temperature and concentration profiles in a laminar boundary layer on a flat plate. Pr value belongs to temperature profile, Sc value to concentration profile. [From J. P. Hartnett and E. R. G. Eckert, Trans. ASME, 79:247–254 (1957).]

Sc = 1 and Pr = 1. For such a two-component fluid, then, the solutions of the three differential equations, $df(\eta)/d\eta = u/u_s$, $\varphi'(\eta)$, $\theta'(\eta)$, must be identical functions. The dashed lines in Fig. 16-3 present these functions for various values of the convective flow parameter $f_w = v_w/u_s \sqrt{Re_x}$ through the surface, according to calculations by a number of investigators.[1] This figure, therefore, presents as dashed lines the velocity, mass-fraction, and temperature profiles throughout the boundary layer. The solid lines are the temperature profiles for a fluid with a Prandtl number equal to 0.7. An inspection of the differential equations shows immediately that the solid curves are also mass-fraction profiles for a two-component gas with a Schmidt number equal to 0.7. By a comparison of the solid and dashed lines, the influence of Prandtl number or Schmidt number, respectively, on the temperature or concentration fields can be estimated. It is seen that the mass-fraction and the temperature profiles are similar when Pr = Sc or when the ratio Sc/Pr = $\alpha/D$, called *Lewis*

[1] J. P. Hartnett and E. R. G. Eckert, *Trans. ASME*, **79**:247–254 (1957).

*number*, is 1. This situation, which is often encountered in gas mixtures, simplifies mass-transfer calculations considerably.

Engineering-wise, a knowledge of the friction, the heat transfer, and the mass transfer at the surface is most important. These parameters are determined by the gradients of the velocity, the temperature, and the

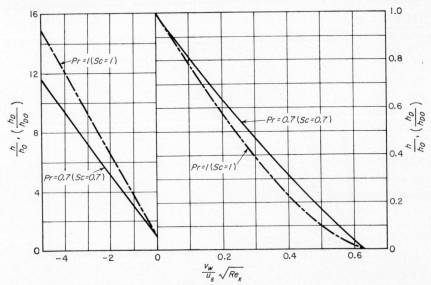

FIG. 16-4. Heat- and mass-transfer parameters for laminar flow over a flat plate. $o$ indicates heat-and mass-transfer coefficients with very small concentration gradients. Pr belongs to $h$, Sc to $h_D$. [*From J. P. Hartnett and E. R. G. Eckert, Trans. ASME,* **79**:247–254 (1957).]

mass-fraction profile at the surface, and Fig. 16-3 shows that, other parameters kept constant, friction, heat transfer, and mass transfer are decreased by a convective flow away from the surface and are increased by a convective flow toward the surface. The heat-transfer rate at the wall surface is characterized by a heat-transfer coefficient defined through the equation

$$q_w = -k \left(\frac{\partial t}{\partial y}\right)_w = h(t_w - t_s) \qquad (16\text{-}23a)$$

A corresponding mass-transfer coefficient can be defined by

$$\dot{m}_{1w} = -\rho D \left(\frac{\partial w_1}{\partial y}\right)_w = \rho h_D (w_{1w} - w_{1s}) \qquad (16\text{-}23b)$$

when $\dot{m}_{1w}$ denotes the mass flow by diffusion of component 1 at the surface. The influence of the convective flow at the surface on both coefficients can be seen in Fig. 16-4 in which the ratio of the actual heat- and mass-transfer coefficients to the coefficients $h_0$, $h_{D0}$ for very small (in the limit zero) convective flow through the surface but for otherwise identical conditions is plotted versus the convective-flow parameter $v_w/u_s \sqrt{Re_x}$. The curves in the figure have been obtained from the respective gradients at $\eta = 0$ in Fig. 16-3. The dashed line applies to a fluid with $Pr = 1$, $Sc = 1$. The solid line gives the ratio $h/h_0$ for a fluid with $Pr = 0.7$ and the ratio $h_D/h_{D0}$ for a fluid with $Sc = 0.7$. The dashed curve in Fig. 16-4 is also proportional to the velocity gradient at the surface. It presents, therefore, also the ratio of friction factors $f_p/f_{p0}$.

It remains to discuss the connection of the convective flow at the surface with the over-all transfer process. This relation will be pointed out in some examples shown in Fig. 16-5. The upper sketch indicates a porous wall through which carbon dioxide is blown, while air flows along the surface of the wall. For the preceding calculations to apply to this case, the blowing would have to start at the leading edge of the wall and the wall porosity be such that the velocity $v_w$ with which the $CO_2$ leaves the surface varies like $1/\sqrt{x}$ along the surface. In this case the velocity $v_w$ will be

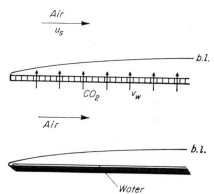

Fig. 16-5. Mass-transfer situations on a flat plate.

known and the velocity, concentration, and temperature profiles can be found from Fig. 16-3. The temperature of the wall surface will also be known or has to be determined from a heat balance of the various heat flows entering and leaving the surface. The mass fraction $w_{1w}$ at the wall surface, however, is generally not known and has to be calculated. If provisions are made that no air passes through the wall, then Eq. (16-11) describes the connection between the blowing velocity $v_w$ and the mass fraction of the carbon dioxide. Written in the mass fraction of the carbon dioxide, the equation reads

$$v_w = -\frac{D}{1 - w_{1w}} \left(\frac{\partial w_1}{\partial y}\right)_w \qquad (16\text{-}24)$$

or introducing the dimensionless parameters and rearranging,

$$\frac{w_{1w} - w_{1s}}{1 - w_{1w}} = \frac{(Sc)f_w}{(d\varphi'/d\eta)_w}$$

Figure 16-6 shows the mass fraction $w_{1w0}$ for the condition $w_{1s} = 0$. Generally, $w_{1w0}$ is obtained from the equation $w_{1w} = w_{1w0} + (1 - w_{1w0})w_s$.

For a prescribed $v_w$ and $w_{1s}$ and for known conditions of the air flow, this figure then allows calculation of the mass fraction $w_{1w}$ at the porous wall surface. Such an arrangement as shown in Fig. 16-5, in which the wall is porous and a cold gas is blown through it, has proved to be an effective means of protecting a wall against a hot gas stream. The reduction in heat-transfer coefficient, obvious from Fig. 16-4, means that with a small amount of cooling gas a very effective reduction of the wall surface temperature can be obtained. This cooling process is referred to in the liter-

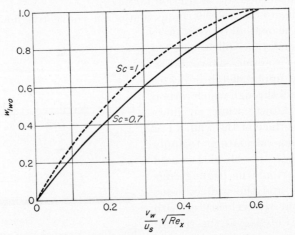

Fig. 16-6. Wall concentration in mass transfer through a laminar boundary layer on a flat plate as a function of mass flow at the surface. [*From J. P. Hartnett and E. R. G. Eckert, Trans. ASME,* **79**:247-254 (1957).]

ature as *transpiration* or *sweat cooling*. Often air is used as a coolant. Then the mass-transfer part of the process as discussed before becomes meaningless. Otherwise Figs. 16-3 and 16-4 are valid for transpiration cooling.[1]

The lower sketch in Fig. 16-5 indicates the transfer of vapor from a water surface into an air stream by evaporation or from the humid air stream to the water surface by condensation. For the calculation in this section to apply, a movement of the water by the friction of the air stream would have to be prevented so that the air velocity at the surface is zero. In addition, the air boundary layer would have to start at the leading edge of the water surface. The convective velocity at the water surface is not known; however, the water surface is impermeable to air, and therefore, Eq. (16-11) holds as well as Fig. 16-6. The temperature of the water surface determines the partial pressure and the mass fraction $w_w$ at the water surface. With a knowledge of the vapor mass fraction $w_s$ in the air stream, the parameter on the ordinate of Fig. 16-6 can be calcu-

[1] For air to air, this process was discussed in Sec. 10-5.

lated. This value and the curve in the figure determine the abscissa value $v_w/u_s \sqrt{\text{Re}_x}$ and thus the convective velocity $v_w$. With this value the profiles and the transfer coefficients can be found in Figs. 16-3 and 16-4.

**16-3. The Integrated Boundary-layer Equations of Mass Transfer.** Approximate procedures based on integrated boundary-layer equations will in many cases lead faster to an answer than an exact solution of the boundary-layer equations. In this paragraph the integrated momentum, mass-flow, and heat-flow equations will be derived for a two-component mixture with variable properties and for two-dimensional forced flow.

Consider a volume element with the dimensions $dx$ and $l$ near the surface from which the mass flow originates (Fig. 16-7). Normal to the plane of the drawing, the dimension is 1. The length $l$ is greater than any of the boundary-layer thicknesses. The subscript 1 will refer to the fluid transported from the surface into the stream. Terms without subscripts refer to the mixture. The state of the fluid outside the boundary layer is

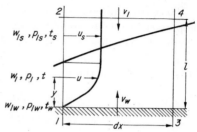

Fig. 16-7. Deriving the boundary-layer equations of mass transfer.

fixed by its velocity $u_s$ (in $x$ direction), the temperature $t_s$, and the partial pressure $p_{1s}$, or the mass fraction $w_{1s}$ of component 1. On the surface the velocity is zero ($u = 0$), the temperature is $t_w$, and the fluid mass fraction $w_{1w}$. At the distance $y$ from the surface, within the boundary layer, the respective values are $u$, $t$, and $w_1$. The total pressure $p$ can again be assumed constant on any normal to the surface within the boundary layer because of the small boundary-layer thickness.

First the continuity equation can be set up for the volume element. The mass flow through the area 1-2 is $\int_0^l \rho u \, dy$. Through the length $dx$ it changes by the amount $\dfrac{d}{dx}\left(\int_0^l \rho u \, dy\right) dx$. In steady state this same amount must enter the volume through the areas 1-3 and 2-4. On the surface (area 1-3) a normal velocity $v_w$ exists whose magnitude was calculated in the previous sections. The velocity through the area 2-4 is $v_1$; therefore continuity demands that

$$\rho_s v_1 = \frac{d}{dx}\int_0^l \rho u \, dy - \rho_w v_w \qquad (16\text{-}25)$$

The equation of momentum (Sec. 6-2) is now

$$\frac{d}{dx}\int_0^l \rho u^2 \, dy - \rho v_1 u_s = -\tau_w - l\frac{dp}{dx}$$

By introducing the continuity equation and using the same procedure as in Sec. 6-2 there is obtained the integrated momentum equation

$$\frac{d}{dx}\int_0^l (u_s - u)\rho u\, dy + \frac{du_s}{dx}\int_0^l (\rho_s u_s - \rho u)\, dy - \rho_w v_w u_s$$
$$= \mu_w \left(\frac{du}{dy}\right)_w \quad (16\text{-}26)$$

The shear stress on the wall is expressed by the velocity gradient from the equation $\tau_w = \mu_w(du/dy)_w$. The equation of momentum has changed as compared with Eq. (6-10) by the summand $\rho_w v_w u_s$.

To write the mass-flow equation for the fluid 1, the fact must be expressed that in steady state the difference between the fluid mass carried by the flow into the volume element through the area 1-2 and out of it through 3-4 must come into the volume through areas 1-3 and 2-4. The mass flux of the fluid on the solid surface is $\dot{m}_{1w}$. Then

$$\dot{m}_{1w} = \frac{d}{dx}\int_0^l \rho_1 u\, dy - v_1 \rho_{1s}$$

Expressing the velocity $v_1$ by the continuity equation (16-25) and using the mass fraction instead of the partial densities give

$$\frac{d}{dx}\int_0^l (w_{1s} - w_1)\rho u\, dy - \rho_w v_w w_{1s} = \dot{m}_{1w}$$

Elimination of the mass velocity $\dot{m}_{1w}$ of the vapor on the solid surface by Eq. (16-8) leads to

$$\frac{d}{dx}\int_0^l (w_{1s} - w_1)\rho u\, dy = \rho_w D_w \left(\frac{dw_1}{dy}\right)_w + \rho_w v_w(w_{1s} - w_{1w}) \quad (16\text{-}27)$$

The heat-flow equation is found in the same way by a heat balance on the volume element. The heat flow per unit area of the solid surface is $q_w$. Then

$$\frac{d}{dx}\int_0^l \rho c_p t u\, dy - \rho_s c_{ps} t_s v_1 = q_w$$

Heat is transferred from the solid surface not only by conduction but also by convection with the component 1. Therefore

$$q_w = -k \left(\frac{dt}{dy}\right)_w + \rho_w c_{pw} t_w v_w$$

Introducing $v_1$ from the continuity equation gives

$$\frac{d}{dx}\int_0^l (c_{ps}t_s - c_p t)\rho u\, dy = k_w \left(\frac{dt}{dy}\right)_w + (c_{ps}t_s - c_{pw}t_w)\rho_w v_w \quad (16\text{-}28)$$

The integrated boundary-layer equations (16-26) to (16-28) for mass and heat transfer differ from the corresponding equation for heat transfer alone by the addition of the summand with the velocity $v_w$. The equations can be solved in the same way as in Sec. 6-4 by introducing proper expressions for the velocity, the temperature, and the mass-fraction profile and by calculating the boundary-layer thicknesses. In order to increase the accuracy of this approximate solution, it is useful to fulfill for the assumed profiles the boundary conditions on the solid surface which are valid for the real profiles. In Sec. 7-3 it was found that the second derivative of the temperature with respect to the wall distance is constant on the solid surface. This follows from the condition that $q_w$ is constant in the immediate proximity of the wall. For the heat flow with mass transfer, the equation preceding Eq. (16-28) holds true. Again, in the immediate proximity of the wall the heat flow $q_w$ must be constant. Therefore

$$\left(\frac{dq}{dy}\right)_w = \rho_w v_w \left(\frac{dc_p t}{dy}\right)_w - \frac{d}{dy}\left(k\frac{dt}{dy}\right)_w = 0 \qquad (16\text{-}29)$$

The mass velocity $\rho v$ must be constant near the wall. For the mass flux of the vapor Eq. (16-8) is valid. By differentiating Eq. (16-8) there is obtained

$$\left(\frac{d\dot{m}}{dy}\right)_w = \rho_w v_w \frac{dw_1}{dy} - \frac{d}{dy}\left(\rho D \frac{dw_1}{dy}\right)_w = 0 \qquad (16\text{-}30)$$

For the shear stress $\tau_w$ without diffusion, Eq. (6-2) holds true. If a mass flow originates from the wall, this equation changes. Imagine a plane in the proximity of the solid surface and parallel to it; then, within this plane a shear stress arises by the viscosity of the fluid flowing along the surface. But in addition, there is a virtual shear stress caused by the transport of momentum with the mass passing through the plane. Such a shear stress has already been considered in Sec. 8-1. The total shear stress therefore is

$$\tau = \mu \frac{du}{dy} - \rho v u \qquad (16\text{-}31)$$

This shear stress must again be constant in the immediate proximity of the wall, so its derivative with respect to $y$ at the wall is zero:

$$\left(\frac{d\tau}{dy}\right)_w = \frac{d}{dy}\left(\mu\frac{du}{dy}\right)_w - \rho_w v_w \left(\frac{du}{dy}\right)_w = 0 \qquad (16\text{-}32)$$

The calculations become much simpler when the partial-pressure differences within the boundary layer are small compared with the average pressure of the fluid. In this case the velocity $v_w$ normal to the wall is small and can be neglected in Eqs. (16-26) to (16-28). The equation of pressure differences. The line $K = K_p = 3$ is valid for a partial pressure of the air equal to zero at the wall, which means for a boiling-water surface. The mass flow from or to the wall can be determined with Fig. 16-9. The value $h_D$ on the ordinate gives the mass-transfer coefficient.

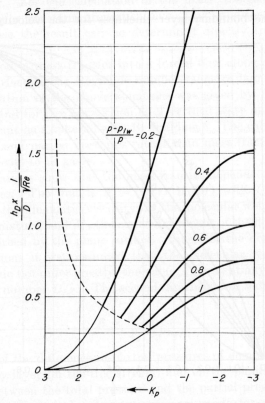

Fig. 16-9. Dimensionless mass-transfer coefficient for a flat surface absorbing or evaporating water into a laminar air flow. [*According to E. Eckert and V. Lieblein, Forsch. Gebiete Ingenieurw.*, **16**:33–42 (1949).]

In situations in which a mass flow occurs at the surface, there exists an ambiguity in the definition of transfer coefficients. Equations (16-23a) and (16-23b) describe the heat flow by conduction and the mass flow by diffusion, respectively. The convective flows have to be added in order to obtain the total heat and mass flow. The definition used in Eq. (16-23) has, on the other hand, the advantage that the transfer coefficients defined in this way depend on a smaller number of parameters. The mass-transfer coefficient in Fig. 16-9 describes the total mass flow at the plate surface as expressed by Eq. (16-8). It will be observed that this coefficient $h_D$ is a function of the pressure ratio $(p - p_{1w})/p$ in addition to the parameter $K_p$, which expresses the convective velocity $v_w$. Only the curves above the dashed line have any physical reality. The values for the mass-transfer coefficient at low partial-pressure differences lie on the vertical line $K_p = 0$. It can be seen that the mass transfer depends considerably on the value $K_p$, which means on the partial-pressure differences. Positive values of $K_p$ hold for evaporation, negative values for absorption of the vapor on the wall.[1]

**16-4. Similarity Relations for Mass Transfer.**[2] As long as the mass fractions of component 1 are small compared with 1, the term with the velocity $v_w$ can be neglected in Eqs. (16-27) and (16-28). Both equations then appear in the same form as the heat-flow equation (7-2) when in addition the properties are considered constant. This can be seen from the following comparison.

Heat-flow equation: $\quad \dfrac{d}{dx}\displaystyle\int_0^l (t_s - t)u\,dy = \alpha \left(\dfrac{dt}{dy}\right)_w$

Mass-flow equation: $\quad \dfrac{d}{dx}\displaystyle\int_0^l (w_{1s} - w_1)u\,dy = D \left(\dfrac{dw_1}{dy}\right)_w$

The latter equation has the same form if written in terms of partial pressure or density. As a consequence of the similar form of the heat-flow and the mass-flow equations, the solutions for both equations must also be similar. Therefore, from the solution of the heat-flow equation, the solution of the mass-transfer equation is obtained when the mass fraction is substituted for the temperature $t$ and the diffusion coefficient $D$ for the thermal diffusivity $\alpha$. An analogy exists also to the general heat-flow equation (without the dissipation term) (7-3) in mass transfer with constant properties:

$$\frac{\partial w_1}{\partial \tau} + u\frac{\partial w_1}{\partial x} + \cdots = D\frac{\partial^2 w_1}{\partial x^2} + \cdots$$

In Sec. 9-1 it was shown that the temperature field in the neighborhood of similar bodies transmitting heat by forced convection with small velocities can be represented in the following way:

$$\vartheta' = f\left(x',y',z',\text{Re},\frac{\nu}{\alpha}\right) = f(x',y',z',\text{Re},\text{Pr}) \tag{16-33}$$

where $\vartheta'$ is a dimensionless value giving the difference between the temperature $t$ of a point at the dimensionless distance $y'$ from a surface and the temperature of any arbitrary reference point divided by any arbitrary

---

[1] Experiments by A. Heyser, *Chem. Ingr. Tech.*, 1956, pp. 161–164, on condensation of vapor from a vapor-air mixture verified the above relation reasonably well.

[2] These relations were first derived simultaneously by Schmidt, *loc. cit.*, and by Nusselt, *loc. cit.*

reference temperature difference, as, for instance, the difference between the temperature $t_s$ of the flowing medium at a great distance from the body and the temperature $t_w$ of the body. Re is the Reynolds number, and $\nu/\alpha$ the Prandtl number. For the same body we get the mass-fraction or partial-pressure field by substituting mass fraction or partial pressures $w_1' = (w_1 - w_{1w})/(w_{1s} - w_{1w})$; $p_1' = (p_1 - p_{1w})/(p_{1s} - p_{1w})$ for the temperatures and the diffusion coefficient $D$ for the thermal diffusivity $\alpha$:

$$w_1' = f\left(x',y',z',\text{Re},\frac{\nu}{D}\right) = f(x',y',z',\text{Re},\text{Sc}) \quad (16\text{-}34)$$

The dimensionless value $\nu/D$ is called the Schmidt number. The function $f$ has the same form in both Eqs. (16-33) and (16-34). Values of the Schmidt number are contained in Table A-9. For gases the Schmidt number is independent of pressure. It varies with temperature and sometimes also with concentration.

In heat-transfer problems, we are primarily interested in the specific heat flow $q_w$ per unit surface area of the body. It can be calculated from the film heat-transfer coefficient $h$ by the equation

$$q_w = h(t_w - t_s)$$

On the other hand there is the equation

$$q_w = -k\left(\frac{\partial t}{\partial y}\right)_w$$

Therefore
$$h(t_s - t_w) = k\left(\frac{\partial t}{\partial y}\right)_w \quad (16\text{-}35)$$

In an analogous manner, one can define a mass-transfer coefficient $h_D$ with a driving potential characteristic of the mass-transfer process. Unfortunately there is no agreement as to which parameter should be used to express the driving potential. We shall use mass fractions and define the mass-transfer coefficient by the equation

$$\dot{m}_{1w} = \rho h_D(w_{1w} - w_{1s}) \quad (16\text{-}36)$$

When, for a fluid with nearly constant properties ($\rho \approx \text{const}$), the mass fractions are converted to partial densities, the equation reads

$$\dot{m}_{1w} = h_D(\rho_{1w} - \rho_{1s})$$

or in mole densities by dividing the last equation by the molecular weight $M_1$

$$\dot{n}_{1w} = h_D(n_{1w} - n_{1s})$$

Assuming in addition to $\rho \approx \text{const}$ that the temperature differences in the field are small as compared with the absolute temperature, one obtains

$$\dot{m}_{1w} = \frac{h_D}{R_1 T}(p_{1w} - p_{1s})$$

This equation was first used by E. Dalton (1788).

The last three equations are most commonly found in the literature. Under the conditions mentioned above, the mass-transfer coefficient is the same in all equations. For large temperature, pressure, and property variations, however, the mass-transfer coefficients in these equations are different quantities.

For a small convective velocity $v$ the mass flow at a solid surface or interface can be expressed by Eq. (16-7).

$$\dot{m}_{1w} = -\rho D \left(\frac{dw_1}{dy}\right)_w$$

Combining this expression with Eq. (16-36) gives

$$h_D(w_{1s} - w_{1w}) = D\left(\frac{\partial w_1}{\partial y}\right)_w \qquad (16\text{-}36a)$$

From Eq. (16-35) the dimensionless film heat-transfer coefficient becomes

$$\frac{hl}{k} = \left(\frac{\partial \vartheta'}{\partial y'}\right)_w$$

with the dimensionless temperature difference $\vartheta'$ and the dimensionless distance $y' = y/l$ from the solid surface. According to Eq. (16-33), for a specific location along the body surface, the last equation becomes

$$\frac{hl}{k} = f(\text{Re}, \text{Pr}) \qquad (16\text{-}37)$$

In the same way, a dimensionless mass-transfer coefficient can be formed by dividing the mass fraction in Eq. (16-36a) by the difference $(w_{1s} - w_{1w})$ and by dividing the coordinate $y$ by the reference length $l$.

$$\frac{h_D l}{D} = \left(\frac{\partial w_1'}{\partial y'}\right)_w \qquad (16\text{-}38)$$

By differentiating Eq. (16-34) with respect to $y'$ and considering a specific location, one obtains at the distance $y = 0$

$$\frac{h_D l}{D} = f\left(\text{Re}, \frac{\nu}{D}\right) = f(\text{Re}, \text{Sc}) \qquad (16\text{-}39)$$

The functions in both Eqs. (16-37) and (16-39) have the same form.

Any of the equations in Part B for heat transfer in laminar flow, therefore, gives also the solution for a corresponding mass-transfer problem if the Nusselt number $hl/k$ is replaced by the dimensionless mass-transfer coefficient $h_D l/D$ and the Prandtl number $\nu/\alpha$ by the dimensionless value $\nu/D$.

This law has also been verified for turbulent flow by a considerable number of experiments on mass transfer for flow through a tube and flow over a flat plate, around cylinders and spheres, and through packed beds. Agreement was found to be within experimental accuracy. Figure 16-10

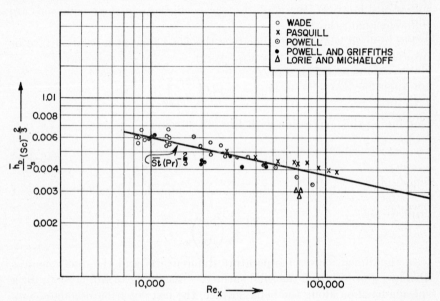

FIG. 16-10. Experimental verification of analogy between heat and mass transfer. *(From J. M. Coulson and J. F. Richardson, "Chemical Engineering," vol. 1, p. 257, McGraw-Hill Book Company, Inc., New York, 1954.)*

shows, as an example, the results of experiments on the evaporation of various liquids from plane surfaces into an air stream.[1] The test points are plotted as a dimensionless group composed of a mass-transfer parameter $(\bar{h}_D/u_s)(\mu/\rho D)^{-2/3}$ which is analogous to Stanton's number in heat transfer and of the Schmidt number over the Reynolds number. The line in the figure illustrates the heat-transfer parameter $\overline{St}\, Pr^{-2/3}$ as obtained from Eq. (8-18). It is seen that the agreement is quite satisfactory.

The mass-transfer coefficients obtained from heat-transfer relations on solid surfaces by the discussed analogy are valid when the mass-fraction or partial-pressure differences causing the mass transfer are sufficiently

---

[1] T. K. Sherwood and R. L. Pigford, "Absorption and Extraction," 2d ed., McGraw-Hill Book Company, Inc., New York, 1952.

small so that the contribution of the convective flow at the surface can be neglected. An approximate correction of the mass-transfer coefficients which takes into account the convective contribution can be made by multiplying the mass-transfer coefficient obtained from the analogy by the factor $p/p_{2m}$ ($p$ total pressure, $p_{2m}$ log mean of the partial pressure of the component 2 at wall and in the freestream). This correction is based on the assumption that the contribution of convection is the same in the situation considered and in the diffusion process illustrated in Fig. 16-2 [Eq. (16-16)]. Experiments by Mickley[1] and coworkers, and by A. Heyser[2] indicate that this correction is satisfactory for laminar and turbulent flow over flat plates as long as Schmidt's number does not differ much from a value 1.

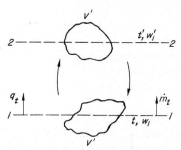

FIG. 16-11. Simplified picture of turbulent mass transfer.

If, for a gas mixture, the diffusion coefficient $D$ equals the thermal diffusivity $\alpha$, the two dimensionless transfer coefficients are equal for a given Reynolds number. By substituting for the diffusion coefficient $D$, the thermal diffusivity $\alpha = k/\rho c_p = k/C_p$ ($C_p$ = specific heat per unit volume at constant pressure) we get

$$h_D = h\frac{D}{k} = h\frac{\alpha}{k} = \frac{h}{C_p} \tag{16-40}$$

This law was derived by *Lewis*[3] and is therefore called the Lewis relation.

For turbulent exchange according to Fig. 16-11, the following conditions apply. $V'$ denotes the gas volume exchanged between the two planes 1 and 2 per unit time and area, and $t$ and $t'$, $w_1$ and $w_1'$, respectively, are the temperatures and mass fractions in the two planes. Then the heat flow per unit area from plane 1 to plane 2 associated with the turbulent exchange is

$$q_t = V'C_p(t - t') \tag{16-41}$$

If this heat flow is characterized by a turbulent film heat-transfer coefficient $h_t$, it can be written that

$$h_t(t - t') = V'C_p(t - t') \tag{16-42}$$

In the same way, the mass flow $\dot{m}_t$ per unit area associated with the

[1] H. S. Mickley, R. C. Ross, A. L. Squyers, and W. E. Stewart, *Natl. Advisory Comm. Aeronaut. Tech. Notes* 3208, 1954.
[2] A. Heyser, *loc. cit.*
[3] W. K. Lewis, *Trans. Am. Inst. Chem. Engrs.*, **20**:9 (1927).

turbulent exchange is

$$\dot{m}_t = \rho V'(w_1 - w_1') \tag{16-43}$$

and, expressed in terms of a turbulent mass-transfer coefficient $h_{Dt}$, is

$$\rho h_{Dt}(w_1 - w_1') = \rho V'(w_1 - w_1') \tag{16-44}$$

Dividing Eq. (16-44) by (16-42) gives

$$h_{Dt} = \frac{h_t}{C_p} \tag{16-45}$$

One can see that the Lewis relation holds true in turbulent flow regardless of whether the ratio $\alpha/D$ equals 1 or not. For the laminar sublayers which always arise in turbulent flow in the immediate neighborhood of solid surfaces, the Lewis relation is valid only for $\alpha/D = 1$.

The turbulent heat flow has been also expressed in Sec. 8-4 as

$$q_t = -\epsilon_q \rho c_p \frac{dt}{dy}$$

where the turbulent diffusivity $\epsilon_q$ for heat replaces the thermal diffusivity in the expression for laminar heat flow. Analogously, a relation for turbulent mass transfer can be obtained by replacing the diffusion coefficient $D$ in Eq. (16-7) by a turbulent diffusivity $\epsilon_D$ for mass:

$$\dot{m}_1 = -\epsilon_D \rho \frac{dw_1}{dy}$$

It is evident that Eq. (16-45) is obtained when $\epsilon_D = \epsilon_q$. It has to be expected from theoretical considerations that this is the case. Experiments give a ratio which is slightly greater than 1.

Heat transfer by free convection can be represented by the following equation, according to Sec. 9-1:

$$\text{Nu} = f(\text{Gr}, \text{Pr}) \tag{16-46}$$

where Nu is the dimensionless heat-transfer coefficient $hl/k$, Gr the Grashof number $[g\beta(t_w - t_s)l^3]/\nu^2$ and Pr the Prandtl number $\nu/\alpha$. In the Grashof number, the product of the thermal-expansion coefficient $\beta$ and the temperature difference $t_w - t_s$ took the place of the ratio of the densities $(\rho_s - \rho_w)/\rho_w$. For mass-transfer processes, where the buoyancy forces are generated by the difference in density of the different mixtures, the Grashof number for mass transfer has the form

$$\text{Gr}_D = \frac{gl^3}{\nu^2}\left(\frac{\rho_s}{\rho_w} - 1\right)$$

where $\rho_w$ is the density of the gas mixture at the solid wall and $\rho_s$ is the density outside the boundary layer. The same deductions used for

MASS TRANSFER

forced convection now lead to the following expression for the mass-transfer coefficient:

$$\frac{h_D l}{D} = f\left(\mathrm{Gr}_D, \frac{\nu}{D}\right) = f(\mathrm{Gr}_D, \mathrm{Sc}) \tag{16-47}$$

The function $f$ is again the same as for the corresponding heat-transfer problem. If heat and mass transfer occur simultaneously on the same surface, it is not possible to calculate the mass-transfer coefficient from the solution of the corresponding problem with heat transfer alone as long as $D \neq \alpha$.†

**Example 16-2.** Along a horizontal water surface an air stream with a velocity $u_s = 10$ fps is flowing. The amount of water evaporated per hour per square foot from the water surface is to be calculated. The temperature of the water on the surface is 59 F, the air temperature is 68 F, the total pressure 14.70 lb$_f$/in.$^2$, and the partial pressure of the water vapor in it is

$$p_{1s} = 0.1130 \text{ lb}_f/\text{in.}^2$$

corresponding to a relative humidity $r = 33.3$ per cent. The water surface in the wind direction has a length of $x = 4$ in.

The Reynolds number for the air flow is

$$\mathrm{Re}_x = \frac{u_s x}{\nu} = \frac{10 \times \frac{4}{12}}{15.99 \times 10^{-5}} = 20{,}850$$

with $\nu = 15.99 \times 10^{-5}$ ft$^2$/sec from the Appendix. The flow therefore is laminar, and the average heat-transfer coefficient is given by Eqs. (7-14) and (7-15): $\overline{\mathrm{Nu}_x} = 0.664 \sqrt[3]{\mathrm{Pr}} \sqrt{\mathrm{Re}_x}$. The mass-transfer coefficient is determined by the analogous equation $\bar{h}_D x/D = 0.664 \sqrt[3]{\nu/D} \sqrt{\mathrm{Re}_x}$. The diffusion coefficient is $D = 1.127$ ft$^2$/hr introduced at a temperature $(59 + 68)/2$ F $= 63.5$ F (see Appendix). Therefore

$$\frac{\nu}{D} = \frac{15.99 \times 10^{-5} \times 3600}{1.127} = 0.511$$

This gives

$$\frac{\bar{h}_D x}{D} = 0.664 \sqrt[3]{0.511} \sqrt{20{,}850} = 76.7$$

and

$$\bar{h}_D = 76.7 \times 1.127 \times 1\tfrac{2}{4} = 259.3 \text{ ft/hr}$$

If it is desired to use the Lewis relation the heat-transfer coefficient must first be determined, with Pr $= 0.710$ for air,

$$\overline{\mathrm{Nu}_x} = 0.664 \sqrt{20{,}850} \times \sqrt[3]{0.710} = 85.5$$

and $\bar{h} = \overline{\mathrm{Nu}_x}(k/x) = 85.5 \times 0.0149 \times 1\tfrac{2}{4} = 3.8$ Btu/hr ft$^2$ F. For air at 68 F and 14.70 lb$_f$/in.$^2$, the specific heat per unit volume is $= 0.0175$ Btu/ft$^3$ F. Therefore $\bar{h}_D = 3.8/0.0175 = 217$ ft/hr. This value is too small by 16 per cent because the assumption $\alpha/D = 1$ under which the Lewis relation holds true is not quite fulfilled here. For approximate calculations on evaporation of water in air, however, the Lewis relation is very useful.

† E. Schmidt, *loc. cit.*

The partial pressure of the water vapor over the water surface is the saturation pressure at the water-surface temperature:

$$p_{1w} = 0.2473 \text{ lb}_f/\text{in.}^2$$

Therefore the amount of water evaporated per square foot per hour is (Eq. (16-36))

$$\dot{m}_{1w} = \frac{h_D}{R_1 T}(p_{1w} - p_{1s}) = \frac{240}{85.74 \times 519}(0.2473 - 0.1130)144 = 0.104 \text{ lb/hr ft}^2$$

**16-5. The Evaporation of Water into Air.** When evaporation of water into air or condensation of water from moist air takes place on a surface, the Lewis relation can be used for approximate calculations, since the ratio of thermal diffusivity to the diffusion coefficient has the value 0.835 at 68 F, which is not greatly different from the value 1. In Sec. 15-2, processes with moist air were studied with the enthalpy–specific-humidity diagram. Therefore, it is useful to transform Eq. (16-36) into a relationship which contains the specific humidities instead of the partial pressures. According to Sec. 15-2

$$s = \frac{M_1}{M_2}\frac{p_1}{p_2}$$

For small partial pressures of the water vapor it can be written that

$$s \approx \frac{M_1}{M_2}\frac{p_1}{p}$$

The range of states represented in the accompanying enthalpy–specific-humidity diagram (see Fig. 15-5) comprises only states with a vapor pressure smaller than 2.2 $\text{lb}_f/\text{in.}^2$ Within this range the above relation gives an error of less than 6 per cent. If we assume this to be tolerable we can write Eq. (16-36) in the form

$$\dot{m}_{1w} = \frac{h_D}{R_1}\frac{p}{T}\frac{M_2}{M_1}(s_w - s_s)$$

The gas constant $R_1$ of the water vapor may be expressed by the universal gas constant $\mathfrak{R}$ and the molecular weight $M_1$ of the water vapor:

$$R_1 = \frac{\mathfrak{R}}{M_1}$$

This gives

$$\dot{m}_{1w} = h_D \frac{p}{R_2 T}(s_w - s_s) = h_D \rho_2 (s_w - s_s)$$

where $\rho_2$ is the density of the dry air. The product $h_D \rho_2$ is called *evaporation coefficient* and is denoted by $\sigma$. The dimension of the evaporation coefficient is pounds per hour per square foot. The above equation assumes the simple form

$$\dot{m}_{1w} = \sigma(s_w - s_s) \tag{16-48}$$

The Lewis relation for the evaporation coefficient becomes

$$\sigma = h_D \rho_2 = \frac{h \rho_2}{C_p} = \frac{h}{c_p} \qquad (16\text{-}49)$$

The numerical value for the specific heat per pound of air can be inserted as 0.24 Btu/lb F. Experiments by different scientists have confirmed the above Lewis relation very well for turbulent flow,[1] but for laminar flow the agreement is not so good (see the above example).

Of special importance for evaporation processes is the temperature which a moist body assumes in an air stream when it exchanges heat only by convection. Such a body assumes a temperature which is lower than the air temperature, since, for the evaporation, heat is used up which must be transferred from the air to the body. This temperature can be calculated by the Lewis relation and is called the *wet-bulb temperature*, since it is assumed by a thermometer whose bulb is covered with a wet cloth.

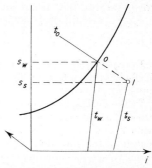

Fig. 16-12. Determining the humidity of air with a dry- and a wet-bulb thermometer.

The mass flow $\dot{m}_w$ and the heat flow $q_w$ per unit surface area are described by the equations

$$\dot{m}_{1w} = \sigma(s_w - s_s)$$
$$q_w = h(t_w - t_s)$$

When the body gets the whole amount of heat necessary for the evaporation from the air by convection, $q_w = \dot{m}_{1w} i_{fg}$ with $i_{fg}$ as heat of evaporation per pound of water. Then

$$\frac{t_s - t_w}{s_s - s_w} = i_{fg} \frac{\sigma}{h} = \frac{i_{fg}}{c_p} \qquad (16\text{-}50)$$

By this equation, the direction of the line is fixed which connects in the enthalpy–specific-humidity diagram the point 0 $(s_w, t_w)$ for the state of the air immediately over the surface with point 1 $(s_s, t_s)$ representing the state of the air at a great distance from the body (Fig. 16-12). R. Mollier[2] showed that the wet-bulb temperature $t_w$ is found in the diagram by finding the isotherm $t_0$ in the supersaturated region whose extension hits point 1.[3]

The moisture content of air is conveniently measured by the psychro-

---

[1] E. Kirschbaum, *Chem. Fabrik*, **14**:171–181 (1941).

[2] R. Mollier, "Stodola-Festschrift," Leipzig, 1929; F. Bosňjaković, "Technische Thermodynamik," vol. 2, p. 43, Theodor Steinkopff, Verlagsbuchhandlung, Dresden, 1937; Kirschbaum, *loc. cit.*

[3] This method is not quite correct, but it agrees very well with experiments.

metric method. A psychrometer consists of an ordinary, or *dry-bulb*, thermometer and a *wet-bulb* thermometer whose bulb is covered with a moist cloth. The air whose moisture is to be measured must be blown with sufficient velocity over the thermometers so that the heat exchange by radiation with the surroundings can be neglected as compared with the heat exchange by convection. By looking up the point in the enthalpy–specific-humidity diagram where the dry-bulb isotherm and the prolonged wet-bulb isotherm intersect, the state point of the air and its humidity are fixed.

**Example 16-3.** A measurement in air at 14.7 $lb_f/in.^2$ with a psychrometer gives a dry-bulb temperature of 68 F and a wet-bulb temperature of 50 F. The specific humidity of the air is to be determined.

In the enthalpy–specific-humidity diagram (Fig. 15-5) the isotherm 50 F in the supersaturated region must be extended until it intersects the isotherm 68 F. The point of intersection gives the state of the moist air. Its specific humidity can be read from the ordinate scale as $s = 0.0037$ *lb* of water per pound of dry air. If the enthalpy–specific-humidity diagram contains lines of constant relative humidity, this value for the investigated air can be read immediately. Otherwise it is necessary to calculate, with Eq. (15-3), the partial pressure of the water vapor of the investigated air and the partial pressure of water vapor of saturated air at 68 F from the specific humidities or to take these values from steam tables or the accompanying *s-i* diagram. The ratio of the two partial pressures is the relative humidity $r = 0.256$.

## PROBLEMS

**16-1.** Develop equations describing the concentration and temperature field as well as the rate of mass and heat transfer for laminar shear flow between a stationary and a moving surface (Couette flow). Assume a two-component gas mixture, one component flowing parallel to the plates only, whereas the other component is released on the stationary surface and absorbed on the moving surface. The mass fractions and temperatures on both surfaces can be prescribed. Otherwise, consider conditions parallel to those in Sec. 16-2.

**16-2.** Derive the boundary-layer mass-transport equation for two-dimensional, steady, laminar flow.

**16-3.** Calculate one of the temperature profiles shown in Fig. 16-3 for $Pr = 1$ by integration of the energy equation (for instance, with the method discussed in Secs. 6-6 and 7-6).

**16-4.** Calculate one of the mass-fraction profiles in Fig. 16-3 for $Sc = 0.7$ by integration of the boundary-layer equations (for instance, with the methods discussed in Secs. 6-6 and 7-6).

**16-5.** Calculate the amount of water vapor evaporated from the surface of a flat pan filled with water at 100 F (180 F) temperature into a stream of dry air flowing with a velocity of 50 fps and a temperature of 150 F parallel to the water surface. The length of the pan in the flow direction is ½ ft. Make calculation using the information in Sec. 16-2 as well as the similarity relations (Sec. 16-4), and compare results. Calculate also the heat flow through the water surface connected with this evaporation process.

PART E

# HEAT EXCHANGERS

Heat exchangers are devices in which heat is transferred from one fluid to another. They can be divided broadly into two classes. In the first class, both fluids are ducted simultaneously through the device and heat is transferred through separating walls. Such a type is called the *transfer-type heat exchanger*. To the second class belong heat exchangers through which the two fluids flow alternately. The device contains a solid matrix with sufficient heat capacity so that it can store the heat received from the hot fluid and transfer it to the cold fluid when it passes through the exchanger. Such a type is called the *storage-type heat exchanger*. Sometimes the matrix in such an exchanger is made to rotate through the two fluid passages which are arranged side by side and in this way to transfer heat from the hot to the cold fluid.

The basic relations for design calculations of transfer-type heat exchangers with simple flow arrangements have been discussed in Sec. 1-4. These relations make use of the logarithmic mean temperature difference. In this part another method based on the thermal effectiveness will be discussed. Relations for heat exchangers with other passage arrangements will also be presented, and calculation procedures for storage-type heat exchangers will be developed.

CHAPTER 17

# HEAT-EXCHANGER CALCULATIONS

**17-1. The Transfer-type Heat Exchanger.** In heat-exchanger calculations two tasks are encountered most frequently. Either the amount of heat to be transferred and the entrance and exit conditions of both fluids are prescribed and the necessary heat-exchanger area is to be determined, or the heat which can be transferred with an exchanger of known area has to be calculated. In the last case the inlet conditions of both fluids must be known. In the following calculations the two fluids will be distinguished by subscripts 1 and 2, inlet conditions will be denoted by a subscript $i$ and exit conditions by $e$, $m$ will denote the mass flow, and $c$ the specific heat. The specific heat is assumed constant. When the entrance conditions for both fluids $(m_1, c_1, t_{1i}, m_2, c_2, t_{2i})$ and the heat $Q$ to be transferred or one of the exit temperatures ($t_{1e}$ or $t_{2e}$) are known, a heat balance results in

$$Q = m_1 c_1 (t_{1i} - t_{1e}) \qquad (17\text{-}1)$$
$$Q = m_2 c_2 (t_{2e} - t_{2i}) \qquad (17\text{-}2)$$

The fluid 1 is assumed to be at the higher temperature. Heat loss of the exchanger will be assumed negligible. From these equations the two gas temperatures at the heat exchanger exit $t_{1e}$ and $t_{2e}$ or the heat flow $Q$ and the missing exit temperature can be calculated. This makes it possible to determine the log mean temperature difference, for instance with Eq. (1-31), and the heat-transferring area required is then obtained from Eq. (1-30).

For this second task, to calculate the amount of heat which is transferred by a heat exchanger with a certain heat-transferring area $A$, the following parameters will be prescribed: $m_1$, $m_2$, $c_1$, $c_2$, $t_{1i}$, $t_{2i}$, and $A$. This calculation cannot be performed in a direct way with the help of the log mean temperature difference because Eq. (1-31) requires a knowledge of the exit temperatures of both fluids. The calculation can be made only by a trial-and-error procedure. If, for instance, the exit temperature of fluid 1 is assumed, then Eq. (17-1) determines the amount $Q$ of heat transferred. Equation (17-2) determines the exit temperature of the second fluid. Equation (1-31) can now be solved for the log mean temperature difference, and Eq. (1-30) for the amount of heat transferred in the heat exchanger. If this heat flow $Q$ is not identical with the one obtained from Eq. (17-1), then the calculation has to be repeated with another assumption on the exit temperature.

A calculation procedure which is based on a parameter called *heat-exchanger effectiveness* avoids this difficulty and permits a direct solution for both problems. It is therefore being used more frequently in design calculations.

The heat-exchanger effectiveness is defined in the following way:

$$\epsilon = \frac{(t_i - t_e)_l}{t_{1i} - t_{2i}} \qquad (17\text{-}3)$$

In this equation the numerator contains the difference between inlet and exit temperature for the one of the two fluids for which this difference is larger. The denominator contains the difference between the two inlet temperatures. In addition, $\epsilon$ is always taken as positive. According to Eq. (17-1) that fluid for which the product $mc$ has the smaller value (indicated by subscript $s$) will have the larger temperature increase in flowing through the heat exchanger. Therefore, the amount of heat transferred per unit time in the exchanger can be expressed as

$$Q = (mc)_s \epsilon (t_{1i} - t_{2i}) \qquad (17\text{-}4)$$

This equation together with Eqs. (17-1) and (17-2) is sufficient for the task of determining the heat flow exchanged in a heat exchanger of prescribed surface area provided the inlet conditions and the heat-exchanger effectiveness are known. Equation (17-4) can be used to calculate the heat flow, and Eqs. (17-1) and (17-2) then determine the exit temperatures of the two fluids. Diagrams and analytic relations have been developed from which the heat-exchanger effectiveness for different flow configurations like parallel or counterflow, crossflow, etc., can be determined. As an example, the effectiveness will be calculated here for the condition of parallel flow.

For this calculation, we shall additionally assume that $m_1 c_1$ is the product with the smaller value. Therefore,

$$m_1 c_1 = (mc)_s \qquad (t_i - t_e)_l = t_{1i} - t_{1e} \qquad (17\text{-}5)$$

Accordingly, the thermal effectiveness is in this case defined by

$$\epsilon = \frac{t_{1i} - t_{1e}}{t_{1i} - t_{2i}}$$

Equation (1-28) can be written as

$$t_{1e} - t_{2e} = (t_{1i} - t_{2i}) e^{-\mu U A}$$

The temperature $t_{2e}$ can be expressed from Eqs. (17-1) and (17-2) as

$$t_{2e} = \frac{m_1 c_1}{m_2 c_2}(t_{1i} - t_{1e}) + t_{2i}$$

If in this equation the temperature difference $t_{1i} - t_{1e}$ is replaced by $\epsilon(t_{1e} - t_{2e})$, then Eq. (1-28) transforms into

$$\epsilon = \frac{1 - \exp\left[-\dfrac{UA}{m_1 c_1}\left(1 + \dfrac{m_1 c_1}{m_2 c_2}\right)\right]}{1 + (m_1 c_1 / m_2 c_2)}$$

This is the equation which expresses the heat-exchanger effectiveness in terms of the known parameters. The calculation can be repeated easily for the situation in which $m_2 c_2$ is smaller than $m_1 c_1$. It is found in this way that, in general, the following equation describes the heat-exchanger effectiveness for parallel-flow conditions:

$$\epsilon = \frac{1 - \exp\left\{-\dfrac{UA}{(mc)_s}\left[1 + \dfrac{(mc)_s}{(mc)_l}\right]\right\}}{1 + [(mc)_s/(mc)_l]} \qquad (17\text{-}6)$$

An analogous calculation determines the effectiveness for counterflow:

$$\epsilon = \frac{1 - \exp\left\{-\dfrac{UA}{(mc)_s}\left[1 - \dfrac{(mc)_s}{(mc)_l}\right]\right\}}{1 - \dfrac{(mc)_s}{(mc)_l}\exp\left\{-\dfrac{UA}{(mc)_s}\left[1 - \dfrac{(mc)_s}{(mc)_l}\right]\right\}} \qquad (17\text{-}7)$$

Results of calculations for other flow arrangements have been summarized by Bosňjaković[1] and Kays and London.[2] Table 17-1 contains relations for the thermal effectiveness of heat exchangers with various flow arrangements. The parameter $UA/(mc)_s$ is sometimes referred to as number of transfer units and denoted by NTU. Diagrams giving the thermal effectiveness of various heat exchangers have been published recently.[3]

**17-2. The Storage-type Heat Exchanger.** Transient heat-conduction processes occur in the storage-type heat exchanger. Warm and cold fluid flow alternately through such a heat exchanger, normally in counterflow. A mass of high heat capacity absorbs and rejects the transferred heat periodically. Storage-type heat exchangers with brick walls as heat

[1] F. Bosňjaković, M. Vilicic, and B. Slipcevic, *VDI-Forschungsheft* 432, p. 5/26, 1951.

[2] W. M. Kays and A. L. London, "Compact Heat Exchangers," National Press, Palo Alto, Calif., 1955.

[3] R. A. Stevens and T. R. Wolf, *ASME Paper* 55-A-90; J. Fernandez and T. R. Wolf, *ASME Paper* 55-A-89.

TABLE 17-1. THERMAL EFFECTIVENESS OF HEAT EXCHANGERS WITH VARIOUS FLOW ARRANGEMENTS

Parallel flow:
$$\epsilon = \frac{1 - e^{-[UA/(mc)_s][1+(mc)_s/(mc)_l]}}{1 + \frac{(mc)_s}{(mc)_l}}$$

Counterflow:
$$\epsilon = \frac{1 - e^{-[UA/(mc)_s][1-(mc)_s/(mc)_l]}}{1 - \frac{(mc)_s}{(mc)_l} e^{-[UA/(mc)_s][1-(mc)_s/(mc)_l]}}$$

Crossflow (both streams unmixed, analytical approximation):
$$\epsilon = 1 - e^{\left\{\left[\exp\left(-\frac{UA}{(mc)_s}\frac{(mc)_s}{(mc)_l}\eta\right) - 1\right]\frac{(mc)_l}{(mc)_s}\frac{1}{\eta}\right\}} *$$

where $\eta = \left[\frac{(mc)_s}{UA}\right]^{0.22}$

Crossflow (both streams mixed):†
$$\epsilon = \frac{\frac{UA}{(mc)_s}}{\frac{\frac{UA}{(mc)_s}}{1 - \exp\left[-\frac{UA}{(mc)_s}\right]} + \frac{\frac{UA}{(mc)_l}}{1 - \exp\left[-\frac{UA}{(mc)_l}\right]} - 1}$$

Crossflow [stream $(mc)_s$ unmixed]:
$$\epsilon = \frac{1 - e^{-[(mc)_s/(mc)_l]\{1-\exp[-UA/(mc)_s]\}}}{\frac{(mc)_s}{(mc)_l}}$$

Crossflow [stream $(mc)_l$ unmixed]:
$$\epsilon = 1 - e^{-[(mc)_l/(mc)_s]\{1-\exp[-UA/(mc)_l]\}}$$

Reverse flow exchanger:
$$\epsilon = \frac{2}{1 + \frac{(mc)_s}{(mc)_l} + \frac{1 + \exp\left[-\frac{UA}{(mc)_s}\sqrt{1 + \frac{(mc)_s^2}{(mc)_l^2}}\right]}{1 + \exp\left[-\frac{UA}{(mc)_s}\sqrt{1 + \frac{(mc)_s^2}{(mc)_l^2}}\right]}\sqrt{1 + \frac{(mc)_s^2}{(mc)_l^2}}}$$

Reverting pipes:

\* According to a communication by R. M. Drake, Jr.
† A stream is considered mixed when no temperature differences exist in any plane normal to the flow on its path through the heat exchanger.

accumulators are frequently used in the steel industry, for instance, as air preheaters for blast furnaces. In steam boiler plants such exchangers made of metal have found a definite place. Also in low-temperature engineering concerned with separation of gases by partial condensation, storage-type heat exchangers are used extensively. An exact calculation procedure for such a heat exchanger is extremely difficult and tedious.[1] Calculation procedures which are used for design purposes are usually simplified by certain assumptions. They can be mainly grouped into two classes. One procedure neglects temperature differences in a direction normal to the surface existing in the solid material in which the heat is stored. This calculation procedure is well adapted to a heat exchanger the matrix of which is constructed from thin-walled material with high heat conductivity. For instance, heat exchangers with matrices made of metal belong to this class. For heat exchangers with matrices of larger wall thickness and lower heat conductivity, the neglect of the temperature differences in the storage walls will lead to large errors, and a calculation procedure which accounts for this fact has to be developed. We shall first consider the calculation of a storage-type heat exchanger under the assumption that we can neglect temperature differences throughout the solid material of the matrix. Figure 17-1 is a sketch of such a heat exchanger. The exchanger can operate with either parallel flow or counterflow. Usually counterflow is employed. Figure 17-1 shows this latter flow arrangement. We start our calculation with a heat balance on a differential length $dx$ of the heat exchanger. The following nomenclature will be used: We call $A_L$ the heat-transferring surface per unit of the heat exchanger length $L$; $V_L$ is the free volume available for the gas flow through the heat exchanger per unit length, and $M_L$ the solid mass of the heat-exchanger matrix per unit length. $c_s$ will be the specific heat of the solid matrix. $\dot{m}$ is the mass flow of the fluid passing through the heat exchanger, $c$ its specific heat, and $\rho$ its density. An index $H$ or $C$ will be used on these terms to indicate whether the flow of the hotter or of the colder fluid through the heat exchanger is considered. $t$ indicates the

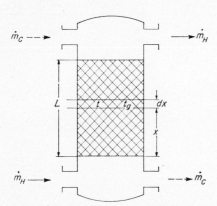

FIG. 17-1. Storage-type heat exchanger.

[1] For a very complete discussion of heat exchangers of both types the reader is referred to the book by H. Hausen, "Wärmeübertragung im Gegenstrom, Gleichstrom und Kreuzstrom," Springer-Verlag, Berlin, 1950.

temperature of the solid at the location $x$, and $t_g$ the temperature of the fluid (gas) at the same position. A heat balance for the solid material within the height $dx$ of the heat exchanger for a time element $d\tau$ results in

$$hA_L(t_g - t)\,dx\,d\tau = c_s M_L\,dx\,\frac{\partial t}{\partial \tau}\,d\tau \qquad (17\text{-}8)$$

$h$ indicates the film heat-transfer coefficient from the fluid to the matrix surface. A similar heat balance performed for the fluid moving through the heat-exchanger element of height $dx$ gives

$$hA_L(t - t_g)\,dx\,d\tau = \rho c V_L\,dx\,\frac{\partial t_g}{\partial \tau}\,d\tau + c\dot{m}\,d\tau\,\frac{\partial t_g}{\partial x}\,dx \qquad (17\text{-}9)$$

The first term on the right-hand side of the last equation describes the heat stored in the fluid which is contained in the height $dx$ of the heat exchanger. The second term describes the difference in the internal energy carried in and out of the heat exchanger with the fluid. The two equations can be simplified.

$$\frac{\partial t}{\partial \tau} = \frac{hA_L}{c_s M_L}(t_g - t) \qquad (17\text{-}10)$$

$$\frac{\partial t_g}{\partial x} + \frac{\rho V_L}{\dot{m}}\frac{\partial t_g}{\partial \tau} = \frac{hA_L}{c\dot{m}}(t - t_g) \qquad (17\text{-}11)$$

In almost all applications, the term $\rho V_L/\dot{m}$ is very small, and accordingly, the second term on the left-hand side of Eq. (17-11) is usually neglected. The two resulting equations constitute two partial differential equations for the two unknown temperatures, of the solid matrix and of the gas flowing through the heat exchanger, as a function of time and position $x$. To obtain a solution, boundary conditions must be prescribed for these differential equations. Heat exchangers of storage type can operate in two ways, either continuously or in a single operation such that a heat exchanger, the matrix of which has been raised to a certain temperature, is used to heat a fluid for a certain period. Heat exchangers of this type are quite frequently used, for instance, in connection with modern wind tunnels which operate only for a short time. A large heat flow is required to heat the air of the wind tunnel to a desired temperature during its operation, and it is difficult to produce this large heat flow, for instance, by electric heating. It is more appropriate to store this heat in a regenerative heat exchanger previous to the operation of the wind tunnel. The boundary conditions which describe this single operation of a regenerator are

At $x = 0$: $\qquad t_g = t_{g0} = \text{const}$
At $\tau = 0$: $\qquad t = t_0 = \text{const}$

In the applications mentioned at the beginning of this paragraph, the heat exchangers are usually in continuous operation in the sense that alternately cold and hot fluid is directed through the heat exchanger and that the temperature variations in it are quasi-steady. The boundary conditions describing this operation for a counterflow regenerator during the heating period of time $\tau_H$ are

At $x = 0$: $\qquad\qquad\qquad t_g = t_{gH}$

During the cooling period with time $\tau_C$, at $x = L$, they are $t_g = t_{gC}$. In addition, the matrix material at the beginning of the heating period must have the same temperature as at the end of the cooling period and vice versa:

$$t \text{ (begin heating)} = t \text{ (end cooling)}$$
$$t \text{ (begin cooling)} = t \text{ (end heating)}$$

The differential equations which describe the temperature field and temperature variation in the heat exchanger can be simplified by the introduction of the following new variables:

$$\xi = \frac{hA_L}{cm} x \qquad \eta = \frac{hA_L}{c_s M_L} \tau \qquad (17\text{-}12)$$

With these variables Eqs. (17-10) and (17-11) become

$$\frac{\partial t}{\partial \eta} = t_g - t \qquad (17\text{-}13)$$

$$\frac{\partial t_g}{\partial \xi} = t - t_g \qquad (17\text{-}14)$$

In Eq. (17-11) the second term on the left-hand side was disregarded. This system of two interconnected partial differential equations together with the boundary conditions described above has been solved only for specific situations. A. Anzelius[1] obtained a solution for single operation of the storage-type heat exchanger, and several scientists have solved the problem by graphical or numerical means for the continuous operation. Figure 17-2 presents the results of the calculation by Anzelius. The temperature of the fluid as well as of the matrix throughout the heat exchanger for any time can be read from this diagram. Of special importance is a knowledge of the temperature with which the fluid leaves the heat exchanger. This temperature can be read from the upper diagram in the figure. It may be observed that the fluid exit temperature is nearly constant for a considerable part of the time of operation. The calculations for the continuous operation of the heat exchanger have been evaluated by Coppage and London[2] in terms of a heat-exchanger effec-

---

[1] A. Anzelius, *Z. angew. Math. Mech.*, **6**:291 (1926).
[2] See Kays and London, *op. cit.*

tiveness $\epsilon$. Figure 17-3 shows as an example the result of this evaluation for a counterflow storage-type exchanger for the condition $m_H c_H = m_C c_C$. The uppermost curve in this diagram is identical with the effectiveness of a steady-type heat exchanger as given by Eq. (17-7). It can be observed that a storage-type heat exchanger has the same effectiveness when the heat capacity of the solid material is very large as compared with the heat

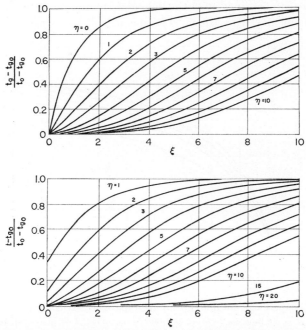

Fig. 17-2. Variation of matrix $t$ and fluid temperature $t_g$ in a storage-type heat exchanger initially at constant temperature $t_0$ and suddenly subject to a fluid flow of different temperature $t_{g0}$. (*According to A. Anzelius, from H. Hausen,* "*Wärmeübertragung, im Gegenstrom, Gleichstrom und Kreuzstrom,*" p. 336, *Springer-Verlag, Berlin* 1950.)

capacity of the fluids. When the ratio $c_s M/c\dot{m}$ is finite, the effectiveness is smaller than for the steady type and can be obtained from the diagram in the figure. It has to be kept in mind that all calculations up to now are based on the assumption that temperature differences at a certain location $x$ within the solid material of the matrix can be neglected.

H. Hausen has developed an elegant method in which the influence of temperature variations in the solid material on the performance of a heat exchanger can be determined.[1] This method will be described now. In a storage-type heat exchanger through which flow alternately hot and

---

[1] H. Hausen, *Arch. Eisenhüttenw.*, **12**:473–480 (1938–1939).

cold gases in opposite directions, the temperatures of the gases as well as of the heat-accumulating walls change not only with respect to location but also with time. The variation of temperatures with time in a cross section within the heat exchanger is given in Fig. 17-4. Figure 17-5 shows the temperature distribution in the wall for various elapsed times after switching from the heating to the cooling period ($\tau_1 < \tau_2 < \tau_3 < \tau_4 < \tau_5$). Apart from the heat-exchanger ends, the temperature $t_{gH}$ of the hot gas rises almost linearly with time, and the temperature $t_{gC}$ decreases practically linearly with time. The mean wall temperature $t_m$ changes

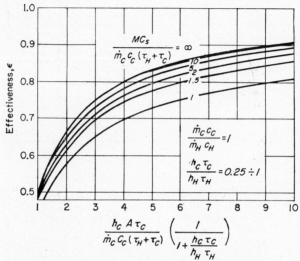

FIG. 17-3. Thermal effectiveness of storage-type heat exchangers in quasi-steady operation. $\tau_c$ is cooling, $\tau_H$ heating period; $M$ is mass; $A$ is heat-transferring area; $C_s$ is specific heat of matrix. (*From W. M. Kays and A. L. London, "Compact Heat Exchangers," National Press, Palo Alto, Calif., 1955.*)

linearly with time. The surface temperature $t_w$ of the wall changes rapidly immediately after switching and soon thereafter approaches a linear variation.

For the quantity of heat $Q_P$† which is transferred to the heat-exchanger wall by the hot gas during the time $\tau_H$ of the heating period, the equation

$$Q_P = h_H A (\bar{t}_{gH} - \bar{t}_{wH}) \tau_H$$

can be written. There $h_H$ is the film heat-transfer coefficient of the gas to the wall, $\tau_H$ is the time of the heating cycle, and $\bar{t}_{gH}$ and $\bar{t}_{wH}$ are the mean values of the temperatures with respect to time. These mean values of the temperatures agree for practical purposes with the instantaneous

† It may be emphasized that $Q_P$ is not the heat flow per unit time, but the heat transferred in the heating (and in the cooling) period.

values in the middle of the heating period. The same quantity of heat is transferred from the wall to the cold gas during the cooling period of duration $\tau_C$. With the film heat-transfer coefficient $h_C$ existing in the period $\tau_C$ and the corresponding mean temperatures $(\bar{t}_{wC} - \bar{t}_{gC})$,

$$Q_P = h_C A (\bar{t}_{wC} - \bar{t}_{gC}) \tau_C$$

In these equations $A$ is the surface of the heat-accumulating walls which is in contact with the gases. Now the surface temperature is to be eliminated from the equations in the same manner as in Sec. 1-3 for the steady-state heat transfer. For this it is necessary to compute the mean

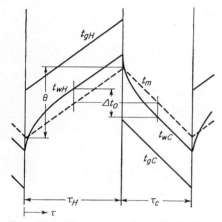

FIG. 17-4. Time-temperature fluctuation of the walls and gases in a heat exchanger.

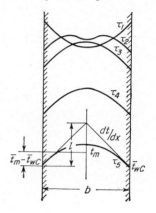

FIG. 17-5. Temperature profiles in the wall of a heat exchanger with time as parameter. [From H. Hausen, Arch. Eisenhüttenw., **12**:473–480 (1938).]

temperature difference $\Delta \bar{t}_w = \bar{t}_{wH} - \bar{t}_{wC}$, which according to Fig. 17-4 is composed of the two components $(\bar{t}_{wH} - t_m)$ and $(\bar{t}_m - \bar{t}_{wC})$. The latter value is plotted in Fig. 17-5. If the heat-accumulating walls are assumed plane, the temperature distribution in them can be found by Eq. (4-10). If the temperature distribution is linear with respect to time as is shown in Fig. 17-4, then $\partial t/\partial \tau = $ const. Thus $\alpha(\partial^2 t/\partial x^2) = $ const, and the double integration leads to the parabolas plotted in Fig. 17-5. A different temperature distribution occurs only immediately after switching, since at this moment the temperature change with respect to time is also different. The slope of the parabolic temperature profile at the surface of the wall can be determined from the fact that the quantity of heat which passes from the surface to the gas per unit of time must equal the heat flowing by conduction from the inside of the wall to the surface, or

$$\frac{Q_P}{\tau_C} = -kA \left(\frac{\partial t}{\partial x}\right)_w$$

The parameter $l$ in Fig. 17-5 is $l = (b/2)(\partial t/\partial x)_w = (Q_P/A)(b/2k\tau_C)$. The height of the parabola is half this value, and the distance $(\bar{t}_m - \bar{t}_{wC})$ is two-thirds the height of the parabola. Therefore,

$$\bar{t}_m - \bar{t}_{wC} = \frac{Q_P}{A} \frac{b}{6k\tau_C}$$

In the same manner for the heating period

$$\bar{t}_{wH} - \bar{t}_m = \frac{Q_P}{A} \frac{b}{6k\tau_H}$$

and consequently for the temperature difference

$$\bar{t}_{wH} - t_{wC} = \frac{Q_P}{A}\left(\frac{1}{\tau_H} + \frac{1}{\tau_C}\right)\frac{b}{6k}$$

If the first two equations of the paragraph are solved for the temperature,

$$t_{gH} - t_{wH} = \frac{Q_P}{A}\frac{1}{h_H \tau_H}$$

$$t_{wC} - t_{gC} = \frac{Q_P}{A}\frac{1}{h_C \tau_C}$$

and by adding the last three equations,

$$t_{gH} - t_{gC} = \frac{Q_P}{A}\left[\frac{1}{h_H \tau_H} + \frac{1}{h_C \tau_C} + \left(\frac{1}{\tau_H} + \frac{1}{\tau_C}\right)\frac{b}{6k}\right]$$

Hence, the result is

$$Q_P = \frac{1}{\frac{1}{h_H \tau_H} + \frac{1}{h_C \tau_C} + \left(\frac{1}{\tau_H} + \frac{1}{\tau_C}\right)\frac{b}{6k}} A(\bar{t}_{gH} - \bar{t}_{gC}) \qquad (17\text{-}15)$$

With this equation the heat $Q_P$ which is transferred from the hot to the cold gas during a full period of the duration $(\tau_H + \tau_C)$ can be computed if the mean values of the gas temperatures with respect to time $\bar{t}_{gH}$ and $\bar{t}_{gC}$ are known. The equation was deduced only for a cross section in the heat exchanger. If it is desired to compute the heat delivery of the total regenerator, then there must be introduced for the temperature difference $(\bar{t}_{gH} - t_{gC})$ the logarithmic mean temperature difference according to Eq. (1-31), which is obtained from the intake and outlet temperatures of the gases. Equation (17-15) shows a great similarity to Eq. (1-16) for steady-state heat flow through a plane wall. The fraction in Eq. (17-15) corresponds to the over-all heat-transfer coefficient in Eq. (1-17).

If the heating period is equal to the cooling period $\tau_H = \tau_C = \tau$, then $Q$, the heat exchanged per unit of time, is given as

$$Q = \frac{Q_P}{2\tau} = \frac{1}{(1/h_H) + (1/h_C) + (b/3k)}\frac{A}{2}(\bar{t}_{gH} - \bar{t}_{gC})$$

If this relation is compared with Eq. (1-17), it must be borne in mind that in the transfer-type heat exchanger only one surface of the wall has to be used for $A$ while here both surfaces have to be employed. The heat delivery of the two transfer heat exchangers becomes equal when the wall thickness of the transfer-type heat exchanger amounts to one-third the wall thickness of the storage-type heat exchanger. The difference is accounted for by the fact that with the transfer-type heat exchanger the total quantity of heat flows through the wall while with the storage-type heat exchanger heat flows into the wall through both surfaces during the heating period and during the cold period leaves the wall in the same way it entered. Hence it is not necessary that the heat flow through the entire wall in the case of the storage-type heat exchanger.

The exchange of heat may be reduced by a layer of dust. If a layer of dust of thickness $b_d$ and thermal conductivity $k_d$ is placed on each surface of the wall, Eq. (17-15) becomes

$$Q_P = \frac{1}{\dfrac{1}{h_H \tau_H} + \dfrac{1}{h_C \tau_C} + \left(\dfrac{1}{\tau_H} + \dfrac{1}{\tau_C}\right)\left(\dfrac{b}{6k} + \dfrac{b_d}{k_d}\right)} A(\bar{t}_{gH} - \bar{t}_{gC})$$

The maximum change $\vartheta$ of the mean wall temperature during a hot or cold period is obtained from the equation

$$Q_P = c\rho b \frac{A}{2} \vartheta$$

in which $c$ is the specific heat and $\rho$ the density of the wall.

In the preceding calculations the processes immediately after switching and at the heat-exchanger entrance and exit of the flow passageways were simplified. Considerable inaccuracies in the results occur only in rare cases by this simplification. H. Hausen[1] has also examined these processes and has published diagrams from which such influences can be seen. According to Hausen's diagram the term $b/6k(1/\tau_H + 1/\tau_C)$ in Eq. (17-15) must be reduced by 10 per cent if $(b^2/2\alpha)(1/\tau_H + 1/\tau_C) = 3$ ($\alpha$ is the thermal diffusivity of the wall). For values lower than 3 the correction is less.

The influences of the heat-exchanger entrance and exit are determined by two characteristic values, $\kappa/c\rho s$ and $\kappa A/(\dot{m}_H c_H \tau_H + \dot{m}_C c_C \tau_C)$, where $\kappa$ is the fraction in Eq. (17-15), that is, the value corresponding to the overall heat-transfer coefficient. $\dot{m}_H$ and $\dot{m}_C$ are the mass flows of the hot and cold gas, and $c_H$ and $c_C$ are their respective specific heats. The exchange of heat according to Eq. (17-15) is correctly determined within 10 per cent as long as $\kappa/c\rho s < 0.3$ at $\kappa A/(\dot{m}_H c_H \tau_H + \dot{m}_C c_C \tau_C) \approx 0$ or $\kappa/c\rho s < 7$ at $\kappa A/(\dot{m}_H c_H \tau_H + \dot{m}_C c_C \tau_C) \approx 50$. The actual value $Q_P$ is somewhat less

[1] H. Hausen, *Z. Ver. deut. Ingr. Beih. Verfahrenstechn.* no. 2, 1942, pp. 31–43.

than the computed value. The heat exchangers which are most frequently used fulfill the above conditions.

## PROBLEMS

**17-1.** Derive the equation for the effectiveness of a counterflow heat exchanger as given in Table 17-1.

**17-2.** A storage-type heat exchanger with 1-ft² cross section is constructed of steel balls ½ in. in diameter. This matrix is heated to a uniform temperature of 500 F, and then a flow of air with atmospheric temperature and pressure is started. The approach velocity of the air is 50 fps. The length of that portion of the heat exchanger containing the matrix is 1 ft. For how long a time can air be blown through the exchanger before its temperature at the exit differs by 50 F from the original matrix temperature?

# APPENDIX OF PROPERTY VALUES

The Appendix is intended primarily to support the text and not necessarily to fulfill the requirements of a handbook; therefore, the accent is on brevity and simplicity. The reader or researcher desiring more detailed or additional material should consult any of the excellent and extensive publications listed here.

Tables A-1 and A-3 were taken from the book "Grundlagen des Wärmeaustausches (Stoffwerte)" by B. Koch (Dissen T. W. Verlagsanstalt H. Bücke und Söhne, 1950), with the kind permission of the author. Some supplementary information has been added to Table A-1 with some corrections of the thermal conductivity values in consideration of newer data. Table A-2 presents data on thermal conductivity of metals and alloys at low temperatures, taken from the paper by C. H. Lees [*Phil. Trans. Roy. Soc. London*, **208A**:381 (1908)]. Table A-3 gives properties of a number of liquids in the saturated state. Excluding the region in the neighborhood of the critical state, the variation of the listed properties with pressure is small so that the table has a universal value. The viscosity is given as the kinematic viscosity.

Table A-4 presents the properties of several important gases at atmospheric pressure. Table A-4 was computed largely on the basis of the National Bureau of Standards *Circular* 564 except for the data on helium and ammonia and supplementary data for air and hydrogen. The supplementary data for air was taken from Stops [*Nature*, **164**:966–967 (1949)] and Glassman and Bonilla [*Chem. Eng. Progr. Symposium Ser.* 5, **49** (1953)]. These data for air are considered to be the best information available at this time. The data for helium were taken from J. B. Ubbink and W. J. de Haas [Thermal Conductivity of Gases, *Physics*, **10** (1943)], W. H. Keesom ("Helium," Elsevier Publishing Company, Amsterdam, 1942), S. W. Akin (The Thermodynamic Properties of Helium, *Trans. ASME*, 1950), and K. K. Kelley (*Bur. Mines Bull.* 371, 1934). The data for ammonia came from B. Koch (see above) except for the thermal conductivity, which came from J. M. Lenoir (*Univ. Ark. Bull.* 18). The supplementary data for hydrogen was taken from Keenan and Kaye ("Gas Tables") and from J. M. Lenoir (*Univ. Ark. Bull.* 18). Excluding again the region of the critical state, the property values for other pressures can be obtained from those given in the following manner. The density can be determined from the ideal gas equation of state $\rho = p/RT$. From this it follows that at any temperature the density is $\rho = \rho_0(p/p_0)$, where $p_0 = 14.696$ lb$_f$/in.$^2$ and $\rho_0$ is the density as given in Table A-4 for the temperature in question. Furthermore, the specific heat $c_p$ varies only a little with pressure for a fairly wide range. This pressure independence is true also for the thermal conductivity $k$, the dynamic viscosity $\mu$, and therefore for the Prandtl number Pr. The kinematic viscosity $\nu$ and the thermal diffusivity $\alpha$ are inversely proportional to the density; therefore, for a given temperature, they are inversely proportional to the pressure. In this way property values for other conditions within the ideal gas range can be found from the ones listed in the table.

The thermal and transport properties of gas mixtures require additional considerations. The properties of primary interest are the density, specific heat, viscosity, and thermal conductivity for binary gas mixtures. The density of the mixture presents no difficulty nor does the specific heat, since these values for nonreactive mixtures can be determined easily from the component properties. For example, the specific heat of such a mixture is given by

$$c_{p_m} = w c_{p_1} + (1 - w) c_{p_2}$$

where $w$ is the mass fraction of component 1. The thermal conductivity and the viscosity, however, present somewhat more difficulty than this. Computation of either of these properties in simple ratio or as molal averages of the components may lead to values that are quite unsatisfactory when compared with experimental measurements. Fortunately calculation methods of reasonable simplicity exist which depend on a knowledge of the pure component properties and provide reasonably accurate results. Wilke[1] gives the following expression for the viscosity of a binary mixture of gases:

$$\mu_m = \frac{\mu_1}{1 + (x_2/x_1)\phi_{12}} + \frac{\mu_2}{1 + (x_1/x_2)\phi_{21}}$$

where
$$\phi_{ij} = \frac{[1 + (\mu_i/\mu_j)^{1/2}(M_j/M_i)^{1/4}]^2}{(4/\sqrt{2})[1 + (M_i/M_j)]^{1/2}}$$

where $M_i$ and $M_j$ = molecular weights of component pure gases
$x_1$ and $x_2$ = mole fractions
$\mu_i$ and $\mu_j$ = viscosities of pure components

Lindsay and Bromley[2] present a method based on the Wassiljewa[3] computation method for thermal conductivities which appears satisfactory, by the relation

$$k_m = \frac{k_1}{1 + A_1(x_2/x_1)} + \frac{k_2}{1 + A_2(x_1/x_2)}$$

where the Wassiljewa constants $A_1$ and $A_2$ are

$$A_1 = \frac{1}{4}\left\{1 + \left[\frac{\mu_1}{\mu_2}\left(\frac{M_2}{M_1}\right)^{3/4}\frac{1 + (S_1/T)}{1 + (S_2/T)}\right]^{1/2}\right\}^2 \frac{1 + (S_{12}/T)}{1 + (S_1/T)}$$

$$A_2 = \frac{1}{4}\left\{1 + \left[\frac{\mu_2}{\mu_1}\left(\frac{M_1}{M_2}\right)^{3/4}\frac{1 + (S_2/T)}{1 + (S_1/T)}\right]^{1/2}\right\}^2 \frac{1 + (S_{12}/T)}{1 + (S_2/T)}$$

where $\mu$ = viscosity of pure component
$M$ = molecular weight
$T$ = absolute temperature, K
$S$ = Sutherland constant

$S_{12} = \sqrt{S_1 S_2}$ except for gases containing water vapor or ammonia, in which case $S_{12} = 0.733\sqrt{S_1 S_2}$. Where Sutherland constants are not available, $S = 1.5 T_B{}^*$ is recommended. The general effect of mixing two gases of different molecular weights is that the resulting thermal conductivity is lower than that calculated by molal averages. Mixtures of gases containing water vapor or ammonia have thermal conductivities higher than those values predicted by molal averages. Lenoir[4] has extended the calculations shown here to polycomponent mixtures and has calculated a number of Wassiljewa constants from the above equations. Lenoir also gives thermal-conductivity values for many technically useful gases at atmospheric pressure.

[1] C. R. Wilke, *J. Chem. Physics*, **18**:517 (1950).
[2] A. L. Lindsay and L. A. Bromley, *Ind. Eng. Chem.*, **42**:1, 508 (1950).
[3] A. Wassiljewa, *Physik Z.*, **5**:737 (1904).
* $T_B$ is the normal boiling temperature of the component in question.
[4] J. M. Lenoir, *Univ. Ark. Eng. Expt. Sta. Bull.* 18, 1953.

APPENDIX OF PROPERTY VALUES     495

The dependence of the property values on pressure and temperature in the neighborhood of the critical state is shown for water in Figs. A-1 to A-8. A similar dependence exists for other materials, and therefore deviations from the values calculated for the ideal gas state can be at least approximately determined from the fact that the percentage variation of any value in the region of the critical state has the same order of magnitude for all materials at corresponding locations with respect to their critical state, i.e., equal values of $p/p_c$ and $T/T_c$, where $p_c$ and $T_c$ represent the critical pressure and critical temperature, respectively. Figures A-1 to A-3 were prepared from steam tables, and Figs. A-4 and A-5 are from measurements by K. Sigwart [*Forsch. Gebiete Ingenieurw.*, **7**:215–230 (1936)] with some changes in the superheat region to bring the values into agreement with other measurements. Figures A-6 and A-7 are from National Bureau of Standards *Circular* 564 and Granet and Gould, *Petroleum Refiner*, May, 1953. Figure A-8 has been calculated from the data shown on the other curves.

Table A-5 is presented to give typical values of nonmetallic substances and stems from several sources notably Marks' "Mechanical Engineers' Handbook" and McAdams' "Heat Transmission." Table A-5 has been supplemented by Table A-6 giving thermal conductivities of insulating materials over their useful temperature range. These data also were taken from L. S. Marks ("Mechanical Engineers' Handbook," 5th ed., McGraw-Hill Book Company, Inc., New York, 1951).

Because of the importance of molten metals in contemporary power-generation systems, Table A-7 has been added giving the properties of several molten metals and molten-metal alloys. The data on which the table was formulated were taken from "Liquid Metals Handbook" (2d ed., Government Printing Office, Washington, D.C., 1952).

Table A-8 gives the critical states of several technically important elements and compounds. These data were taken from "Handbook of Chemistry and Physics," 38th ed., Chemical Rubber Publishing Company, Cleveland, Ohio, 1956–1957.

Table A-9 gives some available diffusion coefficients and the corresponding Schmidt numbers. The supplementary data shown here were taken largely from the data reported in the "Handbook of Chemistry and Physics."

Tables A-10 and A-11 give emissivities for thermal radiation from selected sources which are indicated in the table. The agreement of data on emissivities between investigators is in general not so good, but it is believed that the data shown are reliable.

Table A-12 and A-13 give values of surface tension of various liquids, the data for which were taken completely from data reported in the "Handbook of Chemistry and Physics."

## Table A-1. Property Values

| Metals | Properties at 68 F | | | | k, thermal conductivity, Btu/hr ft F | | | | | | | | | |
|---|---|---|---|---|---|---|---|---|---|---|---|---|---|---|
| | $\rho$ lb/ft³ | $c_p$ $\frac{\text{Btu}}{\text{lb F}}$ | $k$ $\frac{\text{Btu}}{\text{hr ft F}}$ | $\alpha$ $\frac{\text{ft}^2}{\text{hr}}$ | −148 F −100 C | 32 F 0 C | 212 F 100 C | 392 F 200 C | 572 F 300 C | 752 F 400 C | 1112 F 600 C | 1472 F 800 C | 1832 F 1000 C | 2192 F 1200 C |
| **Aluminum:** | | | | | | | | | | | | | | |
| Pure | 169 | 0.214 | 118 | 3.665 | 124 | 117 | 119 | 124 | 132 | 144 | | | | |
| Al-Cu (Duralumin) 94–96 Al, 3–5 Cu, trace Mg | 174 | 0.211 | 95 | 2.580 | 73 | 92 | 105 | 112 | | | | | | |
| Al-Mg (Hydronalium) 91–95 Al, 5–9 Mg | 163 | 0.216 | 65 | 1.860 | 54 | 63 | 73 | 82 | | | | | | |
| Al-Si (Silumin) 87 Al, 13 Si | 166 | 0.208 | 95 | 2.773 | 86 | 94 | 101 | 107 | | | | | | |
| Al-Si (Silumin, copper bearing) 86.5 Al, 12.5 Si, 1 Cu | 166 | 0.207 | 79 | 2.311 | 69 | 79 | 83 | 88 | 93 | | | | | |
| Al-Si (Alusil) 78–80 Al, 20–22 Si | 164 | 0.204 | 93 | 2.762 | 83 | 91 | 97 | 101 | 103 | | | | | |
| Al-Mg-Si 97 Al, 1 Mg, 1 Si, 1 Mn | 169 | 0.213 | 102 | 2.859 | ..... | 101 | 109 | 118 | | | | | | |
| **Lead** | 710 | 0.031 | 20 | 0.924 | 21.3 | 20.3 | 19.3 | 18.2 | 17.2 | | | | | |
| **Iron:** | | | | | | | | | | | | | | |
| Pure | 493 | 0.108 | 42 | 0.785 | 50 | 42 | 39 | 36 | 32 | 28 | 23 | 21 | 20 | 21 |
| Wrought iron (C < 0.5%) | 490 | 0.11 | 34 | 0.634 | ..... | 34 | 33 | 30 | 28 | 26 | 21 | 19 | 19 | 19 |
| Cast iron (C ≈ 4%) | 454 | 0.10 | 30 | 0.666 | | | | | | | | | | |
| Steel (C max ≈ 1.5%) | | | | | | | | | | | | | | |
| Carbon steel C ≈ 0.5% | 489 | 0.111 | 31 | 0.570 | ..... | 32 | 30 | 28 | 26 | 24 | 20 | 18 | 17 | 18 |
| 1.0% | 487 | 0.113 | 25 | 0.452 | ..... | 25 | 25 | 24 | 23 | 21 | 19 | 17 | 16 | 17 |
| 1.5% | 484 | 0.116 | 21 | 0.376 | ..... | 21 | 21 | 21 | 20 | 19 | 18 | 16 | 16 | 17 |
| Nickel steel Ni ≈ 0% | 493 | 0.108 | 42 | 0.785 | | | | | | | | | | |
| 10% | 496 | 0.11 | 15 | 0.279 | | | | | | | | | | |
| 20% | 499 | 0.11 | 11 | 0.204 | | | | | | | | | | |
| 30% | 504 | 0.11 | 7 | 0.118 | | | | | | | | | | |
| 40% | 510 | 0.11 | 6 | 0.108 | | | | | | | | | | |
| 50% | 516 | 0.11 | 8 | 0.140 | | | | | | | | | | |
| 60% | 523 | 0.11 | 11 | 0.182 | | | | | | | | | | |
| 70% | 531 | 0.11 | 15 | 0.258 | | | | | | | | | | |
| 80% | 538 | 0.11 | 20 | 0.344 | | | | | | | | | | |
| 90% | 547 | 0.11 | 27 | 0.452 | | | | | | | | | | |
| 100% | 556 | 0.106 | 52 | 0.892 | | | | | | | | | | |

| Material | | | | | | | | | | | | | | | |
|---|---|---|---|---|---|---|---|---|---|---|---|---|---|---|---|
| Invar Ni = 36% | 508 | 0.11 | 6.2 | 0.108 | 50 | | | | | | | | | 21 | 17 |
| Chrome steel Cr = 0% | 493 | 0.108 | 42 | 0.785 | | 42 | 39 | 36 | 32 | 28 | 23 | 21 | 20 | | |
| 1% | 491 | 0.11 | 35 | 0.645 | | 36 | 32 | 30 | 27 | 24 | 21 | 19 | 19 | | |
| 2% | 491 | 0.11 | 30 | 0.559 | | 31 | 28 | 26 | 24 | 22 | 19 | 18 | 18 | | |
| 5% | 489 | 0.11 | 23 | 0.430 | | 23 | 22 | 21 | 21 | 19 | 17 | 17 | 17 | | |
| 10% | 486 | 0.11 | 18 | 0.344 | | 18 | 18 | 18 | 17 | 17 | 16 | 17 | | | |
| 20% | 480 | 0.11 | 13 | 0.258 | | 13 | 13 | 13 | 13 | 14 | 15 | | | | |
| 30% | 476 | 0.11 | 11 | 0.204 | | | | | | | | | | | |
| Cr-Ni (chrome-nickel): 15 Cr, 10 Ni | 491 | 0.11 | 11 | 0.204 | | 9.4 | 10 | 10 | 11 | 11 | 13 | 15 | 18 | | |
| 18 Cr, 8 Ni (V2A) | 488 | 0.11 | 9.4 | 0.172 | | | | | | | | | | | |
| 20 Cr, 15 Ni | 489 | 0.11 | 8.7 | 0.161 | | | | | | | | | | | |
| 25 Cr, 20 Ni | 491 | 0.11 | 7.4 | 0.140 | | | | | | | | | 13 | | |
| Ni-Cr (nickel-chrome): 80 Ni, 15 Cr | 532 | 0.11 | 10 | 0.172 | | | | | | | | | | | |
| 60 Ni, 15 Cr | 516 | 0.11 | 7.4 | 0.129 | | | | | | | | | | | |
| 40 Ni, 15 Cr | 504 | 0.11 | 6.7 | 0.118 | | 8.1 | 8.7 | 8.7 | 9.4 | 10 | 11 | | | | |
| 20 Ni, 15 Cr | 491 | 0.11 | 8.1 | 0.151 | | | | | | | | | | | |
| Cr-Ni-Al: 6 Cr, 1.5 Al, 0.5 Si (Sicromal 8) | 482 | 0.117 | 13 | 0.237 | | | | | | | | | | | |
| 24 Cr, 2.5 Al, 0.5 Si (Sicromal 12) | 479 | 0.118 | 11 | 0.194 | | | | | | | | | | | |
| Manganese steel Mn = 0% | 493 | 0.118 | 42 | 0.784 | | | | | | | | | | | |
| 1% | 491 | 0.11 | 29 | 0.538 | | 22 | 21 | 21 | 21 | 20 | 19 | | | | |
| 2% | 491 | 0.11 | 22 | 0.376 | | | | | | | | | | | |
| 5% | 490 | 0.11 | 13 | 0.247 | | | | | | | | | | | |
| 10% | 487 | 0.11 | 10 | 0.194 | | | | | | | | | | | |
| Tungsten steel W = 0% | 493 | 0.108 | 42 | 0.785 | | | | | | | | | | | |
| 1% | 494 | 0.107 | 38 | 0.720 | | 36 | 34 | 31 | 28 | 26 | 21 | | | | |
| 2% | 497 | 0.106 | 36 | 0.677 | | | | | | | | | | | |
| 5% | 504 | 0.104 | 31 | 0.591 | | | | | | | | | | | |
| 10% | 519 | 0.100 | 28 | 0.527 | | | | | | | | | | | |
| 20% | 551 | 0.093 | 25 | 0.484 | | | | | | | | | | | |
| Silicon steel Si = 0% | 493 | 0.108 | 42 | 0.785 | | | | | | | | | | | |
| 1% | 485 | 0.11 | 24 | 0.451 | | | | | | | | | | | |
| 2% | 479 | 0.11 | 18 | 0.344 | | | | | | | | | | | |
| 5% | 463 | 0.11 | 11 | 0.215 | | | | | | | | | | | |

TABLE A-1. PROPERTY VALUES (Continued)

| Metals | Properties at 68 F | | | | k, thermal conductivity, Btu/hr ft F | | | | | | | | | |
|---|---|---|---|---|---|---|---|---|---|---|---|---|---|---|
| | $\rho$ lb/ft³ | $c_p$ Btu/lb F | $k$ Btu/hr ft F | $\alpha$ ft²/hr | −148 F −100 C | 32 F 0 C | 212 F 100 C | 392 F 200 C | 572 F 300 C | 752 F 400 C | 1112 F 600 C | 1472 F 800 C | 1832 F 1000 C | 2192 F 1200 C |
| Copper: | | | | | | | | | | | | | | |
| Pure | 559 | 0.0915 | 223 | 4.353 | 235 | 223 | 219 | 216 | 213 | 210 | 204 | | | |
| Aluminum bronze 95 Cu, 5 Al | 541 | 0.098 | 48 | 0.903 | | | | | | | | | | |
| Bronze 75 Cu, 25 Sn | 541 | 0.082 | 15 | 0.333 | | 34 | 41 | | | | | | | |
| Red brass 85 Cu, 9 Sn, 6 Zn | 544 | 0.092 | 35 | 0.699 | | | | | | | | | | |
| Brass 70 Cu, 30 Zn | 532 | 0.092 | 64 | 1.322 | 51 | | 74 | 83 | 85 | 85 | | | | |
| German silver 62 Cu, 15 Ni, 22 Zn | 538 | 0.094 | 14.4 | 0.290 | 11.1 | | 18 | 23 | 26 | 28 | | | | |
| Constantan 60 Cu, 40 Ni | 557 | 0.098 | 13.1 | 0.237 | 12 | | 12.8 | 15 | | | | | | |
| Magnesium: | | | | | | | | | | | | | | |
| Pure | 109 | 0.242 | 99 | 3.762 | 103 | 99 | 97 | 94 | 91 | | | | | |
| Mg-Al (electrolytic) 6–8% Al, 1–2% Zn | 113 | 0.24 | 38 | 1.397 | | 30 | 36 | 43 | 48 | | | | | |
| Mg-Mn 2% Mn | 111 | 0.24 | 66 | 2.473 | 54 | 64 | 72 | 75 | | | | | | |
| Molybdenum | 638 | 0.060 | 71 | 2.074 | 80 | 72 | 68 | 66 | 64 | 63 | 61 | 59 | 57 | 53 |
| Nickel: | | | | | | | | | | | | | | |
| Pure (99.9%) | 556 | 0.1065 | 52 | 0.882 | 60 | 54 | 48 | 42 | 37 | 34 | 32 | 36 | 39 | 40 |
| Impure (99.2%) | 556 | 0.106 | 40 | 0.677 | | 40 | 37 | 34 | 32 | 30 | | | | |
| Ni-Cr 90 Ni, 10 Cr | 541 | 0.106 | 10 | 0.172 | | 9.9 | 10.9 | 12.1 | 13.2 | 14.2 | | | | |
| 80 Ni, 20 Cr | 519 | 0.106 | 7.3 | 0.129 | | 7.1 | 8.0 | 9.0 | 9.9 | 10.9 | 13.0 | | | |
| Silver: | | | | | | | | | | | | | | |
| Purest | 657 | 0.0559 | 242 | 6.601 | 242 | 241 | 240 | 238 | 209 | 208 | | | | |
| Pure (99.9%) | 657 | 0.0559 | 235 | 6.418 | 242 | 237 | 240 | 216 | | | | | | |
| Tungsten | 1208 | 0.0321 | 94 | 2.430 | | 96 | 87 | 82 | 77 | 73 | 65 | 44 | | |
| Zinc, pure | 446 | 0.0918 | 64.8 | 1.591 | 66 | 65 | 63 | 61 | 58 | 54 | | | | |
| Tin, pure | 456 | 0.0541 | 37 | 1.505 | 43 | 38.1 | 34 | 33 | | | | | | |

## Appendix A: Exhaust-gas Analyzer

Any unburned components in the exhaust gas now have another chance to burn as they pass the red-hot wire. They do burn, and this further raises the temperature of the wire. The more unburned components in the exhaust gas, the more combustion, and the higher the temperature of the wire. This increase of temperature causes an increase in resistance and a consequent reduction of current flow in the platinum wire. The ammeter registers this difference of current flow, since it will pass current seeking to flow through its resistance rather than through the wire. The ammeter dial is marked off to read air-fuel ratio directly.

**3. Relative density** The density of the exhaust gas is determined by the proportions of its various components, each of which has a different density. Thus, by determining the density of the exhaust gas, the result can be evaluated in terms of the air-fuel ratio. The type of analyzer that measures the density of the exhaust gas relative to air consists of two fans rotated in two chambers by a motor. Both rotate at the same speed; one in air, the other in exhaust gas. The air or gas movement produced by the fans causes two impulse wheels to move. The two impulse wheels are linked together, and one of them has a pointer registering on a dial. Since the exhaust gas is heavier or denser than air, the impulse wheel in the exhaust-gas chamber will turn more than will the impulse wheel in the air chamber. This causes the pointer to move across the dial and register directly the air-fuel ratio that produced the exhaust gas.

# Appendix B: Glossary

THIS GLOSSARY of automotive terms used in the book is designed to provide a ready reference for the student. The definitions may differ somewhat from those given in a standard dictionary; they are not intended to be all-inclusive, but have the purpose of serving as reminders so that the student can quickly refresh his memory on automotive terms of which he may be unsure. More complete definitions and explanations of the terms are found in the text.

**Abrasive** In automobile service, a substance used for cutting, grinding, or polishing metal.

**Accelerator** The foot-operated pedal linked to the carburetor throttle valve.

**Accelerator pump** In the carburetor, a pump linked to the accelerator, which momentarily enriches the mixture when the accelerator pedal is depressed.

**Air bleed** An opening into a gasoline passage through which air can pass (or bleed) into the gasoline as it moves through the passage.

**Air cleaner** The device mounted on the carburetor, through which air must pass on its way into the carburetor air horn. It filters out dirt and dust particles and also silences the intake noise.

**Air-cooled engine** An engine that is cooled by passage of air around the cylinders and not by passage of a liquid through water jackets.

**Air-fuel mixture** Mixture delivered to engine by carburetor.

**Air horn** In the carburetor, the tubular passage through which the incoming air must pass.

**Air line** A hose or pipe through which air passes.

**Air pressure** Atmospheric pressure (14.7 psi at sea level) or pressure of air produced by pump, by compression in engine cylinder, etc.

**Antifriction bearing** Type of bearing in which moving parts are in rolling contact; ball, roller, or tapered roller bearing.

**Antiknock** In engine fuels, the property that opposes knocking.

**Antipercolator** The device in the carburetor that opens a vent when the throttle is closed, to permit release of fuel vapors in the high-speed circuit so that fuel vapor will not push fuel out of the high-speed nozzle.

**Atmospheric pressure** Pressure of the atmosphere, or air, due to its weight pressing downward. Average is 14.7 psi at sea level.

## Appendix B: Glossary

**Atomization**  The spraying of a liquid that makes it a very fine mist.
**Automatic choke**  A choke that positions the choke valve automatically in accordance with engine temperature.
**Backfiring**  Pre-explosion of air-fuel mixture so that explosion passes the still-opened intake valve and flashes back through the intake manifold.
**Balanced carburetor**  Carburetor in which the float bowl is vented into upper air horn, below air cleaner, to eliminate effects of clogged cleaner.
**Ball check valve**  A valve consisting of a ball and seat. Fluid can pass in one direction only; when it attempts to flow the other way, it is checked by the ball seating on the seat.
**Barrel**  The air horn in the carburetor; used particularly to refer to that part of the air horn in which the throttle valve is located.
**BDC**  Bottom dead center, which see.
**Bearing**  Generally, the curved surface on a shaft or in a bore, or the part assembled onto one or into the other to permit relative rotation with minimum wear and friction.
**Bellows**  A device, usually metal, that can lengthen or shorten much like an accordian. The thermostat in the cooling system is usually a bellows.
**Bimetal**  Referring to the thermostatic bimetal element made up of two different metals with different heat-expansion rates; temperature change produces a bending or distorting movement.
**Blow-by**  Leakage of compressed air-fuel mixture or burned gases from the combustion chamber, past the piston rings, and into the crankcase.
**Body**  The assembly of sheet-metal sections, together with windows, doors, seats, and other parts, that provides an enclosure for the passengers, engine, etc.
**Borderline knock test**  A test used to establish octane rating, or knock resistance, of different fuels.
**Borderline lubrication**  Type of lubrication resulting when greasy friction exists. Moving parts are coated with a very thin film of lubricant.
**Bore**  Diameter of engine cylinder hole; also may be diameter of any hole, as, for instance, the hole in which a bushing fits.
**Bottom dead center**  The piston position at which the piston has moved to the bottom of the cylinder and the center line of the connecting rod is parallel to the cylinder walls.
**Brake horsepower**  The power delivered by the engine which is available for driving the vehicle.

## Automotive Fuel, Lubricating, and Cooling Systems

**Breather**   The opening that allows air to circulate in the crankcase and that is part of the *crankcase ventilator,* which see.

**Burnisher**   A cylindrical tool with integral collars that increase in diameter from one end of the tool to the other.

**Burr**   A featheredge of metal left on a part being cut with a file or other cutting tool.

**Bushing**   A sleeve placed in a bore to serve as a bearing surface.

**Butane**   One form of liquefied petroleum gas, which see.

**Butterfly**   The choke or throttle valve.

**Bypass filter**   Type of oil filter in which only some of the oil from the oil pump flows through the filter. The remainder of the oil bypasses the filter on its way to engine parts.

**Caliper**   A measuring tool that can be set to measure the thickness of a block, diameter of a shaft, or bore of a hole (inside caliper).

**Cam**   An irregularly shaped moving part designed to move or alter the motion of another part.

**Cam ground**   Refers to oval-shaped piston, so ground as to permit piston to expand and assume a round shape when hot.

**Camshaft**   The shaft in the engine which has a series of cams for operating the valve mechanisms. It is driven by gears or sprockets and chain from the crankshaft.

**Carbon**   A substance deposited on engine parts by the combustion of the fuel. Carbon forms on pistons, rings, valves, etc., inhibiting their action.

**Carbon dioxide**   A gas resulting from burning of fuel.

**Carbon monoxide**   A poisonous gas produced by a running gasoline engine.

**Carburetion**   The actions that take place in the carburetor: converting liquid fuel to vapor and mixing it with air to form a combustible mixture.

**Carburetor**   The device in the fuel system which mixes air and gasoline (vaporizing the gasoline as it does so) in varying proportions to suit engine operating conditions.

**Cetane**   Ignition quality of diesel fuel. A high-cetane fuel ignites more easily (at lower temperature) than a low-cetane fuel.

**CFR Uniontown road test**   A road test used to establish octane rating of different fuels.

**Change of state**   Changing of a substance from solid to liquid, or from liquid to vapor, or vice versa.

**Chassis**   The assembly of mechanisms that make up the major operating part of the vehicle. It is usually assumed to include everything except the car body.

## Appendix B: Glossary

**Choke** In the carburetor, a device that chokes off the air flow through the air horn, producing a partial vacuum in the air horn for greater fuel delivery and a richer mixture.

**Clearance** The space between two moving parts or between a moving and a stationary part, such as a journal and a bearing. Clearance is considered to be filled with lubricating oil when engine is running.

**Coil spring** A spring made up of an elastic metal, such as steel, formed into a wire or bar and wound into a coil.

**Combination fuel pump** A fuel pump with which a vacuum pump for operating the windshield wipers has been combined.

**Combustion** In the engine, the rapid burning of air-fuel mixture in cylinder.

**Combustion chamber** The space at the top of the cylinder and in the head in which combustion of the air-fuel mixture takes place.

**Compensating circuit** A special circuit in some carburetors to compensate for variations in fuel discharge from main nozzle. Compensating-circuit fuel nozzle discharges more fuel when nozzle discharges less, and vice versa, so that a balanced air-fuel mixture is delivered at all times.

**Compression gauge** A device for testing the amount of pressure developed in the engine cylinder during cranking.

**Compression ratio** The ratio between the volume in the cylinder with the piston at BDC and the volume with the piston at TDC.

**Compression rings** The upper ring or rings on a piston, designed to hold the compression in the cylinder and prevent blow-by.

**Compression stroke** The piston stroke from BDC to TDC during which both valves are closed and the air-fuel mixture is compressed.

**Condensation** The changing of a vapor to a liquid due to temperature, pressure, or other changes.

**Connecting rods** In the engine, linkages between the cranks on the crankshaft and the pistons.

**Cooling system** In the engine, the system that removes heat from the engine and thereby prevents overheating. It includes water jackets, water pump, radiator, and thermostat.

**Crank** A device for converting reciprocating motion into rotary motion, or vice versa.

**Crankcase** The lower part of the engine, in which the crankcase rotates. The upper part of the crankcase is lower section of the cylinder block, while the lower part is made up of the oil pan.

**Crankcase dilution** Dilution of the lubricating oil in the oil pan by the seepage of liquid gasoline down the cylinder walls.

## Automotive Fuel, Lubricating, and Cooling Systems

**Crankcase ventilator**  The device that permits air to flow through engine crankcase when engine is running.

**Crankpin**  The bearing surface on a crank of the crankshaft, to which the connecting rod is attached.

**Crankshaft**  The main rotating member, or shaft, of the engine, with cranks to which the connecting rods are attached.

**Cycle**  In the engine, the four piston strokes (or two piston strokes) that complete the working process and produce power.

**Cylinder**  In the engine, the tubular opening in which the piston moves up and down.

**Cylinder block**  The basic framework of the engine in and on which the other engine parts are attached. It includes the engine cylinders and the upper part of the crankcase.

**Cylinder head**  The part of the engine that encloses the cylinder bores. Contains water jackets and, on I-head engine, the valves.

**Degree**  1/360 part of a circle.

**Detergent**  A chemical sometimes added to the engine oil, designed to help keep the internal parts of the engine clean by preventing the accumulation of deposits.

**Detonation**  In the engine, excessively rapid burning of the compressed charge which results in *knock*, which see.

**DG oil**  Lubricating oil for average, or normal, diesel-engine service.

**Dial indicator**  A gauge that has a dial face and needle to register movement. Used to measure variations in size, movements too small to be measured conveniently by other means, etc.

**Diaphragm**  A flexible membrane, in automotive components usually made of fabric and rubber, clamped at the edges, and usually spring-loaded; used in fuel pump, vacuum pump, distributor, etc.

**Diesel cycle**  An engine cycle of events in which air alone is compressed and fuel oil injected at the end of the compression stroke. The heat produced by compressing the air ignites the fuel oil.

**Dip stick**  The oil-level indicator stick.

**Downdraft carburetor**  A carburetor in which the air horn is so arranged that the air passes down through it on its way to the intake manifold.

**Drill**  Also called *twist drill*. A cylindrical bar with helical grooves and a point, for cutting holes in material. Also refers to the device that rotates the drill.

**Dry friction**  The friction between two dry solids.

**DS oil**  Lubricating oil for severe, or heavy-duty, diesel-engine service.

**Dual carburetors**  Carburetors with two air horns, fuel nozzles, throttle valves, idle circuits, etc.

*Appendix B: Glossary*

**Dynamometer**   A device for measuring power output of an engine.
**Eccentric**   Off center.
**Economizer valve**   The mechanism in the carburetor that permits a rich mixture for full-load engine operation but leans the mixture for more economical operation on part throttle.
**Efficiency**   Ratio between the effect produced and the energy expended.
**Electric system**   In the automobile, the system that electrically cranks the engine for starting, furnishes high-voltage sparks to the engine cylinders to fire the compressed air-fuel charges, lights the lights, operates the heater motor, radio, etc. Consists, in part, of starting motor, wiring, battery, generator, regulator, ignition distributor, ignition coil.
**Energy**   Energy is the capacity or ability to do work.
**Engine**   The assembly that burns fuel to produce power, sometimes referred to as the power plant.
**Engine tune-up**   The procedure of checking and adjusting various engine components so that engine is restored to top operating condition.
**Ethyl**   Tetraethyllead, which see.
**Evaporation**   The change of a liquid to a vapor, or gas.
**Exhaust-gas analyzer**   A device for analyzing exhaust gases to determine carburetor action.
**Exhaust manifold**   The part of the engine that provides a series of passages through which burned gases from the engine cylinders can flow.
**Exhaust muffler**   The device in the exhaust line that muffles the sound of exhaust.
**Exhaust stroke**   The piston stroke from BDC to TDC during which the exhaust valve is open so that the burned gases are forced from the cylinder.
**Exhaust valve**   The valve that opens to allow the burned gases to exhaust from the engine cylinder during the exhaust stroke.
**Expansion plug**   A plug that is slightly dished out. When driven into place, it is flattened and expanded to fit tightly.
**Extreme-pressure lubricant**   A special lubricant for use in hypoid-gear differentials.
**Fan**   The device on the front of the engine that rotates to draw a blast of cooling air through the radiator.
**Fast idle**   The mechanism on the carburetor that holds the throttle valve slightly open when the engine is cold so that the engine will idle at a higher rpm when cold.

## Automotive Fuel, Lubricating, and Cooling Systems

**Feeler stock** Strips of metal of accurately known thickness, used to measure clearances.

**File** A cutting tool with a large number of cutting edges arranged along a surface.

**Filter** That part in the lubricating or fuel system through which fuel, air, or oil must pass so that dust or dirt is removed.

**Firing order** The order in which the engine cylinders fire, or deliver, their power strokes.

**Float bowl** In the carburetor, the reservoir from which gasoline feeds into the passing air.

**Float circuit** The circuit in the carburetor that controls entry of fuel and fuel level in the float bowl.

**Float level** The float position at which the needle valve closes the fuel inlet to the carburetor to prevent further delivery of fuel.

**Flywheel** The rotating metal wheel, attached to the crankshaft, that helps even out the power surges from the power strokes and also serves as part of the clutch and engine-cranking system.

**Four-barrel carburetor** A carburetor with four air horns. In effect, two two-barrel, or dual, carburetors in a single assembly. Used on several V-8 engines.

**Four cycle** Short for four-stroke cycle, which see.

**Four-stroke cycle** The four operations of intake, compression, power, and exhaust, or four piston strokes, that make up the complete cycle of events in the four-stroke-cycle engine.

**Friction** The resistance to motion between two bodies in contact with each other.

**Friction bearing** Type of bearing in which moving parts are in sliding contact; sleeve, guide, or thrust bearing.

**Fuel gauge** The gauge that indicates to the driver the height of the fuel level in the fuel tank.

**Fuel injector** A device in a diesel-engine fuel system for injecting fuel oil into the cylinder.

**Fuel jet** See *jet*.

**Fuel line** The pipe or tube through which fuel travels from the tank to the fuel pump and from the pump to the carburetor.

**Fuel nozzle** The tube in the carburetor through which gasoline feeds from the float bowl to the passing air.

**Fuel system** In the automobile, the system that delivers to the engine cylinders the combustible mixture of vaporized fuel and air. It consists of fuel tank, lines, gauge, carburetor, manifold.

**Fuel tank** The metal tank that serves as a storage place for gasoline.

*Appendix B: Glossary*

**Full-flow filter**  Type of oil filter in which all the oil from the oil pump flows through the filter.

**Full throttle**  Wide-open throttle position with accelerator pressed all the way down to floor board.

**Gasket**  A flat strip, of cork or other material, placed between two surfaces to provide a tight seal between them.

**Gasket cement**  An adhesive material used to apply gaskets.

**Gasoline**  A hydrocarbon suitable as an engine fuel, obtained from petroleum.

**Gear lubricant**  A type of grease or oil designed especially to lubricate gears.

**Gear-type pump**  A pump using a pair of matching gears that rotate; meshing of the gears forces oil (or other liquid) from between the teeth through the pump outlet.

**Generator**  The part of the electric system that converts mechanical energy into electric energy for lighting lights, charging the battery, operating the ignition system, etc.

**Goggles**  Special glasses worn over the eyes to protect them from flying chips, dirt, or dust.

**Governor**  A device, often installed under the carburetor, that prevents engine speed from exceeding a preset maximum.

**Gravity**  The attractive force between objects that tends to bring them together. A stone dropped from the hand falls to the earth because of gravity.

**Grease**  Lubricating oil to which thickening agents have been added.

**Greasy friction**  The friction between two solids coated with a thin film of oil.

**Grinding wheel**  An abrasive wheel used for grinding metal objects held against it.

**Heat-control valve**  In the engine, a thermostatically operated valve in the exhaust manifold for varying heat to intake manifold with engine temperature.

**Heat of compression**  Increase of temperature brought about by compression.

**Heptane**  A reference fuel that knocks very easily, used in various proportions with iso-octane for comparative test of knock characteristics of fuels.

**High compression**  A term used to refer to the increased compressions of modern automotive engines.

**High-speed circuit**  The circuit in the carburetor that supplies fuel to the air passing through the air horn during medium- and high-speed, part- to full-throttle operation.

*Automotive Fuel, Lubricating, and Cooling Systems*

**High-test gasoline**  A term referring to the octane rating of a fuel. A high-test fuel has a high octane rating.

**Hone**  An abrasive stone that is rotated in a bore or bushing to remove material.

**Horsepower**  A horsepower is a measure of a definite quantity of power; 33,000 ft-lb of work per minute.

**Hydrocarbon**  A compound made of the elements hydrogen and carbon; gasoline is a hydrocarbon.

**Hydrometer**  A device to determine the specific gravity (roughly the heaviness) of a liquid. This determination indicates the freezing point of the coolant in the cooling system, for example.

**Idle circuit**  In the carburetor, the passage through which fuel is fed when the engine is idling.

**Idle mixture**  The air-fuel mixture supplied to the engine during idle.

**Idle-mixture adjustment screw**  The adjustment screw that can be turned in or out to lean or enrich the idle mixture.

**Idle port**  The opening into the air horn through which the idle circuit in the carburetor discharges.

**Idling speed**  The speed at which the engine runs without load when the accelerator pedal is released.

**Ignition coil**  That part of the ignition system which acts as a transformer to step up the battery voltage to many thousands of volts; the high-voltage surge then produces a spark at the spark-plug gap.

**Ignition distributor**  That part of the ignition system which closes and opens the circuit to the ignition coil with correct timing and distributes to the proper spark plugs the resulting high-voltage surges from the ignition coil.

**Ignition system**  In the automobile, the system that furnishes high-voltage sparks to the engine cylinders to fire the compressed air-fuel charges. Consists of battery, ignition coil, ignition distributor, ignition switch, wiring, spark plugs.

**Indicated horsepower**  A measurement of engine power based on power actually developed in the engine cylinders.

**Inertia**  Property of objects that causes them to resist any change in speed or direction of travel.

**Intake manifold**  The part of the engine that provides a series of passages from the carburetor to the engine cylinders through which air-fuel mixture can flow.

**Intake stroke**  The piston stroke from TDC to BDC during which the intake valve is open and the cylinder receives a charge of air-fuel mixture.

## Appendix B: Glossary

**Intake valve**  The valve that opens to permit air-fuel mixture to enter the cylinder on the intake stroke.

**Iso-octane**  A reference fuel that shows great resistance to knocking, used in various proportions with heptane for comparative test of knock characteristics of various fuels.

**Jackets**  The water jackets that surround the cylinders, through which the cooling water passes.

**Jet**  A fuel nozzle or calibrated fuel passage in the carburetor.

**Knock**  In the engine, a rapping or hammering noise resulting from excessively rapid burning of the compressed charge.

**Liquefied petroleum gas**  A hydrocarbon suitable as an engine fuel obtained from petroleum and natural gas, a vapor at atmospheric pressure but liquefied if put under sufficient pressure.

**Lock nut**  A second nut turned down on a holding nut to prevent loosening.

**Low-speed circuit**  The circuit in the carburetor that supplies fuel to the air passing through the air horn during low-speed, part-throttle operation.

**LPG**  Liquefied petroleum gas, which see.

**Lubrication system**  The system in the engine that supplies moving engine parts with lubricating oil.

**Main fuel nozzle**  The fuel nozzle in the carburetor that supplies fuel when the throttle is partially to fully open.

**Manifold**  The intake or exhaust manifold, which see.

**Manifold vacuum**  The vacuum in the intake manifold that develops as a result of the vacuum in the cylinders on their intake strokes.

**Mechanical efficiency**  In an engine, the ratio between brake horsepower and indicated horsepower.

**Mechanical octane**  Octane needs of an engine, resulting from the mechanical design, or shape and relation of parts.

**Mechanism**  A system of interrelated parts that make up a working agency.

**Metering rod**  A device in the carburetor that enlarges or decreases the fuel passage to the fuel nozzle, varying fuel delivery for various throttle openings.

**Micrometer**  A measuring device that measures accurately such dimensions as shaft or bore diameter or thickness of an object.

**Mike**  A slang term for micrometer, which see.

**Missing**  In the engine, the failure of a cylinder to fire when it should.

**ML oil**  Oil for light automotive service.

**MM oil**  Oil for medium, or average, automotive service.

**MS oil**  Oil for severe automotive service.

## Automotive Fuel, Lubricating, and Cooling Systems

**Muffler**  In the exhaust, a device, through which the exhaust gases must pass, that muffles the sound.

**Nozzle**  Fuel nozzle, or jet, through which fuel passes when it is discharged into the carburetor air horn.

**Octane**  A measure of antiknock value of engine fuel.

**Oil cleaner**  The filtering device through which oil passes, which filters dirt and dust from the oil.

**Oil-control rings**  The lower ring or rings on a piston designed to prevent excessive amounts of oil from working up into the combustion chamber.

**Oil dilution**  Dilution of oil in the crankcase, caused by leakage of liquid gasoline from the combustion chamber past the pistons.

**Oil filter**  That part of the lubricating system that removes dirt and dust from the oil circulated through it.

**Oil-level indicator**  The indicator, usually a "stick," that can be removed to determine the level of oil in the crankcase.

**Oil pan**  The detachable lower part of the engine, usually made of sheet metal, that encloses the crankcase and acts as an oil reservoir.

**Oil-pressure indicator**  Oil gauge that reports to the driver the oil pressure in the engine lubricating system.

**Oil pump**  In the lubrication system, the device that delivers oil from the oil pan to the various moving engine parts.

**Oil pumping**  Passing of oil past the piston rings into the combustion chamber because of defective rings, piston, worn cylinder walls, etc.

**Oil seal**  A seal placed around a rotating shaft, etc., to prevent escape of oil.

**Oilstone**  A block of abrasive material bonded together, used for removing metal.

**Orifice**  A small opening, or hole, into a cavity.

**Otto cycle**  The four operations of intake, compression, power, and exhaust; so named for the inventor of the four-stroke cycle engine, Dr. Nikolaus Otto.

**Petroleum**  The crude oil extracted from the ground from which gasoline, lubricating oil, and other products are refined.

**Ping**  A metallic rapping sound from engine cylinder, caused by detonation.

**Piston**  In the engine, the cylindrical part that moves up and down in the cylinder.

**Piston pin**  Also called *wrist pin*. The cylindrical, or tubular, metal object that attaches the piston to the connecting rod.

## Appendix B: Glossary

**Piston-pin bearings**  The bearings or bushings in the piston and upper end of the connecting rod, in which the piston pin rides.

**Piston rings**  The rings fitted into grooves in the piston. There are two types, compression rings (for sealing the compression into the combustion chamber) and oil rings (to scrape excessive oil off the cylinder walls and thereby prevent it from working up into and burning in the combustion chamber).

**Piston skirt**  The lower part of the piston.

**Piston slap**  Hollow, muffled, bell-like sound made by excessively loose piston slapping cylinder wall.

**Poppet valve**  A mushroom-shaped valve, widely used in automotive engines.

**Port**  In the carburetor, an opening or jet through which fuel is discharged into the air horn.

**Power**  The rate of doing work.

**Power jet**  The fuel nozzle that discharges additional fuel into the high-speed circuit of the carburetor when the throttle is opened wide.

**Power piston**  The vacuum-operated piston in carburetors that releases at wide-open throttle to permit delivery of a richer air-fuel mixture.

**Power stroke**  The piston stroke from TDC to BDC during which the air-fuel mixture burns and forces the piston down, so that the engine produces power.

**Power train**  The group of mechanisms that carries the rotary motion developed in the engine to the car wheels; it includes clutch, transmission, propeller shaft, differential, and axles.

**Preignition**  Ignition of the air-fuel mixture in the engine cylinder (by any means) before the ignition spark occurs at the spark plug.

**Press fit**  A fit so tight, as a piston pin in a pin bushing, for example, that the pin has to be pressed into place (usually with an arbor press).

**Pressure cap**  The type of radiator cap used with pressure cooling systems; it contains a pressure relief valve and a vacuum valve.

**Pressure-feed**  A type of engine lubricating system that makes use of an oil pump to force oil through tubes and passages to the various engine parts requiring lubrication.

**Pressure regulator**  The device in an LPG fuel system that reduces the pressure on the LPG, permitting the fuel to vaporize in readiness for mixing with air in the carburetor.

**Pressure relief valve**  A valve in the oil line that opens to relieve excessive pressures that the oil pump might develop.

*...ive Fuel, Lubricating, and Cooling Systems*

...vice for measuring power output of an engine (brake ...*r*).

...One type of LPG, which see.

...ounds per square inch; usually used to indicate pressure of a liquid or gas.

**Puller**  Generally, a service tool that permits removal of one part from another without damage. Contains a screw or screws that can be turned to apply gradual pressure.

**Quadrijet carburetor**  Four-barrel carburetor, which see.

**Radiator**  In the cooling system, the device that removes heat from water passing through it; it takes hot water from the engine and returns cooled water to the engine.

**Radiator cap**  The cap placed on the radiator filler tube.

**Ribbon-cellular radiator core**  One type of radiator core consisting of ribbons of metal soldered together along their edges.

**Rocker arm**  A part in the fuel pump, linked to the diaphragm, that rocks back and forth as its end rides on a cam on the camshaft.

**Rotor pump**  A type of pump using a pair of rotors, one inside the other, to produce the oil pressure required to circulate oil to engine parts.

**Rpm**  Revolutions per minute.

**Scored**  Scratched or grooved, as a cylinder wall may be scored by abrasive particles moved up and down by the piston rings.

**Scraper**  A device used in engine service to scrape carbon, etc., from engine block, pistons, etc.

**Screen**  A fine-mesh screen in the fuel line that prevents the passage of dirt or dust into the carburetor.

**Shim**  A strip of copper or similar material used, for example, under a bearing cap to increase bearing clearance.

**Sludge**  Accumulation in oil pan, containing water, dirt, and oil; sludge is very viscous and tends to prevent lubrication.

**Soldering**  The uniting of pieces of metal with solder, flux, and heat.

**Spark plug**  The assembly, which includes a pair of electrodes and an insulator, that has the purpose of providing a spark gap in the engine cylinder.

**Splash-feed**  A type of engine lubricating system that depends on splashing of the oil for lubrication to moving engine parts.

**Spring**  An elastic device that yields under stress or pressure but returns to its original state or position when the stress or pressure is removed.

**Storage battery**  The part of the electric system which acts as a reservoir for electric energy, storing it in chemical form.

## Appendix B: Glossary

**Stroke**  In an engine, the distance that the piston moves from BDC to TDC.
**Stud**  A headless bolt threaded on both ends.
**Tachometer**  A device for measuring engine speed, or rpm.
**Tank unit**  The unit of the fuel-indicating system that is mounted in the fuel tank.
**Tap**  A special cutting tool for cutting threads in a hole.
**Taper**  A decrease in diameter from one to another place as taper in a cylinder, taper of a shaft.
**TDC**  Top dead center, which see.
**Tel**  Tetraethyllead, which see.
**Temperature indicator**  A gauge that indicates to the driver the temperature of the coolant in the cooling system, thus giving warning of impending damage if the temperature goes too high.
**Tetraethyllead**  A chemical put into engine fuel which increases octane rating, or reduces knock tendency.
**Thermal efficiency**  Relationship between the power output and the energy in the fuel burned to produce the output.
**Thermostat**  A device that operates on temperature changes. Several thermostats are used in engines. There is one in the cooling system, another in the manifold heat control, etc.
**Thermosiphon cooling**  Cooling by natural circulation of water, resulting from fact that a given volume of hot water is lighter than an identical volume of cold water.
**Throttle cracker**  Linkage from the starting motor switch to the throttle, which opens the throttle slightly when the engine is being cranked.
**Throttle-return check**  A device on the carburetor that prevents excessively sudden closing of the throttle.
**Throttle valve**  The round disk in the lower part of the carburetor air horn that can be turned to admit more or less air.
**Timing**  In the engine, refers to timing of valves and also timing of ignition.
**Top dead center**  The piston position at which the piston has moved to the top of the cylinder and the center line of the connecting rod is parallel to the cylinder walls.
**Torque**  Turning or twisting effort, measured in pound-feet.
**Torque wrench**  A special wrench with a dial that indicates the amount of torque being applied to a nut or bolt.
**Trouble-shooting**  The detective work necessary to run down the cause of a trouble; implies the correction of the trouble by elimination of cause.

## Automotive Fuel, Lubricating, and Cooling Systems

**Tube-and-fin radiator core** One type of radiator core, consisting of tubes to which cooling fins are attached; water flows through the tubes between the upper and lower radiator tanks.

**Turbulence** In the engine, the rapid swirling motion imparted to the air-fuel mixture entering the cylinder.

**Two-barrel carburetor** A dual carburetor, which see.

**Two cycle** Short for two-stroke cycle, which see.

**Two-stroke cycle** The series of events taking place in a two-stroke-cycle engine, which are intake, compression, power, and exhaust, all of which take place in two piston strokes.

**Unbalanced carburetor** Carburetor in which the float bowl is vented into the open air, as opposed to a *balanced carburetor*, which see.

**Updraft carburetor** Carburetor in which air horn and other parts are so arranged that the air passes up through the air horn on its way to the intake manifold. Used on engines where there is not enough headroom for a downdraft carburetor.

**Vacuum** An absence of air or other substance.

**Vacuum gauge** In automotive-engine service, a device that measures intake-manifold vacuum and thereby indicates actions of engine components.

**Valve** A device that can be opened or closed to allow or stop the flow of a liquid, gas, or vapor from one place to another.

**Valve clearance** The clearance between the adjusting screw on the valve lifter and the valve stem (in L-head engines) or between the rocker arm and the valve stem (in I-head engines).

**Valve lifter** Also called *valve tappet*. The cylindrical part of the engine that rests on a cam of the camshaft and is lifted by the cam action so that the valve is opened. There is a valve lifter for each valve.

**Vapor lock** A condition in the fuel system, in which gasoline has vaporized, as in the fuel line, so that fuel delivery to the carburetor is blocked or retarded.

**Vent** An opening from an enclosed chamber, through which air can pass.

**Venturi** In the carburetor, the constriction in the air horn that produces the vacuum responsible for the movement of gasoline into the passing air.

**Vibration** A complete rapid motion back and forth; oscillation.

**Viscosity** The term used to describe a liquid's resistance to flow. A thick oil has greater viscosity than a thin oil.

**Viscous** Thick, tending to resist flowing.

**Viscous friction** Friction between layers of a liquid.

**Vise** A gripping device for holding a piece while it is being worked on.

## Appendix B: Glossary

**Volatility**  A measurement of the ease with which a liquid vaporizes.

**Volumetric efficiency**  Ratio between amount of air-fuel mixture that actually enters an engine cylinder to the amount that could enter under ideal conditions.

**Water-distributing tube**  In the engine cooling system, a tube that improves water circulation around exhaust valves and other areas that might overheat.

**Water jacket**  The space between the inner and outer shells of the cylinder block or head, through which cooling water can circulate.

**Water pump**  In the cooling system, the device that maintains circulation of the water between the engine water jackets and the radiator.

**Water sludge**  A black, viscous substance that forms in the engine crankcase due to water's collecting and being whipped into the oil by the crankshaft.

**Work**  Work is the changing of the position of a body against an opposing force, measured in foot-pounds.

# Index

Accelerator-pump circuit, 84–87
Air bleed, 100, 101
Air cleaner, 51–53
  servicing of, 212
Air-cooled engines, 355, 356
Air-fuel ratios, 67, 68
Antifreeze solutions, 370, 371
  testing of, 381, 382
Antifriction bearings, 298
Anti-icing in carburetor, 97, 98
Antipercolator, 100
Atmospheric pressure, 37, 38
Atomization, 55
Atoms, 29–32

Backfiring of engine, 205
Ball bearings, 298
Bearings, 294–298
Body of oil, 301, 302
Borderline knock test, 169, 170
Borderline lubrication, 292
Buick carburetors, overhaul of, 258–267, 276–289
Bypass oil filter, 323

Carburetors, 54–61, 67–140, 239–289
  accelerator pump in, 84–87
  adjustments of, 241–246
  air bleed in, 100, 101
  antipercolator in, 100
  checks of, 205, 206
  choke in, 87, 94, 213–217
  circuits in, 68–94
  compensating system in, 101–103
  dual type of, 124–126
  float-bowl vent in, 71, 72
  float circuit in, 68–72
  four-barrel type of, 126–134
  fundamentals of, 45–61, 67–103
  installation of, 289

Carburetors, removal of, 246
  repair of, 239–289
  servicing of, 239–289
  troubles in, 239–241
Cetane number, 177, 178
CFR Uniontown road test, 169
Change of state, 35
Chassis dynamometers, 190
Chevrolet carburetors, overhauling of, 249–255
Chevrolet oil-pump service, 346, 347
Chevrolet water-pump service, 389–391
Choke, 87–94
  adjustment of, 213–216
  servicing of, 216, 217
Circuits in carburetor, 68–94
Combustion, 33, 34, 176
Combustion analyzers, 186, 187
Compensating system, 101–103
Compression ratio, 163, 164
Cooling systems, engine (see Engine cooling systems)
Crankcase ventilation, 328–330

Detergents in oil, 303, 304
Detonation, 171, 172
DG oil, 309
Diesel engines, 145–154
  fuels for, 176–179
Dry friction, 291, 292
DS oil, 309
Dual carburetors, 124–126
  overhaul of, 258–267
Dynamometers, 190

Elements, 29, 30
Engine, 3–25
  diesel type of, 145–154
  operation of, 4–12

[419]

Engine cooling systems, 354–395
  air-cooled type of, 355, 356
  care of, 382
  cleaning of, 383–388
  flushing of, 383–388
  forced circulation in, 357
  liquid-cooled type of, 356–360
  purpose of, 354, 355
  servicing of, 376–395
  testing of, 376–383
  trouble-shooting of, 382, 383
Engine fan, 361, 362
Engine lubricating system, 291–351
  checks of, 336–340
  combination splash and pressure, 315–317
  pressure feed, 313, 314
  purpose of, 299, 300
  servicing of, 335–351
  splash feed, 313, 314
  trouble-shooting of, 337–340
  types of, 313–317
Evaporation, 55
Exhaust-gas analyzers, 186, 187
Exhaust system, 61, 62
  dual, 63
Expansion due to heat, 36

Fan, engine, 361, 362
Fast idle, 99, 100
Float bowl, 59–61
  vents in, 71, 72
Float circuits, 68–72
Fluidity of oil, 301, 302
Flywheel, 14
Ford carburetors, 134–140
  overhaul of, 267–276
Ford oil-pump service, 349–351
Ford water-pump service, 394, 395
Four-barrel carburetor, 126–134
  overhaul of, 276–289
Friction, 291–294
Friction bearings, 295–298
Fuel filters, 42
Fuel gauge, 42–44
  servicing of, 220, 221
Fuel-injection system, 145–150
Fuel-line service, 218–220
Fuel-mileage testers, 185, 186
Fuel-nozzle action, 58

Fuel pump, 44–51, 187–189, 225–235
  electric, 50, 51
  inspection of, 222, 223
  installation of, 235
  removal of, 225
  servicing of, 225–235
  testers for, 187–189
  troubles in, 223, 224
Fuel systems, 28–289
  diesel type of, 147–154
  operation of, 41–63
  servicing of, 211–289
  troubles in, 184–206
Fuel tank, 41, 42
  servicing of, 217, 218
Full-flow oil filter, 323–325
Full-power circuit, 80–83

Gasoline, 159–176
  antiknock value of, 167–175
  octane rating of, 167–175
  origin of, 159, 160
  volatility of, 160–163
Governor, 112–115
Greases, 311–313
Greasy friction, 292

Heat, 34, 35
  of compression, 164
Heater, car, 365, 366
High-speed circuits, 77–83

Idling circuits, 72–77
Ignition system, 18–25
  advance mechanisms in, 20–25
  controls of, 107–109
Intake manifold, 53, 54

Knocking, cause of, 165–167
  control of, 172–175

Liquefied petroleum gas, 154–156, 179, 180
Low-speed circuits, 72–77
LPG (*see* Liquefied petroleum gas)
Lubricants, automotive, 311–313
Lubricating systems, engine (*see* Engine lubricating systems)
Lubrication, theory of, 293, 294

Manifold heat control, 94–97
  in V-8, 96, 97
Missing of engine, 205

[420]

## Index

ML oil, 308
MM oil, 308
Molecules, 32, 33
MS oil, 308
Muffler, 62, 63

Octane, 167–175
Oil, changing of, 309, 310, 340–342
    consumption of, 310, 311
    function of, 299, 300
    properties of, 301, 305
    service ratings of, 308, 309
    sources of, 300, 301
    viscosity of, 301, 302
Oil coolers, 325, 326
Oil filter, 322–325
    servicing of, 344–345
Oil-level indicators, 330
Oil pan, 313–318
    servicing of, 342–344
Oil-pressure indicators, 326–328
    servicing of, 351
Oil pumps, 318–320
    servicing of, 345–351

Piston rings, 12, 13
Plymouth carburetors, overhaul of, 255–258
Plymouth oil-pump service, 347–349
Plymouth water-pump service, 391–394
Preignition, 171, 172
Pressure, gas, 36, 37
Pressure tester for lubricating system, 335, 336

Quadrijet carburetor, 126–134
    overhaul of, 258–267, 276–289

Radiator, 363–365
    flushing of, 386, 387
    repairing of, 388, 389
    testing of, 377, 378
Radiator pressure cap, 368–370

Relief valve in lubricating system, 320–322
    servicing of, 344
Rings, piston, 12, 13
Roller bearings, 298

Sludge, crankcase, 306–308
Stalling of engine, 204, 205
Starting-control switches, 109, 110

Tachometers, 189, 190
Temperature indicators, 371–373
Thermosiphon cooling system, 356, 357
Thermostat in cooling system, 366–368
    testing of, 377
Throttle cracker, 98, 99
Throttle-return checks, 110–112
Throttle valve, 58, 59
Two-barrel carburetors, 124–126
    overhaul of, 258–267

Updraft carburetors, 116–122

Vacuum, 38
Vacuum gauges, 189
Vacuum pump, 48–50
Valves, engine, 7–9
    arrangements of, 16–18
Vapor lock, 202
Vapor-pressure temperature indicator, 371, 372
Venturi, 56–58, 78
VI (viscosity index), 304, 305
Viscosity, 292, 293, 301, 302
Viscosity index, 304, 305
Viscous friction, 292, 293
Volatility, 160

Water-distributing tubes, 358–360
Water jackets, engine, 358–360
Water pumps, 360
    servicing of, 389–395
    testing of, 378